Plant Science: New Frontiers in Botany

Plant Science: New Frontiers in Botany

Editor: Edward Ackles

STATES
ACADEMIC PRESS
www.statesacademicpress.com

States Academic Press,
109 South 5th Street,
Brooklyn, NY 11249, USA

Visit us on the World Wide Web at:
www.statesacademicpress.com

ISBN: 978-1-63989-424-6 (Hardback)

Cataloging-in-Publication Data

Plant science : new frontiers in botany / edited by Edward Ackles.
p. cm.
Includes bibliographical references and index.
ISBN 978-1-63989-424-6
1. Botany. 2. Plants. I. Ackles, Edward.
QK45.2 .P53 2022
580--dc23

Table of Contents

Preface

I am honored to present to you this unique book which encompasses the most up-to-date data in the field. I was extremely pleased to get this opportunity of editing the work of experts from across the globe. I have also written papers in this field and researched the various aspects revolving around the progress of the discipline. I have tried to unify my knowledge along with that of stalwarts from every corner of the world, to produce a text which not only benefits the readers but also facilitates the growth of the field.

Botany is a sub-discipline of biology which deals with the study of life of plants. They are crucial for the sustenance of life on Earth. The primary areas of study in this discipline include reproduction, primary metabolism, development, chemical products, evolutionary relationships and systematics. The techniques used in this domain include electron microscopy, live cell imaging and optical microscopy. The discipline of botany further branches out into plant anatomy, plant ecology, plant systematics, phytosociology, paleobotany, plant morphology, plant reproduction, plant physiology, etc. It has applications in horticulture, forestry and agriculture, breeding and genetic modification, environmental management and biodiversity. The book is a compilation of chapters that discuss the most vital concepts and emerging trends in the field of botany. The topics included herein are of utmost significance and bound to provide incredible insights to the readers. This book presents the complex subject of botany in the most comprehensible and easy to understand language.

Finally, I would like to thank all the contributing authors for their valuable time and contributions. This book would not have been possible without their efforts. I would also like to thank my friends and family for their constant support.

Editor

1

Microarray meta-analysis to explore abiotic stress-specific gene expression patterns in Arabidopsis

Po-chih Shen[1], Ai-ling Hour[2] and Li-yu Daisy Liu[1*]

Abstract

Background: Abiotic stresses are the major limiting factors that affect plant growth, development, yield and final quality. Deciphering the underlying mechanisms of plants' adaptations to stresses using few datasets might overlook the different aspects of stress tolerance in plants, which might be simultaneously and consequently operated in the system. Fortunately, the accumulated microarray expression data offer an opportunity to infer abiotic stress-specific gene expression patterns through meta-analysis. In this study, we propose to combine microarray gene expression data under control, cold, drought, heat, and salt conditions and determined modules (gene sets) of genes highly associated with each other according to the observed expression data.

Results: By analyzing the expression variations of the Eigen genes from different conditions, we had identified two, three, and five gene modules as cold-, heat-, and salt-specific modules, respectively. Most of the cold- or heat-specific modules were differentially expressed to a particular degree in shoot samples, while most of the salt-specific modules were differentially expressed to a particular degree in root samples. A gene ontology (GO) analysis on the stress-specific modules suggested that the gene modules exclusively enriched stress-related GO terms and that different genes under the same GO terms may be alternatively disturbed in different conditions. The gene regulatory events for two genes, *DREB1A* and *DEAR1*, in the cold-specific gene module had also been validated, as evidenced through the literature search.

Conclusions: Our protocols study the specificity of the gene modules that were specifically activated under a particular type of abiotic stress. The biplot can also assist to visualize the stress-specific gene modules. In conclusion, our approach has the potential to further elucidate mechanisms in plants and beneficial for future experiments design under different abiotic stresses.

Keywords: Abiotic stress, Gene module, The coefficient of intrinsic dependence, Analysis of variance, The weighted gene co-expression network analysis, Singular value decomposition, Biplot

Background

Facing the challenge of climate change, raising crop production to feed enough people indicates to increase the tolerance of plants to severe environments (Ronald 2011). Plant is a sessile organism and must maintain a complex system of genetic expression to accommodate the impacts of different environments in order to survive with success (Trewavas 2003). When the plant is subjected to a stress, genes coded on its DNA usually take the initial actions required to trigger proper self-defensive mechanisms (Sachs and Ho 1986). It is therefore straightforward to monitor gene expression patterns and their interactions as the first step to deciphering the underlying mechanisms of a plant subjected to stresses. Embraced by rapidly developed biotechnologies, it has become very convenient to accurately monitor global gene expression under different circumstances in living organisms (Ritchie et al. 2015). Although the sequencing

*Correspondence: lyliu@ntu.edu.tw
[1] Biometrics Division, Department of Agronomy, National Taiwan University, Taipei, Taiwan
Full list of author information is available at the end of the article

of messenger RNA by the latest generation of sequencing technology (also known as RNA-Sequencing or RNA-Seq) is more straightforward, sensitive, and accurate in terms of the quantification of gene expressions, the systematic error rates and costs of said technology remain high compared to those of microarray technology, which has been in use for more than two decades (Mantione et al. 2014).

After collecting a global set of gene expressions, finding differentially expressed genes is the first step in deciphering the underlying mechanisms of a plant that copes with stress. In addition, biologists have recently been asking more about the systematic explanations of gene expression patterns. (e.g., Atkinson and Urwin 2012; Hahn et al. 2013; Priest et al. 2014). Such inquiries have motivated the advancement of gene set analysis and the utilization of microarray data to make inferences regarding genetic networks. Gene set analysis concerns the disturbed gene sets instead of individual genes whereas the gene sets of interest are predetermined (e.g., the co-expressed genes, the genes in the same category of the gene ontology, the genes involved in the same metabolic pathway, etc.) (Kaever et al. 2014, 2015; Rest et al. 2016). Network inference, which is the focus of this study, links genes with edges that indicate potential associations to depict the possible interactions among the chosen set of genes (Todaka et al. 2012; Rasmussen et al. 2013; Nakashima et al. 2014).

The construction of a correlation network is one approach that can be taken after a microarray experiment, while conducting a weighted gene co-expression network analysis (WGCNA) is another. The former can be accomplished by computing the pairwise Pearson correlation coefficients of genes and connecting a given gene pair if the Pearson correlation coefficient exceeds a user-specified threshold (Song et al. 2012). The choice of the threshold, however, might be very subjective, and different thresholds result in networks with different topologies (Borate et al. 2009). A WGCNA, however, preserves all possible edges in the network but assigns different weights to them instead. By clustering the genes with high-weight edges, WGCNA is able to determine the modules (gene sets) of genes highly associated with each other according to the observed expression data. In this study, we further extend the application of WGCNA in three directions:

1. Instead of using the correlation coefficient as the measure of the association between genes, we choose the coefficient of intrinsic dependence (CID).
2. We explore stress-specific modules by conducting the analysis of variance on the expression levels of the Eigen genes representing the gene modules and by comparing the edge weights of the stress-specific modules.
3. A biplot is introduced to visualize the stress-specific modules for convenience in further interpretation.

Along with long-term development of the technology, a huge amount of results from a wide range of microarray experiments has been accumulated. As of August 18, 2016, data from a total of 179 microarray experiments had been included in the Arabidopsis Information Resource (TAIR) database. Typical experiments have consisted of several treated samples under a particular condition and several controlled samples as the background of comparison for the reasons of purification and simplicity. However, analyzing the expression patterns under one stressed condition versus those under control conditions can only reveal a corner of a huge puzzle. One is not able to depict an overview of the entire system or of the interactions between the impacts caused by different stress sources on the living organisms. Therefore, in this study, we combine all possible samples from different stress conditions and perform a meta-analysis of the gene regulatory network on the combined dataset.

To that end, the coefficient of intrinsic dependence (CID), instead of the typical Pearson correlation coefficient, is used to measure the association between genes because the Pearson correlation coefficient only measures the linearity of the gene associations. However, past studies (Liu 2005; Liu et al. 2009) have shown that a nonlinear relationship between the expressions of two associated genes might occur in some cases. The CID does not require distributional and functional assumptions regarding the data and is useful for analyzing noisy microarray data. Relatedly, while systematic errors are well controlled in highly developed microarray technology, samples from different experiments contribute noise to each other when a meta-analysis is conducted due to the fact that the expression patterns from different experiments have a wide range of variation (Ramasamy et al. 2008; Campain and Yang 2010). The CID had been applied to investigate gene regulatory events incorporating the Galton–Pearson correlation coefficient (Liu et al. 2009, 2012), to identify associations among multivariate variables (Liu and Tsai 2013), and to select relevant features on a step-by-step basis according to their importance in relation to the target variable (Hsiao and Liu 2016). In this study, we strictly followed the definition and methodology of CID described in Hsiao and Liu (2016) and focus on utilizing the CID in measuring the magnitude of association in general between genes based on microarray gene expression data.

A CID matrix is used to construct the weighted gene co-expression network produced by a WGCNA.

According to the weighted network, gene modules containing genes with similar expression patterns are then identified. The WGCNA further performs a principle component analysis (Pearson 1901; Hotelling 1933; Zhang and Horvath 2005) on the expression matrix of each gene module and uses the first principle component (designated as "Eigen gene") as the representative of the gene module. The expression of the Eigen gene is the linear combination of the expressions from all genes in the gene module, which has been utilized to identify quantitative-trait-associated gene modules by computing the Pearson correlation coefficient between the observed values of the quantitative trait and the expression levels of the Eigen gene (Zhang and Horvath 2005; Langfelder and Horvath 2008). In this study, it is the qualitative variable (i.e., the treated conditions of the sample) that is of interest. We propose to utilize the analysis of variance for identification of stress-related gene modules by using the expression levels of the Eigen gene as the dependent variable and the treatment conditions of the sample as the independent variable.

Finally, we present the stress-related gene module using a biplot (Fig. 1), which unifies the information from both the gene module (shown as a gray arrow) and treatment conditions (shown as black arrows) in a single two-dimension plot. The biplot was introduced by Gabriel (1971); Yan and Kang (2002) subsequently described various methods for visualizing and interpreting a biplot. In our case, the length and direction of the projection of a specified condition vector on the vector of the gene module represents the magnitude of the impact of that condition on the gene module. A biplot can also assist in the interpretation of stress-specific modules. For example, an illustration of a biplot shown in Fig. 1, gene module M10 is up-regulated under the heat condition (as indicated by the acute angle between the M10 and HEAT vectors) but down-regulated under the cold and salt conditions (as indicated by the obtuse angles between the M10 and the COLD and SALT vectors). In addition, M10 is heat-specific, but it cannot tell the difference between the cold and salt conditions (the M10 vector is close to the red line, which is perpendicular to the line connecting two condition points; see Materials and Methods).

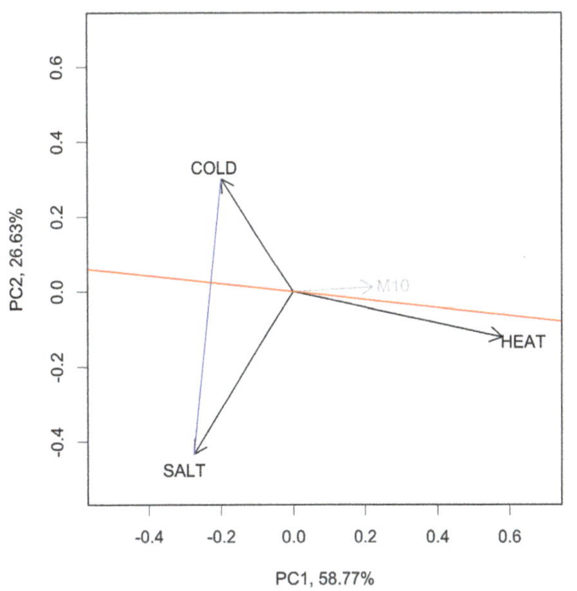

Fig. 1 An illustration of a biplot. The *blue line* represents the connection between two stresses, and the *red line* passes through the origin and runs perpendicular to the *blue line*. The *gray* and *black arrows* are vectors of each gene module and stress, respectively

Methods

The microarray expression data were downloaded from TAIR database (Huala et al. 2001). Our dataset consisted of 24 cold-treated, 24 salt-treated, 28 drought-treated, 16 heat-treated, and 36 controlled time-series arrays (Additional file 1: Table S1). All 128 arrays were hybridized using the Affymetrix ATH1 chip, which contains 22,810 probe sets (genes). The downloaded microarray raw data was preprocessed using the robust multi-array average (RMA) method and log2 transformed by RMAExpress version 1.0.5 (Bolstad et al. 2003; Irizarry et al. 2003a, b). The output of RMAExpress was taken log2 again to make the expression data more normally distributed (Additional file 2: Figure S1) in order to fulfill the essential assumption of following the analysis of variance. The processed microarray data were subjected to the analyses of variance (ANOVA) to identify the genes differentially expressed using the design models:

$$R_{ijkl} = \mu_{Ri} + \alpha_{Rij} + \alpha\tau_{Rijk} + \epsilon_{Rijkl}$$

for samples of root tissues, or

$$S_{ijkl} = \mu_{Si} + \alpha_{Sij} + \alpha\tau_{Sijk} + \epsilon_{Sijkl}$$

for samples of shoot tissues,

where R_{ijkl} (S_{ijkl}) is the normalized expression level of the lth replicate for the ith gene under the jth condition at kth time point in the root (shoot) samples; μ_{Ri} (μ_{Si}) is the overall average expression level for the ith gene in the root (shoot) samples; α_{Rj} (α_{Sj}) is the condition effect observed in the root (shoot) samples; $\alpha\tau_{Rjk}$ ($\alpha\tau_{Sjk}$) is the time effect under the jth condition in the root (shoot) samples; and ε_{Rijkl} (ε_{Sijkl}) is the sampling error. The analysis of variance which we performed to determine a differential expression gene considered the alternative hypothesis that the gene's expression was significantly different in at least one of the cold, heat, drought, salted, and controlled conditions. Four p values (two for testing the conditional effects and the other two for testing time effects under different conditions) can be obtained from the above model for each gene. Combining the two p values from the root samples and the two p values from the shoot samples, the smallest one was designated as the overall p value of a gene. The most significant 2281 genes yielding the smallest 10% of overall p values were claimed to be differentially expressed in this study.

The following steps to construct the weighted gene co-expression network were analogue to those of the well-known WGCNA package in R (Langfelder and Horvath 2008). The pairwise CID values for the expression levels of g DE genes were computed to form the g*g CID matrix, $C = \{c_{i,j}\}$, where $c_{i,j} = CID(g_i|g_j)$ for two DE genes, g_i and g_j, by setting the subgroup size of g_j to be 20 (Liu et al. 2009). Unlike the correlation matrix, the CID matrix is asymmetric; that means $c_{i,j}$ is not necessarily equal to $c_{j,i}$. The g*g adjacency matrix, $A = \{a_{i,j}\}$, was set as C taken to the power of three to make the connectivity of the DE genes follow the scale-free property of the biological network (Albert 2005; Pavlopoulos et al. 2011). Specifically, in this study, the connectivity of gene g_i for the asymmetric adjacency matrix was defined as $\Sigma_j a_{i,j} = \Sigma_{j \neq i}(c_{i,j})^3$, $i,j = 1, \ldots, 2281$. The adjacency matrix, A, was analyzed by WGCNA with minModuleSize = 60 to obtain the gene modules (Langfelder and Horvath 2008).

The Eigen gene of each gene module was defined as the first principle component of the genes in the module (Langfelder and Horvath 2008). We performed the singular value decomposition (SVD) toward the s by m expression matrix $M = \{m_{ij}\}$, where m_{ij} is the average expression level of the Eigen gene of jth ($j = 1, \ldots, m$) module under the ith ($i = 1, \ldots, s$) condition. More specifically, the SVD decomposed the matrix M into

$$\mathbf{M} = \mathbf{USV}^T = \left(\mathbf{US}^{1/2}\right)\left(\mathbf{VS}^{1/2}\right)^T = \mathbf{U}^*\mathbf{V}^{*T}$$

where $\mathbf{U}$ and $\mathbf{V}$ are s by r left and m by r right singular matrices, respectively (Golub and Van Loan 1996), $\mathbf{S}$ is an r by r diagonal matrix having the singular values, and r is the number of non-zero values of the diagonal of S. The first two columns of $\mathbf{U}^*$ and $\mathbf{V}^*$ are referred as the first two principle components (PCs) that most explain the variability of the numbers in the rows (conditions) and the columns (modules) of $\mathbf{M}$, respectively.

By overlaying the values of the first two PCs in $\mathbf{U}^*$ and $\mathbf{V}^*$ together in a two-dimension plot (known as the "biplot"), we can visualize the effects of the conditions and the modules and their interactions more easily (Gabriel 1971; Yan and Kang 2002). According to the inner-product property of a biplot (Yan and Kang 2002), the projection of a module vector on a condition vector represents the magnitude of the effect of the gene module on the condition, or vice versa. If the angle between two vectors is less than 90° (greater than 90°), the gene module is up-regulated (down-regulated) in the condition. We further connect two points of two conditions (for example, the blue line in Fig. 1 represents such a connection) on the biplot and draw a straight line passing through the origin and running perpendicular to the line connecting two condition points (for example, the red line in Fig. 1). Any module vector having a small angle to the perpendicular line (the red line in Fig. 1) cannot effectively distinguish the two conditions, meaning that similar expression levels (i.e., the projection lengths) on the gene module have been observed in the two conditions. Therefore, we claim that the gene module is specific for a condition if it fulfills two criteria: (1) the length of the gene module vector projected on the condition vector is relatively long, and (2) the gene module vector is almost perpendicular to the connected line (e.g., the blue line in Fig. 1) between any other two conditions. The genes in the stress specific gene modules were verified through gene ontology (GO) analysis using agriGO analytic tools (Du et al. 2010).

Results

Arabidopsis thaliana (Arabidopsis) microarray expression data consisting of 22,810 probe sets (genes) and 216 samples under five conditions (control, cold, drought, heat, and salt) were analyzed. The data were first preprocessed in order to fulfill the normality assumption of ANOVA (Additional file 2: Figure S1), which were run separately on the root and shoot samples. The most significant 2281 genes under different conditions were collected for further analysis. We had observed the tissue-specific responses of the genes under different conditions; only 554 genes were included in the top 10% lists for both the root and shoot samples. Furthermore, there were 813 genes that were differentially expressed only in the shoot samples and 914 genes that were differentially expressed only in the root samples.

From the 2281 DE genes, the WGCNA resulted in 19 gene modules (Additional file 3: Table S4). The sizes of the gene modules ranged from 60 to 295 genes, with an average of 120.05 genes (Additional file 4: Table S2). The Eigen gene (ME) for each gene module was designated as the first principle component after conducting the singular value decomposition (SVD) on the expression matrix of the gene module. Figure 2 shows the patterns of average expressions for 19 Eigen genes of all conditions in shoot/root samples. Fourteen out of the 19 Eigen genes (ME1-ME4, ME6, ME8-ME12, ME14, ME16-ME17, and ME19) were mostly highly expressed in root samples. Three Eigen genes (ME5, ME7, and ME15) were expressed more in shoot samples, but the other two Eigen

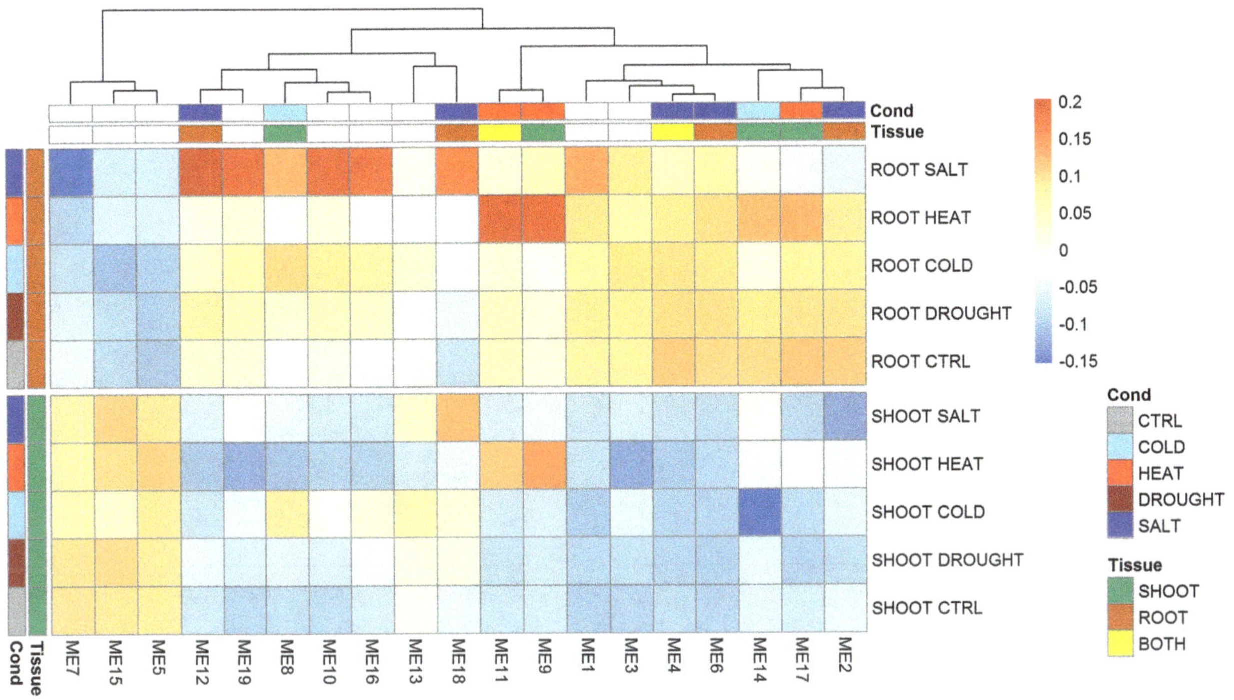

Fig. 2 A heatmap consisting of 19 gene modules, 5 stress conditions, and 2 tissue types. This heatmap was consisting of 19 gene modules (M1 to M19), 5 stress conditions (cold, heat, salt, drought, and control), and 2 tissue types (root and shoot tissues)

genes (ME13 and ME18) had similar patterns of expression in both root and shoot samples.

By showing both the diversities of the Eigen genes and the experiments in one biplot, we aimed to link the expression patterns and the abiotic stresses to find the so-called tissue-specific or abiotic stress-specific modules. To simplify the biplot, we used arrows to present the Eigen gene vectors and text to indicate the ends of the abiotic-stress vectors in different tissues starting from the origins (Fig. 3). The SALT and HEAT vectors are the longest among those of all five conditions in the root and shoot samples, respectively, indicating that the expressions of the Eigen genes have relatively large variation under the salt and heat conditions in the root and shoot samples, respectively. The DROUGHT vectors in both the root and shoot samples were very short, meaning that the expressions of the Eigen genes were close to their average values under the drought condition, which can also be observed through the heatmap (Fig. 2). Also, the DROUGHT vectors were very close to the CTRL vectors (the two vectors in root/shoot samples pointed in the same direction and were almost parallel to each other). This means that the expression patterns from the drought-treated samples were very similar to those from the control samples (Fig. 2). Therefore, the drought condition was ignored when searching for stress-specific modules.

Discussion

By examining the angles between the Eigen gene vectors and the condition vectors in root or shoot samples, we identified the abiotic stress-responsive modules. A smaller angle implies a stronger interaction between the Eigen gene and the condition. We subjectively selected the Eigen genes having angles of less than 15° to any condition vector for further inspection (italics in Tables 1, 2). Under the cold condition, five of the Eigen gene vectors (ME3 and ME9 in the root samples and ME7, ME8, and ME14 in the shoot samples) had angles of less than 15° to the cold vectors; six Eigen gene vectors (ME3, ME11, and ME13 in the root samples and ME9, ME11, ME17, and ME19 in the shoot samples) had angles of less than 15° to the heat vectors. For salt vectors, six Eigen gene vectors (ME2, ME4, ME5, ME6, ME12, and ME18 in the root samples and only ME2 and ME4 in the shoot samples) had angles of less than 15° to them. Note that most of the salt-responsive Eigen genes were present in the root samples and that there were slightly more temperature- (cold- or heat-) responsive Eigen genes detected in the shoot samples. The expressions of the Eigen genes were also tissue-specific because the responsive modules in root and shoot samples were mostly different for a selected condition.

We further identified Eigen genes which were specifically regulated only under the cold, heat, or salt condition

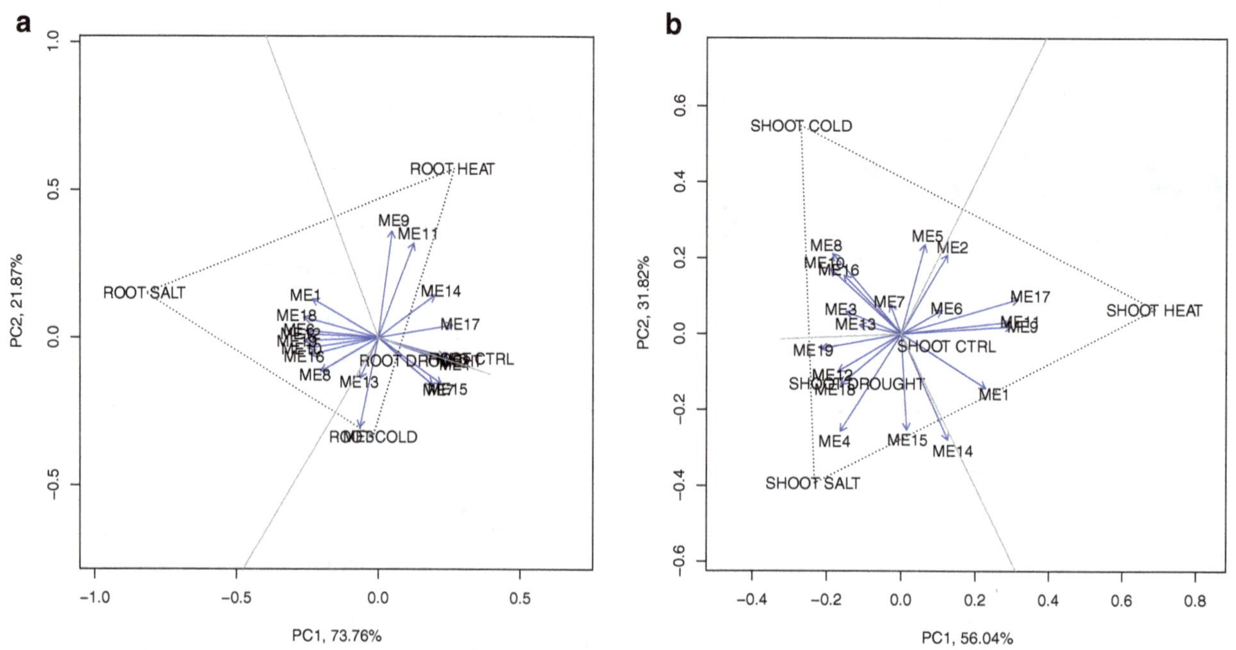

Fig. 3 Biplots of 19 Eigen genes and 5 stress conditions. Those biplots were consisting of 19 Eigen genes (ME1 to ME19) and 5 stress conditions (cold, heat, salt, drought, and control) for **a** root tissues, and **b** shoot tissues, respectively

Table 1 Angles between each gene module and stress for root samples

Module	Length	C	H	S	P(C,S)	P(H,S)	P(C,H)
ME9	0.364	*−4.444*	17.070	87.317	23.562	−28.981	−80.381
ME11	0.344	−17.980	*3.535*	−79.148	*10.026*	−42.517	86.084
ME3	0.312	*8.814*	*−12.700*	88.313	−19.192	33.352	84.751
ME18	0.273	−78.822	−79.664	*4.051*	−73.172	−54.284	*−2.885*
ME15	0.269	57.671	−79.185	−25.201	−85.677	33.134	18.265
ME1	0.268	−64.099	85.613	18.774	−87.896	−39.561	*−11.838*
ME19	0.264	84.335	−62.821	*12.793*	−56.329	−71.128	−19.728
ME17	0.262	−77.256	55.742	−19.871	49.250	78.206	26.807
ME6	0.257	−87.773	−70.713	*4.900*	−64.222	−63.235	*−11.836*
ME4	0.254	74.502	83.984	*−8.370*	77.492	49.965	*1.434*
ME2	0.253	76.637	81.849	*−6.236*	75.357	52.100	*0.700*
ME12	0.253	89.584	−68.070	*7.543*	−61.579	−65.878	−14.479
ME10	0.251	78.739	−57.225	18.388	−50.733	−76.723	−25.324
ME7	0.248	53.104	−74.618	−29.768	−81.110	28.567	22.832
ME16	0.248	73.152	−51.638	23.975	−45.146	−82.311	−30.911
ME14	0.245	−50.562	29.048	−46.565	22.556	−75.100	53.501
ME5	0.239	77.480	81.006	*−5.393*	74.514	52.943	*1.543*
ME8	0.234	57.615	−36.101	39.512	−29.609	82.153	−46.448
ME13	0.151	21.791	*−0.277*	75.337	*−6.215*	46.328	−82.272

The italic underline gene modules had length greater than 0.25

The italics angles were less than 15 degrees

by drawing reference lines on the biplot. A line connecting the ends of any two vectors of conditions, C_1 and C_2, was made; this line was called $L(C_1,C_2)$ for convenience. Then we made another line, called $P(C_1,C_2)$, passing through the origin and running perpendicular to $L(C_1,C_2)$. A "stress-specific" module for the condition C_3 would be

Table 2 Angles between each gene module and stress for shoot samples

Module	Length	C	H	S	P(C,S)	P(H,S)	P(C,H)
ME17	0.329	−79.780	*10.677*	−43.221	*13.794*	79.689	47.246
ME14	0.306	*−1.953*	71.056	55.047	67.939	*2.044*	−51.022
ME4	0.305	−58.792	−52.105	*1.792*	−55.222	58.883	*−5.817*
ME9	0.299	−67.399	*1.704*	−55.601	*1.414*	67.308	59.626
ME11	0.293	−69.798	*0.695*	−53.203	*3.812*	69.706	57.228
ME8	0.281	*14.763*	−54.340	−71.763	−51.223	*−14.672*	67.737
ME1	0.269	−31.729	37.374	88.729	34.256	31.638	−84.704
ME10	0.255	21.774	−47.329	−78.773	−44.212	−21.683	74.748
ME15	0.254	−22.970	−87.927	34.029	88.956	23.062	−30.004
ME2	0.244	57.258	53.639	*−0.258*	56.756	−57.349	*4.283*
ME5	0.244	41.448	69.449	−15.551	72.566	−41.540	*11.526*
ME19	0.220	73.898	*−4.795*	49.102	*−7.912*	−73.807	−53.127
ME16	0.218	18.926	−50.177	−75.925	−47.060	−18.834	71.900
ME18	0.210	−77.204	−33.693	20.204	−36.811	77.295	−24.229
ME12	0.196	−86.352	−24.545	29.353	−27.662	86.443	−33.378
ME3	0.160	42.199	−26.904	80.802	−23.787	−42.107	−84.827
ME6	0.127	86.898	23.999	−29.899	27.116	−86.990	33.924
ME13	0.116	52.040	−17.063	70.961	*−13.946*	−51.948	−74.986
ME7	0.083	*5.531*	−74.634	−51.469	−71.516	*−5.622*	47.444

The italic underline gene modules had length greater than 0.25

The italics angles were less than 15 degrees

almost parallel to both P(C_1,C_2) and the vector of C_3. The angles between the Eigen gene vectors and all three perpendicular lines, P(C[OLD], S[ALT]), P(H[EAT], S[ALT]), P(C[OLD], H[EAT]) are also shown in Tables 1 and 2. When searching for the stress-specific Eigen genes, their lengths were also taken into consideration to identify the Eigen genes that were more affected by the condition of interest; a shorter Eigen gene vector implied that the expression levels of that Eigen gene did not change much under different conditions. The stress-specific Eigen gene vectors with a length greater than 0.25 are marked in italics and underlining in Tables 1 and 2.

In shoot samples, two Eigen gene vectors (ME8 and ME14) under the cold condition (Fig. 3b) had very sharp angles (less than 15°) to P(HEAT, SALT). This implies that the expression patterns of these two Eigen genes are similar under the heat and salt conditions but very different under the cold condition in shoot samples (Fig. 2). Similarly, ME9 and ME17 were heat-specific Eigen genes only in shoot samples. Four Eigen genes (ME2, ME6, ME12, and ME18) were identified as salt-specific only in root samples. Two Eigen genes, ME4 and ME11, were identified as salt-specific and heat-specific, respectively, in both root and shoot tissues.

We validated our findings through gene ontology (GO) enrichment analysis using the genes in the selected gene modules. The lists of the enriched GO are provided in S3 Table. There were 39 GO terms enriched by two cold-specific modules, M8 and M14 (Fig. 4a), with only one cellular component term, "plasma membrane" (GO: 0005886), in common. According to the enriched GO terms, M8 and M14 are complementary to each other and specify different aspects of cold-responsive mechanisms. In particular, the genes in M8 largely participate in the biological processes of stress response (GO: 0006950), temperature stimulus response (GO: 0009266), and cold response (GO: 0009409) (Additional file 5: Table S3A). Cold may also stimulate chemical and endogenous responses in the cells that cause them to accumulate carbohydrate (GO: 0009743) and chitin (GO: 0010200). However, the GO terms enriched by the genes in M14 mostly belong to molecular function and cellular component categories.

Of three heat-specific gene modules, M9 and M17 were specifically identified in shoot samples, while M11 was identified in both root and shoot samples. Similarly, we expected that shoot-only M9 and M17 exclusively enriched heat-related GO terms. The three common GO terms among the 55 enriched in either M9 or M17 were more general GO terms including "response to chemical stimulus" (GO: 0042221), "response to stimulus" (GO: 0050896), and "plasma membrane" (GO: 0005886) (Fig. 4b). Combining the results from three modules, there were several heat-related GO terms that drew our attention, including "response to heat" (GO: 0009408), "protein folding" (GO: 0006457), "response to high light intensity" (GO: 0009644), "response to oxidative

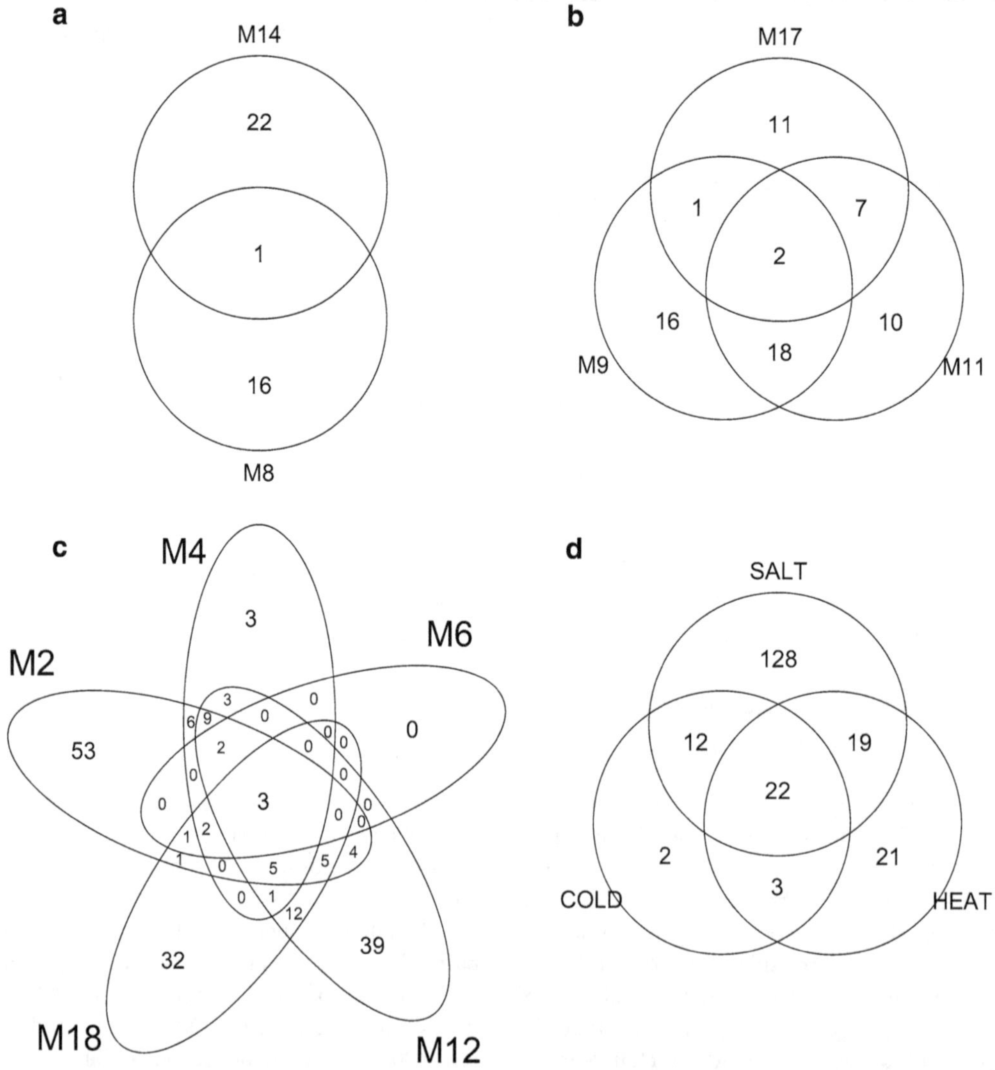

Fig. 4 Venn diagrams of enriched gene ontology items. Those venn diagrams were produced for **a** cold-specific, **b** heat-specific, **c** salt-specific, and **d** all stress-specific gene modules. The *numbers* in the venn diagrams were the numbers of enriched GO items in the stress-specific gene modules. For example, in **a**, there were 22, 16, and 1, GO items enriched in M14 only, M8 only, and in both M8 and M14

stress" (GO: 0006979), "response to radiation" (GO: GO: 0009314), and "response to cadmium ion" (GO: 0046686) (Additional file 5: Table S3B). This implied that the plant may simultaneously suffer adversities from light, radiation, and cadmium toxicity when under heat stress.

Under the salt stress, genes in the M4 module were specifically disturbed in both root and shoot samples. The three GO terms enriched only in M4 were "response to inorganic substance" (GO: 0010035), "response to metal ion" (GO: 0010038), and "response to cadmium ion" (GO: 0046686) (Fig. 4c; Additional file 5: Table S3C). The module M6 contained genes in response to osmotic stress (GO: 0006970), especially salt stress (GO: 0009651). This meant that the enriched GO terms of M6 mostly overlapped with those enriched in the other salt-specific modules. Genes in the M2 module were more involved in the processes of localization (GO: 0051179), cellular component biogenesis (GO: 0044085), macromolecule biosynthesis (GO: 0009059), and vitamin biosynthesis (GO: 0009110). A large portion of genes in the M12 module were expressed in response to wounding (GO: 0009611), biotic stimulus (GO: 0009607), organic acid metabolism (GO: 0006082), and cellular amino acid metabolism (GO: 0006520). The M18 module contains genes related to reproduction (GO: 00000003), fruit and seed development (GO: 0010154 and GO: 0048316), and embryonic/post-embryonic development (GO: 0009790 and GO: 0009791).

Although the stress-specific modules were composed of different genes, they eventually enriched the same 22

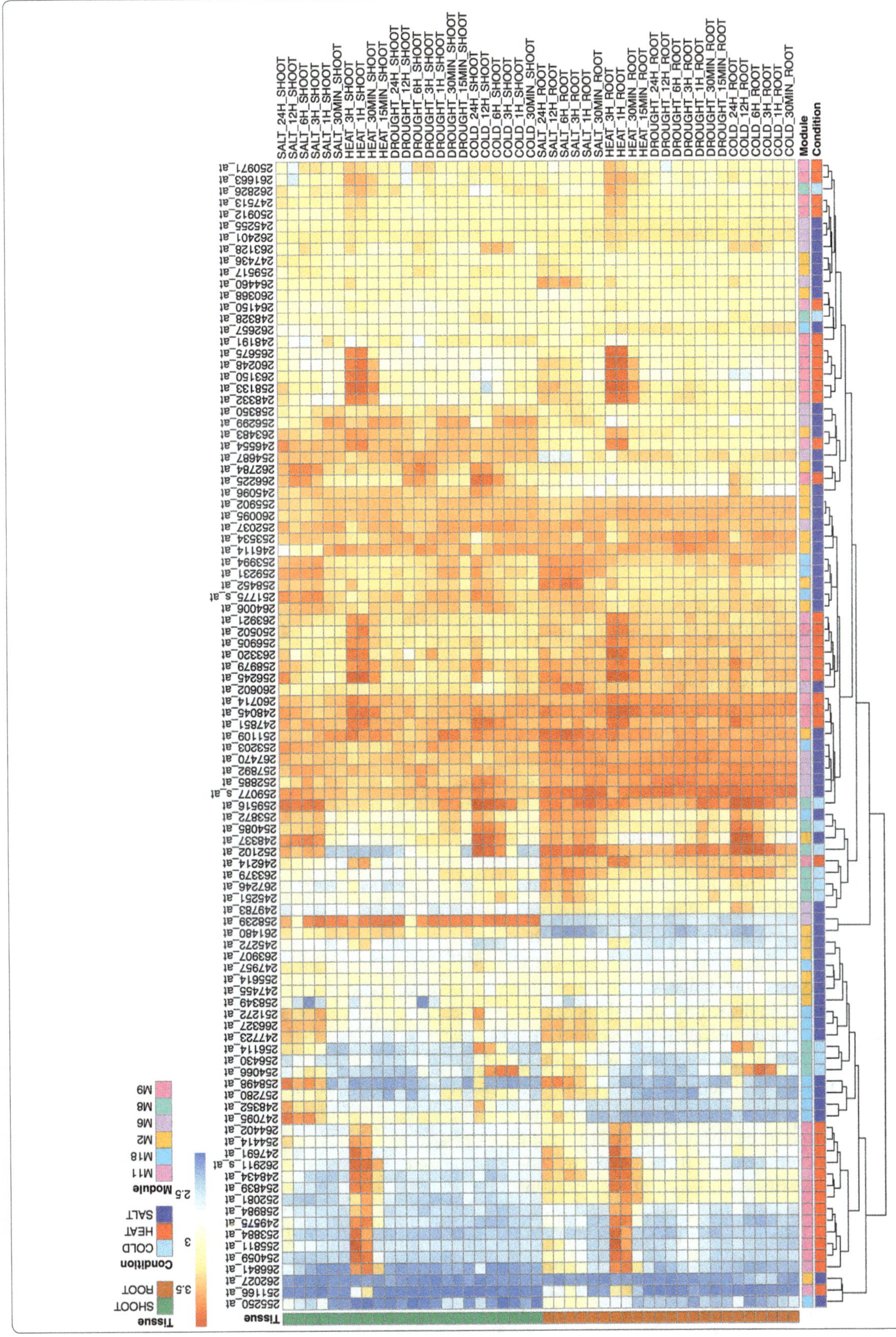

Fig. 5 A heatmap of the enriched GO term, "response to abiotic stimulus" (GO: 0009628). This heatmap was produced using the 99 genes corresponding to the enriched GO term, "response to abiotic stimulus" (GO: 0009628) in the four stress conditions

GO terms (Fig. 4d), including "primary metabolic process" (GO: 0044238), "response to chemical stimulus" (GO: 0042221), "response to stimulus" (GO: 0050896), "response to temperature stimulus" (GO: 0009266), "response to stress" (GO: 0006950), "response to hormone stimulus" (GO: 0009725), "response to endogenous stimulus" (GO: 0009719), and "response to abiotic stimulus" (GO: 0009628) (Additional file 5: Table S3D). That means that different genes in the same GO may be disturbed in different ways under different conditions. For example, the expressions for the 99 genes corresponding to the enriched "response to abiotic stimulus" (GO: 0009628) in different modules presented tissue-specific and/or stress-specific patterns (Fig. 5).

The inferred stress-specific gene regulatory events were partly supported by experimental results through literature search. For instance, we compared the AGI locus identifiers of the 98 genes in the cold-specific gene module M8 to those known as cold-temperature responsive genes in the Gramene Pathway Browser (http://plantreactome.gramene.org/PathwayBrowser/#SPECIES=9079025&DIAGRAM=9085214&PATH=9085019,9085215&DTAB=MT) and only two genes, *DREB1A* (AT4G25480) and *DEAR1* (AT3G50260), were matched. They belong to CBF1 homolog and RAP2 homolog, respectively, in the Gramene pathway modules. *DREB1A* was the intermediate between the well-known cold induced transcription factor, ICE1, and the cold responsive genes, *RD29A* and *COR15A* (Yamaguchi-Shinozaki and Shinozaki 2006), while *DEAR1* would repress the expression of *RD29A* and *COR15A* under the cold stress (Tsutsui et al. 2009).

Conclusion

In this study, we identified abiotic stress-specific modules after conducting a weighted gene correlation network analysis (WGCNA) using the analysis of variance and biplot visualization. Our first step in doing so was to differentiate the relevant gene module(s) according to different categorical traits (the different stress conditions, in our case) by the analysis of variance, something which might not have been accomplished by an analysis of correlation. Furthermore, the geometric interpretation of a biplot aims to utilize mRNA levels to point out to plant physiologists a plausible direction for further in vitro validation of tissue-specific and/or stress-specific mechanisms. The readers need to be aware of the fact that the heatmaps and biplots presented in this study were constructed based on the expression levels of the "pseudo" Eigen genes, which represent the diverse gene expressions in the gene modules specified by the WGCNA. It is thus possible that none of the genes in a given module exactly matches the expression patterns of the Eigen gene. The results of the analyses only provide hints about the underlying biological processes, which need to be further confirmed, as for example, by the gene ontology analysis in this study. In conclusion, our approach has the potential to further elucidate stress-specific mechanisms in plants via meta-analysis of massive amounts of microarray data. It can be used to complement the conventional bioinformatics analyses associated with the studied phenotypes.

Abbreviations

GO: gene ontology; RNA-Seq: RNA-sequencing; WGCNA: the weighted gene co-expression network analysis; CID: the coefficient of intrinsic dependence; TAIR: The Arabidopsis Information Resource; RMA: robust multi-array average; ANOVA: analysis of variance; SVD: singular value decomposition; PC: principle component.

Authors' contributions

PCS performed most of the analysis, and drafted the manuscript under the supervision of LYDL. ALH performed the gene ontology analysis and validated the inferred gene regulatory events. LYDL initiated the research, revised the manuscript, and provided funding for the study. All authors read and approved the final manuscript.

Author details

[1] Biometrics Division, Department of Agronomy, National Taiwan University, Taipei, Taiwan. [2] Department of Life Science, Fu-Jen Catholic University, Xinbei, Taiwan.

Acknowledgements

We would like to thank Dr. Men-Chi Chang from Department of Agronomy at National Taiwan University for reviewing the paper and providing valuable comments regarding to this research.

Competing interests

The authors declare that they have no competing interests.

References

Albert R (2005) Scale-free networks in cell biology. J Cell Sci 118:4947–4957. doi:10.1242/jcs.02714

Atkinson NJ, Urwin PE (2012) The interaction of plant biotic and abiotic stresses: from genes to the field. J Exp Bot. doi:10.1093/jxb/ers100

Bolstad BM, Irizarry RA, Åstrand M, Speed TP (2003) A comparison of normalization methods for high density oligonucleotide array data based on variance and bias. Bioinformatics 19:185–193. doi:10.1093/bioinformatics/19.2.185

Borate BR, Chesler EJ, Langston MA et al (2009) Comparison of threshold selection methods for microarray gene co-expression matrices. BMC Res Notes 2:240. doi:10.1186/1756-0500-2-240

Campain A, Yang YH (2010) Comparison study of microarray meta-analysis methods. BMC Bioinform 11:408. doi:10.1186/1471-2105-11-408

Du Z, Zhou X, Ling Y et al (2010) agriGO: a GO analysis toolkit for the agricultural community. Nucleic Acids Res 38:W64–W70. doi:10.1093/nar/gkq310

Gabriel KR (1971) The biplot graphic display of matrices with application to principal component analysis. Biometrika 58:453–467. doi:10.1093/biomet/58.3.453

Golub GH, Van Loan CF (1996) Matrix computations, 3rd edn. Johns Hopkins University Press, Baltimore

Hahn A, Kilian J, Mohrholz A et al (2013) Plant core environmental stress response genes are systemically coordinated during abiotic stresses. Int J Mol Sci 14:7617–7641. doi:10.3390/ijms14047617

Hotelling H (1933) Analysis of a complex of statistical variables into principal components. J Educ Psychol 24:417

Hsiao Y-C, Liu L-YD (2016) A stepwise approach of finding dependent variables via coefficient of intrinsic dependence. J Comput Biol 23:42–55. doi:10.1089/cmb.2015.0150

Huala E, Dickerman AW, Garcia-Hernandez M et al (2001) The Arabidopsis Information Resource (TAIR): a comprehensive database and web-based information retrieval, analysis, and visualization system for a model plant. Nucleic Acids Res 29:4

Irizarry RA, Bolstad BM, Collin F et al (2003a) Summaries of Affymetrix GeneChip probe level data. Nucleic Acids Res 31:e15

Irizarry RA, Hobbs B, Collin F et al (2003b) Exploration, normalization, and summaries of high density oligonucleotide array probe level data. Biostatistics 4:249–264. doi:10.1093/biostatistics/4.2.249

Kaever A, Landesfeind M, Feussner K et al (2014) Meta-analysis of pathway enrichment: combining independent and dependent omics data sets. PLoS ONE 9:e89297. doi:10.1371/journal.pone.0089297

Kaever A, Landesfeind M, Feussner K et al (2015) MarVis-pathway: integrative and exploratory pathway analysis of non-targeted metabolomics data. Metabolomics 11:764–777. doi:10.1007/s11306-014-0734-y

Langfelder P, Horvath S (2008) WGCNA: an R package for weighted correlation network analysis. BMC Bioinform 9:559. doi:10.1186/1471-2105-9-559

Liu L-Y (2005) Coefficient of intrinsic dependence: a new measure of association. Thesis

Liu L-Y, Tsai C-A (2013) Identifying gene set association enrichment using the coefficient of intrinsic dependence. PLoS ONE. doi:10.1371/journal.pone.0058851

Liu L-Y, Chen C-Y, Chen M-J et al (2009) Statistical identification of gene association by CID in application of constructing ER regulatory network. BMC Bioinform 10:85

Liu L-Y, Chang L-Y, Kuo W-H et al (2012) Major functional transcriptome of an inferred center regulator of an ER(−) breast cancer model system. Cancer Inform. doi:10.4137/cin.s8633

Mantione KJ, Kream RM, Kuzelova H et al (2014) Comparing bioinformatic gene expression profiling methods: microarray and RNA-Seq. Med Sci Monit Basic Res 20:138–141. doi:10.12659/MSMBR.892101

Nakashima K, Yamaguchi-Shinozaki K, Shinozaki K (2014) The transcriptional regulatory network in the drought response and its crosstalk in abiotic stress responses including drought, cold, and heat. Front Plant Sci. doi:10.3389/fpls.2014.00170

Pavlopoulos GA, Secrier M, Moschopoulos CN et al (2011) Using graph theory to analyze biological networks. BioData Min 4:1–27. doi:10.1186/1756-0381-4-10

Pearson K (1901) LIII. On lines and planes of closest fit to systems of points in space. Lond Edinb Dublin Philos Mag J Sci 2:559–572

Priest HD, Fox SE, Rowley ER et al (2014) Analysis of global gene expression in *Brachypodium distachyon* reveals extensive network plasticity in response to abiotic stress. PLoS ONE 9:e87499. doi:10.1371/journal.pone.0087499

Ramasamy A, Mondry A, Holmes CC, Altman DG (2008) Key issues in conducting a meta-analysis of gene expression microarray datasets. PLOS Med 5:e184. doi:10.1371/journal.pmed.0050184

Rasmussen S, Barah P, Suarez-Rodriguez MC et al (2013) Transcriptome responses to combinations of stresses in Arabidopsis. Plant Physiol 161:1783–1794. doi:10.1104/pp.112.210773

Rest JS, Wilkins O, Yuan W et al (2016) Meta-analysis and meta-regression of transcriptomic responses to water stress in Arabidopsis. Plant J Cell Mol Biol 85:548–560. doi:10.1111/tpj.13124

Ritchie ME, Phipson B, Wu D et al (2015) limma powers differential expression analyses for RNA-sequencing and microarray studies. Nucleic Acids Res 43:e47. doi:10.1093/nar/gkv007

Ronald P (2011) Plant genetics, sustainable agriculture and global food security. Genetics 188:11–20. doi:10.1534/genetics.111.128553

Sachs MM, Ho THD (1986) Alteration of gene expression during environmental stress in plants. Annu Rev Plant Physiol 37:363–376. doi:10.1146/annurev.pp.37.060186.002051

Song L, Langfelder P, Horvath S (2012) Comparison of co-expression measures: mutual information, correlation, and model based indices. BMC Bioinform 13:328. doi:10.1186/1471 2105 13 328

Todaka D, Nakashima K, Shinozaki K, Yamaguchi-Shinozaki K (2012) Toward understanding transcriptional regulatory networks in abiotic stress responses and tolerance in rice. Rice 5:6. doi:10.1186/1939-8433-5-6

Trewavas A (2003) Aspects of plant intelligence. Ann Bot 92:1–20. doi:10.1093/aob/mcg101

Tsutsui T, Kato W, Asada Y et al (2009) DEAR1, a transcriptional repressor of DREB protein that mediates plant defense and freezing stress responses in Arabidopsis. J Plant Res 122:633. doi:10.1007/s10265-009-0252-6

Yamaguchi-Shinozaki K, Shinozaki K (2006) Transcriptional regulatory networks in cellular responses and tolerance to dehydration and cold stresses. Annu Rev Plant Biol 57:781–803. doi:10.1146/annurev.arplant.57.032905.105444

Yan W, Kang MS (2002) GGE biplot analysis: a graphical tool for breeders, geneticists, and agronomists. CRC Press, Boca Raton

Zhang B, Horvath S (2005) A general framework for weighted gene co-expression network analysis. Stat Appl Genet Mol Biol 4:1–45

2

Characterization of LhSorTGA2, a novel TGA2-like protein that interacts with LhSorNPR1 in oriental hybrid lily Sorbonne

Le Wang[1,2], Zhihong Guo[1], Yubao Zhang[1], Yajun Wang[1], Li Wang[3], Guo Yang[1], Wenmei Li[1,2], Ruoyu Wang[1] and Zhongkui Xie[1*]

Abstract

Background: Non-expressor of pathogenesis-related genes 1 (NPR1) regulates expression of *pathogenesis-related (PR)* genes by interacting with TGA family proteins during systemic acquired resistance (SAR). However, no TGA-like proteins or their interacting partners have been characterized in lily.

Results: In the present study, LhSorTGA2, a novel TGA-like protein, was identified as an interacting partner of LhSorNPR1 (an NPR-like protein) by bimolecular fluorescence complementation (BIFC) and yeast two-hybrid assay (Y2H). Subcellular localization of GFP-tagged proteins targeted LhSorTGA2 to the nucleus, whereas GFP-labeled LhSorNPR1 was observed both in the nucleus and at the cytomembrane. Sequence alignment revealed that LhSorTGA2 was featured with a basic leucine zipper (bZIP) domain and two glutamine rich acid domains (QI and QII). Further phylogenetic analysis showed that TGA family proteins can be grouped into three subclades, within which LhSorTGA2 was clustered into subclade I, together with AtTGA2/5/6. Expression of *LhSorTGA2* was investigated in different tissues by qPCR, and the highest expression level was observed in stem. Besides, when treated with phytohormones (SA, MeJA, ETH and ABA) or fungal pathogen *Botrytis elliptica*, *LhSorTGA2* expression was also induced at different time points post treatments.

Conclusions: Collectively, these results suggested that LhSorTGA2 was an interacting partner of LhSorNPR1, which might function in regulating expression of *PR* genes in lily during SAR.

Keywords: Non-expressor of pathogenesis-related genes 1, TGA-like proteins, Protein interaction, Gene expression, Subcellular localization, *Lilium* spp.

Background

Infections at local sites often induce resistance against further invasions of pathogenic organisms in the distal uninfected parts of plants. This primed resistance, which protects plants from further invasion of a broad spectrum of pathogens at the whole-plant level, is referred to as systemic acquired resistance (SAR) (Fu and Dong 2013). The establishment of SAR is accompanied by accumulation of salicylic acid (SA) (Gao et al. 2015), and induction of *pathogenesis-related (PR)* genes. *PRs* encode for small proteins, of which some are endowed with direct anti-microbial activities. Since SA level, PR accumulation, and SAR are tightly linked, SAR-conferred resistance could be severely compromised either by interfering the level of SA synthesis or by disrupting signaling pathways control *PR* gene expression (Gaffney et al. 1993). Mutation of the *non-expressor of PR genes 1 (NPR1)* gene, a central node in SA-mediated defense signaling, blocks the priming of SAR (Cao et al. 1997). NPR1 lacks a DNA binding domain itself, but features with an ankyrin repeat domain and a BTB/POZ domain, both of which mediate protein–protein interaction (Pieterse and Van Loon 2004). The ankyrin repeat at C-terminus interacts with TGA2

*Correspondence: wxhcas@lzb.ac.cn
[1] Gaolan Station of the Agricultural and Ecological Experiment, Northwest Institute of Eco-Environment and Resources, Chinese Academy of Sciences, Lanzhou 730000, China
Full list of author information is available at the end of the article

to activate *PR1* transcription, which is critical to plant defense against biotrophic pathogens in the state of SAR (Fan and Dong 2002; Zhang et al. 1999).

TGA proteins are members of the group D basic region/leucine zipper (bZIP) transcription factors (TFs) (Jakoby et al. 2002), which recognize the TGACG motif (also known as *activation sequence 1; as-1*) in the promoter region (Gatz 2013). With the identification of as-1 elements in promoters of a number of genes, the crucial roles TGA TFs played in gene expression regulation were unveiled. Mutagenesis of *LS7* element, which contained a TGA-binding site, resulted in complete abolishment of *PR1* expression. On the contrary, mutation of *LS5*, another TGACG motif containing element, augmented *PR1* expression (Kesarwani et al. 2007). The opposite effects exerted by these two TGACG motifs to *PR1* expression implied that transcription of genes containing *as-1* elements in the adjacent promoter, might subject to complex regulation by different TGAs in the genome.

Lilies are herbaceous perennials that renowned for their elegant flowers. However, bulb and cut flower production in genus *Lilium*, are both under severe threat of pathogens in different types. For instance, lily bulbs can be infected by some soil-borne fungal pathogens (*Fusarium oxyporum*, *Penicillium albocoremium*, and *Penicillium tulipae*), causing rot of scales, bulb bases, and buds (Lecomte et al. 2016; Wang et al. 2017a). Furthermore, infection of air-borne fungi, such as *Botrytis cinerea* and *Botrytis elliptica*, could lead to fire blight disease, which is destructive to cut flower production (Huang et al. 2012; Wang et al. 2017d). In addition, multiplication of viruses (lily symptomless virus, cucumber mosaic virus, and lily mottle virus) also affects plant growth, resulting in leaf mottle and contortion, as well as stunted growth of lily plants (Zhang et al. 2016). To counteract pathogenic organisms, a comprehensive understanding of defense related genes that govern plant immunity in lily would be of great necessity.

In the present study, LhSorTGA2, a TGA2-like protein that interacted with LhSorNPR1 (Wang et al. 2017b) was characterized. Subcellular localization revealed that presence of GFP-tagged LhSorNPR1 was detected both in the nucleus and at the cytomembrane, but the GFP-tagged LhSorTGA2 was predominantly detected in the nucleus. *LhSorTGA2* expression in response to various phytohormone treatments and fungal pathogen *B. elliptica* infection was also explored.

Methods

Plant material and treatments

Bulbs (10–12 cm in diameter) of oriental hybrid cultivar 'Sorbonne' were planted in plastic pots containing peat moss as a growing substrate. All pots were then transferred to the growth chamber of Gaolan Station of Agricultural and Ecological Experiment, to keep plants under controlled conditions (16 h light/8 h dark; 22 °C). *Nicotiana benthamiana* plants were raised under the same conditions for observation of protein localization in planta.

Full-length cloning of *LhSorTGA2*

Total RNA was extracted from the leaves of 1-month-old seedlings using the RNAprep Pure Kit for plants (TIANGEN Corporation, Beijing, China). Partial coding sequence of *LhSorTGA2*, which was annotated in our previous RNA-seq data, was used to design gene specific primers (5RACE-1, 5RACE-2, 3RACE-1, 3RACE-2; Table 1) for rapid amplification of cDNA ends (RACE). 5′ RACE and 3′ RACE were conducted according to the instructions of the SMARTer® RACE 5′/3′ Kit (Takara, China). The cDNA full-length sequence of *LhSorTGA2* was obtained by assembling the 5′-RACE sequence, the partial coding sequence, and the 3′-RACE sequence.

In silico analysis of LhSorTGA2

The open reading frame (ORF) of the assembled *LhSorTGA2* full-length cDNA was predicted by ORF finder. The theoretical isoelectric point and molecular weight were calculated by using the Compute pI/Mw tool of ExPASy. Sequences of TGA proteins were aligned with the Clustal W program. Construction of phylogenetic tree was carried out by Mega 5 software using the neighbor-joining method (Tamura et al. 2011).

Subcellular localization and protein–protein interaction validated by BIFC

The ORF of *LhSorTGA2* without stop codon was amplified using primers LhSorTGA2-sub-F and LhSorTGA2-sub-R, and inserted into *Spe*I digested pCAMBIA1302 vector to express LhSorTGA2 in frame with GFP at the N-terminus. For the vector construction of LhSorNPR1, the ORF of *LhSorNPR1* was amplified with primers LhSorNPR1-sub-F and LhSorNPR1-sub-R, and cloned into *Nco*I and *Spe*I digested pCAMBIA1302 vector after sequencing. The recombinant plasmids pCAMBIA1302-LhSorTGA2 and pCAMBIA1302-LhSorNPR1 were each introduced into *Agrobacterium tumefaciens* (strain GV3101) competent cells via the freeze-thaw method. GV3101 cells harbouring pCAMBIA1302-LhSorTGA2 or pCAMBIA1302-LhSorNPR1 construct were infiltrated into leaf epidermal cells of 4-week-old *N. benthamiana* plants using a 1-ml needleless syringe. The pCAMBIA1302 empty vector was used as the control for protein localization.

To validate the protein–protein interaction between LhSorTGA2 and LhSorNPR1, the bimolecular

Table 1 List of primers used in the experiment

Experiment	Primer name	Primer sequence (5′–3′)
LhSorTGA2 5′ RACE	LhSorTGA2-5RACE-1	CAGGTCAGCAGCATCCTCAATTCGG
	LhSorTGA2-5RACE-2	ATAGCCCCGAAGTCAGAAACCCGGG
LhSorTGA2 3′ RACE	LhSorTGA2-3RACE-1	GAACCTCCAGCAGTCCTCCCAGCAG
	LhSorTGA2-3RACE-2	GGGTCAGATGGCTATGGCGATGGGA
BIFC	LhSorTGA2-BIFC-F	CGTCTAGAATGCACTCCTCCCGGGTTTCT
	LhSorTGA2-BIFC-R	CGGGTACCTTCCCGTGGACGGGCTAGCCA
	LhSorNPR1-BIFC-F	CGGGGCCCATGGCCGACGCCGCCGAGTG
	LhSorNPR1-BIFC-R	CGGGATCCTCATTCTTCCATATCTACCAGACA
Y2H	LhSorTGA2-GADT7-F	CGGAATTCATGCACTCCTCCCGGGTTTCT
	LhSorTGA2-GADT7-R	CGGGATCCCTCATTCCCGTGGACGGGCTAG
	LhSorNPR1-GBKT7-F	CGCCATGGAGATGGCCGACGCCGCCGA
	LhSorNPR1-GBKT7-R	CGGGATCCCTCATTCTTCCATATCTACCAGACAG
Subcellular localization	LhSorNPR1-sub-F	ATATCCATGGTAATGGCCGACGCCGCCGA
	LhSorNPR1-sub-R	CGACTAGTTTCTTCCATATCTACCAGACAGACA
	LhSorTGA2-sub-F	CGACTAGTATGCACTCCTCCCGGGTTTCTGA
	LhSorTGA2-sub-R	CGACTAGTTTCCCGTGGACGGGCTAGCCA
LhSorTGA2 qPCR	LhSor TGA2 qPCR-F	TCAGATGGCTATGGCGATGG
	LhSor TGA2 qPCR-R	CGGATTGGCGAGTAGTGAGT

Restriction sites for vector construction are underlined

fluorescence complementation (BiFC) assay was performed using vectors described previously. *LhSorNPR1* was constructed into the pSPYNE(R)173 vector to express LhSorNPR1 in fusion with the N-terminus of YFP (LhSorNPR1-YFP^{N}), and *LhSorTGA2* was inserted into the pSPYCE(M) vector to express LhSorTGA2 in fusion with the C-terminus of YFP (LhSorTGA2-YFP^{C}). The resulting recombinant plasmids, pSPYNE-LhSorNPR1 and pSPYCE-LhSorTGA2, were transformed into GV3101 and co-infiltrated into *N. benthamiana* leaf epidermal cells using a protocol described previously (Kerppola 2013). Fluorescence of YFP or GFP was observed 72 h post infiltration using a confocal laser scanning microscope (Leica TCS SP8, Leica Microsystems, Germany).

Yeast two-hybrid assay

To further confirm the protein interaction between LhSorTGA2 and LhSorNPR1, the ORF of *LhSorNPR1* was amplified (LhSorNPR1-GBKT7-F, LhSorNPR1-GBKT7-R; Table 1) and inserted into pGBKT7 vector to construct the pGBKT7-LhSorNPR1 bait plasmid. The prey plasmid, pGADT7-LhSorTGA2, was generated by cloning the *LhSorTGA2* ORF into the pGADT7 vector after amplification with primers (LhSorTGA2-GADT7-F, LhSorTGA2-GADT7-R; Table 1). Constructed vectors were co-transformed into AH109 competent yeast cells in pairs, and transformants were plated on DDO and QDO mediums to test protein interactions. Transformants were also streaked on QDO medium supplemented with 40 μg/ml 5-bromo-4-chloro-3-indoxyl-α-D-galactopyranoside (X-α-Gal) to further confirm interaction of different co-transformants.

Gene expression analysis by qPCR

For tissue-specific expression of *LhSorTGA2*, root, stem, leaf, petal, and scale were sampled from 2-month-old fully bloomed plants and subjected to total RNA isolation as described above. For *LhSorTGA2* expression in respond to hormone or pathogen treatment, lily seedlings of 1-month-old were subjected to 20 mM methyl jasmonate (MeJA), 10 mM sodium salicylate (SA), 5 mM ethephon (ETH), 2 μM abscisic acid (ABA) or *B. elliptica* (8×10^5 conidia/ml) inoculation treatments. Preparation of phytohormone solutions, the *B. elliptica* inoculum, and the corresponding mock treatments were performed according our previous report (Wang et al. 2017a). Leaves were sampled at 0, 2, 4, 8, 12, and 24 h post treatments, and froze immediately in liquid nitrogen and kept in a refrigerator at − 80 °C) until use.

After RNA extraction, 500 ng of total RNA was reversely transcribed to cDNA with the HiScript II Q RT SuperMix for qPCR kit (Vazyme Biotech, Nanjing, China). *LhSorTGA2* specific primers (LhSorTGA2 qPCR-F, LhSorTGA2 qPCR-R; Table 1) were used for qPCR amplification, and the housekeeping gene *polyubiquitin 4* was used as an internal reference (Wang et al. 2017c). All PCR reactions were conducted in a MX3000P qPCR thermocycler system (Stratagene Corp., USA). The following protocol was used for amplification: pre-denaturation for 10 min at 95 °C, followed by 40 cycles of 94 °C for 15 s, 55 °C for 15 s, and 72 °C for 30 s. The relative expression software tool (REST) was used for gene expression data analysis, and we applied P(H1) testing for statistical analysis. The Origin Pro 8.1 software was used for data graphing.

Results

Molecular cloning and sequence analysis of *LhSorTGA2*

The partial coding sequence of *LhSorTGA2* was retrieved from sequences annotated in our previous RNA-seq data. Gene specific primers were designed for 5′ RACE and 3′ RACE PCR to obtain the *LhSorTGA2* full-length sequence. DNA fragments amplified during RACE cloning were shown in Additional file 1: Figure S1. The *LhSorTGA2* cloned was 2010 bp in length, comprising a 5′ untranslated region (UTR) of 337 bp, a predicted open reading frame (ORF) of 1278 bp, and a 3′ UTR of 395 bp. The ORF encoded a protein of 425 amino acids with a theoretical molecular mass of 47.34 kDa and a predicted isoelectric point (pI) of 6.74. The *LhSorTGA2* full-length cDNA sequence has been submitted to GenBank and deposited under the Accession Number MF685037.

Alignment of LhSorTGA2 protein sequence to TGAs in *Arabidopsis thaliana* revealed that LhSorTGA2 was highly similar to members of the TGA family protein (Fig. 1). LhSorTGA2 contained the feature structures of bZIP transcription factors: a DNA binding domain (DBD) and a leucine zipper that enables protein dimerization. Within the DBD, glutamine, alanine, and serine residues, which were highly conserved for all TGA homologues were identified. Phosphorylation of the serine residue within the DBD led to DNA binding impediment, suggesting that phosphorylation might involve in the fine tune of transcription control of LhSorTGA2-regulated downstream genes (Kirchler et al. 2010). Leucine zippers of TGA proteins were formed by four heptads, within which three conserved leucine residues and one glycine residue were identified (Fig. 1). Two glutamine rich acid domains (QI and QII), which functioned in transcription activation, were also found in LhSorTGA2. The signature sequence for group D bZIP transcription factors, Yx_2RL[RQ]ALSS[LS]W, was identified at the C-terminus of all TGA homologues. A phylogenetic tree was built to analyze the evolutionary relationship of LhSorTGA2 with other TGA family proteins in plants. As shown in Fig. 2, TGA family proteins were subdivided into three clades, and LhSorTGA2 was most closely related to GhTGA2 in subclade I.

Subcellular localization of LhSorTGA2 and LhSorNPR1, and protein–protein interaction validated by BIFC

LhSorTGA2 and LhSorNPR1 (GenBank Accession: KY073343) were fused to the N-terminus of GFP respectively, and the fusion proteins were transiently expressed in *N. benthamiana* leaf epidermal cells under the control of CaMV 35S promoter to examine their subcellular localization. As shown in Fig. 3a, fluorescence of GFP-tagged LhSorTGA2 was predominantly detected in the nucleus. Different from LhSorTGA2, fluorescence of the LhSorNPR1-GFP infusion protein was observed both in the nucleus and at the cytomembrane. Strong green fluorescence was detected throughout the entire cell for the pCAMBIA1302-GFP control as expected. Our observation of subcellular localization of LhSorTGA2 and LhSorNPR1 was in agreement with their corresponding orthologues in *Gladiolus hybridus* from previous report (Zhong et al. 2015).

Protein–protein interaction between NPR1 and TGA2 was first identified in *Arabidopsis*. To test whether the interaction between LhSorNPR1 and LhSorTGA2 exists or not, we performed BIFC experiment by co-expressing LhSorNPR1-YFP^N and LhSorTGA2-YFP^C transiently in *N. benthamiana* leaf epidermal cells. As shown in Fig. 3b, strong fluorescence was detected in the LhSorNPR1-YFP^N and LhSorTGA2-YFP^C co-infiltration treatment, indicating reconstitution of YFP fluorophore by the interaction between LhSorNPR1 and LhSorTGA2. In contrast, co-infiltration of LhSorNPR1-YFP^N and YFP^C or YFP^N and LhSorTGA2-YFP^C, did not produce any detectable fluorescent signal. These findings indicate that LhSorNPR1 interacts with LhSorTGA2.

Confirmation of LhSorNPR1–LhSorTGA2 interaction by Y2H

To further confirm the interaction between LhSorTGA2 and LhSorNPR1, the prey plasmid pGADT7-LhSorTGA2 and the bait plasmid pGBKT7-LhSorNPR1 were co-transformed into AH109 competent yeast cells. Transformants were selected on DDO and QDO mediums, and protein–protein interaction between LhSorTGA2 and LhSorNPR1 was identified (Fig. 4a). In contrast, no positive interaction was detected in a series of co-transformants used as negative controls. All co-transformants were streaked on QDO medium supplemented with X-α-Gal chromogenic substrate, and blue colonies were only observed for the pGADT7-LhSorTGA2 and pGBKT7-LhSorNPR1 co-transformant and the positive control (pGADT7-T and pGBKT7-53 co-transformant) (Fig. 4b ,c).

Expression analysis of *LhSorTGA2* in different tissues and in response to pathogen infection and various hormone treatments

Expression of *LhSorTGA2* was detected in all the five tissues explored, but the transcript abundance varied from tissue to tissue (Fig. 5). The transcript levels of *LhSorTGA2* observed in leaf, petal, root, and scale, were

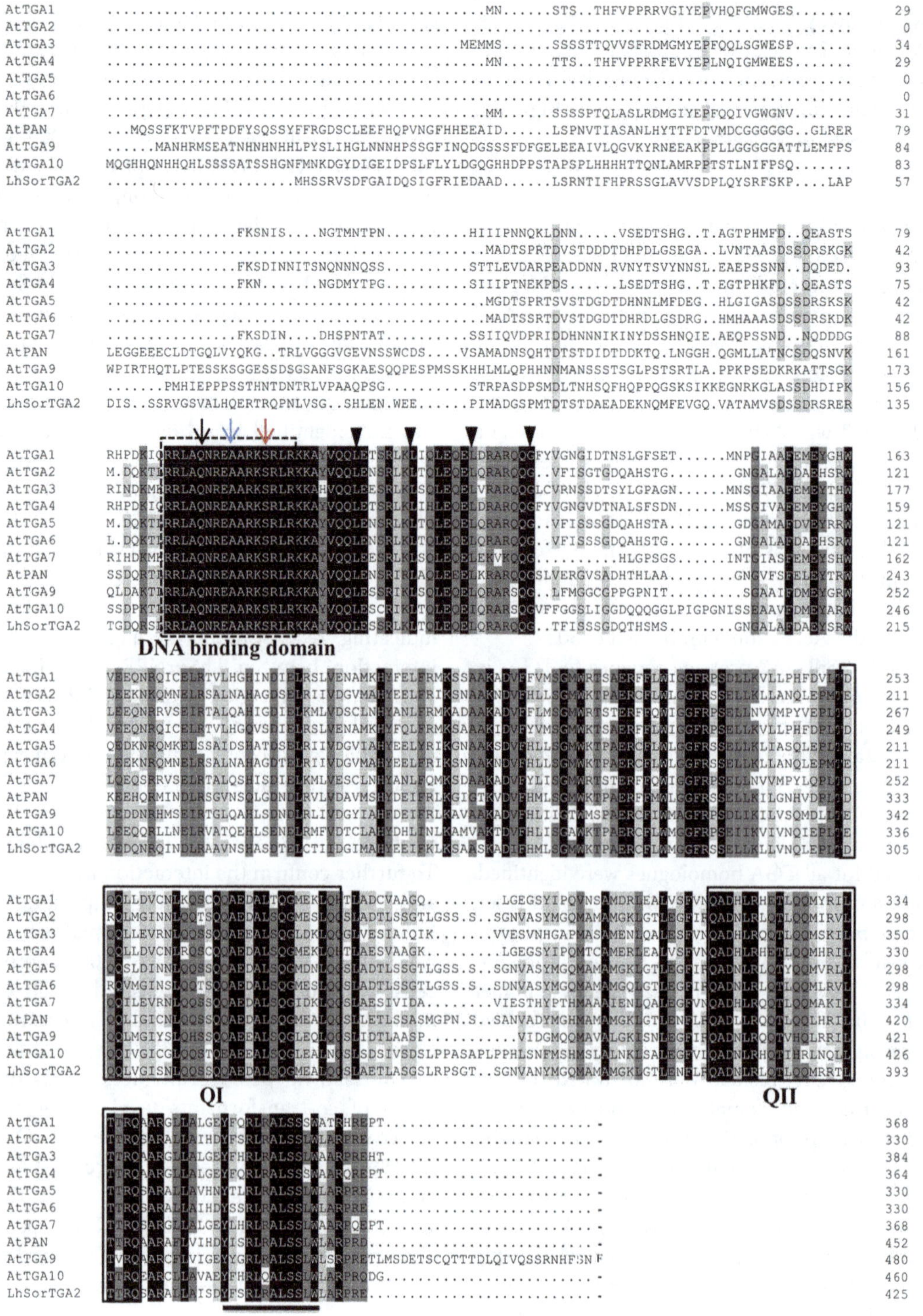

Fig. 1 Sequence alignment of LhSorTGA2 with proteins of the TGA family in *Arabidopsis*. Sequences were aligned using Clustal W. Identical residues are shaded in black, highly similar residues are shaded in dark gray, and similar residues are shaded in gray. The DNA binding domain (DBD) is indicated by a dashed line box. The conserved glutamine residues (black arrow), alanine residues (blue arrow) and serine residues (red arrow), are indicated by arrows. Three leucine residues and one glycine residue for the four heptad repeats in leucine zipper are highlighted by triangles. Two glutamine rich acid domains (QI and QII) that mediate transcription activation are highlighted by solid line boxes. The signature sequence for group D bZIP transcription factors, $Yx_2RL[RQ]ALSS[LS]W$, is underlined. The accession numbers of all the TGA proteins used for sequence alignment are listed in Additional file 2: Table S1

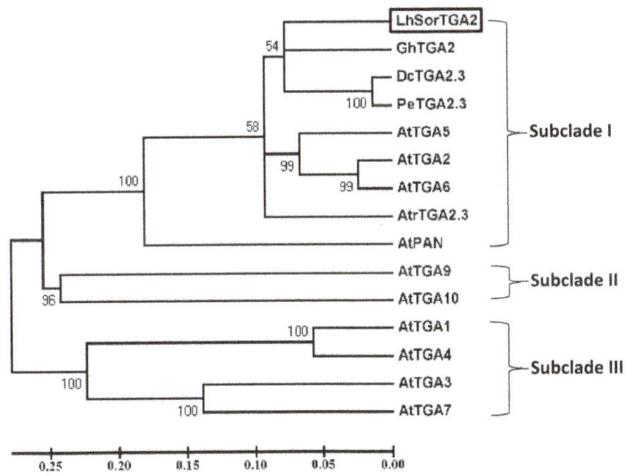

Fig. 2 Phylogenetic relationship of LhSorTGA2 with TGA family proteins from other plants. The phylogram was constructed using the neighbor-joining method in Mega 5, and tested with 1000 bootstrap iterations. The Accession Numbers of all the TGA proteins used for phylogram construction are listed in Additional file 2: Table S1

all lower than that in stem, making stem the most abundant tissue expressing *LhSorTGA2*.

To investigate *LhSorTGA2* expression in respond to pathogen infection or hormone treatments, transcript levels of *LhSorTGA2* were quantified after subjected to different treatments (MeJA, SA, ETH, ABA, and *B. elliptica* infection). As shown in Fig. 6a, when subjected to MeJA treatment, *LhSorTGA2* transcription was significantly induced, reaching its peak level 4 h after treatment. Foliar spray of SA also stimulated the accumulation of *LhSorTGA2*, resulting in a 4.15 times increase of *LhSorTGA2* transcript level 24 h after treatment (Fig. 6b). Following the ETH treatment, *LhSorTGA2* was up-regulated 4.02 times the control level (Fig. 6c). For the ABA treatment, *LhSorTGA2* expression was rapidly induced 2 h post treatment, reaching a level 2.17 times that of in the control. When exposed to pathogenic fungus *B. elliptica*, the expression of *LhSorTGA2* was also up-regulated, reaching its highest level 12 h after treatment (by a factor of 2.77 compared to the control). Upregulation of *LhSorTGA2* in respond to fungal infection and hormone treatments suggested that LhSorTGA2 might involve in diverse defense pathways regulated by multiple phytohormones.

Discussion

Transcriptional reprogramming during SAR is predominantly mediated by NPR1. The interactions between NPR1 and TGA family proteins modulate expression of *PR* genes during SAR. bZIP TFs in *Arabidopsis* are subdivided into 10 groups (A–I and S), among which group

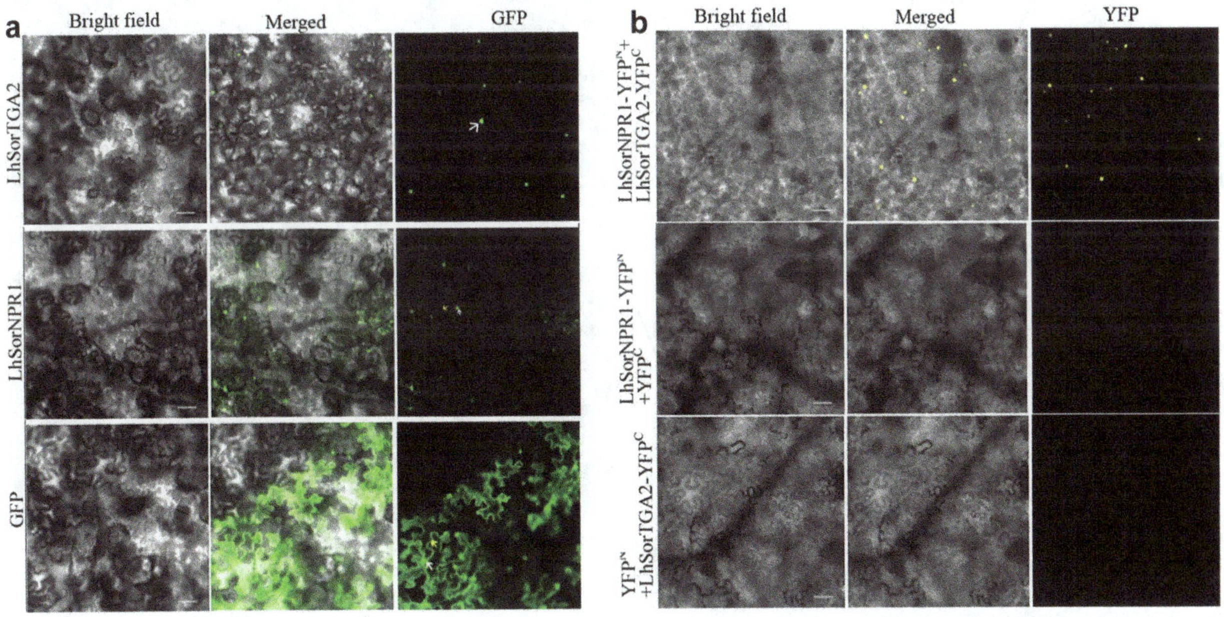

Fig. 3 Subcellular localization of LhSorTGA2 and LhSorNPR1, and their interaction validated by BIFC in planta. **a** Localization of LhSorTGA2 and LhSorNPR1 in *N. benthamiana* leaf epidermal cells. **b** BIFC analysis to validate the protein interaction between LhSorTGA2 and LhSorNPR1. LhSorNPR1 was fused to the N-terminal, while LhSorTGA2 was fused to the C-terminal half of YFP protein. Yellow fluorescence indicated protein interaction between LhSorTGA2 and LhSorNPR1 in the nuclei. Fluorescence was observed by confocal microscopy, and bright field and merged images were also shown for each observation. White arrows indicate nuclei of epidermal cells, and yellow arrows indicate the cytomembrane. Bar = 100 μm

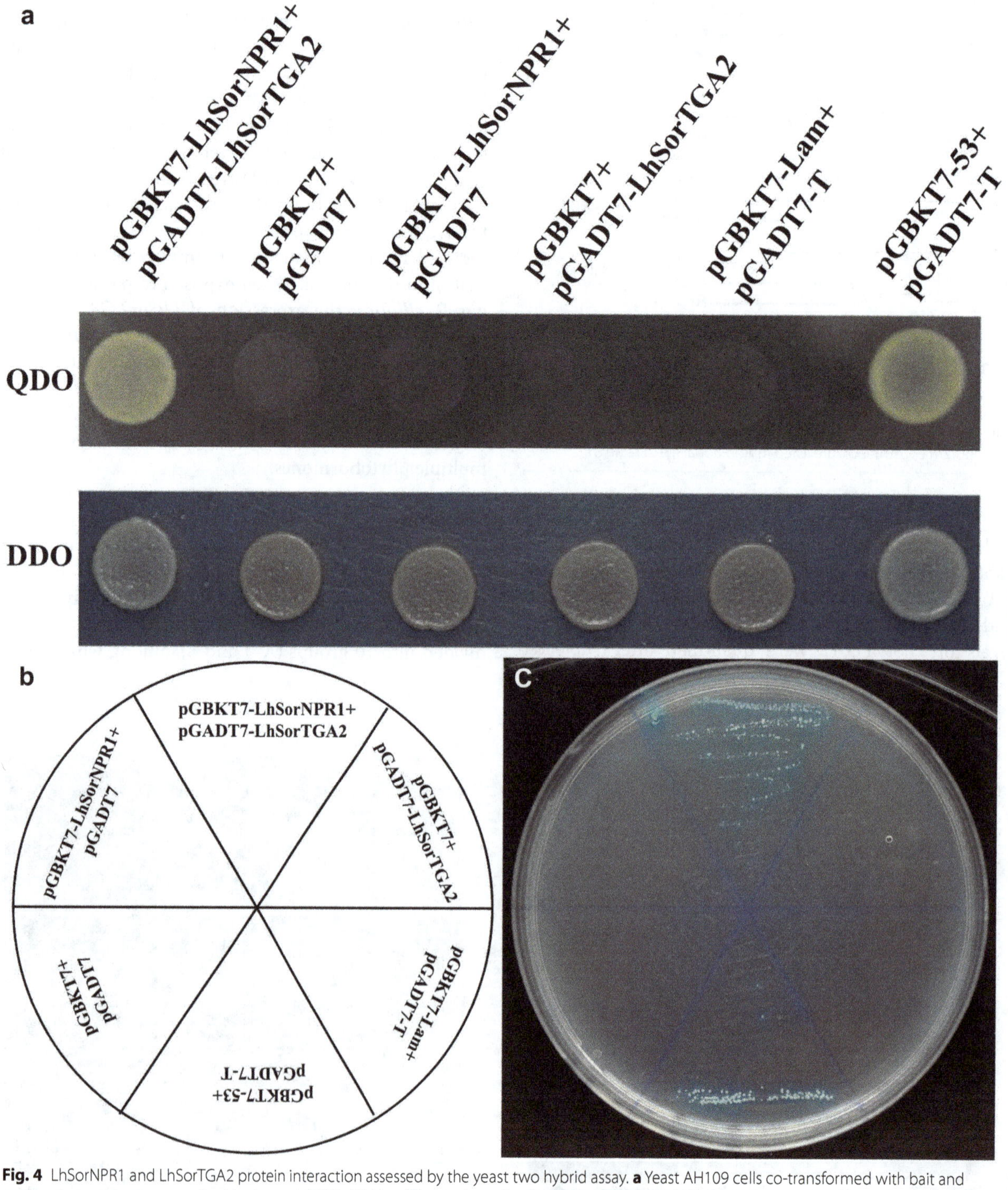

Fig. 4 LhSorNPR1 and LhSorTGA2 protein interaction assessed by the yeast two hybrid assay. **a** Yeast AH109 cells co-transformed with bait and prey vectors were grown on double dropout (DDO, SD/-Leu/-Trp) and quadruple dropout (QDO, SD/-Ade/-His/-Leu/-Trp) mediums. **b** The diagram indicates the corresponding co-transformants for the assay in **c**. **c** Transformants were streaked on QDO medium supplemented with 40 μg/ml X-α-Gal. pGBKT7-53 and pGADT7-T co-transformant was used as the positive control. Yeast cells co-transformed with pGBKT7 and pGADT7, pGBKT7-LhSorNPR1 and pGADT7, pGBKT7 and pGADT7-LhSorTGA2, or pGBKT7-Lam and pGADT7-T were used as negative controls

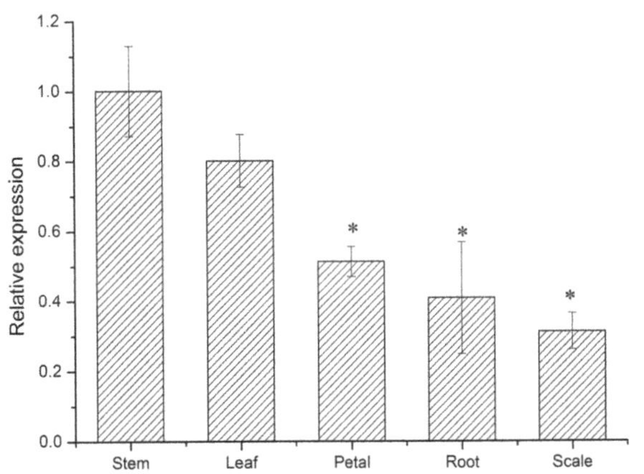

Fig. 5 Tissue-specific expression of *LhSorTGA2* in oriental hybrid lily cv. Sorbonne. The expression of *LhSorTGA2* in stem was used as calibrator (designated as 1.0) to determine the relative expression of the target gene in different tissues. Relative expression was calculated using REST 2009 software, and the values were presented as mean ± standard deviation (SD) for three replicates. Bars labeled with an asterisk indicate significant differences from the calibrator at $P < 0.05$ (the REST statistical randomization test)

D proteins are crucial participants in response to pathogen infection (Jakoby et al. 2002). Previous survey of the *Arabidopsis* genome identified 10 members of the group D bZIP TFs (TGA family proteins), of which seven have been confirmed to exist interaction with NPR1 (Kesarwani et al. 2007).

Lilies are one of the most important floriculture crops in cut flower market and are valued for their elegant flowers emitting fragrance. However, a lack of genome information hinders the genetic improvement targeting genes of interest, for instance, genes related to ornamental traits or disease resistance. Species in the genus *Lilium* have huge genomes (around 36 GB), which makes sequencing and assembly of their genomes a technical challenge (Du et al. 2017). In light of the constraint in availability of the genomic data, RNA-seq seems to be the most practical approach to profile the transcriptional change of genes of interest in lily. In the present study, the partial coding sequence of *LhSorTGA2* was retrieved from our previous transcriptome sequencing data, and the *LhSorTGA2* full-length cDNA was obtained by RACE. Subsequent in silicon analysis confirmed the homology of LhSorTGA2 to TGA-like proteins in other plants, such as AtTGA2 (72.23% similarity) in *Arabidopsis* and GhTGA2 (85.55% similarity) in *Gladiolus* hybrid cultivar. Sequence alignment of LhSorTGA2 with TGA proteins in *Arabidopsis* revealed the structural basis for TGA2-mediated transcription control. The leucine zipper domain determines the dimerization specificity of bZIP TFs (Vinson et al. 2006), forming homo- or heterodimeric coiled-coil structures (zipper). The contiguous basic region, which contains a predicted nuclear localization signal (amino acids 143–159 in LhSorTGA2), contacts the DNA. Whereas the glutamine rich acid domains (QI and QII) have been proven to transactivate *PR1* expression in a NPR1-dependent way during SAR (Fan and Dong 2002).

A phylogenetic tree was built to characterize the evolutionary relationship of TGA-like proteins originated from different species. LhSorTGA2 was grouped in subcladed I, together with AtTGA2, AtTGA5 and AtTGA6. These three closely related TGAs have been proposed as transcriptional repressors of *PR1*, which are required for the basal repression of *PR1*. However, in the state of SAR, their biological functions are both redundant and essential, since only the *tga2 tga5 tga6* triple mutant is blocked from induction of *PR1* during SAR (Zhang et al. 2003).

To further investigate the function of LhSorTGA2 and LhSorNPR1, subcellular localization of GFP-tagged proteins was examined using confocal microscopy. As mentioned above, a NLS was predicted in the middle of LhSorTGA2, which probably making it a nucleus-localized protein. Consistent with the prediction, GFP-tagged LhSorTGA2 was predominately detected in the nucleus. A NLS was also predicted in LhSorNPR1 at the C terminus, but presence of GFP-tagged LhSorNPR1 was detected both in the nucleus and at the cytomembrane. Consecutive expression of GFP-tagged AtNPR1 or TcNPR1 in *Arabidopsis* has proven that NPR1 localized in the cytoplasma if the seedlings were not treated with SA (Mou et al. 2003; Shi et al. 2010). However, under SA treatment (in the state of SAR), both AtNPR1-GFP and TcNPR1-GFP fusion proteins were translocated into the nucleus. Agroinfiltration causes immune response. Hence, SAR might be triggered during subcellular localization of GFP-labeled proteins using tabacco transient expression system. Our observation of LhSorNPR1-GFP in the nucleus was consistent with previous reports (GhNPR1 and VvNPR1), which might be caused by the translocation of LhSorNPR1-GFP into the nucleus in the state of SAR during agroinfiltration (Le Henanff et al. 2009; Zhong et al. 2015).

In *Arabidopsis*, interactions between NPR1 and TGA2 clade proteins modulated the expression of *PR1*. To validate the possible interaction between LhSorTGA2 and LhSorNPR1, vectors were constructed to test interaction in two different systems. As shown in Figs. 3 and 4, both BIFC and Y2H tests confirmed the interaction between LhSorTGA2 and LhSorNPR1. However, whether transcription of *PR* genes, especially *PR1*, is regulated by LhSorTGA2–LhSorNPR1 interaction in lily still needs to be confirmed.

Expression of *LhSorTGA2* was detected in all the five tissues (leaf, stem, root, petal and scale) explored, but

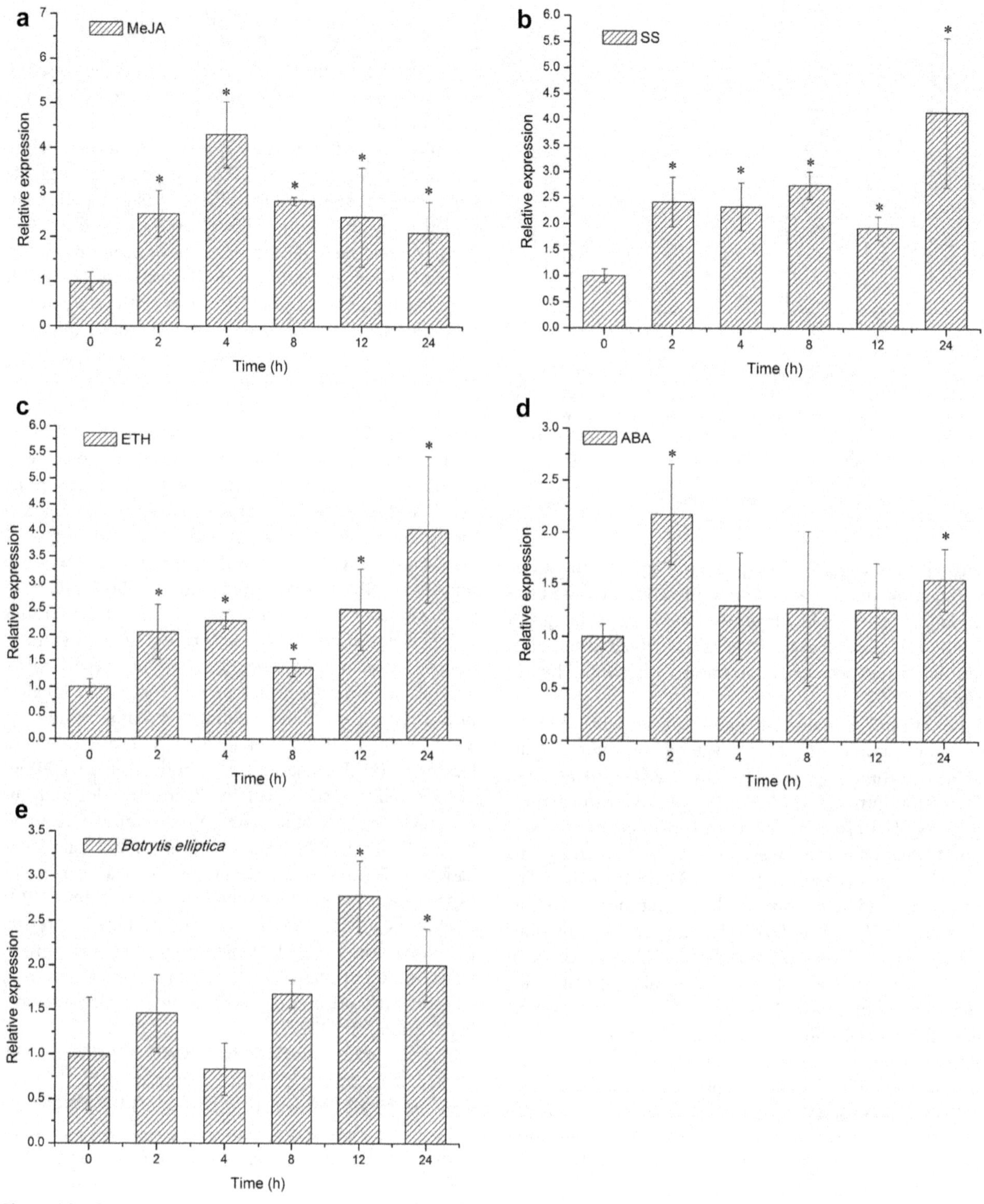

Fig. 6 *LhSorTGA2* expression in respond to **a** methyl jasmonate (MeJA), **b** sodium salicylate (SS), **c** ethephon (ETH), **d** abscisic acid (ABA) and **e** *Botrytis elliptica* inoculation treatments. Foliar spray of MeJA, SS, ETH, ABA, or Spores of *B. elliptica* (8×10^5 conidia/ml), was applied to 1-month-old lily seedling. Leaves were sampled 0, 2, 4, 8, 12, 24 h post treatments to explore *LhSorTGA2* expression. Expression at 0 h was used as the calibrator (designated as 1.0) to determine the relative expression of the target gene at different time points. Values are shown as mean ± standard deviations (SD) for three replicates. Bars labeled with an asterisk indicate significant differences from the control (0 h) at $P < 0.05$ (the REST statistical randomization test)

the transcription levels varied substantially (Fig. 5). TGA2 participated in SA-regulated SAR induction, and the expression of *TGA2* itself has been shown to be responsive to SA in *Arabidopsis* (Fan and Dong 2002). In our case, *LhSorTGA2* expression was responsive to SA, MeJA, ETH, ABA, and *B. elliptica* inoculation treatments, implying diverse roles it might play in defense pathways regulated by different hormones. Phytohormones are tightly linked to physiological processes in plant (Weyers and Paterson 2001). Induction of *LhSorTGA2* in respond to multiple hormones suggested that LhSorTGA2 might relate to physiological processes contributing to plant defense. Although there is still no direct evidence linking expression of *TGA2-like* genes to physiological changes under pathologic conditions, AtTGA10 and AtTGA9, two other members of the TGA family in *Arabidopsis*, has been shown to be involved in reactive oxygen species (ROS)-mediated responses to bacterial PAMP flg22 (Noshi et al. 2016). Accumulation of ROS is associated with hypersensitive response (HR) that restricts the spread of pathogens (Zurbriggen et al. 2010). However, whether TGA family proteins in lily modulate disease resistance through ROS accumulation is unknown. Among these TGA proteins, which one is the key regulator involved in ROS-mediated HR response to pathogen? Answers to these questions are mysteries awaited to be uncovered in future research.

Conclusions

LhSorTGA2, a novel TGA-like protein, was identified as an interacting partner of LhSorNPR1 in lily. Sequence alignment revealed that LhSorTGA2 was a member of the group D bZIP transcription factors, which featured with a conserved bZIP domain and two glutamine rich acid domains. *LhSorTGA2* was differentially expressed in various tissues, and its expression was responsive to SA, MeJA, ETH, ABA, and *B. elliptica* inoculation treatments. However, whether transcription of *PR* genes is modulated by the interaction between LhSorTGA2 and LhSorNPR1 still needs to be tested.

Abbreviations

SAR: systemic acquird resistance; NPR1: non-expressor of pathogenesis-related genes 1; bZIP: basic leucine-zipper; RACE: rapid amplification of cDNA ends; BIFC: bimolecular fluorescence complementation; Y2H: yeast two-hybrid; QDO: SD/-Ade/-His/-Leu/-Trp; DDO: SD/-Leu/-Trp; ABA: abscisic acid; SA: salicylic acid; SS: sodium salicylate; MeJA: methyl jasmonate; ETH: ethephon.

Authors' contributions

ZX and YW conceived and designed the experiments; LW and ZG performed the experiments; YZ, LW and WL analyzed the data; LW, GY and RW wrote the manuscript. All authors read and approved the final manuscript.

Author details

[1] Gaolan Station of the Agricultural and Ecological Experiment, Northwest Institute of Eco-Environment and Resources, Chinese Academy of Sciences, Lanzhou 730000, China. [2] University of Chinese Academy of Sciences, Beijing 100049, China. [3] The Forest Tree Seedling Station of the Alxa League, Alxa League 750300, China.

Competing interests

The authors declare that they have no competing interests.

Funding

This work was supported by the Ningxia Agricultural Comprehensive Development Office (NTKJ2015-05-01), the National Natural Science Foundation of China (No. 31101190 and No. 31201651), the CAS Key Technology Talent Program (No. 2016-65), the "West Light" project of the Chinese Academy of Sciences in 2014, the China Postdoctoral Science Foundation project (No. 2015M580894), and the Hundred Talents Programs of Chinese Academy of Sciences (No. Y629721002 and No. 27Y127L41002).

References

Cao H, Glazebrook J, Clarke JD, Volko S, Dong XN (1997) The *Arabidopsis* NPR1 gene that controls systemic acquired resistance encodes a novel protein containing ankyrin repeats. Cell 88:57–63

Du YP, Bi Y, Zhang MF, Yang FP, Jia GX, Zhang XH (2017) Genome size diversity in *Lilium* (Liliaceae) is correlated with karyotype and environmental traits. Front Plant Sci 8:1303

Fan WH, Dong XN (2002) In vivo interaction between NPR1 and transcription factor TGA2 leads to salicylic acid-mediated gene activation in *Arabidopsis*. Plant Cell 14:1377–1389

Fu ZQ, Dong X (2013) Systemic acquired resistance: turning local infection into global defense. Annu Rev Plant Biol 64:839–863

Gaffney T, Friedrich L, Vernooij B, Negrotto D, Nye G, Uknes S, Ward E, Kessmann H, Ryals J (1993) Requirement of salicylic-acid for the induction of systemic acquired-resistance. Science 261:754–756

Gao QM, Zhu S, Kachroo P, Kachroo A (2015) Signal regulators of systemic acquired resistance. Front Plant Sci 6:228

Gatz C (2013) From pioneers to team players: TGA transcription factors provide a molecular link between different stress pathways. Mol Plant Microbe Interact 26:151–159

Huang CJ, Liu YH, Yang KH, Chen CY (2012) Physiological response of *Bacillus cereus* C1L-induced systemic resistance in lily against *Botrytis* leaf blight. Eur J Plant Pathol 134:1–12

Jakoby M, Weisshaar B, Droge-Laser W, Vicente-Carbajosa J, Tiedemann J, Kroj T, Parcy F (2002) bZIP transcription factors in *Arabidopsis*. Trends Plant Sci 7:106–111

Kerppola TK (2013) Bimolecular fluorescence complementation (BiFC) analysis of protein interactions in live cells. Cold Spring Harbor Protoc 2013:727–731

Kesarwani M, Yoo J, Dong X (2007) Genetic interactions of TGA transcription factors in the regulation of pathogenesis-related genes and disease resistance in *Arabidopsis*. Plant Physiol 144:336–346

Kirchler T, Briesemeister S, Singer M, Schutze K, Keinath M, Kohlbacher O, Vicente-Carbajosa J, Teige M, Harter K, Chaban C (2010) The role of phosphorylatable serine residues in the DNA-binding domain of *Arabidopsis* bZIP transcription factors. Eur J Cell Biol 89:175–183

Le Henanff G, Heitz T, Mestre P, Mutterer J, Walter B, Chong J (2009) Characterization of *Vitis vinifera* NPR1 homologs involved in the regulation of pathogenesis-related gene expression. BMC Plant Biol 9:54

Lecomte C, Alabouvette C, Edel-Hermann V, Robert F, Steinberg C (2016) Biological control of ornamental plant diseases caused by *Fusarium oxysporum*: a review. Biol Control 101:17–30

Mou Z, Fan WH, Dong XN (2003) Inducers of plant systemic acquired resistance regulate NPR1 function through redox changes. Cell 113:935–944

Noshi M, Mori D, Tanabe N, Maruta T, Shigeoka S (2016) *Arabidopsis* clade IV TGA transcription factors, TGA10 and TGA9, are involved in ROS-mediated responses to bacterial PAMP flg22. Plant Sci 252:12–21

Pieterse CM, Van Loon L (2004) NPR1: the spider in the web of induced resistance signaling pathways. Curr Opin Plant Biol 7:456–464

Shi Z, Maximova SN, Liu Y, Verica J, Guiltinan MJ (2010) Functional analysis of the *Theobroma cacao* NPR1 gene in *Arabidopsis*. BMC Plant Biol 10:248

Tamura K, Peterson D, Peterson N, Stecher G, Nei M, Kumar S (2011) MEGA5: molecular evolutionary genetics analysis using maximum likelihood, evolutionary distance, and maximum parsimony methods. Mol Biol Evol 28:2731–2739

Vinson C, Acharya A, Taparowsky EJ (2006) Deciphering B-ZIP transcription factor interactions in vitro and in vivo. BBA Gene Struct Expr 1759:4–12

Wang L, Li LJ, Zhang YB, Guo ZH, Wang YJ, Yang G, Wang L, Wang RY, Xie ZK (2017a) Characterization of LhSorPR5, a Thaumatin-like protein from the oriental hybrid lily Sorbonne. Int J Agric Biol 19:865–872

Wang L, Guo ZH, Zhang YB, Wang YJ, Yang G, Yang L, Wang L, Wang RY, Xie ZK (2017b) Overexpression of LhSorNPR1, a NPR1-like gene from the oriental hybrid lily 'Sorbonne', conferred enhanced resistance to *Pseudomonas syringae* pv. tomato DC3000 in *Arabidopsis*. Physiol Mol Biol. https://doi.org/10.1007/s12298-017-0466-3

Wang L, Guo ZH, Zhang YB, Wang YJ, Yang G, Yang L, Wang RY, Xie ZK (2017c) Characterization of LhSorP5CS, a gene catalyzing proline synthesis in oriental hybrid lily Sorbonne: molecular modelling and expression analysis. Bot Stud 58:10

Wang L, Guo ZH, Zhang YB, Wang YJ, Yang G, Yang L, Wang RY, Xie ZK (2017d) Isolation and characterization of two distinct class II PR4 genes from the oriental lily hybrid Sorbonne. Russ J Plant Physiol. 64:707–717

Weyers JDB, Paterson NW (2001) Plant hormones and the control of physiological processes. New Phytol 152:375–407

Zhang YL, Fan WH, Kinkema M, Li X, Dong XN (1999) Interaction of NPR1 with basic leucine zipper protein transcription factors that bind sequences required for salicylic acid induction of the PR-1 gene. Proc Natl Acad Sci USA 96:6523–6528

Zhang YL, Tessaro MJ, Lassner M, Li X (2003) Knockout analysis of *Arabidopsis* transcription factors TGA2, TGA5, and TGA6 reveals their redundant and essential roles in systemic acquired resistance. Plant Cell 15:2647–2653

Zhang YB, Wang YJ, Xie ZK, Wang RY, Guo ZH, van der Werf W, Wang L (2016) Purification and immuno-gold labeling of lily mottle virus from lily leaves. J Virol Methods 232:33–38

Zhong XH, Xi L, Lian QL, Luo X, Wu Z, Seng SS, Yuan X, Yi MF (2015) The NPR1 homolog GhNPR1 plays an important role in the defense response of Gladiolus hybridus. Plant Cell Rep 34:1063–1074

Zurbriggen MD, Carrillo N, Hajirezaei MR (2010) ROS signaling in the hypersensitive response: when, where and what for? Plant Signal Behav 5:393–396

3

Tobacco plants expressing the maize nitrate transporter ZmNrt2.1 exhibit altered responses of growth and gene expression to nitrate and calcium

Abubakar Ibrahim[1,2], Xiao-Lu Jin[1], Yu-Bin Zhang[1,3], Jessica Cruz[1], Paveena Vichyavichien[1], Nwadiuto Esiobu[1] and Xing-Hai Zhang[1*]

Abstract

Background: Nitrate uptake is a highly regulated process. Understanding the intricate interactions between nitrate availability and genetically-controlled nitrate acquisition and metabolism is essential for improving nitrogen use efficiency and increasing nitrate uptake capacity for plants grown in both nitrate-poor and nitrate-enriched environments. In this report, we introduced into tobacco (*Nicotiana tabacum*) the constitutively expressed maize high-affinity transporter ZmNrt2.1 gene that would bypass the tight control for the endogenous nitrate-responsive genes. By using calcium inhibitors and varying levels of NO_3^-, Ca^{2+} and K^+, we probed how the host plants were affected in their nitrate response.

Results: We found that the ZmNrt2.1-expressing plants had better root growth than the wild type plants when Ca^{2+} was deficient regardless of the nitrate levels. The growth restriction associated with Ca^{2+}-deficiency can be alleviated with a high level of K^+. Furthermore, the transgenic plants exhibited altered expression patterns of several endogenous, nitrate-responsive genes, including the high- and low-affinity nitrate transporters, the Bric-a-Brac/Tramtrack/Broad protein BT2 and the transcription factor TGA-binding protein TGA1, in responding to treatments of NO_3^-, K^+ or inhibitors for the calcium channel and the cytosolic Ca^{2+}-regulating phospholipase C, as compared to the wild type plants under the same treatments. Their expression was not only responsive to nitrate, but also affected by Ca^{2+}. There were also different patterns of gene expression between roots and shoots.

Conclusion: Our results demonstrate the ectopic effect of the maize nitrate transporter on the host plant's overall gene expression of nitrate sensing system, and further highlight the involvement of calcium in nitrate sensing in tobacco plants.

Keywords: Calcium signaling, Maize high-affinity nitrate transporter ZmNrt2.1, Nitrate sensing, Nitrate responsive genes, Potassium, Tobacco

Background

Nitrogen is an essential element for all forms of life. Nitrate (NO_3^-) is a major form of nitrogen in the environment taken up by plants (Guerrero et al. 1981). Because of high demand for nitrate by crop plants to achieve high yield, over 100 million tons of nitrogen-fertilizers are being applied annually to farm land worldwide in recent years (FAO 2015). Yet due to low affinity between nitrate and most types of soils, as much as 50–70% of the applied nitrogen may be leached from the soil (Peoples et al. 1995), causing pollution and eutrophication of natural water systems, especially near agricultural and urban areas (Unkles et al. 2004; Howarth and Marino 2006).

*Correspondence: xhzhang@fau.edu
[1] Department of Biological Sciences, Florida Atlantic University, Boca Raton, FL 33431, USA
Full list of author information is available at the end of the article

Better understanding of how plants respond to nitrate availability in soil and regulate nitrate uptake is important in finding suitable solutions to over-fertilization as well as remedy for nitrogen deficiency.

Plants take up NO_3^- from soil by transporting it across the plasma membrane of the root epidermal and cortical cells. This process is carried out by various nitrate transporters through a proton-coupled gradient $2H^+/1NO_3^-$ symport mechanism (Meharg and Blatt 1995; Crawford and Glass 1998). Once inside the cell, most nitrate is first reduced to nitrite by nitrate reductases, and then to ammonium by nitrite reductases. Ammonium is metabolized into glutamine, a universal precursor for many metabolic processes (Guerrero et al. 1981; Good et al. 2004; Wang et al. 2012). Plants have developed multiple and independently operating systems to adjust for changing nitrate availability in the environment. There are at least three known nitrate uptake systems: a constitutive high-affinity transport system (cHATS), a NO_3^- inducible high-affinity transport system (iHATS), and a low-affinity constitutively expressed transport system (LATS) (Crawford and Glass 1998; Forde 2000). Previous studies have shown that the LATS operates when NO_3^- levels are high (> 1 mM), whereas the cHATS and iHATS operate at lower NO_3^- levels (< 1 mM) (Crawford and Glass 1998; Quaggiotti et al. 2003; Katayama et al. 2009). Although the cHATS has a higher affinity for NO_3^-, the iHATS is more active with a V_{max} 30 times that of cHATS and is responsible for a greater NO_3^- uptake (Kronzucker et al. 1996; Crawford and Glass 1998). Upon exposed to nitrate, nitrogen starved plants increase nitrate uptake several folds largely due to the iHATS (Tsay et al. 2007).

In Arabidopsis, at least four families of nitrate-transporting proteins have been studied: low-affinity nitrate transporter 1/peptide transporter (NRT1 or NPF), high-affinity nitrate transporter 2 (NRT2), chloride channel and slow anion channel associated homologues (Forde 2000; Quaggiotti et al. 2003; Krapp et al. 2014). Members of NRT1 family have been shown to play a variety of important physiological roles such as sustaining NO_3^- nitrate uptake and influx, transporting plant hormones, regulating NRT2 gene expression and HATS activity, and nitrate and calcium sensing (Muños et al. 2004; Remans et al. 2006; Chiba et al. 2015; Undurraga et al. 2017). Interestingly, unlike NRT2, NRT1.1 gene is not induced by nitrate deficiency nor suppressed by reduced nitrogen metabolism (Lejay et al. 1999). But Loqué et al. (2003) reported repression of NTR1.1 gene expression by the reduced NO_3^- product NO_2^- in Arabidopsis roots.

The NRT2 family is a part of the iHATS responsible for the influx of nitrate into the plant cell when nitrate becomes available in the environment. For example, the Arabidopsis NRT2.1 is crucial for NO_3^- influx especially when nitrate is deficient (Wirth et al. 2007). Structurally, NRT2 nitrate transporters belong to the nitrate-nitrite transporter family, a part of the Major Facilitator Superfamily of membrane transporters (Pao et al. 1998). Thus, the NRT2 membrane topology is made of two sets of six transmembrane domains connected by a cytosolic loop (Pao et al. 1998; Forde 2000). For several plants investigated, the transport activity of NRT2 protein is shown to require the direct interaction with another protein, NAR2 (Quesada et al. 1994; Zhou et al. 2000; Orsel et al. 2006). For example, in Arabidopsis both the NRT2.1 and NAR2.1 proteins were shown to localize and interact at the plasma membrane in the roots, where the active nitrate transporter is presumably formed as a NRT2.1/NAR2.1 heterooligomer (Wirth et al. 2007; Yong et al. 2010). In a study carried out in barley, nitrate and nitrite transport activity was only observed when HvNRT2.1 and HvNAR2.3 genes were both injected into the *Xenopus* oocytes expression system, and no transport activity was detected when HvNRT2.1 gene was injected alone or coupled with HvNAR2.1 or HvNAR2.2 genes (Tong et al. 2005). In maize roots, the accessory protein ZmNAR2.1 (ZmNRT3.1A) was important for iHATS activity and appeared to form an oligomer with ZmNRT2.1 (Pii et al. 2016). Although the two-component system in several species tested seems to be specific for NRT and NAR proteins, the rice OsNAR2.1 had a broader range of interactions with OsNRT2.1, OsNRT2.2 and OsNRT2.3a (Yan et al. 2011). Furthermore, the NRT2 genes in grass species are distinct from the Arabidopsis NRT2 by the absence of gene family clusters in the genomes and the lack of an intron, suggesting that the divergence of the NRT2 family occurred after the evolutionary split between dicots and monocots (Plett et al. 2010).

The iHATS including NRT2 is part of nitrate sensing system tightly controlled to maintain nitrogen homeostasis, where its activity dramatically increases upon first provision of NO_3^- and is quickly repressed after NO_3^- exposure (Crawford and Glass 1998; Quaggiotti et al. 2003; Medici and Krouk 2014). Down-regulation occurs through mRNA stability and with influx of other nitrogen metabolites such as ammonium, glutamate, glutamine, asparagine, and arginine (Imsande and Touraine 1994; Forde and Clarkson 1999). Growth and development is another signal for NRT2 regulation. For example, in Arabidopsis NRT2.1 protein levels remain stable in older plants and are not affected by environmental cues such as nutrient availability or darkness, while in younger (8-day old) seedlings the amount of NRT2.1 protein is decreased after 24 h of darkness (Laugier et al. 2012). Another study showed that light, sucrose or nitrogen treatments strongly affect both NRT2.1 transcription and HATS activity in Arabidopsis, but NRT2.1 protein level remains largely steady in response to these treatments (Wirth et al. 2007). Yet a different study reported that cellular glucose elevates NRT2.1 protein levels and transport activity in Arabidopsis, independent of NRT2.1 transcription (de Jong et al. 2014). Furthermore, post-transcriptional control was reported to be important for NRT2.1/NAR2.1 transport system in Arabidopsis roots (Laugier et al. 2012) and NRT2.1-nitrate influx in *Nicotiana plumbaginifolia* (Fraisier et al. 2000). Regardless of the various points of view and some seemingly discrepancies in literature, it is clear that NRT2.1 is actively regulated at various levels of transcription and translation, and there is an intricate crosstalk between plant metabolism and nitrate gene expression throughout growth and development.

In plants, hundreds of genes, including the aforementioned NO_3^- uptake systems and nitrate transporters (NRT), respond to nitrate as a regulatory signal (Wang et al. 2004; Krapp et al. 2014; Medici and Krouk 2014). However, increasing evidence has shown that calcium is another essential player in the nitrate signaling network. For example, Ca^{2+} and calcium-binding proteins such as CIPKs are important in modulating NRT gene expression in response to cellular and environmental nitrate levels (Albrecht et al. 2001; Hu et al. 2009). The universal calcium-mobilizing second messenger, inositol 1,4,5-trisphosphate, is produced by phosphoinositide-specific phospholipase C (PLC) enzymes from hydrolyzing the highly phosphorylated lipid phosphatidylinositol 4,5-bisphosphate (Streb et al. 1983; Hunt et al. 2004). Changes in cellular Ca^{2+} levels through the actions of PLC and membrane-bound calcium-permeable channels can significantly affect the expression of nitrate-responding genes (Sakakibara et al. 1997; Riveras et al. 2015). Thus, inhibitors such as U73122 for PLC (Franklin-Tong et al. 1996) and La^{3+} for several calcium channels (White 2000) have become a useful tool for analyzing Ca^{2+}- or nitrate-responsive genes.

It is obvious that nitrate uptake and metabolism in plants is tightly regulated by various signals at different levels (Wang et al. 2012; Krapp et al. 2014; Medici and Krouk 2014). Studies of the high-affinity nitrate transporter NRT2, a major nitrate uptake avenue for plants, and other nitrate responsive and regulatory genes will help better understand the intricate interactions between nitrate availability in the environment and genetically-controlled nitrate acquisition and metabolism. This knowledge is needed for achieving high nitrogen use efficiency and high capacity of nitrate uptake for plants in both nitrate-poor and anthropologically nitrate-enriched environments, in order to aim for an optimal balance between fertilizer usage, plant productivity and environmental protection (Good et al. 2004). As part of the effort to investigate plant nitrate response and regulation, we introduced in tobacco plants a maize high-affinity transporter ZmNrt2.1 gene driven by the constitutive CaMV 35S promoter. Therefore, this ZmNrt2.1 transgene would bypass the tight transcriptional regulation exerted on host plant's endogenous nitrate-responsive genes. Thus, we hoped to use ZmNrt2.1, along with calcium inhibitors and varying availability of NO_3^-, Ca^{2+} and K^+, to probe changes of nitrate responses in the host plant. Our studies reveal the ectopic effect of ZmNrt2.1 on growth and nitrate sensing system involving several endogenous nitrate-responsive genes.

Materials and methods

Plant material

As described previously (Hill et al. 2016), plants of tobacco (*Nicotiana tabacum* L., cv. Havana petit) were grown in MS medium solidified with 2.7 g/L phytagel in a growth chamber at 24°C under 8 h of darkness and 16 h of cool white light of 150 μmol photons m^{-2} s^{-1}. The seedlings were used for various treatments as described below.

Construction of transformation vector expressing maize nitrate transporter

The coding sequence of the high-affinity nitrate transporter gene ZmNrt2.1 from maize (*Zea mays*, Genbank accession: NM_001111725; Quaggiotti et al. 2003), fused with the FLAG tag (DYKDDDDK) coding sequence, was synthesized by GenScript (USA). The 1.6-kb *Sma*I/*Sac*I-cut fragment was ligated into the *Sma*I/*Sac*I digested binary plasmid E1492 (Zhang et al. 2004) that is derived from pBI101 (Clontech, USA). The resultant plasmid p35S-ZmNrt (Additional file 1: Figure S1) was introduced to *Agrobacterium tumefaciens* strain EHA105.

Agrobacterium-mediated plant transformation

Plant transformation was carried out as previously described (Zhang et al. 2004; Tsai et al. 2010; Hill et al. 2016) using kanamycin (100 mg/L) as selection reagent. Eight independent, homozygous lines of transgenic plants were identified by screening T1 and T2 seeds germinated on kanamycin-containing medium. Based on the segregation ratios, plants with a single copy ZmNrt2.1 transgene were identified. Three independent lines of these plants were used for this study.

DNA, RNA and protein analysis

Genomic DNA was extracted from leaves of the wild type and transgenic plants using Plant DNeasy kit (Qiagen, USA). Polymerase chain reaction (PCR) was carried out to confirm the presence of ZmNrt2.1 in the transgenic plants.

Total RNA was extracted from shoots or roots of the wild type and transgenic plants in similar physiological conditions or under different treatments using Qiagen Plant RNeasy kit, including a DNase treatment to remove residual DNA. The first-strand cDNA was synthesized using M-MLV Reverse Transcriptase RNase H^- kit (Solis BioDyne, Estonia) with random primers and oligo dT_{18}. Quantitative real-time PCR (qPCR) using SYBR Green (Life Technologies, USA) was carried out to examine the transcription of the transgene ZmNrt2.1 and several endogenous genes with gene-specific primers. Tobacco elongation factor 1α and ribosomal large subunit protein L25 (rpL25) cDNAs were used as reference for qPCR, according to Schmidt and Delaney (2010). The information for all the primers used in this study is listed in Additional file 1: Table S1.

To detect the presence of the ZmNrt protein, proteins were extracted from same leaf area of the wild type and transgenic plants according to Zhang et al. (2011) and Hill et al. (2016). Western blot analysis was done as described previously (Zhang et al. 2011, 2014), using monoclonal FLAG tag antibody (Cell Signaling, USA).

qPCR analysis of tobacco endogenous nitrate-responsive genes

To examine changes in expression of the endogenous nitrate-responsive genes in the wild type and transgenic tobacco plants, the homologous genes in tobacco were identified by analyzing the *Arabidopsis* and *Nicotiana* sequence databases. Gene specific primers (Additional file 1: Table S1) were designed for qPCR analysis of tobacco Bric-a-Brac/Tramtrack/Broad gene family member BT2 (GenBank Accessions: XM_016603770 and XM_016606879), leucine-zipper transcription factor TGA-binding protein TGA1 (KJ808740 and M62855), NAR2.1 (XM_009799779 and XM_009612602), NRT1.1 (AB102805 to AB102808) and NRT2.1 (AJ557583 and AJ557584).

Nitrate and protein determination

NO_3^- concentrations were determined from an aqueous extraction of 0.2 g frozen leaves and roots using the method described in Liu et al. (2014), spectrophotometrically measured at 410 nm using $NaNO_3$ as standard. Protein extraction and measurement was done according to Hill et al. (2016).

Nutrient treatments

Seven-day-old seedlings grown on MS medium were transferred to the medium with various NO_3^-, Ca^{2+} and K^+ concentrations, and grown for 14 days in a growth chamber at 24 °C under 8 h of darkness and 16 h of light at 150 μmol photons m^{-2} s^{-1}. Root length and biomass were measured. Samples of roots and shoots were also collected for gene expression assays.

Plant nutrient solutions were purchased from LaMotte (USA). The medium nutrient compositions are listed in Additional file 1: Table S2.

Inhibitor treatments

Twenty-one-day-old seedlings of the wild type and ZmNrt2.1-expressing plants grown on MS medium were removed from the medium, gently rinsed with water and incubated with gentle shaking for 1 h in water, 5 mM $LaCl_3$, 10 μM U73122 or 10 μM U73343, followed by incubation for 1 h in either 5 mM KNO_3 or 5 mM KCl, according to Riveras et al. (2015). Roots and shoots were collected, used immediately for RNA extraction or flash frozen in liquid N_2 and stored at − 80 °C. $LaCl_3$, U73122 and U73343 were purchased from Sigma-Aldrich (USA). $LaCl_3$ was dissolved in water. U73122 and U73343 were dissolved in a minimal amount of dimethyl sulfoxide and diluted in water to the final concentration. All the solutions were adjusted to pH 6.5.

Statistical analysis

At least three replicates for each assay were performed. The differences were detected using one way ANOVA coupled with Games-Howell (equal variance not assumed) and the Least Significant Difference (LSD) methods (equal variance assumed) with the corresponding patterns of experimental data. Statistical analysis was performed using SPSS software (ver. 16.0). The significant level was set at $P < 0.05$.

Results

Generation of transgenic plants expressing the maize high-affinity nitrate transporter ZmNrt2.1

The maize high-affinity nitrate transporter ZmNrt2.1 is composed of 524 amino acids, and shares 69% residue identity and 82% similarity with the tobacco 530-amino acid NtNrt2.1 (Additional file 1: Figure S2). Kanamycin-resistant transgenic plants generated via *Agrobacterium*-mediated transformation were initially screened by PCR, an example of which is shown in Additional file 1: Figure S3. All the transgenic plants grew and reproduced normally, with no obvious phenotypic differences as compared to the untransformed wild type plants. Homozygous lines of transgenic plants were used in this study.

While RT-PCR analyses indicate an absence of ZmNrt2.1 transcription in the wild type plants, various levels of ZmNrt2.1 mRNA were detected in leaves and roots of the transgenic ZmN plants grown in the nutrient-rich MS medium (Additional file 1: Figures S4 and S5). As expected, the expression of the tobacco endogenous nitrate transporter genes NtNrt2.1 (high-affinity) and NtNrt1.1 (low-affinity) was also detected in the leaf and root of both the wild type plants and transgenic plants (Additional file 1: Figures S4 and S5). Likewise, Western blot analyses detected the presence of FLAG-fused ZmNrt2.1 protein in the transgenic plants but not in the wild type plants (Additional file 1: Figure S5). These results demonstrate that as expected, both the wild type and transgenic plants expressed the endogenous genes such as NtNrt2.1 and NtNrt1.1, and the transgenic ZmN plants also expressed the maize ZmNrt2.1. Three independent ZmN lines with a single transgene, along with the untransformed wild type plants, were studied in parallel. Similar patterns of changes and responses were observed in all these ZmNrt2.1-expressing transgenic plant lines examined. Here, we present the data on the plant line ZmN6.

Growth response to various nitrate treatments

For the growth analysis, our control ("normal") medium contained NO_3^-, Ca^{2+} and K^+ at 10, 5 and 5 mM, respectively (see medium compositions in Additional file 1: Table S2). The nitrate contents of leaves or roots were largely similar between the wild type plants and the transgenic plants (Additional file 1: Figure S6). This is consistent with previous observations of similar NO_3^- contents between the wild type and CaMV 35S-NpNrt2.1 transgenic *Nicotiana plumbaginifolia* plants regardless of NO_3^- levels supplied (Fraisier et al. 2000). When grown under relatively low nitrate (1 mM) but normal Ca^{2+} and K^+ conditions, both the wild type and ZmNrt2.1-expressing plants had similar primary root growth. However, under low levels (both 1 mM) of NO_3^- and Ca^{2+}, the transgenic ZmN6 plants grew longer primary roots than the wild type (Fig. 1A, B), which is consistent with previous reports that constitutively expressing the NRT2.1 gene is associated with increased growth (Fraisier et al. 2000; Laugier et al. 2012). On the other hand, both the wild type and ZmN6 plants grew longer primary roots under 10 mM NO_3^- with normal levels of Ca^{2+} (5 mM) and K^+ (5 mM) than when either Ca^{2+} or K^+ was low. However, under 10 mM nitrate, Ca^{2+}-restricted root growth can be compensated with a high level (50 mM) of K^+ for both plants, and ZmN6 plants grew significantly longer primary roots than the wild type (Fig. 1A, B and Additional file 1: Figure S7).

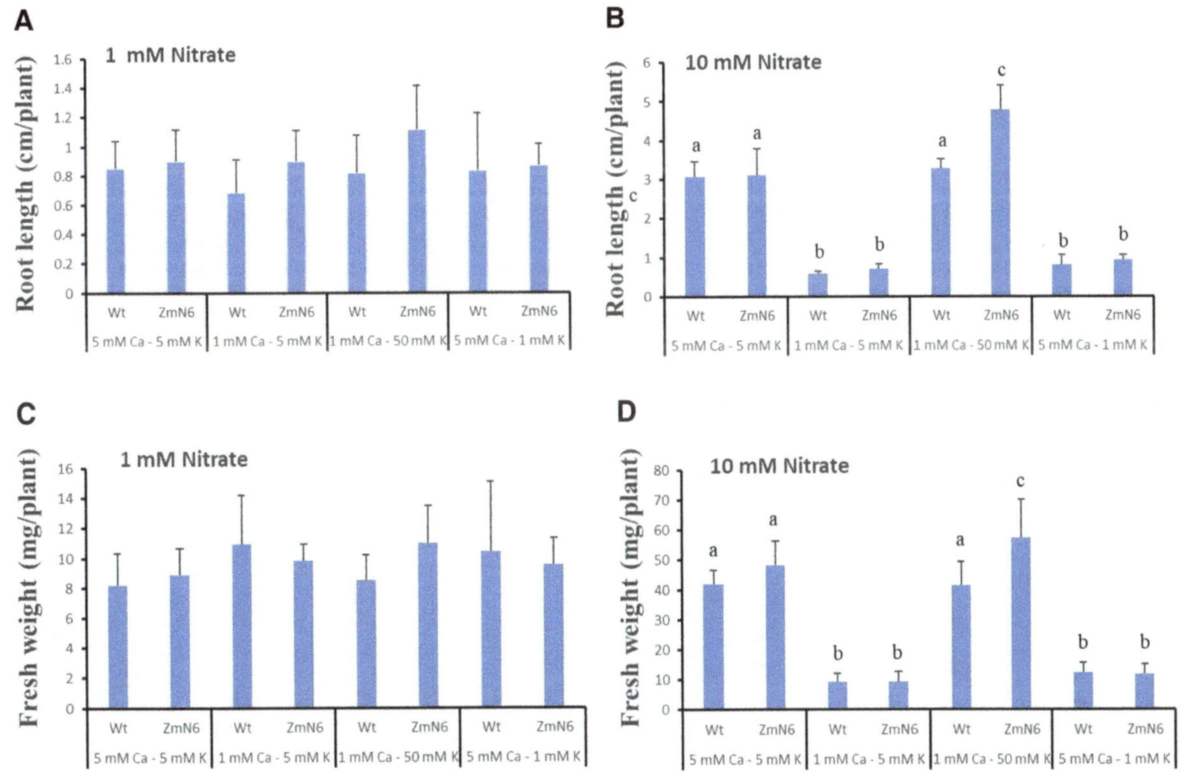

Fig. 1 Growth of 7 days-seedlings under 1 mM nitrate (**A, C**) or 10 mM nitrate (**B, D**) and various Ca^{2+} and K^+ concentrations for 14 days. Primary root length (**A, B**) and fresh biomass (**C, D**) of six to ten plants were measured. Wt: wild type, ZmN6: transgenic plant expressing ZmNrt2.1. Error bars indicate the standard deviation. Different letters indicate significant differences ($P < 0.05$) between plants-treatments of each panel, whereas no letter or same letters indicate insignificant differences

Likewise, under 1 mM nitrate, regardless of Ca^{2+} or K^+ levels, both the wild type and ZmN6 plants had similar biomass (Fig. 1C, D). As expected, under 10 mM nitrate both plants had higher biomass when Ca^{+2} and K^+ were sufficient (Fig. 1D). Again, high (50 mM) K^+ level was able to overcome Ca^{2+}-deficient (but not low NO_3^-) caused growth restriction for both the ZmN and the wild type plants, when compared to the growth under 5 mM K^+, and the ZmN plants grew better than the wild type (Fig. 1C, D and Additional file 1: Figure S7). Overall, these observations point to a role of Ca^{2+} and K^+ in nitrate response and also hint at altered nitrate sensing in the ZmN plants involving both Ca^{2+} and K^+, indicating a consequence of ectopic expression of the maize nitrate transporter gene. This prompted us to further examine expression in roots and shoots of several endogenous genes in response to NO_3^-, Ca^{2+} and K^+.

Altered responses in expression of the endogenous, nitrate-responsive genes in roots

Under 1-h KNO_3 treatment, the calcium channel blocker $LaCl_3$ had largely no effect on the transcription of the endogenous nitrate transporter genes NtNrt2.1 and NtNrt1.1 in the roots of the ZmN transgenic plants, but $LaCl_3$ inhibited these genes in the wild type plants (Fig. 2A, C). By contrast, no inhibitory effect of $LaCl_3$ was observed for both plants when nitrate was absent (Fig. 2B, D). On the other hand, the phospholipase C (PLC) inhibitor U73122 that blocks increases in the cytosolic Ca^{2+} levels had no effect on NtNrt2.1 gene expression regardless of whether nitrate was present, similar as its nonfunctional analog U73343 (Fig. 2A, B). For NtNrt1.1, U73122 and U73343 treatments exhibited largely similar expression levels (Fig. 2C, D), suggesting a lack of inhibition. In stark contrast to the NtNrt2.1 gene, the transcription of NtNAR, a protein functionally associated with NtNrt2.1, was inhibited by $LaCl_3$ in the transgenic ZmN roots, but not the wild type roots, in the presence of either KNO_3 or KCl. U73122, on the other hand, variously inhibited NtNAR genes for both plants (Additional file 1: Figure S7). These results suggest that (1) calcium, nitrate and potassium all influenced the expression of these genes, and (2) the ZmNrt2.1-expressing plants showed altered responses of the native genes NtNrt2.1 and NtNAR to changes in the calcium channels and cytosolic Ca^{2+} levels, as compared to the wild type plants.

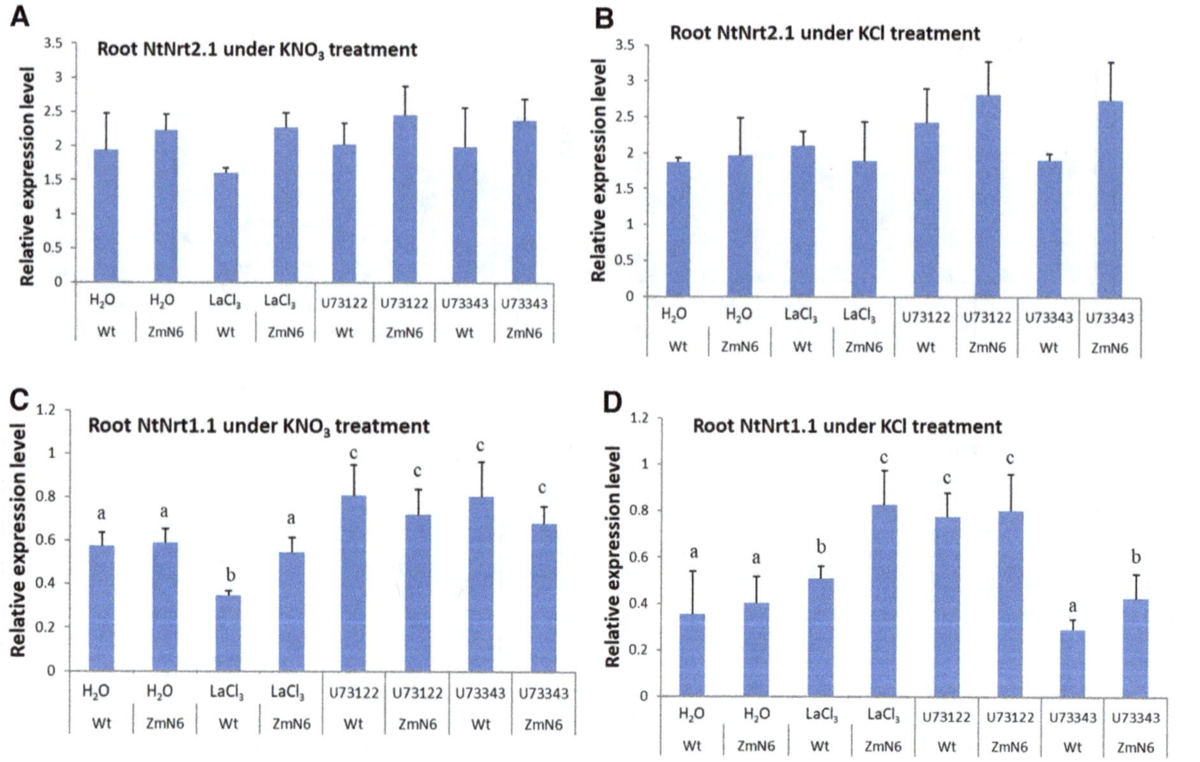

Fig. 2 Expression of root high-affinity (NtNrt2.1; **A, B**) and low-affinity (NtNrt1.1; **C, D**) nitrate transporter genes in response to calcium inhibitor and nitrate treatments. Three-week-old plants were incubated for 1 h in H_2O, 5 mM $LaCl_3$, 10 μM U73122 or 10 μM U73343, followed by 1 h in 5 mM KNO_3 or 5 mM KCl. The significant differences ($P < 0.05$) for multiple comparisons are shown with different letters separately for each panel, whereas no letter or same letters indicate insignificant variation

In Arabidopsis, a member of Bric-a-Brac/Tramtrack/Broad gene family, BT2, has been shown to be regulated by multiple signals including nitrogen nutrients and to suppress nitrate uptake (Mandadi et al. 2009; Araus et al. 2016). We examined the responses of the tobacco BT2 gene to calcium and nitrate. When treated with $LaCl_3$ or U73122 followed by KNO_3, the BT2 gene expression was elevated for both the wild type and ZmN plants (Fig. 3A), implicating a down-regulatory role by Ca channels and cytosolic $[Ca^{2+}]$. Under KCl without nitrate, there was also a significant increase in BT2 expression with $LaCl_3$ and U73122 treatments as compared to H_2O only treatment, but the wild type exhibited a stronger response than the ZmN plants (Fig. 3B). This indicates a potassium effect on BT2 genes when nitrate was absent and calcium signaling pathway was impaired, reflecting a complex crosstalk among NO_3^-, Ca^{2+} and K^+ in regulating genes for nitrate nutrition. Again, the ZmN transgenic plants exhibited altered responses different from the wild type plants.

Furthermore, the transcription factor TGA-binding protein, TGA1, is reported to be responsive to nitrate and may play a regulatory role in nitrate sensing (Ho and Tsay 2010; Alvarez et al. 2014). We found that the endogenous TGA1 expression in the ZmN roots was different from that in the wild type roots. Under KNO_3 treatment, TGA1 transcription in the ZmN plants was increased after incubation with $LaCl_3$ or U73122, whereas any enhancement of TGA1 expression was abolished by these inhibitors in the wild type plants (Fig. 3C). On the other hand, without NO_3^-, $LaCl_3$ and U73122 did not show inhibitory effect on TGA1 transcription in both plants (Fig. 3D). This suggests that while K^+ may not be directly involved in TGA1 gene expression, Ca^{2+} is important in mediating the nitrate-sensing TGA1 gene expression. What is revealing is that the ZmNrt2.1-expression plants under nitrate induction exhibited an altered pattern of TGA1 expression after treatment with Ca inhibitors $LaCl_3$ and U73122 (Fig. 3C), again highlighting the effect of the heterologous ZmNrt2.1 gene on the overall nitrate-related gene expression in the host plant.

Altered responses in expression of the endogenous, nitrate-responsive genes in shoots

Similarly, the shoots of the wild type and ZmN plants also exhibited some differences in gene expression as well

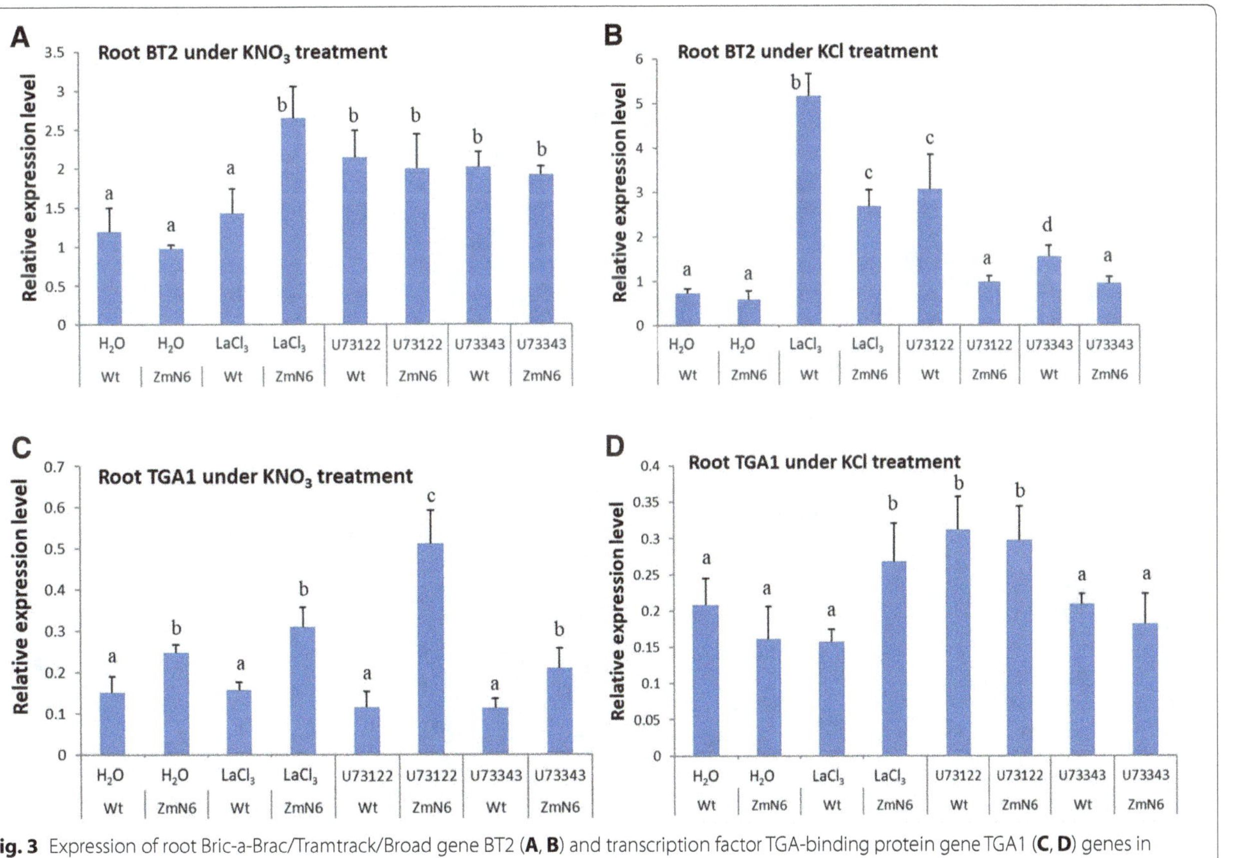

Fig. 3 Expression of root Bric-a-Brac/Tramtrack/Broad gene BT2 (**A**, **B**) and transcription factor TGA-binding protein gene TGA1 (**C**, **D**) genes in response to calcium inhibitor and nitrate treatments, as described in Fig. 2

as from the roots. In the presence of KNO_3, $LaCl_3$ did not inhibit NtNrt2.1 for both the wild type and ZmN plants (Fig. 4A), similar as in roots (Fig. 2A). U73122, however, almost completely abolished the NtNrt2.1transcription in the ZmN plants, whereas only partial inhibition was measured for the wild type (Fig. 4A). Without NO_3^-, $LaCl_3$ and U73122 almost completely inhibited NtNrt2.1 expression for both plants as compared to H_2O only treatment (Fig. 4B), which is in stark contrast to the roots where no obvious inhibition was observed (Fig. 2B). Likewise, the shoot NtNrt1.1 gene in both plants was completely inhibited by $LaCl_3$ or U73122, under either KNO_3 or KCl (Fig. 4C, D), suggesting that the calcium channel and PLC-regulated Ca^{2+} signaling play a significant role in the expression of nitrate transporter genes in shoots. These results also highlight root/shoot differences in nitrate transporter gene expression. This organ differential expression was observed for NtNAR genes as well, which in shoots exhibited almost identical patterns as NtNrt2.1 but was quite different from the roots (Fig. 4A, B; Additional file 1: Figure S8).

Unlike in roots, BT2 gene expression in shoots of the wild type plants under KNO_3 treatment was significantly inhibited by both $LaCl_3$ and U73122, whereas for the ZmN plants this inhibition was either not observed (for $LaCl_3$) or much less severe than the wild type (for U73122) (Fig. 5A). Without nitrate, on the other hand, BT2 expression was nearly completely abolished by these two inhibitors for both plants (Fig. 5B). Similarly, the TGA1 expression in shoots was significantly inhibited by both $LaCl_3$ and U73122 regardless of whether KNO_3 or KCl was present, although subtle difference can be detected between these two plants (Fig. 5C, D). All these results implicate the involvement of the Ca^{2+} signaling pathway in nitrate-related gene expression, thus nitrate sensing. Meanwhile, compared to the wild type plants, the ZmNrt2.1-expressing plants exhibited certain altered patterns of nitrate-related gene expression in response to nitrate availability and Ca^{2+} signaling inhibitors, reflecting the ectopic effect of the maize nitrate transporter on nitrate sensing in the transgenic plants.

Discussion

A main goal for this study was to investigate what would happen to the host plant's nitrate signaling if a foreign nitrate transporter gene, a member of the NO_3^- inducible high-affinity transport system (iHATS), was introduced and expressed. Our results show that the plants expressing the maize high-affinity nitrate transporter ZmNrt2.1 had an overall better primary root growth and higher biomass than the wild type when grown under high nitrate and K^+ but Ca^{2+} deficient conditions,

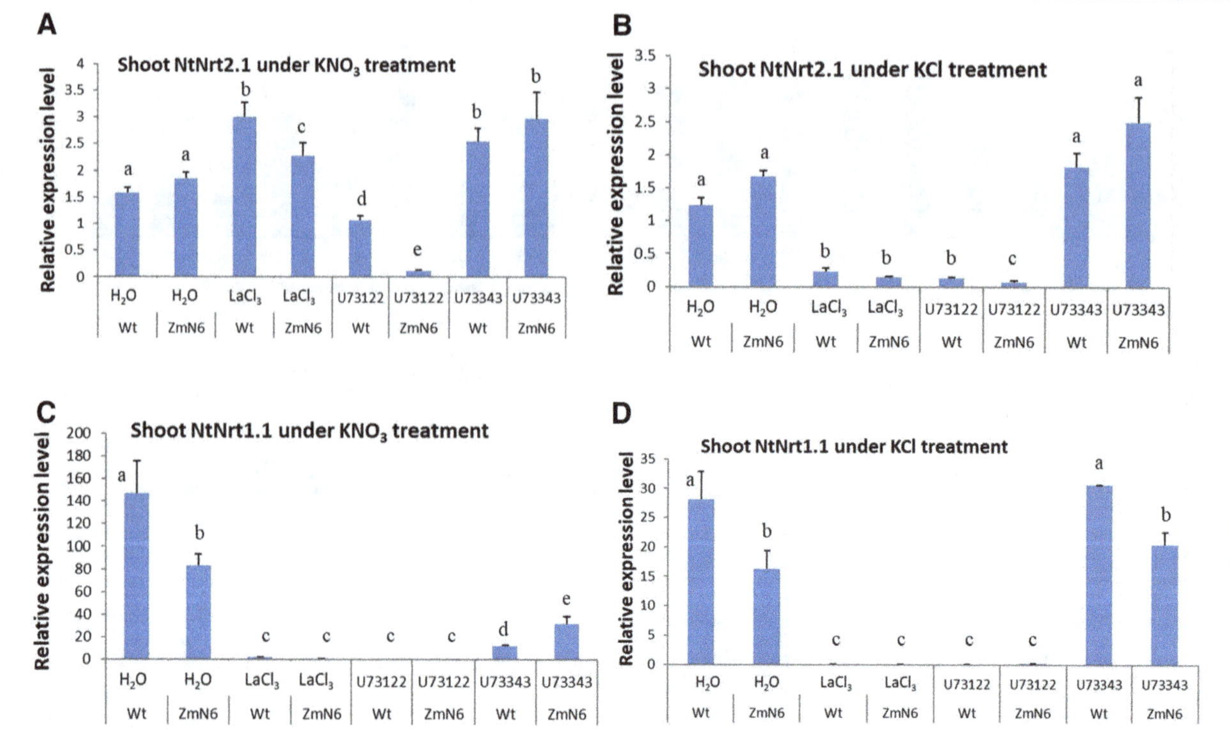

Fig. 4 Expression of shoot nitrate transporter genes NtNrt2.1 (**A, B**) and NtNrt1.1 (**C, D**) in response to calcium inhibitor and nitrate treatments, as described in Fig. 2

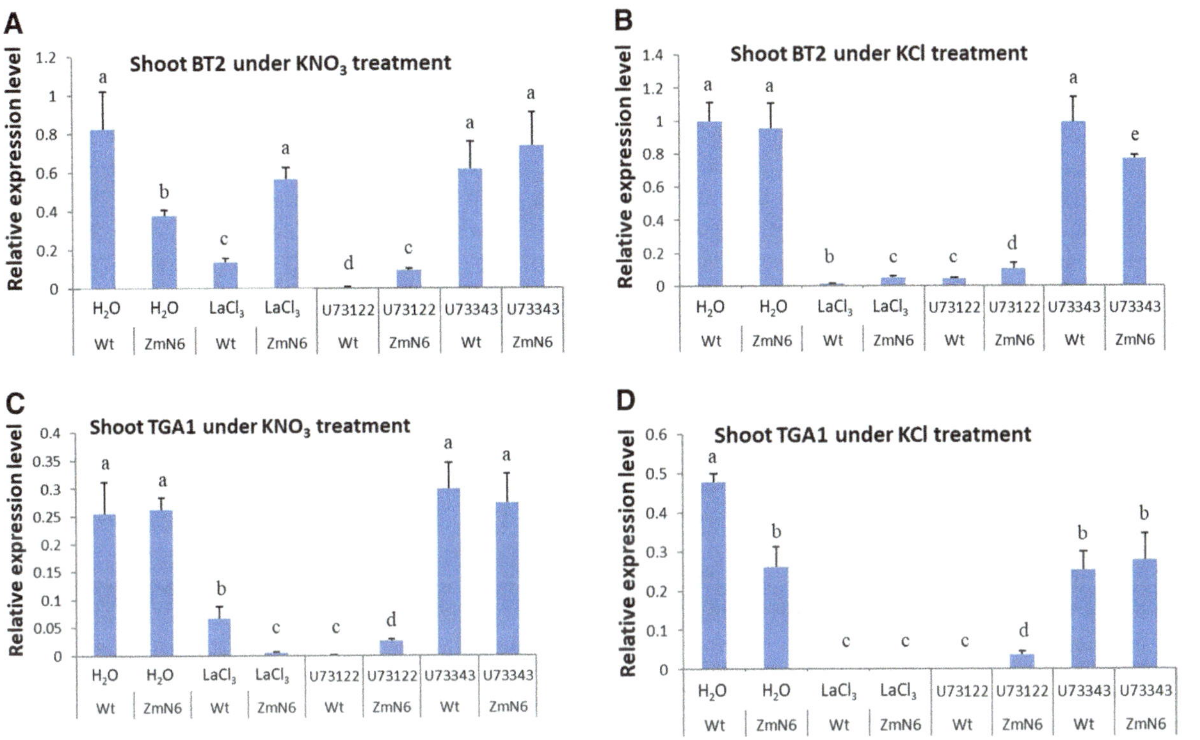

Fig. 5 Expression of shoot BT2 (**A**, **B**) and TGA1 (**C**, **D**) genes in response to calcium inhibitor and nitrate treatments, as described in Fig. 2

whereas there was no significant difference when nitrate was deficient (Fig. 1). One explanation may be that the additional nitrate transporter could somewhat enhance plants' capacity for nitrate uptake and assimilation, particularly under nitrogen eutrophication where plant's endogenous iHATS tends to be suppressed by high levels of nitrate. Plant growth responds to soil environmental cues such as the availability of minerals and nutrients. For example, root growth is generally enhanced by nutrient-poor conditions and inhibited by nutrient-rich conditions (Ericsson 1995). Root architecture is affected by nitrate transporters, such as the repression of lateral root initiation by NRT2.1 in Arabidopsis (Little et al. 2005). Therefore, lateral roots and primary roots may respond differently to NO_3^- ion as a signaling molecule (Zhang and Forde 2000; Little et al. 2005). We observed an enhanced primary root growth, particularly in our ZmN plants, when Ca^{2+} was low but K^+ was high (Fig. 1), suggesting that Ca^{2+} and K^+ may be just as important as nitrate for root development. On the other hand, other nitrate and nitrite transporters such as Nrt1.1 are shown to be important for K^+ translocation in Arabidopsis (Drechsler et al. 2015), reflecting an overlapping role for these genes and a complex network of $NO_3^-/Ca^{2+}/K^+$ interactions. Any phenotypic changes observed under different nitrate nutrient conditions must be the results of interactions among these nitrogen-responsive genes, including ZmNrt2.1 in the ZmN plants.

It is somewhat puzzling that overexpression of Nrt2.1 does not seem to significantly change NO_3^- contents in plants regardless of nitrate provision, as reported by Fraisier et al. (2000) and observed by us (Additional file 1: Figure S6). It is possible that in our study expression of ZmNrt2.1 may not be functionally compatible with the native NtNAR2, therefore limiting its nitrate transportation activity. However, Fraisier et al. (2000) overexpressed a copy of native NpNrt2.1 gene in *Nicotiana plumbaginifolia* plants, therefore Nrt2.1-NAR2 compatibility is not an issue. Lack of clear cause-effect association between Nrt2.1 expression and cellular NO_3^- contents under various NO_3^- availabilities reflects the complex regulation of nitrate homeostasis and nitrogen metabolism during plant growth and development.

It should be pointed out that our study only examined the young plants (up to 3 weeks old) grown in the nutrient-defined medium. It remains to assess how our transgenic plants would compare to the wild type plants when grown under nitrate-challenging (too much or too little) conditions in soil throughout a life cycle. Nonetheless, our study demonstrates a feasibility that elevated high-affinity nitrate transporter transcription that does not respond to the typical suppression by high nitrate levels could be used as a strategy for enhanced nitrate sequestration by plants from the nitrate-polluting environment.

It is also worth noting that in our study, and in literature (e.g. Fraisier et al. 2000; Quaggiotti et al. 2003; Okamoto et al. 2003; Wirth et al. 2007), 1 mM was used as "low" nitrate treatment, which appears to exceed the NRT2's operation limit of < 1 mM (Crawford and Glass 1998). In fact, due to plant growth, the nitrate content in the medium decreases rapidly. For example, in our experiments, the nitrate level on the 1 mM medium dropped below 0.5 mM after 1 week and was near 0.1–0.2 mM after 3 weeks (data not shown), with relatively healthy and reasonable-sized plants. Therefore, 1 mM nitrate should be an acceptable treatment for moderately low nitrate in nitrate transporter studies.

Calcium is a well-known important signal for a wide variety of physiological and biochemical processes in the cell (Dodd et al. 2010; Carafoli and Krebs 2016), as is an essential element for plant growth. However, its importance for nitrate signaling and nitrate-responsive gene regulation has not been fully understood or appreciated. Our results demonstrated a Ca^{2+} requirement for several nitrate-responsive genes, so that their transcription in shoots could be severely inhibited by inhibitors for calcium channel or Ca^{2+} level-regulating PLC, particularly when nitrate was limiting (Figs. 4 and 5). Thus, it is likely that Ca^{2+} and NO_3^-, perhaps K^+ as well, are all important part of the transcriptional regulation network for many nitrate-responsive genes. Overall, our studies are consistent with the notion that Ca^{2+} acts as a second messenger, beside NO_3^-, in the nitrate signaling pathway (Riveras et al. 2015; Undurraga et al. 2017). Regardless of the role of Ca^{2+} or K^+, nitrate availability seems to be a primary and essential signal for the cascade of gene expression for nitrogen nutrition in plants.

Using a combination of transgenic plants, inhibitors and nutrient variation, we were able to begin to dissecting, to a certain extent, the effect of a maize nitrate transporter gene on the expression of several native nitrate-responsive genes. The ZmNrt2.1 is moderately similar to the tobacco NtNrt2.1 with a similar codon usage (Additional file 1: Figure S2 and Table S3). It is not known how post-transcriptional and/or post-translational processing occurs on NRT2.1 protein in either maize or tobacco plants, as reported to take place for the Arabidopsis NRT2.1 (Wirth et al. 2007; Laugier et al. 2012). It is not clear how much, if any, ZmNrt2.1 could oligomerize with NtNAR2 to form an active nitrate transporter complex, as suggested for other plants (Tong et al. 2005; Yong et al. 2010; Pii et al. 2016). It will be important to investigate the genetic or biochemical mechanisms of how the ZmNrt2.1 interacts (or interferes) with the tobacco nitrate transporters and other native nitrate-involving genes/proteins in the host plant grown under various NO_3^- or K^+ conditions. However, regardless of whether ZmNrt2.1 was present as a monomer or functionally complexed with an accessory partner, this study clearly showed that compared to the untransformed plants, the ZmNrt2.1-expressing plants exhibited a different response in gene transcription when nitrate was limited or the Ca^{2+} signal system was impaired. For example, the fact that ZmN plants were less sensitive to $LaCl_3$ inhibition for BT2 and TGA1 transcription than the wild type (Figs. 3 and 5) suggests that being part of the host's genetic system, the maize ZmNrt2.1 somehow altered or interfered with the endogenous gene regulatory network for nitrogen nutrition in the host plant. Further exploration of this type of ectopic effects imposed by a foreign gene may be worthwhile not only for better understanding plants' nitrogen homeostasis, but also for creating desirable plants suitable for various nitrate environments.

In summary, the transgenic tobacco plants expressing the maize nitrate transporter gene ZmNrt2.1 exhibited altered expression patterns of a set of genes that are modulated not only by nitrate, but also by calcium. Introduction of selective foreign nitrate transporter genes into crop plants, along with managing Ca^{2+} or K^+ availability in soil, may prove to be a feasible approach to improving nitrogen use efficiency and nitrate uptake capacity, and dealing with environmental eutrophication caused by fertilizer overuse.

Authors' contributions

XHZ, AI and NE conceived and designed the experiments. AI, XLJ, JC, PV and XHZ performed the experiments. XLJ, YBZ and XHZ analyzed the data. XHZ wrote the manuscript. All commented on the manuscript. All authors read and approved the final manuscript.

Author details

[1] Department of Biological Sciences, Florida Atlantic University, Boca Raton, FL 33431, USA. [2] Department of Soil Sciences, Modibbo Adama University of Technology, Yola, Nigeria. [3] Monitoring Center for Marine Resources and Environments, Guangdong Ocean University, Zhanjiang 524088, China.

Acknowledgements

This work was supported in part by a fund from Modibbo Adama University of Technology, Nigeria, and Florida Atlantic University, USA.

Competing interests

The authors declare that they have no competing interests.

References

Albrecht V, Ritz O, Linder S, Harter K, Kudla J (2001) The NAF domain defines a novel protein-protein interaction module conserved in Ca^{2+}-regulated kinases. EMBO J 20:1051–1063

Alvarez JM, Riveras E, Vidal EA, Gras DE, Contreras-López O, Tamayo KP, Aceituno F, Gómez I, Ruffel S, Lejay L, Jordana X, Gutiérrez RA (2014) Systems approach identifies TGA1 and TGA4 transcription factors as important regulatory components of the nitrate response of *Arabidopsis thaliana* roots. Plant J 80:1–13

Araus V, Vidal EA, Puelma T, Alamos S, Mieulet D, Guiderdoni E, Gutiérrez RA (2016) Members of BTB gene family of scaffold proteins suppress nitrate uptake and nitrogen use efficiency. Plant Physiol 171:1523–1532

Carafoli E, Krebs J (2016) Why calcium? How calcium became the best communicator. J Biol Chem 291:20849–20857

Chiba Y, Shimizu Y, Miyakawa S, Kanno Y, Koshiba T, Kamiya Y, Seo M (2015) Identification of *Arabidopsis thaliana* NRT1/PTR FAMILY (NPF) proteins capable of transporting plant hormones. J Plant Res 128:679–686

Crawford NM, Glass AD (1998) Molecular and physiological aspects of nitrate uptake in plants. Trends Plant Sci 3:389–395

de Jong F, Thodey K, Lejay LV, Bevan MW (2014) Glucose elevates NITRATE TRANSPORTER2.1 protein levels and nitrate transport activity independently of its hexokinase1-mediated stimulation of *NITRATE TRANSPORTER2.1* expression. Plant Physiol 164:308–320

Dodd AN, Kudla J, Sanders D (2010) The language of calcium signaling. Annu Rev Plant Biol 61:593–620

Drechsler N, Zheng Y, Bohner A, Nobmann B, von Wirén N, Kunze R, Rausch C (2015) Nitrate-dependent control of shoot K homeostasis by the nitrate transporter1/peptide transporter family member NPF7.3/NRT1.5 and the Stelar K^+ Outward Rectifier SKOR in Arabidopsis. Plant Physiol 169:2832–2847

Ericsson T (1995) Growth and shoot:root ration of seedlings in relation to nutrient availability. Plant Soil 168–169:205–214

FAO, Food and Agriculture Organization of the United Nations (2015) World fertilizer trends and outlook to 2018. Rome. ISBN 978-92-5-108692-6

Forde BG (2000) Nitrate transporters in plants: structure, function and regulation. Biochim Biophys Acta 1465:219–235

Forde BG, Clarkson DT (1999) Nitrate and ammonium nutrition of plants: physiological and molecular perspectives. Adv Bot Res 30:1–90

Fraisier V, Gojon A, Tillard P, Daniel-Vedele F (2000) Constitutive expression of a putative high-affinity nitrate transporter in *Nicotiana plumbaginifolia*: evidence for post-transcriptional regulation by a reduced nitrogen source. Plant J 23:489–496

Franklin-Tong VE, Drobak BK, Allan AC, Watkins P, Trewavas AJ (1996) Growth of pollen tubes of *Papaver rhoeas* is regulated by a slow-moving calcium wave propagated by inositol 1,4,5-trisphosphate. Plant Cell 8:1305–1321

Good AG, Shrawart AK, Muench DG (2004) Can less yield more? Is reducing nutrient input into the environment compatible with maintaining crop production? Trends Plant Sci 9:597–605

Guerrero MG, Vega JM, Losada M (1981) The assimilatory nitrate-reducing system and its regulation. Annu Rev Plant Physiol 32:169–204

Hill W, Jin X-L, Zhang X-H (2016) Expression of an arctic chickweed dehydrin, CarDHN, enhances tolerance to abiotic stress in tobacco plants. Plant Growth Regul 80:323–334

Ho CH, Tsay YF (2010) Nitrate, ammonium, and potassium sensing and signaling. Curr Opin Plant Biol 13:604–610

Howarth RW, Marino R (2006) Nitrogen as the limiting nutrient for eutrophication in coastal marine ecosystems: evolving views over three decades. Limnol Oceanography 51:364–376

Hu HC, Wang YY, Tsay YF (2009) AtCIPK8, a CBL-interacting protein kinase, regulates the low-affinity phase of the primary nitrate response. Plant J 57:264–278

Hunt L, Otterhag L, Lee JC, Lasheen T, Hunt J, Seki M, Shinozaki K, Sommarin MD, Gilmour DJ, Pical C, Gray JE (2004) Gene-specific expression and calcium activation of *Arabidopsis thaliana* phospholipase C isoforms. New Phytol 162:643–654

Imsande J, Touraine B (1994) N demand and the regulation of nitrate uptake. Plant Physiol 105:3–7

Katayama H, Mori M, Kawamura Y, Tanaka T, Mori M, Hasegawa H (2009) Production and characterization of transgenic rice plants carrying a high-affinity nitrate transporter gene (OsNRT2. 1). Breed Sci 59:237–243

Krapp A, David LC, Chardin C, Girin T, Marmagne A, Leprince A-S, Chaillou S, Ferrario-Méry S, Meyer C, Daniel-Vedele F (2014) Nitrate transport and signalling in *Arabidopsis*. J Exp Bot 65:789–798

Kronzucker HJ, Siddiqi MY, Glass A (1996) Kinetics of NH_4^+ influx in spruce. Plant Physiol 110:773–779

Laugier E, Bouguyon E, Mauries A, Tillard P, Gojon A, Lejay L (2012) Regulation of high-affinity nitrate uptake in roots of Arabidopsis depends predominantly on posttranscriptional control of the NRT2.1/NAR2.1 transport system. Plant Physiol 158:1067–1078

Lejay L, Tillard P, Lepetit M, Olive F, Filleur S, Daniel-Vedele F, Gojon A (1999) Molecular and functional regulation of two NO_3^- uptake systems by N- and C-status of *Arabidopsis* plants. Plant J 18:509–519

Little D, Rao H, Oliva S, Daniel-Vedele F, Krapp A, Malamy JE (2005) The putative high-affinity nitrate transporter NRT2.1 represses lateral root initiation in response to nutritional cues. Proc Natl Acad Sci USA 102:13693–13698

Liu C, Wang Y, Pan K, Zhu T, Li W, Zhang L (2014) Carbon and nitrogen metabolism in leaves and roots of dwarf bamboo (*Fargesia denudata* Yi) subjected to drought for two consecutive years during sprouting period. J Plant Growth Regul 33:243–255

Loqué D, Tillard P, Gojon A, Lepetit M (2003) Gene expression of the NO_3^- transporter NRT1.1 and the nitrate reductase NIA1 is repressed in Arabidopsis roots by NO_2^-, the product of NO_3^- reduction. Plant Physiol 132:958–967

Mandadi KK, Misra A, Ren S, McKnight TD (2009) BT2, a BTB protein, mediates multiple responses to nutrients, stresses, and hormones in Arabidopsis. Plant Physiol 150:1930–1939

Medici A, Krouk G (2014) The primary nitrate response: a multifaceted signalling pathway. J Exp Bot 65:5567–5576

Meharg AA, Blatt MR (1995) NO_3^- transport across the plasma membrane of *Arabidopsis thaliana* root hairs: kinetic control by pH and membrane voltage. J Membrane Biol 145:49–66

Muños S, Cazettes C, Fizames C, Gaymard F, Tillard P, Lepetit M, Lejay L, Gojon A (2004) Transcript profiling in the *chl1-5* mutant of Arabidopsis reveals a role of the nitrate transporter NRT1.1 in the regulation of another nitrate transporter, NRT2.1. Plant Cell 16:2433–2447

Okamoto M, Vidmar JJ, Glass ADM (2003) Regulation of *NRT1* and *NRT2* gene families of *Arabidopsis thaliana*: responses to nitrate provision. Plant Cell Physiol 44:304–317

Orsel M, Chopin F, Leleu O, Smith SJ, Krapp A, Daniel-Vedele F, Miller AJ (2006) Characterization of a two-component high-affinity nitrate uptake system in *Arabidopsis*. Physiology and protein-protein interaction. Plant Physiol 142:1304–1317

Pao SS, Paulsen IT, Saier MH (1998) Major facilitator superfamily. Microbiol Mol Biol Rev 62:1–34

Peoples MB, Freney JR, Mosier AR, Bacon PE (1995) Minimizing gaseous losses of nitrogen. In: Bacon PE (ed) Nitrogen Fertilization in the Environment. Marcel Dekker, Inc, New York, pp 565–602

Pii Y, Alessandrini M, Dall'Osto L, Guardini K, Prinsi B, Espen L, Zamboni A, Varanini Z (2016) Time-resolved investigation of molecular components involved in the induction of NO_3^- high affinity transport system in maize roots. Front Plant Sci 7:1657. https://doi.org/10.3389/fpls.2016.01657

Plett D, Toubia J, Garnett T, Tester M, Kaiser BN, Baumann U (2010) Dichotomy in the NRT gene families of dicots and grass species. PLoS ONE 5:e15289

Quaggiotti S, Ruperti B, Borsa P, Destro T, Malagoli M (2003) Expression of a putative high-affinity NO_3^- transporter and of an H^+-ATPase in relation to whole plant nitrate transport physiology in two maize genotypes differently responsive to low nitrogen availability. J Exp Bot 54:1023–1031

Quesada A, Galvan A, Fernandez E (1994) Identification of nitrate transporter genes in *Chlamydomonas reinhardtii*. Plant J 5:407–419

Remans T, Nacry P, Pervent M, Filleur S, Diatloff E, Mounier E, Tillard P, Forde BG, Gojon A (2006) The *Arabidopsis* NRT1.1 transporter participates in the signaling pathway triggering root colonization of nitrate-rich patches. Proc Natl Acad Sci USA 103:19206–19211

Riveras E, Alvarez JM, Vidal EA, Oses C, Vega A, Gutiérrez RA (2015) The calcium ion is a second messenger in the nitrate signaling pathway of Arabidopsis. Plant Physiol 169:1397–1404

Sakakibara H, Kobayashi K, Deji A, Sugiyama T (1997) Partial characterization of the signaling pathway for the nitrate-dependent expression of genes for nitrogen-assimilatory enzymes using detached maize leaves. Plant Cell Physiol 38:837–843

Schmidt GW, Delaney SK (2010) Stable internal reference genes for normalization of real-time RT-PCR in tobacco (*Nicotiana tabacum*) during development and abiotic stress. Mol Genet Genomics 283:233–241

Streb H, Irvine RF, Berridge MJ, Schulz I (1983) Release of Ca^{2+} from a nonmitochondrial intracellular store in pancreatic acinar cells by inositol-1,4,5-trisphosphate. Nature 306:67–69

Tong Y, Zhou JJ, Li Z, Miller AJ (2005) A two-component high-affinity nitrate uptake system in barley. Plant J 41:442–450

Tsai F-Y, Zhang X-H, Ulanov A, Widholm JM (2010) The application of the yeast

N-acetyltransferase MPR1 gene and the proline analogue L-azetidine-2-carboxylic acid as a selectable marker system for plant transformation. J Exp Bot 61:2561–2573

Tsay YF, Chiu CC, Tsai CB, Ho CH, Hsu PK (2007) Nitrate transporters and peptide transporters. FEBS Lett 581:2290–2300

Undurraga SF, Ibarra-Henríquez C, Fredes I, Miguel Álvarez JM, Gutiérrez RA (2017) Nitrate signaling and early responses in Arabidopsis roots. J Exp Bot 68:2541–2551

Unkles SE, Rouch D, Wang Y, Siddiqi MY, Glass ADM, Kinghorn JR (2004) Two perfectly conserved arginine residues are required for substrate binding in a high-affinity nitrate transporter. Proc Natl Acad Sci USA 101:17549–17554

Wang R, Tischner R, Gutiérrez RA, Hoffman M, Xing X, Chen M, Coruzzi G, Crawford NM (2004) Genomic analysis of the nitrate response using a nitrate reductase-null mutant of Arabidopsis. Plant Physiol 136:2512–2522

Wang YY, Hsu PK, Tsay YF (2012) Uptake, allocation and signaling of nitrate. Trends Plant Sci 17:458–467

White PJ (2000) Calcium channels in higher plants. Biochim Biophys Acta 1465:171–189

Wirth J, Chopin F, Santoni V, Viennois G, Tillard P, Krapp A, Gojon A (2007) Regulation of root nitrate uptake at the NRT2.1 protein level in *Arabidopsis thaliana*. J Biol Chem 282:23541–23552

Yan M, Fan X, Feng H, Miller AJ, Shen Q, Xu G (2011) Rice OsNAR2.1 interacts with OsNRT2.1, OsNRT2.2 and OsNRT2.3a nitrate transporters to provide uptake over high and low concentration ranges. Plant Cell Environ 34:1360–1372

Yong Z, Kotur Z, Glass AD (2010) Characterization of an intact two-component high-affinity nitrate transporter from *Arabidopsis* roots. Plant J 63:739–748

Zhang H, Forde BG (2000) Regulation of *Arabidopsis* root development by nitrate availability. J Exp Bot 51:51–59

Zhang X-H, Takagi H, Widholm JM (2004) Expression of a novel yeast gene that detoxifies the proline analog azetidine-2-carboxylate confers resistance during tobacco seed germination, callus and shoot formation. Plant Cell Rep 22:615–622

Zhang X-H, Webb J, Huang Y-H, Lin L, Tang R-S, Liu A (2011) Hybrid Rubisco of tomato large subunits and tobacco small subunits is functional in tobacco plants. Plant Sci 180:480–488

Zhang X-H, Keating P, Wang X-W, Huang Y-H, Martin J, Hartmann JX, Liu A (2014) Production of functional native human interleukin-2 in tobacco chloroplasts. Mol Biotechnol 56:369–379

Zhou JJ, Fernández E, Galván A, Miller AJ (2000) A high affinity nitrate transport system from *Chlamydomonas* requires two gene products. FEBS Lett 466:225–227

4

Nostoc calcicola extract improved the antioxidative response of soybean to cowpea aphid

Van-Chung Mai[1*], Ba-Hoanh Nguyen[1], Duc-Dien Nguyen[2] and Le-Ai-Vinh Nguyen[2]

Abstract

Background: Soybean (*Glycine max* (L.) Merr. cv. "Nam Dan") is one of the most valuable crops in agricultural production in Nghe An province (Vietnam). Our previous study revealed that extract of the cyanobacterium strain *Nostoc calcicola* HN9 expressed positive effect on growth and development, and raised soybean productivity (Tran et al. in Proceeding of Vietnam national conference of research on biology, Da Nang, 2016). We hypothesized that *N. calcicola* HN9 would improve the defense responses of *G. max* cv. "Nam Dan" to cowpea aphid (*Aphis craccivora* Koch)-a serious pest of leguminous crops.

Results: Infestation of *A. craccivora* caused oxidative stress in leaves of *G. max* cv. "Nam Dan". A strong generation of endogenous reactive oxygen species (ROS) such as superoxide anion radical ($O_2^{\cdot-}$) and hydrogen peroxide (H_2O_2) resulted in the cellular damages in the aphid-infested leaves through high levels of injury percentage and lipid peroxidation. To protect from aphid attack themselves, soybean plants triggered the antioxidant defense systems, in which, enzymatic antioxidants such as superoxide dismutase (SOD, 1.15.1.1), catalase (CAT, 1.11.1.6) and GPx (EC 1.11.1.9) were strongly accumulated to reduce the toxic effects of ROS. Components of *N. calcicola* HN9 extract might strengthen the defensive capability of *G. max* cv. "Nam Dan" to cowpea aphid infestation via establishing the chemical constraints on oxidative stress. Under effect of cyanobacteria extract, generation of $O_2^{\cdot-}$ and H_2O_2 was strictly limited, activities of SOD, CAT and GPx were remarkably accumulated in the aphid-infested leaves leading to a significant reduction of oxidative damages.

Conclusions: *Nostoc calcicola* HN9 extract probably not only controlled the generation and effects of $O_2^{\cdot-}$ and H_2O_2 but also augmented the accumulated activity of SOD, CAT and GPx in soybean leaves that allowed them to control oxidative stress, contributed to increase the resistance of *G. max* cv. "Nam Dan" to *A. craccivora*. The improvement of cyanobacteria extract on the antioxidative response of soybean "Nam Dan" to cowpea aphid can be a novel aspect to contribute to current knowledge regarding the soybean-aphid interaction.

Keywords: *Aphis craccivora*, *Glycine max*, *Nostoc calcicola* HN9, Hydrogen peroxide, Superoxide anion radical, Catalase, Glutathione peroxidase, Superoxide dismutase

Background

Cyanobacteria (or blue-green algae) are the large group of prokaryotic, photosynthetic microorganisms, which are a rich source of potentially various bioactive compounds effecting on living organisms (Zulpa et al. 2003). The endogenous substances in cyanobacteria, e.g., vitamins, enzymes, carbohydrates, amino acids, have been known to induce crop growth and raise yields (Shariatmadari et al. 2013; Prasanna et al. 2013). Other natural products from cyanobacteria like carotenoids, phenolic compounds, etc., exhibit various biological effects such as cytotoxic, antibacterial and antifungal activities (Kulik 1995; Zulpa et al. 2003). Therefore, an inoculation with cyanobacteria not only promotes plants growth but

*Correspondence: chungmv@vinhuni.edu.vn
[1] Department of Plant Physiology, School of Natural Sciences Education, Vinh University, str. Le Duan 182, Vinh, Nghe An Province, Vietnam
Full list of author information is available at the end of the article

also enhances their defense responses (Rana et al. 2012; Prasanna et al. 2013).

Nostoc, an unbranched filamentous cyanobacterium genus (Komárek 2010), has been regarded as good candidate for producing secondary substances that can effect on living organisms (Hirata et al. 2003). Treating *Solanum lycopersicum, Cucurbita maxima, Cucumis sativus* plants with *Nostoc calcicola* extracts revealed a significant positive effect in most physiological examined factors (Shariatmadari et al. 2011). The similar results were also confirmed by studies on other crops such as rice (Saadatnia and Riahi 2009), maize (Jäger et al. 2010). Several substances in *Nostoc* spp. expressed the antifungal, antibacterial, antiviral or cytotoxic activities on crops (Nowruzi et al. 2012). Nostocine A exhibited strong inhibitory activity to the root elongation of barnyard grass, high antifeedant activity to cotton ballworm, and acute toxicity to mice resulting in its classification as a dangerous poison (Hirata et al. 2003). Extract of *Nostoc* sp. induced oxidative stress by accumulating reactive oxygen species (ROS), causing massive cell death in *Mimosa pigra* (Sukkhaeng et al. 2015). Contrary, an in vitro study of Li and co-workers (2007) demonstrated that *Nostoc* species have their ability to scavenge ROS and inhibit lipid peroxidation.

Reactive oxygen species such as superoxide anion radical ($O_2^{\cdot-}$), hydrogen peroxide (H_2O_2) are produced as a normal product of plant cellular metabolism. Despite their destructive activity, they are well-described second messengers in a variety of cellular processes, including conferment of tolerance to various stresses. Whether ROS would serve as signaling molecules or could cause oxidative damage to the tissues depends on the delicate equilibrium between ROS productions, and their scavenging (Sharma et al. 2012). In plants tissues affected by aphid attack, $O_2^{\cdot-}$ and H_2O_2 are often enhanced in apoplastic as well as in symplastic regions, besides their main concentration in exocellular matrix, peroxisomes/mitochondria and plasma membrane (Smith and Boyko 2007; Maffei et al. 2007). High level of those ROS can exert toxic effects, and their uncontrolled productions can damage cellular components and associated loss of function leads to cell death (Ahmad et al. 2008). Plants have the efficient complex antioxidant defense systems to avoid the toxic effects of ROS, including enzymatic antioxidants such as superoxide dismutase (SOD), catalase (CAT), peroxidases (POX) and others (Maffei et al. 2007; Ahmad et al. 2008). Functions of those enzymatic antioxidants in plant defense responses have been reported in several plant-aphid interactions. Enhanced activity of SOD and CAT was suggested as one of the most essential elements of defense responses in pea seedling to oxidative stress caused by infestation of pea aphid (Mai et al. 2013). Accumulation of POX was strongly enhanced in wheat after feeding of the Russian wheat aphid (Moloi and van der Westhuizen 2006; Ni et al. 2001), in response of soybean to cowpea aphid (Mai et al. 2016). In cabbage plants, activities of polyphenol peroxidase and polyphenol oxidase were stimulated by cabbage aphid infestation (Khattab 2007). However, up to date, lack information has been mentioned about effects of cyanobacteria on plant antioxidant defense systems against aphid impacts.

Soybean (*Glycine max* (L.) Merr. cv. "Nam Dan") is an important crop in the agricultural production in Vietnam for its edible seed as the foremost material of vegetative protein and oil used in the food processing. During its life, soybean plant has to face to attack from cowpea aphid (*Aphis craccivora* Koch)—one of the most destructive aphid species of leguminous crops. Our previous study (Tran et al. 2016) revealed that inoculation extract of the cyanobacterium strain *N. calcicola* HN9 presented the positive effect on growth, development and productivity of soybean "Nam Dan". The aim of present work was to investigate effects of *N. calcicola* HN9 extract on the antioxidative responses of *G. max* cv. "Nam Dan" when soybean plants were cultured in Hoagland medium in phytotron, infested by different density of cowpea aphid in different time intervals. Cytochemical localization and the alteration in contents of $O_2^{\cdot-}$, H_2O_2 as well as changes in activity of antioxidant enzymes such as SOD (EC 1.15.1.1), CAT (EC 1.11.1.6), and GPx (EC 1.11.1.19) in soybean leaves were carefully assessed. As a secondary objective, degree of cellular oxidative damages was investigated on measurement of the electrolyte leakage and levels of thiobarbituric acid reactive substances (TBARS)-a final product of lipid peroxidation. It is worth noting that the aforementioned aspects will provide the additionally convinced evidences to clarify whether *G. max* cv. "Nam Dan" is an aphid resistant cultivar of soybean.

Methods

Cyanobacteria culture and extraction

The cyanobacterium strain, *N. calcicola* HN9 (voucher specimen is deposited at herbarium of the Phycology lab, Vinh University), was collected from the rice field in Hung Nguyen district (Nghe An province, Vietnam) and isolated by the streak plate method. *Nostoc* cells were cultured in BG11 medium, pH 6.5 at ambient temperature (25 ± 2 °C), light intensity of 110–130 μM photons m^{-2} s^{-1}, light period of 14 light/10 dark (Rippka et al. 1979).

The cyanobacteria biomass in the cultured day of 23th (in the stationary phase) was harvested, centrifuged at 6000×*g* for 15 min, and subsequently dried at 55 °C for 72 h. The dried cells were ground to powder, and extracted with 80% methanol for 24 h at ambient temperature. Supernatant was collected after centrifuging by 10,000×*g* for 15 min, then evaporated to obtain a crude brown gum. This gum was dissolved in the deionized water to the concentration of 0.03%, which expressed the positive effect to growth of soybean "Nam Dan" (Tran et al. 2016).

Plant and inoculation experiment

Experiments were performed on soybean (*Glycine max* (L.) Merr. cv. "Nam Dan") plants whose seeds were exclusively supplied by Nghe An seed center (Vietnam). Seeds were surface-sterilized by $HgCl_2$ 0.01% for 10 min and then were imbibed in the incubator at 23 ± 1 °C for 48 h. Germinating seeds were cultured in 10 cm-diameter plastic pots (one plant per pot) containing Hoagland medium placed in phytotron with temperature of 23 ± 1 °C, relative humidity of 70–75%, light intensity of 110–130 μM m^{-2} s^{-1} and light period 14 light/10 dark hours.

As growing in the V3 stage (first two trifoliate leaves fully developed, third trifoliate leaf unrolled), soybean plants were divided into two equal groups. The first one included soybean plants that were sprayed on the surface of leaves by 1 mL solution *N. calcicola* HN9 extract, concentration 0.03%, per plant. The second one was soybean without cyanobacteria inoculation.

Aphid and infestation experiment

The aphid species used to treat soybean "Nam Dan" is cowpea aphid (*Aphis craccivora* Koch), which is obtained from Department of Applied Entomology (Vietnam Academy of Science and Technology). The virus free individuals were cultured on host, *G. max*, in phytotron in temperature of 23 ± 1 °C, relative humidity of 70–75%, light intensity of 110–130 μM m^{-2} s^{-1} and light period of 14 light/10 dark hours (Mai et al. 2016).

After spraying of *N. calcicola* HN9 extract 24 h, all soybean plants in two mentioned groups were used for setting up experiments of infestation. Each soybean plant was separately treated by 10, 20 or 30 wingless adults of *A. craccivora*. The similarly wingless adults of cowpea aphid were carefully selected and transferred to soybean leaves with a fine paintbrush. The newly produced nymphs and winged individuals were monitored daily to prevent density-dependent effects on aphid performance; therefore, number of wingless adults of *A. craccivora* in experiments was constant. Control was soybean plants without aphid infestation. The control and aphid-infested variants were separately put in glass boxes (50 cm × 50 cm × 50 cm) covered by nylon gauze; all were placed in the phytotron with the environmental factors such as temperature, related humidity, light intensity and light period controlled.

Material for analysis

Mature leaves of soybean plants in all variants were carefully collected at 0, 6, 12, 24, 48, 72 and 96 h post-infestation (hpi) of cowpea aphid. Aphid individuals all were removed; leaves were weighed, frozen in nitrogen liquid and kept at − 20 °C for subsequent analyses of lipid peroxidation and activity of antioxidant enzymes. Content and cytochemical localization of superoxide anion radical ($O_2^{\cdot-}$), hydrogen peroxide (H_2O_2) and injury percentage were determined in fresh materials.

Determination of superoxide anion radical content

Content of superoxide anion radical ($O_2^{\cdot-}$) was determined following the nitro blue tetrazolium (NBT) assay (Doke 1983). Soybean leaves (0.30 g fresh weight-FW) were incubated in 3 mL of 10 mM phosphate buffer (pH 7.8) containing 0.05% NBT and 10 mM NaN_3 at ambient temperature for 1 h. After incubation, the reacted solution was heated at 85 °C for 15 min, then rapidly cooled. Measurements were carried out in the UV–Vis CARY 60 spectrophotometer (Agilent, USA) connected with a computer; the spectral data were analyzed by using the UV-Win 5.0 application software. Content of $O_2^{\cdot-}$ was expressed as absorbance at 580 nm per 1 g fresh materials (A_{580} g^{-1} FW).

Detection of superoxide anion radical localization

The cytochemical localization of $O_2^{\cdot-}$ was detected following method using dihydroethidium (DHE), the specific fluorescent dye, described by Morkunas and Bednarski (2008) with minor modification. The fresh soybean leaves were submerged in 50 M DHE in 5 mM DMSO after immersing 12 h in darkness at ambient temperature. After rinsing with the 100 M $CaCl_2$ solution pH 5.0, the DHE dying leaves were observed by using a Zeiss Axiovert 200 M fluorescence microscope (model LSM 510, filter set no. 9, excitation 450–490 nm, emission 520 nm or more), magnified 5× and photographed by a digital camera (AxioCam, Zeiss). An argon laser with excitation at 488 nm and emission at 565–615 nm was used. Microscope, laser and photomultiplier settings were held stably during experiments to obtain comparable data. Images were analyzed by the LSM Image Browser software, version 4.2.

Determination of hydrogen peroxide concentration

Concentration of endogenous hydrogen peroxide (H_2O_2) was determined following the spectrophotometric method (Becana et al. 1986) that was modified by Mai et al. (2013). 0.50 g fresh material were homogenized at 4 °C with 3 mL of 5% trichloroacetic acid (TCA) and 0.10 g activated charcoal. The homogenate was centrifuged at 12,000×*g* for 30 min at 4 °C to select supernatant using as extract. The reagent was prepared of 0.6 mM 4-(-2 pyridylazo)resorcinol, 0.6 mM of potassium-titanium oxalate in 1:1 proportion. Total 3 mL the reaction mixture contained extract, 100 mM phosphate buffer (pH 8.4) and reagent. The decrease of absorbance was measured at wavelength of 508 nm in the UV–Vis CARY 60 spectrophotometer (Agilent, USA) system. The spectral data were analyzed by using the UV-Win 5.0 application software. Amount of hydrogen peroxide in soybean leaves was expressed as M H_2O_2 g^{-1} FW.

Detection of hydrogen peroxide localization

Cytochemical localization of endogenous H_2O_2 generated in soybean leaves was detected by staining with specific fluorescence, the 2′,7′-dichlorofluorescein diacetate (DCFH-DA) following methods of Małecka et al. (2009) with minor modification. The fresh soybean leaves were submerged in 4 M DCFH-DA dissolved in 50 mM potassium phosphate buffer (pH 7.5) for 12 h. Leaves were washed twice with the loading buffer and then were observed with the Zeiss LSM 510 confocal microscopy system. Fluorescence was excited using 488 nm of an argon laser with emission at 500–550 nm. All images were obtained at the same depth, and were analyzed by the LSM Image Browser software, version 4.2.

Electrolyte leakage

Electrolyte leakage was the conductometrical method used for assessing the injury percentage of cell membrane (Bajji et al. 2002). Five fresh, complete leaves of each sample were washed in deionized water for a few seconds, embedded in test-tubes containing 20 mL of deionized water and took the first conductometric measurement (EC_i); then incubated on a shaking platform at ambient temperature for 3 h and measured again (EC_f) by a conductivity meter (Orion Star A212, Thermo Scientific, USA). After autoclaving the solution at 100 °C for 20 min to rupture completely the membranes allowing ion leakage of the entire cell contents and cooling to ambient temperature, the third measurement was done (EC_t). The membrane damage of soybean leaves' cells was evaluated as the injury percentage compared with control and was calculated as $I_d = [(R_s - R_c)/(1 - R_c)] \times 100\,(\%)$; where R_s and R_c represent $(EC_f - EC_i)/(EC_t - EC_i)$ for aphid and/or cyanobacteria treated tissues and control, respectively.

Lipid peroxidation

The thiobarbituric acid reactive substances (TBARS) assay is the most generally used test in the appreciation of lipid peroxidation (Heath and Packer 1968). 0.50 g of frozen soybean leaves was homogenized in 3 mL of 0.5% thiobarbituric acid (TBA) in 20% TCA and then centrifuged at 12,000×*g* for 20 min at 4 °C to select the extract. The reaction mixture [extract, phosphate buffer (pH 7.0) and reagent (0.5% TBA in 20% TCA, w/v)] were incubated at 95 °C for 30 min and then quickly cooled in an ice bath. After that, the mixtures were centrifuged at 10,000×*g* for 10 min. The specific absorbance at 532 nm and the non-specific absorbance at 600 nm of the clear supernatant were measured by using the UV–Vis CARY 60 spectrophotometer. Level of lipid peroxidation was calculated and expressed as μM TBARS g^{-1} FW.

Enzyme assay

Briefly, 0.50 g frozen soybean leaves were homogenized at 4 °C in 5.0 mL of 50 mM phosphate buffer (pH 7.0) containing 1.0 mM EDTA, 2% NaCl and 1% polyvinyl pyrrolidone and centrifuged at 15,000×*g* for 15 min to select supernatant as the enzymatic extract.

Activity of SOD (EC 1.15.1.1) was spectrophotometrically assayed by measuring its ability in inhibition the photochemical reduction of NBT to a blue solution (Beauchamp and Fridovich 1971). The reacted mixture contained 50 mM phosphate buffer (pH 7.8), 13 mM methionine, 75 mM NBT, 0.1 mM EDTA, enzymatic extract and 2 mM riboflavin. The reaction was started by switching on a 30 W fluorescent lamp placed 30 cm above test-tubes and proceeded for 15 min. Samples without the enzymatic extract were selected so that the absorption difference between blank and examined tests was about 50%. Measurements were done by using the UV–Vis CARY 60 spectrophotometer. Amount of enzyme that caused the inhibition of NBT reduction by 50% was taken as a unit of SOD activity.

Activity of CAT (EC 1.11.1.6) was determined by measuring hydrogen peroxide removal (Dhindsa et al. 1981). The reacted mixture contained 100 mM phosphate buffer (pH 7.0), 3% H_2O_2 and enzymatic extract. Absorbance was assessed by measuring at 240 nm against a calibration curve using the UV–Vis CARY 60 spectrophotometer.

Activity of GPx (EC 1.11.1.19) was assayed according to the procedure of Flohé and Günzler (1984) with minor modifications. The reaction mixture consisting of 0.5 mL of each: 0.4 M sodium phosphate buffer (pH 7.0), 10 mM sodium azide, 4 mM reduced glutathione, 5 mM H_2O_2, and enzyme extract was incubated at 0, 30, 60, 90 s respectively. The reaction was terminated with 0.5 mL of 10% TCA and centrifuged; after that 2 mL of the supernatant was added to 3 mL of phosphate buffer and 1 mL of DTNB reagent (0.04% DTNB in 1% sodium citrate). Absorbance was measured at a wavelength of 412 nm in the UV–Vis CARY 60 spectrophotometer.

Activity of SOD, CAT and GPx was expressed as nanokatal per 1 mg of protein (nkat mg^{-1} protein). The protein concentration in the samples was determined according to Bradford (1976) method by using bovine serum albumin as a standard.

Statistical analysis

Analyses all were performed at least in three independent experiments. Analysis of variance (ANOVA) was applied to verify whether means from independent experiments within each given variant were significant at level $P < 0.05$. Data shown in the figures are means and standard errors (s.e.), using asterisks to present the significant differences.

Results

Generation of ROS in leaves of soybean "Nam Dan"

Generally, a strongly release the reactive oxygen species (ROS) productions such as superoxide anion radical ($O_2^{\cdot-}$) and hydrogen peroxide (H_2O_2) was observed in aphid-infested leaves of soybean (*Glycine max* cv. "Nam Dan") in stage V3.

Without inoculating *N. calcicola* HN9 extract, superoxide anion radical in the aphid-infested leaves was remarkably released and reached to the highest content at 12 h post-infestation (hpi) of 10, 20, or 30 aphids, having values of 1.64-, 2.48-, and 2.63-fold higher than at the beginning, respectively. This ROS molecule then slightly reduced to lower levels until 96 hpi. Whereas, this free radical in control plants maintained at low level throughout the experiment (Fig. 1; the dash lines). ANOVA results showed the significant difference in levels of $O_2^{\cdot-}$ in aphid-infested variants and control within 6–48 hpi.

Contrary, a small amount of $O_2^{\cdot-}$ was generated in leaves inoculated with *N. calcicola* HN9 extract, and content of this free radical was slightly increased from the beginning up to 96 hpi (Fig. 1; the solid lines). In most of investigated time points, content of $O_2^{\cdot-}$ in soybean treated by *N. calcicola* HN9 extract was in lower than that in plant without cyanobacteria inoculating.

The applying confocal microscopy results revealed the relative release of $O_2^{\cdot-}$ in soybean leaves via DHE-fluorescence staining in epidermis cells surrounding the aphid feeding sites. DHE can passively cross the cells membrane, and is oxidized by $O_2^{\cdot-}$ to ethidium bromide that is detected by using the confocal laser scanning microscope. The fluorescence is proportional to the intracellular $O_2^{\cdot-}$ level. Figure 2 presented an example about the presence of $O_2^{\cdot-}$ at 12 hpi, in which time a burst of $O_2^{\cdot-}$ in aphid-infested/non cyanobacteria inoculated leaves was recorded by the spectrophotometric method. Cytochemical localization of $O_2^{\cdot-}$ presented that, a stronger intensity and larger surface area of the DHE-derived fluorescence

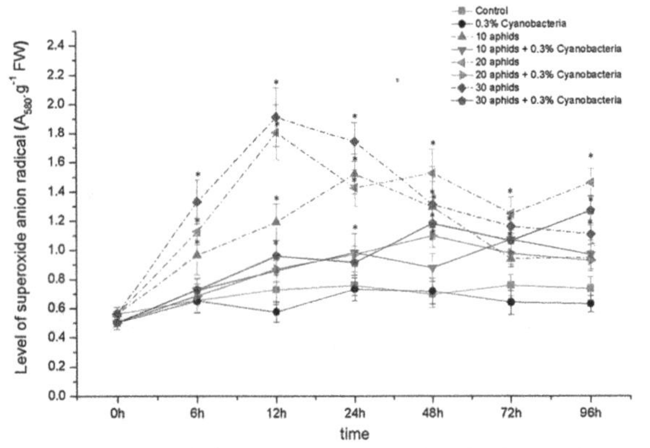

Fig. 1 Generation of superoxide anion radical ($O_2^{\cdot-}$) in leaves of *Glycine max* cv. "Nam Dan" under infestation of *Aphis craccivora* and inoculation of *Nostoc calcicola* HN9 extract. Values represent means and s.e. from three independent experiments. (*) In figure shown the significant difference between the experimental variants and control at level $P < 0.05$

appeared in leaves only infested by *A. craccivora*; while a small fluorescent stain was observed in the additionally inoculated cyanobacteria leaves. The fluorescent intensity also provided an important evidence that the release of $O_2^{\cdot-}$ was limited in soybean leaves inoculated by *N. calcicola* HN9 extract.

A strong generation and continuous increase of H_2O_2 in aphid-infested leaves was observed in soybean plants without cyanobacteria inoculating within 0–48 hpi (Fig. 3; the dash lines). The maximum levels of this ROS production were obtained at 24 hpi in 20- and 30-aphid infested leaves, whereas content of H_2O_2 in 10-aphid infested variant was continuously increased until 72 hpi. Infestation of 30 cowpea aphids per plant caused to release the highest amount of H_2O_2, obtaining 16.97 mM g^{-1} FW, having by 2.36- and 3.63-fold higher than in control and from beginning, respectively. The most remarkable difference was recorded between content of H_2O_2 in aphid-infested and control plants within 6–48 hpi. Similar to $O_2^{\cdot-}$, an accumulation of H_2O_2 in leaves inoculated with *N. calcicola* HN9 was much lower than that observed in soybean without inoculating cyanobacteria (Fig. 3; the solid lines).

In addition, observations of the relative generation H_2O_2 in soybean leaves using confocal microscopy revealed similar trends to the spectrophotometric results. After staining with 2′,7′-dichlorofluorescein diacetate (DCFH-DA) there was observed DCFH-DA-derived fluorescence in soybean leaves. DCFH-DA is a nonpolar dye, converted into the polar derivative DCFH that are non-fluorescent but switched to highly fluorescent DCF when oxidized by intracellular H_2O_2. DCF can be

Soybean leaves without treatment *Nostoc calcicola* HN9 extract | Soybean leaves treated by *Nostoc calcicola* HN9 extract

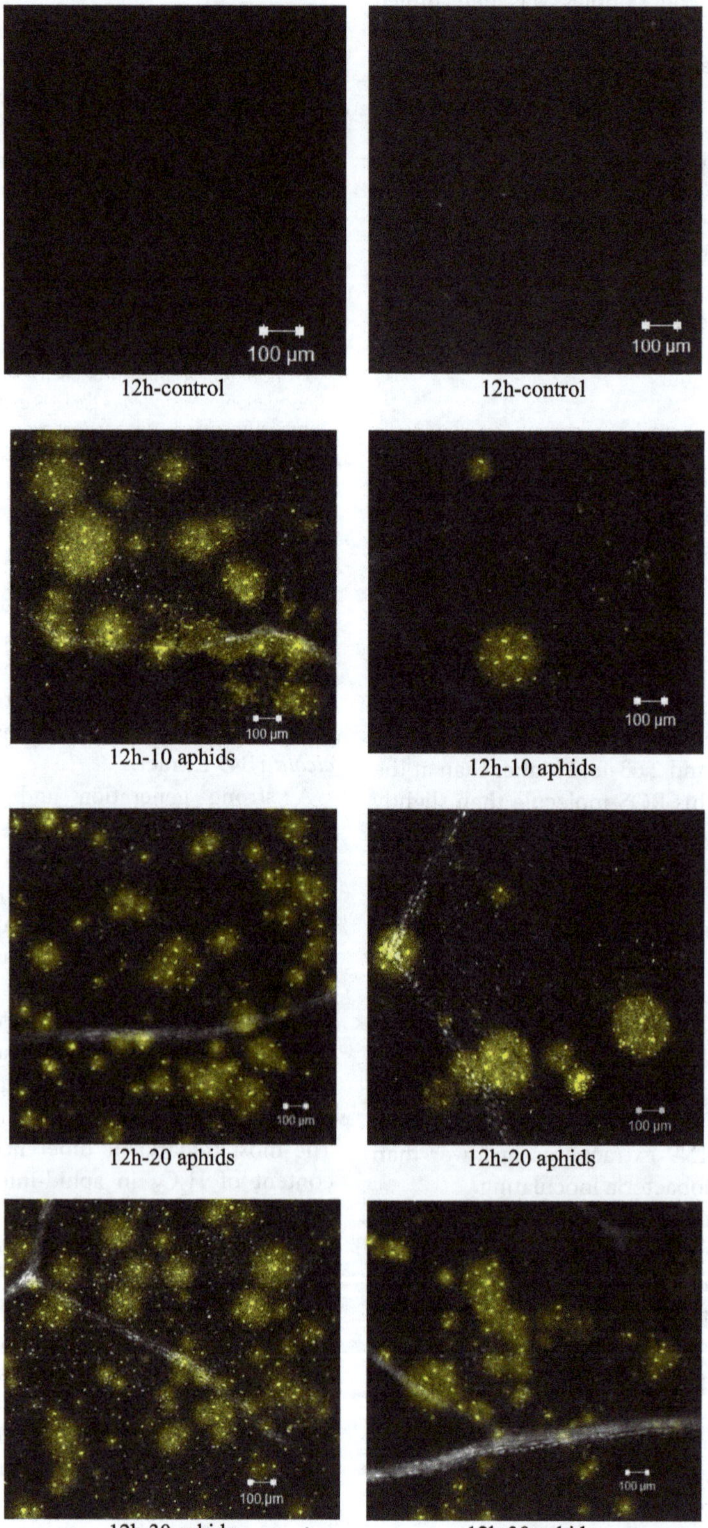

12h-control | 12h-control

12h-10 aphids | 12h-10 aphids

12h-20 aphids | 12h-20 aphids

12h-30 aphids | 12h-30 aphids

(See figure on previous page.)
Fig. 2 Cytochemical localization of superoxide anion generation (O_2^{-}) in leaves of *Glycine max* cv. "Nam Dan" under infestation of *Aphis craccivora* and inoculation of *Nostoc calcicola* HN9 extract at 12 hpi. Yellow fluorescence came from DHE (dihydroethidium), which is observed by a Zeiss LSM 510 confocal microscope, with excitation at 488 nm, emission at 565–615 nm, objective magnification of 5×, scale bar 100 μm

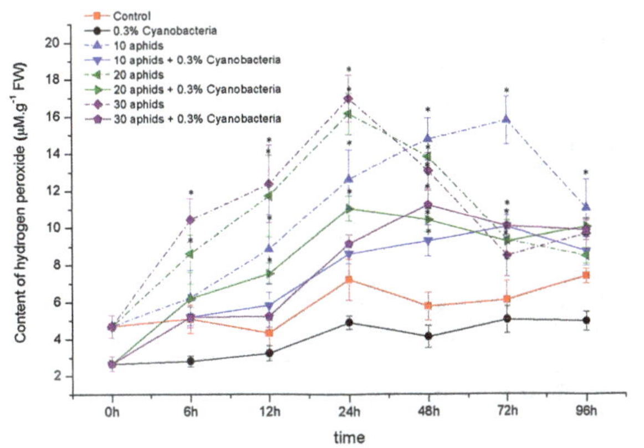

Fig. 3 Generation of hydrogen peroxide (H_2O_2) in leaves of *Glycine max* cv. "Nam Dan" under infestation of *Aphis craccivora* and inoculation of *Nostoc calcicola* HN9 extract. Values represent means and s.e. from three independent experiments. (*) In figure shown the significant difference between the experimental variants and control at level P < 0.05

observed as green color under the fluorescent microscope and intensity of fluorescence is proportional to the endogenous H_2O_2 level. A high intensity of fluorescence was observed in the aphid-infested leaves and covered a large uniform area of tissue, whereas no or small trace of fluorescent staining was detected in control (e.g., at 24 hpi, Fig. 4). Intensity of green color in *N. calcicola* HN9-inoculated leaves was remarkably lower in comparing with that recorded in soybean without cyanobacteria treatment.

Cell membrane injury

The membrane injury degrees in *G. max* cells were estimated through assessment of electrolyte leakage and lipid peroxidation.

The electrolyte leakage of soybean leaves' cells was evaluated as the injury percentage compared with control. Without inoculating with *N. calcicola* HN9 extract, the injury percentage in all aphid-infested leaves increased from 0 to 48 hpi and maintained in high levels until 96 hpi (Fig. 5; the dash lines). The highest degree of cellular injury was recorded at 72 hpi in leaves infested by 30 aphids, having 22.45% higher than in control.

Inoculation of *N. calcicola* HN9 extract seemed to reduce injury rate in the aphid-infested leaves. The injury percentage in soybean inoculated with *N. calcicola* HN9 extract was always significant lower than that observed in plants without treating cyanobacteria. The maximum degree of injury recorded in 30-aphid infested leaves at 96 hpi was only 9.72% (Fig. 5; the solid lines).

Infestation of *A. craccivora* induced the lipid peroxidation process in *G. max* cv. "Nam Dan" leaves (Fig. 6; the dash lines). A small but progressive increase in content of TBARS, the final products of lipid peroxidation, reached the highest level at 96 hpi, e.g., 19.87, 21.04 and 23.02 M TBARS g^{-1} FW under effect from 10, 20 and 30 aphids, respectively. ANOVA results showed that, content of TBARS in aphid-infested leaves was significantly higher than in control, and an appreciable proportion between levels of TBARS and the infestation intensity of cowpea aphid was exposed since 24 hpi.

Similar to effect on injury percentage of cellular membrane, *N. calcicola* HN9 extract might control lipid peroxidation. This process in plants inoculated with cyanobacteria extract was minor changed during the investigated time; TBARS content was mostly smaller than that observed in soybean without inoculating cyanobacteria (Fig. 6; the solid lines). The maximum content of TBARS obtained in 30-aphid infested/*N. calcicola* HN9 inoculated leaves at 48 hpi was 17.96 M TBARS g^{-1} FW that was only 78.02% in comparing with that in 30-aphid infested leaves.

Activity of enzymatic antioxidants

Analyses of spectrophotometric assays showed that infestation of *A. craccivora* and inoculation *N. calcicola* HN9 extract differently induced activity of enzymatic antioxidants such as SOD, CAT and GPx in leaves of *G. max* cv. "Nam Dan".

Activity of SOD in soybean leaves infested by different densities of cowpea aphid increased in the first 24 h of infestation. The highest value of SOD activity recorded in 20 aphid-infested leaves at 24 hpi was 18.08 nkat mg^{-1} protein, which was 288.9 and 351.3% higher than SOD activity in control and at the beginning, respectively. After reaching the peak within 24–48 hpi, activity of this enzyme was strongly reduced to low levels but always higher than in control (Fig. 7; the dash lines). ANOVA analyses confirmed the significant differences between the SOD activities in the aphid-infested and control plants during experimental time. However, no appreciable proportion between the infestation intensity and the enhanced activity of SOD was recorded.

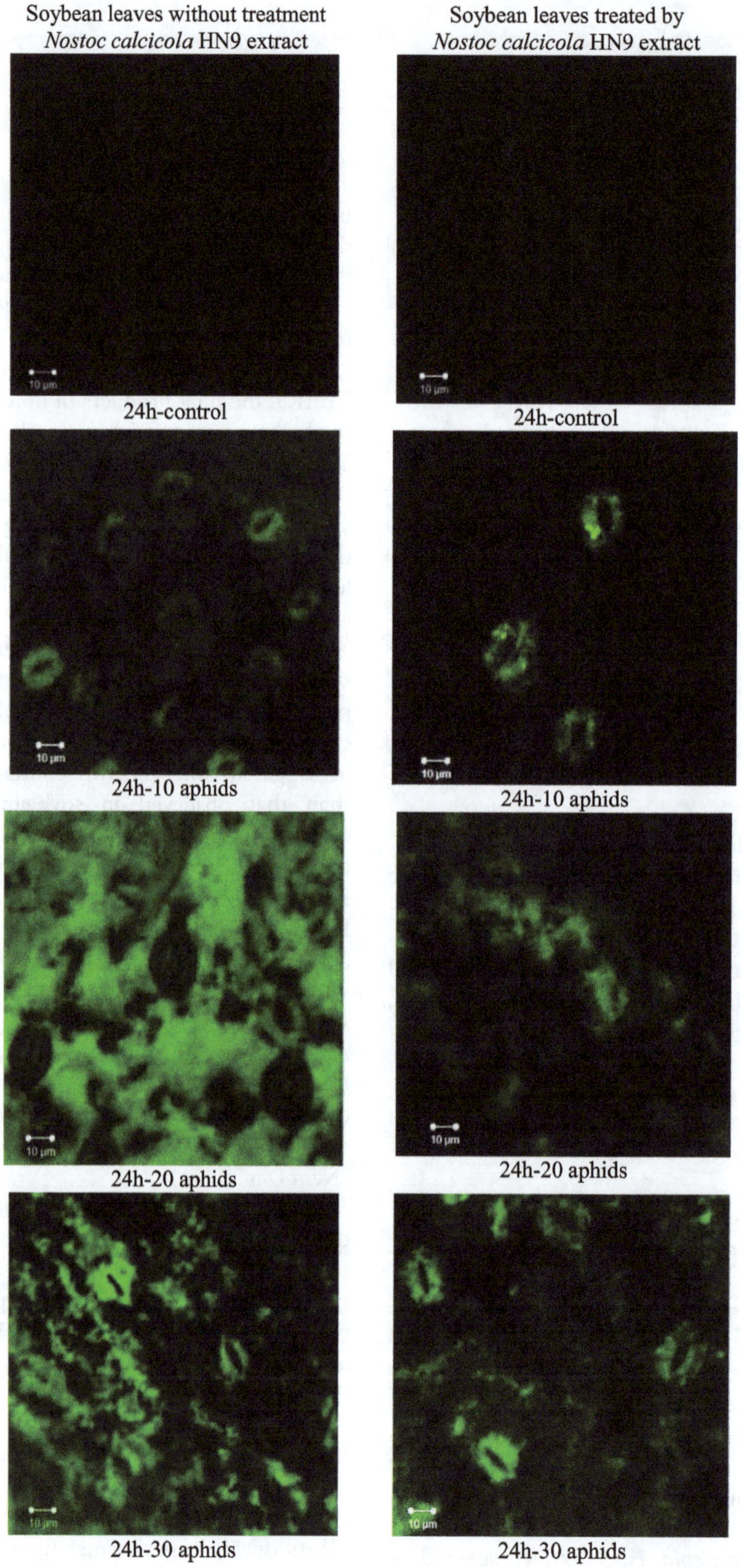
Soybean leaves without treatment
Nostoc calcicola HN9 extract
Soybean leaves treated by
Nostoc calcicola HN9 extract
24h-control
24h-control
24h-10 aphids
24h-10 aphids
24h-20 aphids
24h-20 aphids
24h-30 aphids
24h-30 aphids

(See figure on previous page.)

Fig. 4 Cytochemical localization of hydrogen peroxide (H_2O_2) in leaves of *Glycine max* cv. "Nam Dan" under infestation of *Aphis craccivora* and inoculation of *Nostoc calcicola* HN9 extract at 24 hpi. Green fluorescence came from DCFH-DA (dichlorodihydro-fluorescein diacetate), which is observed by a Zeiss LSM 510 confocal microscope, with excitation at 488 nm, emission at 500–550 nm, objective magnification of 63×, scale bar 10 µm

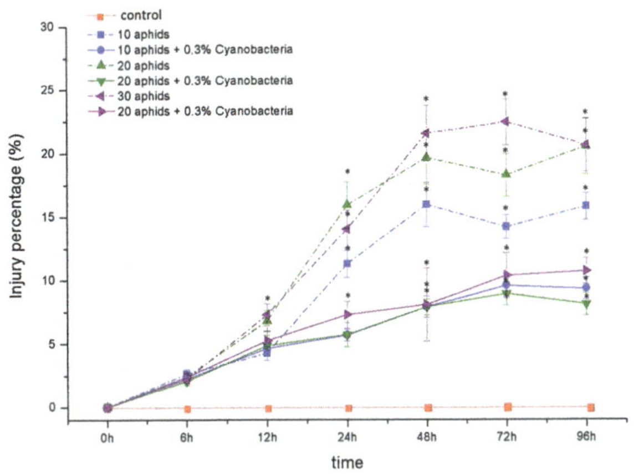

Fig. 5 The percentage of injury (in comparing with control) in *Glycine max* cv. "Nam Dan" leaves under infestation of *Aphis craccivora* and inoculation of *Nostoc calcicola* HN9 extract. Values represent means and s.e. from three independent experiments. (*) In figure shown the significant difference between the experimental variants and control at level P < 0.05

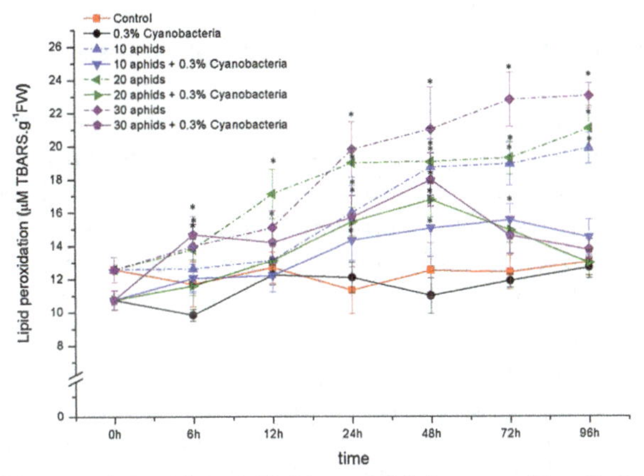

Fig. 6 Lipid peroxidation in leaf tissues of *Glycine max* cv. "Nam Dan" leaves under infestation of *Aphis craccivora* and inoculation of *Nostoc calcicola* HN9 extract. Values represent means and s.e. from three independent experiments. (*) In figure shown the significant difference between the experimental variants and control at level P < 0.05

Inoculation soybean with *N. calcicola* HN9 extract sooner accumulated action of SOD in the aphid-infested leaves. Activity of this enzyme was strongly increased to peak within 12–24 hpi, and maintained in high level up to 96 hpi. There was an appreciable proportion between the accumulated activity of SOD and intensity of aphid infestation within 6–72 hpi. The highest activity of SOD enhanced by infestation of 30 aphid individuals at 12 hpi was 20.79 nkat mg^{-1} protein, having by 2.61- and 3.32-fold higher than that in control and at the beginning, respectively (Fig. 7; the solid lines). ANOVA analyses confirmed that SOD activities in the aphid-infested/cyanobacteria inoculated soybean leaves were significant higher than that in control plants within 6–96 hpi.

Similar to SOD, enzyme CAT in leaves of soybean "Nam Dan" was also induced after cowpea aphid infestation. An accumulated activity of CAT was observed from the beginning, reached to peak at 48 hpi. The highest activity of CAT obtained in 30-aphid infested leaves were 26.54 nkat mg^{-1} protein, having by 195.49% higher than in the control (Fig. 8; the dash lines). ANOVA results revealed a significant difference between the activities of CAT in 20- and 30-aphid infested soybean plants and control within 12–72 hpi.

Under effect of *N. calcicola* HN9 extract, an explosive expression of CAT was observed in aphid-infested leaves. Activity of CAT was remarkably increased and reached to peak within 24–48 hpi. The highest activity of CAT in 30-aphid infested soybean was 42.31 nkat mg^{-1} protein at 24 hpi, having by 1.59-folder higher than the maximum activity of CAT in plants without inoculating cyanobacteria (26.54 nkat mg^{-1} protein). Interestingly, an appreciable proportion between activity of CAT and the infestation intensity of cowpea aphid was also recorded (Fig. 8; the solid lines).

Cowpea aphid infestation also enhanced activity of GPx in leaves of soybean "Nam Dan". Activity of this enzyme was accumulated since 12 hpi, increased to maximum at 48 hpi, and then slightly reduced to ending of experiments (Fig. 9; the dash lines). The highest activity of GPx obtained in 20 aphid-infested leaves at 48 hpi was 6.91 nkat mg^{-1} protein, having by 9.24- and 3.32-fold-higher than that at the beginning and control, respectively.

Inoculating soybean "Nam Dan" by *N. calcicola* HN9 extract seemed to improve action of GPx in plants infested by cowpea aphid. This antioxidant enzyme in the inoculated leaves was soon induced since 6 hpi and prolonged high activity until 96 hpi. Under effect of *N. calcicola* HN9 extract, the maximum activity of GPx recorded in 30-aphid infested plants at 48 hpi was 12.13 nkat mg^{-1} protein, having by 175.54% that observed in soybean without inoculating cyanobacteria. There was a

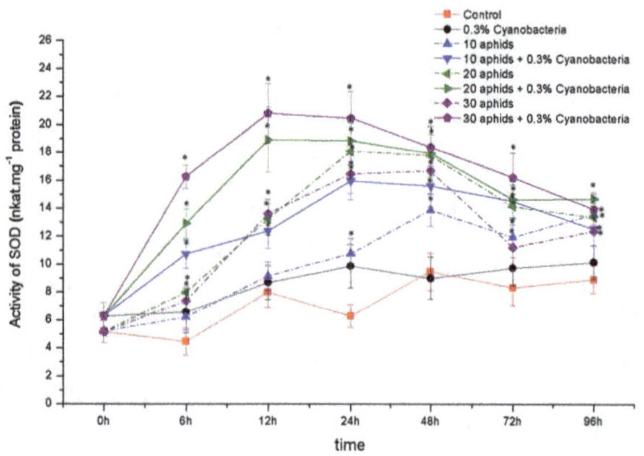

Fig. 7 Activity of superoxide dismutase (SOD) in *Glycine max* cv. "Nam Dan" leaves under infestation of *Aphis craccivora* and inoculation of *Nostoc calcicola* HN9 extract. Values represent means and s.e. from three independent experiments. (*) In figure shown the significant difference between the experimental variants and control at level $P < 0.05$

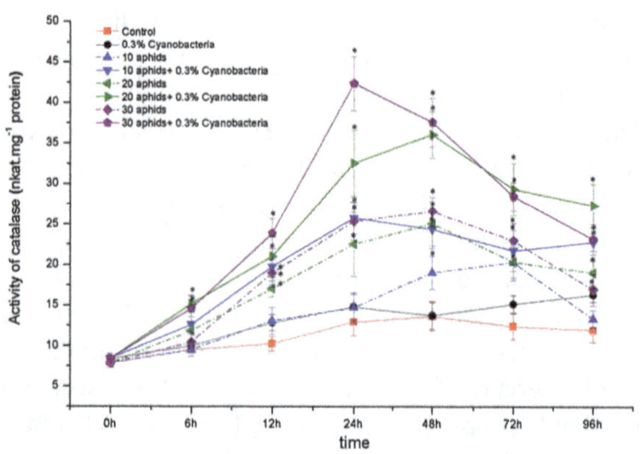

Fig. 8 Activity of catalase (CAT) in *Glycine max* cv. "Nam Dan" leaves under infestation of *Aphis craccivora* and inoculation of *Nostoc calcicola* HN9 extract. Values represent means and s.e. from three independent experiments. (*) In figure shown the significant difference between the experimental variants and control at level $P < 0.05$

significant difference between activity of GPx in the aphid-infested plants and control since 6 hpi, however, the relationship between the accumulated enzymatic activity and intensity of infestation was unappreciable (Fig. 9; the solid lines).

Discussion

Feeding of insects, including aphids, has been known to induce generation of the endogenous ROS such as $O_2^{\cdot-}$ and H_2O_2 in plants (Orozco-Cárdenas and Ryan 1999; Ni et al. 2001; Argandona et al. 2001; Zhu-Salzman et al. 2004; Moloi and van der Westhuizen 2006; Ślesak et al. 2007; Kuśnierczyk et al. 2008; Rangasamy et al. 2009; Liu et al. 2010; Radville et al. 2011; Morkunas et al. 2011; Mai et al. 2013, 2016). Similar to previous studies, infestation of *A. craccivora* accumulated a strong release of $O_2^{\cdot-}$ and H_2O_2 in leaves of *G. max* cv. "Nam Dan" (Figs. 1, 3). The induced generation of $O_2^{\cdot-}$ constitutes an important aspect of plant potential defense (Mai et al. 2013). Induction of $O_2^{\cdot-}$ within a short time after stress mediated rapid local resistance mechanisms in plants against insect larvae, leading to cell death (Bown et al. 2002), might play an important function in protection against infection by opportunistic pathogens at the site of the larval footprint damage (Hall et al. 2004). A large amount of $O_2^{\cdot-}$ in cell wall may limit the pathogen invasion that can be directly transmitted from aphid saliva during its penetration into plant tissues (Capinera 2001). Different from $O_2^{\cdot-}$ functions, enhancement of H_2O_2 stimulates a cascade of many physiological and molecular responses (Maffei et al. 2007). H_2O_2 causes to release of Ca^{2+} flux into the cellular matrix, which is the control point for signal transduction processes or activates some members of the mitogen-activated protein kinase (MAPK) family-central for mediating cellular responses to stresses (Neill et al. 2002). H_2O_2 can tighten the cell wall by driving lignification as well as enabling the oxidative coupling of polysaccharide-bound phenolics, which strengthen barriers to aphid probing within plant tissues (Vreeburg and Fry 2005; Gapper and Dolan 2006), leading to prevent or minimize aphid attack (Argandona et al. 2001), reflecting a defensive role on protection plant tissues from wounding.

However, high levels of ROS can exert toxic effects and caused "oxidative stress" to plant cells. The uncontrolled productions of ROS can result in damages of cellular components such as proteins, lipids, and nucleic acids (Ahmad et al. 2008), often associated cell membrane injury leading to electrolyte leakage and lipid peroxidation. As aphid probing into plant tissues, components of sheath and watery saliva may establish the physical and chemical properties to cause the extensive cellular disruption. An increased electrolyte leakage is used to indicate loss of membrane integrity, and soybean cells membranes may be subject to changes under cowpea aphid infestation. Lipid peroxidation involves the formation of lipid radicals, the eventual destruction of membrane lipids, is also considered as the oxidative damage to cell structures that may lead to cell death (Dianzani and Barrera 2008). Under effects from different densities of *A. craccivora*, an extension of electrolyte leakage and a progressive increase in content of TBARS-the final products of lipid peroxidation were observed in leaves of *G. max* cv. "Nam Dan" (Figs. 5, 6; the dash lines). That was certain symptoms of cellular damage resulting from oxidative stress caused by cowpea aphid infestation.

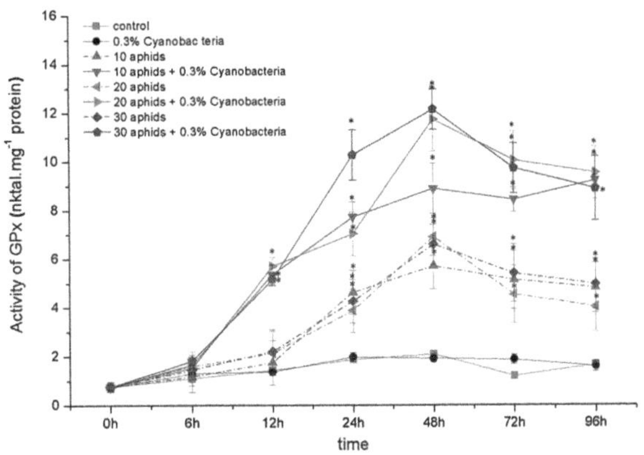

Fig. 9 Activity of glutathione peroxidase (GPx) in *Glycine max* cv. "Nam Dan" leaves under infestation of *Aphis craccivora* and inoculation of *Nostoc calcicola* HN9 extract. Values represent means and s.e. from three independent experiments. (*) In figure shown the significant difference between the experimental variants and control at level $P < 0.05$

Regarding effects of inoculation cyanobacteria to crops, several previous reports showed that cyanobacterial materials could induce productions of oxidants. *Nostoc* sp. extract caused a burst of $O_2^{\cdot-}$ and H_2O_2 generation leading to the cellular damages in root of *Mimosa pigra* (Sukkhaeng et al. 2015). Lipid peroxidation in *M. pigra* and *Vallisneria natans* plants also significantly increased after treatment of cyanobacterial extract (Jiang et al. 2011; Sukkhaeng et al. 2015). The cyanobacterial bloom products caused an oxidative damage in *Medicago sativa* seedlings (Pflugmacher et al. 2006). Furthermore, some bioactive substances from cyanobacteria extract can be toxic to plant tissues, even result in cell death. For example, Nostocine A produced by *N. spongiaeforme* TISTR 8169 exhibited an inhibitory activity on root growth of barnyard grass (Hirata et al. 2003). Microcystin-RR from *Microcystis aeruginosa* decreased cell viability of tobacco after inoculation (Yin et al. 2005). Different from the above results, *Nostoc* species have several antioxidant properties, due to their ability to scavenge free radicals such as $O_2^{\cdot-}$ and $OH^{\cdot}$ and inhibit lipid peroxidation (Li et al. 2007). In this light, our study also revealed that, inoculation of *N. calcicola* HN9 extract depressed generation of $O_2^{\cdot-}$ and H_2O_2; levels of those ROS remarkably decreased in leaves of *G. max* cv. "Nam Dan" infested by *A. craccivora*. Extract of this cyanobacterium strain may exhibit potential adaptation strategies against oxidative stress in soybean "Nam Dan". The extract could activate plant antioxidant defense system that comprises various enzymatic antioxidants that mainly caused a reduction of ROS content (Bharanidharan et al. 2013; Guedes et al. 2013). Moreover, cyanobacteria extract also contains phenolic acids, flavonoids and other non-enzymatic antioxidants that improve plant adaptive responses to oxidative stress (Honga et al. 2008; Krishnaraj et al. 2012). Phenolic acids and flavonoids are able to donate hydrogen atom to the free radical thus stopping the propagation chain reaction during lipid peroxidation process (Abd El-Baky et al. 2009). It is known that prevention of the chain initiation step by scavenging ROS products is considered to be an important mode of antioxidative mechanism (Ruberto et al. 2001). Moreover, most phenolic compounds found in the cyanobacteria extract contained the protonated form with the high chelating ability (Goh et al. 2010). The ability to chelate off the ferrous ion is important to avoid a Fenton reaction, which would produce the harmful free radicals such as $O_2^{\cdot-}$. The presence of phenolic acids and flavonoids were supposed to impart prominent antioxidant properties that detoxify ROS products (Perron and Brumaghim 2009; Reuter et al. 2010; Singh et al. 2017).

Parallel action of non-enzymatic antioxidants, plant cells can scavenge toxicity of ROS products via enzymes such as superoxide dismutase (SOD), catalase (CAT), and other enzymatic antioxidants (Maffei et al. 2007; Ahmad et al. 2008). Potential roles of SOD and CAT in signaling and control of ROS production have been implicated in plant defense response to insect herbivores, including aphids (Ni et al. 2001; Gomez et al. 2004; Heng-Moss et al. 2004; Wei et al. 2007; Moloi and van der Westhuizen 2008; Rangasamy et al. 2009; Ferry et al. 2011; Mai et al. 2013). In agreement with the above results, our study revealed that the accumulated activity of SOD and CAT under infestation of *A. craccivora* is an aspect of antioxidative responses of *G. max* cv. "Nam Dan" to cowpea aphid. Within living cells, SOD constitutes the first line of defense against ROS (Alscher et al. 2002), because this enzyme can counteract oxidative damage caused by regulating levels of $O_2^{\cdot-}$ (Abassi et al. 1998). The strong accumulated activity of CAT protects the plant cells against an excess of H_2O_2 and thus against considerable membrane damage (Rani and Jyothsna 2010). The potential defensive roles of SOD, CAT to control of ROS products have been involved in plant resistance to aphids (Rangasamy et al. 2009; Ferry et al. 2011).

Up to date, limited information has been mentioned about effects of cyanobacteria species on expression of the enzymatic antioxidants in plant defense mechanism. Inoculation cyanobacterial crude extract on alfalfa enhanced activities of SOD, CAT and other enzymes in response of alfalfa to oxidative stress (Pflugmacher et al. 2006). CAT activity significantly increased in *M. pigra* plant since 24 h after treating by *Nostoc* sp. extract (Sukkhaeng et al. 2015). Extract of *N. commune*, *Anabaena flos-acquae* and *Westiellopsis* sp. was found to enhance activity of CAT and POX of medicinal plant flax (Naresh et al. 2013). In soybean-cowpea aphid interaction, inoculation the aphid-infested leaves with *N. calcicola* HN9 extract improved activity of SOD and CAT (Figs. 7, 8). Because most cyanobacterial species contain a wide range of enzymes such as SOD, CAT, and other enzymatic antioxidants (Bharanidharan et al. 2013; Guedes et al. 2013; Gunes et al. 2015; Ghalab et al. 2016), SOD and CAT from *N. calcicola* HN9 extract probably augmented to the antioxidant defense system of soybean. The accumulated activity of SOD and CAT in soybean defense mechanism might be resulted from the acted combination or resonance of those enzymes originated from both soybean plant and cyanobacteria extract. The presence of high activity of enzymatic antioxidants in cyanobacteria extract may also be considered as a part of the key functions which contribute to protect of plants from effect of unfavorable factors, including aphids (Aydas et al. 2013; Singh et al. 2014).

Another important enzyme in H_2O_2-scavenging is GPx from the non-heme containing peroxidase family (Bela et al. 2015). Different from SOD and CAT, GPx was not detected in cyanobacteria (Tel-Or et al. 1986); therefore, the accumulated activity of this enzyme recorded in our study was originated only from soybean plant. An accumulation of GPx was as an aspect of soybean defense mechanism. GPx can regulate diverse defensive signaling pathways during oxidative stress and other unfavorable factors (Milla et al. 2003; Navrot et al. 2006; Diao et al. 2014; Gao et al. 2014). Typically, GPx is part of an alternative pathway that scavenges peroxides such as H_2O_2, phospholipid hydroperoxides, thereby protects cell membranes from peroxidative damage (Gueta-Dahan et al. 1997). It should be stressed that, inoculation of *N. calcicola* HN9 extract strongly improved activity of GPx in soybean leaves during cowpea aphid attack. This enzyme was accumulated since 6 hpi, reached to maximum activity at 48 hpi and maintained in high level up to ending of experiment; GPx activity was always significantly higher than that in soybean without inoculating cyanobacteria extract (Fig. 9).

Taken together, the aforementioned results provided the persuasive evidence that components of *N. calcicola* HN9 extract may strengthen the defensive capability of *G. max* cv. "Nam Dan" to cowpea aphid infestation via establishing the chemical constraints on oxidative stress. Cyanobacteria extract probably not only directly controlled the generation of $O_2^{\cdot-}$ and H_2O_2 but also developed a range of chemical properties to limit aphid feeding. Furthermore, *N. calcicola* HN9 extract might augment activity of SOD and CAT, enhance action of GPx, allowed those enzymatic antioxidants to play crucial functions in defense mechanism of soybean "Nam Dan" to strictly reduce the toxic effects of ROS products.

Additionally, a high generation of $O_2^{\cdot-}$ and H_2O_2, an accumulated activity of enzymatic antioxidants, and low cellular damages in leaves of *G. max* cv. "Nam Dan" after *A. craccivora* infestation suggest that it may be a aphid resistant variety of soybean.

In conclusion, the enzymatic antioxidants such as SOD, CAT and GPx are important elements of soybean defense responses against cowpea aphid infestation. The accumulation of those enzymes can be thought of a reflection of intrinsic strategies to reduce damages caused due to ROS generated during aphid attack. Inoculation soybean plant with *N. calcicola* HN9 extract, which contains various enzymatic and non-enzymatic antioxidants, strengthened the adaptive potential to protect soybean from oxidative stress, contributed to increase the resistance of *G. max* cv. "Nam Dan" to *A. craccivora*. The improved antioxidative responses of *G. max* cv. "Nam Dan" to *A. craccivora* infestation under effect of *N. calcicola* HN9 extract can be a novel aspect of bio-protection, which is an integral component of the friendly strategies for sustainable agriculture.

Abbreviations

CAT: catalase; FW: fresh weight; GPx: glutathione peroxidase; H_2O_2: hydrogen peroxide; hpi: hours post infestation; nkat: nanokatal; $O_2^{\cdot-}$: superoxide anion radical; SOD: superoxide dismutase.

Authors' contributions

LAVN was responsible for cyanobacteria culture. DDN supplied all laboratory and chemical substances used in this study. VCM and BHN designed the experiment. VCM and LAVN determined, analyzed the data, and drafted the manuscript. All authors read and approved the final manuscript.

Author details

[1] Department of Plant Physiology, School of Natural Sciences Education, Vinh University, str. Le Duan 182, Vinh, Nghe An Province, Vietnam. [2] Department of Environmental Sciences, School of Chemo-Biology and Environment Technology, Vinh University, str. Le Duan 182, Vinh, Nghe An Province, Vietnam.

Acknowledgements

We are thankful to Vietnam National Foundation for Science and Technology Development (NAFOSTED) under Grant Number: 106-NN.03-2014.2 for financial support.

Competing interests

The authors declare that they have no competing interests.

Funding

This research is funded by Vietnam National Foundation for Science and Technology Development (NAFOSTED) under Grant Number: 106-NN.03-2014.2.

References

Abassi NA, Kushad MM, Endress AG (1998) Active oxygen-scavenging enzymes activities in developing apple flowers and fruits. Sci HorticAmst 74:183–194

Abd El-Baky HH, El-Baz FK, El-Barot GS (2009) Production of phenolic compounds from *Spirulina maxima* microalgae and its protective effects. Afr J Biotechnol 8:7059–7067

Ahmad P, Sarwat M, Sharma S (2008) Reactive oxygen species, antioxidants and signaling in plants. J Plant Biol 51:167–173

Alscher RG, Erturkm N, Heathm LS (2002) Role of superoxide dismutases (SODs) in controlling oxidative stress in plants. J Exp Bot 53:1331–1341

Argandona VH, Chaman M, Cardemil L, Munoz O, Zuniga GE, Corcuera LJ (2001) Ethylene production and peroxidase activity in aphid-infested barley. J Chem Ecol 27:53–68

Aydas SB, Ozturk S, Aslim B (2013) Phenylalanine ammonia lyase (PAL) enzyme activity and antioxidant properties of some cyanobacteria isolates. Food Chem 136:164–169

Bajji M, Kinet J-M, Lutts S (2002) The use of the electrolyte leakage method for assessing cell membrane stability as a water stress tolerance test in durum wheat. Plant Growth Regul 36(1):61–70

Beauchamp C, Fridovich I (1971) Superoxide dismutase: improved assays and an assay applicable to acrylamide gels. Anal Biochem 44:276–287

Becana M, Aparicio-Tejo P, Irigoyen JJ, Sanchez-Diaz M (1986) Some enzymes of hydrogen peroxide metabolism in leaves and root nodules of *Medicago sativa*. Plant Physiol 82:1169–1171

Bela K, Horváth E, Gallé Á, Szabados L, Tari I, Csiszár J (2015) Plant glutathione peroxidases: emerging role of the antioxidant enzymes in plant development and stress responses. J Plant Physiol 176:192–201

Bharanidharan M, Sivasubramanian V, Raja SR, Nayagam V (2013) Evaluation of antioxidant and antimicrobial potential of cyanobacteria, *Chroococcus turgidus* (Kützing) Nägeli. Int J Curr Microbiol App Sci 2:300–305

Bown AW, Hall DE, MacGregor KB (2002) Insect footsteps on leaves stimulate the accumulation of 4-aminobutyrate and can be visualized through increased chlorophyll fluorescence and superoxide production. Plant Physiol 129:1430–1434

Bradford M (1976) A rapid and sensitive method for quantitation of microgram quantities of protein utilizing the principle of protein-dye binding. Anal Biochem 72:248–254

Capinera JL (2001) Handbook of vegetable pests. Academic Press, New York

Dhindsa RS, Plumb-Dhindsa P, Thorpe TA (1981) Leaf senescence: correlated with increased levels of membrane permeability and lipid peroxidation, and decrease levels of superoxide dismutase and catalase. J Exp Bot 32:93–101

Dianzani M, Barrera G (2008) Pathology and physiology of lipid peroxidation and its carbonyl products. In: Álvarez S, Evelson P (eds) Free radical pathophysiology. Transworld Research Network, Thiruvananthapuram, pp 19–38

Diao Y, Xu H, Li G, Yu A, Yu X, Hu W et al (2014) Cloning a glutathione peroxidase gene from *Nelumbo nucifera* and enhanced salt tolerance by overexpressing in rice. Mol Biol Rep 41:4919–4927

Doke N (1983) Involvement of superoxide anion generation in the hypersensitive response of potato tuber tissues to infection with an incompatible race of *Phytophthora infestans* and to the hyphal wall components. Physiol Mol Plant Pathol 23:345–355

Ferry N, Stavroulakis S, Guan W, Davison GM, Bell HA, Weaver RJ, Down RE, Gatehouse JA, Gatehouse AMR (2011) Molecular interactions between wheat and cereal aphid (*Sitobion avenae*): analysis of changes to the wheat proteome. Proteomics 11:1985–2002

Flohé L, Günzler WA (1984) Assays of glutathione peroxidase. Methods Enzymol 105:114–121

Gao F, Chen J, Ma T, Li H, Wang N, Li Z et al (2014) The glutathione peroxidase gene family in *Thellungiella salsuginea*: genome-wide identification, classification, and gene and protein expression analysis under stress conditions. Int J Mol Sci 15:3319–3335

Gapper C, Dolan L (2006) Control of plant development by reactive oxygen species. Plant Physiol 141:341–345

Ghalab NM, Tantawy EA, Khalil HMA, Shaban KA (2016) Plant growth promoters substances that excreting from bacteria and cyanobacteria as essential factors for alleviation soil salt stress on rice plant. J Microbiol Res 6(5):103–110

Goh S, Yusoff FM, Loh SP (2010) A comparison of the antioxidant properties and total phenolic content in a diatom, *Chaetoceros* sp. and a green microalga, *Nannochloropsis* sp. J Agric Sci 2:123–130

Gomez SK, Oosterhuis D, Rajguru SN, Johnson DR (2004) Foliar antioxidant enzyme responses in cotton after aphid herbivory. J Cotton Sci 8:99–104

Guedes AC, Gião MS, Seabra R, Ferreira AC, Tamagnini P, MoradasFerreira P, Malcata FX (2013) Evaluation of the antioxidant activity of cell extracts from microalgae. Mar Drugs 11:1256–1270

Gueta-Dahan Y, Yaniv Z, Zilinskas BA, Ben-Hayyim G (1997) Salt and oxidative stress: similar and specific responses and their relation to salt tolerance in *Citrus*. Planta 203:460–469

Gunes S, Tamburaci S, Imamoglu E, Dalay MC (2015) Determination of superoxide dismutase activities in different cyanobacteria for scavenging of reactive oxygen species. J Biol Act Prod Nat 5(1):25–32

Hall DE, MacGregor KB, Nijsse J, Bown AW (2004) Footsteps from insect larvae damage leaf surfaces and initiate rapid responses. Eur J Plant Pathol 110:441–447

Heath RL, Packer L (1968) Photoperoxidation in isolated chloroplasts. I. Kinetics and stoichiometry of fatty acid peroxidation. Arch Biochem Biophys 125:189–198

Heng-Moss T, Sarath G, Baxendale F, Novak N, Bose S, Ni X, Quisenberry S (2004) Characterization of oxidative enzyme changes in buffalograsses challenged by *Blissus occiduus*. J Econ Entomol 97:1086–1095

Hirata K, Yoshitomi S, Dwi S, Iwabe O, Mahakhant A, Plochai J, Miyamoto K (2003) Bioactivities of Nostocine-a produced by a freshwater cyanobacterium *Nostoc spongiaeforme* TISTR 8169. J Biosci Bioeng 95:512–517

Honga Y, Hua HY, Xiea X, Li FM (2008) Responses of enzymatic antioxidants and non-enzymatic antioxidants in the cyanobacterium *Microcystis aeruginosa* to the allelochemical ethyl 2-methyl acetoacetate (EMA) isolated from reed (*Phragmites communis*). J Plant Physiol 165:1264–1273

Jäger K, Bartók T, Ördög V, Barnabás B (2010) Improvement of maize (*Zea mays* L.) anther culture responses by algae-derived natural substances. S Afr J Bot 76:511–516

Jiang J, Gu X, Song R, Wang X, Yang L (2011) Microcystin-LR induced oxidative stress and ultrastructural alterations in mesophyll cells of submerged macrophyte *Vallisneria natans* (Lour.) Hara. J Hazard Mater 190:188–196

Khattab H (2007) The defense mechanism of cabbage plant against phloem-sucking aphid (*Brevicoryne brassicae* L.). Aust J Basic Appl Sci 1:56–62

Komárek J (2010) Modern taxonomic revision of planktic nostocacean cyanobacteria: a short review of genera. Hydrobiologia 639:231–243

Krishnaraj RN, Babu SV, Ashokkumar B, Malliga P, Varalakshmi P (2012) Antioxidant property of fresh and marine water cyanobacterial extracts in Swiss mice. J Biopestic 5:250–254

Kulik MM (1995) The potential for using cyanobacteria (blue-green algae) and algae in the biological control of plant pathogenic bacteria and fungi. Eur J Plant Pathol 101(6):585–599

Kuśnierczyk A, Winge P, Jørstad TM, Troczyńska J, Rossiter JT, Bones AM (2008) Towards global understanding of plant defense against aphids-timing and dynamics of early *Arabidopsis* defense responses to cabbage aphid (*Brevicoryne brassicae*) attack. Plant Cell Environ 31:1097–1115

Li AH, Cheng K, Wong C, King-Wai F, Feng C, Yue J (2007) Evaluation of antioxidant capacity and total phenolic content of different fractions of selected microalgae. Food Chem 102:771–776

Liu X, Williams CE, Nemacheck JA, Wang H, Subramanyam S, Zheng C, Chen MS (2010) Reactive oxygen species are involved in plant defense against a gall midge. Plant Physiol 152:985–999

Maffei ME, Mithöfer A, Boland W (2007) Insects feeding on plants: rapid signals and responses preceding the induction of phytochemical release. Phytochemistry 68:2946–2959

Mai VC, Bednarski W, Borowiak-Sobkowiak B, Wilkaniec B, Samardakiewicz S, Morkunas I (2013) Oxidative stress in pea seedling leaves responses to *Acyrthosiphon pisum* infestation. Phytochemistry 93:49–62

Mai VC, Tran NT, Nguyen DS (2016) The involvement of peroxidases in soybean seedlings' defense against infestation of cowpea aphid. ArthropodPlant Interact 10(4):283–292

Małecka A, Piechalak A, Tomaszewska B (2009) Reactive oxygen species production and antioxidative defense system in pea root tissues treated with lead ions: the whole roots level. Acta Physiol Plant 31:1053–1063

Milla MAR, Maure A, Huete AR, Gustafson JP (2003) Glutathione peroxidase genes in *Arabidopsis* are ubiquitous and regulated by abiotic stresses through diverse signaling pathways. Plant J 36:602–615

Moloi MJ, van der Westhuizen AJ (2006) The reactive oxygen species are involved in resistance responses of wheat to the Russian wheat aphid. J Plant Physiol 163:1118–1125

Moloi MJ, van der Westhuizen AJ (2008) Antioxidative enzymes and the Russian wheat aphid (*Diuraphis noxia*) resistance response in wheat (*Triticum aestivum*). Plant Biol 10:403–407

Morkunas I, Bednarski W (2008) *Fusarium oxysporum*-induced oxidative stress and antioxidative defenses of yellow lupine embryo axes with different sugar levels. J Plant Physiol 165:262–277

Morkunas I, Mai VC, Gabrys B (2011) Phytohormonal signaling in plant responses to aphid feeding. Acta Physiol Plant 33:2057–2073

Naresh L, Alex BK, Koshy EP (2013) Effect of different cyanobacterial species on growth, photosynthetic activity and antioxidant system of flax plant. Int J Pharm Bio Sci 4(4):(B) 446–(B) 455

Navrot N, Collin V, Gualberto J, Gelhaye E, Hirasawa M, Rey P, Knaff DB, Issakidis E, Jacquot JP, Rouhier N (2006) Plant glutathione peroxidases are functional peroxiredoxins distributed in several subcellular compartments and regulated during biotic and abiotic stresses. Plant Physiol 142:1364–1379

Neill SJ, Desikan R, Clarke A, Hurst RD, Hancock JT (2002) Hydrogen peroxide and nitric oxide as signalling molecules in plants. J Exp Bot 53:1237–1247

Ni X, Quisenberry SS, Heng-Moss T, Markwell JP, Sarath G, Klucas R, Baxendale FP (2001) Oxidative responses of resistant and susceptible cereal leaves to symptomatic and nonsymptomatic cereal aphid (Hemiptera: Aphididae) feeding. J Econ Entomol 94:743–751

Nowruzi B, Khavari-Nejad R-A, Sivonen K, Kazemi B, Najafi F, Nejadsattari T (2012) Identification and toxigenic potential of a *Nostoc* sp. Algae 27:303–313

Orozco-Cárdenas M, Ryan CA (1999) Hydrogen peroxide is generated systemically in plant leaves by wounding and systemin via the octadecanoid pathway. PNAS USA 96:6553–6557

Perron NR, Brumaghim JL (2009) A review of the antioxidant mechanisms of polyphenol compounds related to iron binding. Cell Biochem Biophys 53:75–100

Pflugmacher S, Jung K, Lundvall L, Neumann S, Peuthert A (2006) Effects of cyanobacterial toxins and cyanobacterial cell-free crude extract on germination of alfalfa (*Medicago sativa*) and induction of oxidative stress. Environ Toxicol Chem 25(9):2381–2387

Prasanna R, Chaudhary V, Gupta V, Babu S, Kumar A, Singh R, Shivay YS, Nain L (2013) Cyanobacteria mediated plant growth promotion and bioprotection against *Fusarium* wilt in tomato. Eur J Plant Pathol 136:337–353

Radville L, Chaves A, Preisser EL (2011) Variation in plant defense against invasive herbivores: evidence for a hypersensitive response in Eastern Hemlocks (*Tsuga canadensis*). J Chem Ecol 37:592–597

Rana A, Joshi M, Prasanna R, Shivay YS, Nain L (2012) Biofortification of wheat through inoculation of plant growth promoting rhizobacteria and cyanobacteria. Eur J Soil Biol 50:118–126

Rangasamy M, Rathinasabapathi B, McAuslane HJ, Cherry RH, Nagata RT (2009) Oxidative responses of St. Augustinegrasses to feeding of southern chinch bug, *Blissus insularis* Barber. J Chem Ecol 35:796–805

Rani PU, Jyothsna Y (2010) Biochemical and enzymatic changes in rice plants as a mechanism of defense. Acta Physiol Plant 32:695–701

Reuter S, Gupta SC, Chaturvedi MM, Aggarwal BB (2010) Oxidative stress, inflammation and cancer: how are they linked? Free Radic Biol Med 49:1603–1616

Rippka R, Deruelles J, Waterbury JB, Herdman M, Stanier RY (1979) Generic assignments, strain histories and properties of pure cultures of cyanobacteria. J Gen Microbiol 111:1–61

Ruberto G, Baratta MT, Biondi DM, Amico V (2001) Antioxidant activity of extracts of the marine algal genus *Cystoseira* in a micellar model system. J Appl Physiol 13:403–407

Saadatnia H, Riahi H (2009) Cyanobacteria from paddy fields in Iran as a biofertilizer in rice plants. Plant Soil Environ 55:207–212

Shariatmadari Z, Riahi H, Shokravi S (2011) Study of soil blue-green algae and their effect on seed germination and plant growth of vegetable crops. Rostaniha 12:101–110

Shariatmadari Z, Riahi H, Hashtroudi MS, Ghassempour AR, Aghashariatmadary Z (2013) Plant growth promoting cyanobacteria and theirdistribution in terrestrial habitats of Iran. Soil Sci Plant Nutr 59(4):535–547

Sharma P, Jha AB, Dubey RS, Pessarakli M (2012) Reactive oxygen species, oxidative damage, and antioxidative defense mechanism in plants under stressful conditions. J Bot 2012:26. https://doi.org/10.1155/2012/217037

Singh DP, Prabha R, Meena KK, Sharma L, Sharma AK (2014) Induced accumulation of polyphenolics and flavonoids in cyanobacteria under salt stress protects organisms through enhanced antioxidant activity. Am J Plant Sci 5:726–735

Singh DP, Prabha R, Verma S, Meena KK, Yandigeri M (2017) Antioxidant properties and polyphenolic content in terrestrial cyanobacteria. Biotech 7:134. https://doi.org/10.1007/s13205-017-0786-6

Ślesak I, Libik M, Karpinska B, Karpinski S, Miszalski Z (2007) The role of hydrogen peroxide in regulation of plant metabolism and cellular signalling in response to environmental stresses. Acta Biochim Pol 54:39–50

Smith CM, Boyko EV (2007) The molecular bases of plant resistance and defense responses to aphid feeding: current status. Entomol Exp Appl 122:1–16

Sukkhaeng S, Sanevas N, Suwanwong S (2015) *Nostoc* sp. extract induces oxidative stress mediated root cell destruction in *Mimosa pigra* L. Bot Stud 56:3. https://doi.org/10.1186/s40529-014-0081-3

Tel-Or E, Huflejt ME, Packer L (1986) Hydroperoxide metabolism in cyanobacteria. Arch Biochem Biophys 246(1):396–402

Tran VH, Nguyen DD, Nguyen LAV, Mai VC (2016) Effect of *Nostoc calcicola* extract on soybean "Nam Dan" in the vegetative stages. In: Proceeding of Vietnam national conference of research on biology, Da Nang, 20 May 2016; pp 1044–1051 **(In Vietnamese)**

Vreeburg RAM, Fry SC (2005) Reactive oxygen species in cell walls. In: Smirnoff N (ed) Antioxidants and reactive oxygen species in plants. Blackwell Publishing Ltd., Oxford, pp 215–249

Wei H, Zhikuan J, Qingfang H (2007) Effects of herbivore stress by *Aphis medicaginis* Koch on the malondialdehyde contents and the activities of protective enzymes in different alfalfa varieties. Acta Ecol Sin 27(6):2177–2183

Yin L, Huang J, Huang W, Li D, Wang G, Liu Y (2005) Microcystin-RR-induced accumulation of reactive oxygen species and alteration of antioxidant systems in tobacco BY-2 cells. Toxicon 46:507–512

Zhu-Salzman K, Salmon RA, Ahn JE, Koiwa H (2004) Transcriptional regulation of sorghum defense determinants against a phloem-feeding aphid. Plant Physiol 134:420–431

Zulpa G, Zaccaro MC, Boccazzi F, Parada JL, Storni M (2003) Bioactivity of intra and extracellular substances from cyanobacteria and lactic acid bacteria on "wood blue stain" fungi. Biol Control 27:345–348

5

Identification and cytochemical immunolocalization of acetyl-CoA acetyltransferase involved in the terpenoid mevalonate pathway in *Euphorbia helioscopia* laticifers

Meng Wang, Dou Wang, Qing Zhang, Jia Chai, Yong Peng and Xia Cai*

Abstract

Background: Terpenoids, the largest class of natural products in the plant kingdom, have been widely used in medicine. The precursors of terpenoids, isoprene phosphate (IPP) and dimethylallyl pyrophosphate (DMAPP), were synthesized from a mevalonate (MVA) pathway and a 2-C-methyl-D-erythritol-4-phosphate (MEP) pathway respectively. The acetyl-CoA acetyltransferase (AACT) is the initial enzyme in MVA pathway and is considered presently to be essential for terpenoid backbone biosynthesis. The basic research on cytochemistry of terpenoid metabolic enzymes is important for understanding the mechanisms underlying major metabolic processes. However, compartmentalization of AACT in plants is in controversy. *Euphorbia helioscopia* L. containing laticifers in the whole plant is a famous ancient folk medicine for tumor treatment, and the terpenoid is an active ingredient. Furthermore, the laticifer cell is the main synthesizing and storing site for terpenoids.

Results: The gene of *AACT* was cloned successfully from *E. helioscopia*, and named as *EhAACT*. The *EhAACT* expression has no significant difference among roots, stems and leaves. However, compared with the roots and stems, the *EhAACT* expression level is slightly higher in leaves. In addition, EhAACT recombinant protein was expressed by procaryotic expression system and anti-EhAACT antibody was prepared, the molecular weight is about 43 kDa. Western blotting results illustrated that the EhAACT antibodies specifically recognized the endogenous proteins in *E. helioscopia* laticifers. At last, the subcellular localization of EhAACT in *E. helioscopia* laticifers was observed by using colloidal gold immune-electron microscopy. EhAACT was found to mainly distribute in the endoplasmic reticulum (ER), vacuoles originated from ER and cytosol aound vacuoles originated from ER.

Conclusions: As a result, we speculated that in *E. helioscopia* laticifers, EhAACT located in cytosol would be transferred to small vacuoles dilated from ER, and the precursors of terpenoids were synthesized in these small vacuoles, then terpenoids were further synthesized into latex particles. This result would provide theoretical basis for regulating and controlling of terpenoid biosynthesis in laticifers.

Keywords: Terpenoid, Acetyl-CoA acetyltransferase, Gene clone, Prokaryotic expression, Expression, Immunolocalization

*Correspondence: caix@nwu.edu.cn
Key Laboratory of Resource Biology and Biotechnology in Western China, (Northwest University), Ministry of Education, Xi'an 710069, China

Introduction

Terpenoids are the largest class of natural products in the plant kingdom. Their structural unit is isoprene, and the general formula is $(C_5H_8)_n$. Terpenoids have potential medicinal value and are widely used in the medicine, for example, paclitaxel, β-elemene, beta ionone and geraniol are used as anticancer drugs (Eisenreich et al. 1998; Hua and Cai 2006; Duncan et al. 2004). The precursors of terpenoids are five carbon compounds which are isoprene phosphate (IPP) and dimethylallyl pyrophosphate (DMAPP), these two kinds of precursors were synthesized from a mevalonate (MVA) pathway (Chappell 1995) and a 2-C-methyl-D-erythritol-4-phosphate (MEP) pathway (Lange et al. 1998) respectively. Compartmentalization of MVA pathway enzymes has been extensively studied in mammalian (Kovacs et al. 2002; Hogenboom et al. 2004; Wanders and Waterham 2006). In plants, there only exists fragmented experimental data regarding the localization of the MVA pathway enzymes (Sapir-Mir et al. 2008). Rodríguez-Concepción and Boronat (2015) pointed out that future efforts to address the regulation of isoprenoid metabolism must also take into account the spatial distribution of the biosynthetic pathways as well as the targeting and subcellular location of the enzyme isoforms involved, the intracellular transport of metabolic intermediates, and the different storage structures for end products (Rodríguez-Concepción and Boronat 2015). The acetyl-CoA acetyltransferase (AACT), or named acetoacetyl-CoA thiolase (ACAT) is a kind of enzyme that catalyzes hydrolysis and condensation. In the MVA pathway, AACT is the initial enzyme to catalyze two acetyl CoA to produce acetyl acetyl CoA and is essential for terpenoid backbone biosynthesis. Soto et al. (2011) found that acetoacetyl-CoA thiolase is a regulatory enzyme in isoprenoid biosynthesis involved in abiotic stress adaptation (Soto et al. 2011). Jin et al. (2012) considered that the AACT2-derived acetoacetyl-CoA pool generates isoprenoids which are required for normal growth and development, and these cannot be generated from other compensatory isoprenoid biosynthetic pathways (Jin et al. 2012). Therefore, AACT is highly significant for basic researches on the cytochemistry of terpenoid synthesis and the mechanisms underlying major metabolic processes. Furthermore, the localization of AACT in plants has been also in controversy.

Euphorbia helioscopia L. is a biennial herb of genus *Euphorbia* (Euphorbiaceae). The whole plant contains laticifers from which the white latex flowed out when the plant was broken. As an ancient folk medicine for tumor treatment, its active ingredients are terpenoids (Cai et al. 1999; Wang et al. 2012). The laticifer cell is the main synthesizing and storing site for terpenoids in *E. helioscopia*. In the mature laticifers, latex particles of various sizes were accumulated in the latex (Cai et al. 2009).

As a result, in order to further understand the compartmentalization of AACT, synthesis site, transportation and accumulation of the terpenoids in laticifers, we prepared AACT antibodies and analyzed the immunolocalization of EhAACT on the levels of proteins, cells and organelles in laticifers by using techniques of western blotting and colloidal gold immunoelectron microscopy. The results above would provide theoretical foundations for biosynthesis and regulation of terpenoids.

Materials and methods

Plant material

The healthy 3-month-old seedlings of *Euphorbia helioscopia* collected from the field at the Botanical Garden of Northwest University in Shaanxi Province (Shaanxi, People's Republic of China) were used for RNA extraction.

Molecular cloning of full length cDNA of *EhAACT*

Total RNA was extracted from stems, leaves and roots of three seedlings by the General Plant Total RNA Extraction Kit (BioTeke, China) according to the manufacturer's instructions. The quality and concentration of RNA were determined by agarose gel electrophoresis and spectrophotometric analysis (Eppendorf, Germany). For cloning the conserved fragment of *EhAACT*, the cDNA was synthesized with a PrimeScript™ 1st Strand cDNA Synthesis Kit (Takara, China) following the manufacturer's instructions. The primers of P1-S and P1-A were designed according to the conserved nucleotide sequences of *AACT* genes shared by other species (Table 1). A fragment of *EhAACT* was amplified by PCR using the cDNA as templates under the following conditions: 94 °C for

Table 1 Sequences of PCR primers used in this study

Target code	Sequence (5′ → 3′)	Purpose
P1-S	AATGACTTTGGAATGGGAGTTTG	Conserved fragment cloning
P1-A	TAAATAGTTCAGGAGCCTGAGC	Conserved fragment cloning
P2-S	GGCTTAGGAAAGTTTGATGCTG	3′-RACE outer PCR
P3-S	CTGTTACTGCTGGAAATGCCTCTA	3′-RACE inner PCR
P2-A	CCGCATCACCATATCCACG-TATTCTAG	5′-RACE outer PCR
P3-A	GTGCAGATATTCCACGCTCAAAGC	5′-RACE inner PCR
P4-S	ATTCCCCTTTCCTTCAATCTCAG	Full-length cDNA cloning
P4-A	TGTAACAGACAGAACAGGATGGC	Full-length cDNA cloning
Actin-S	GGTAACATTGTGCTCAGTGGTGG	Reference gene
Actin-A	AACGACCTTAATCTTCATGCTGC	Reference gene
P5-S	GCAGGACGAGGAAAATCATC	Quantitative real-time PCR
P5-A	CCAGCAGTAACAGAACCACC	Quantitative real-time PCR

3 min followed by 30 cycles of amplification (94 °C for 30 s, 55 °C for 30 s, and 72 °C for 1 min), and a final elongation at 72 °C for 10 min. The amplified product was purified (Tiangen, China), ligated into a pMD 19-T Vector (Takara, China) and cloned in *Escherichia coli* strain DH5α followed by sequencing. For 3′-RACE of *EhAACT*, the two round 3′-RACE was carried out with a 3′-Full RACE Core Set (Takara, China). According to the manufacturer's suggestion, an aliquot of 1 μg of total RNA was reverse transcribed to get the 3′ RACE-Ready cDNA with 3′-RACE Adaptor (provided in the kit). The Outer 3′-RACE PCR was first performed with primer P2-S (Table 1) and 3′-RACE Outer Primer (provided in the kit) under the conditions of 94 °C for 3 min followed by 35 cycles of amplification (94 °C for 30 s, 56 °C for 30 s, and 72 °C for 2 min), and a final elongation at 72 °C for 10 min. Using the Outer PCR products as templates, the Inner 3′-RACE PCR was performed with primer P3-S (Table 1) and 3′-RACE Inner Primer (provided in the kit) at the same reaction condition described above. The amplified product was purified, ligated into a pMD 19-T Vector and cloned in *E. coli* strain DH5α followed by sequencing. For 5′-RACE of *EhAACT*, the 5′ RACE-Ready cDNA was synthesized by reverse transcripting 2 μg of total RNA with 5′-CDS primer A and SMART II Aoligo supplied in the SMARTer™ RACE cDNA Amplification Kit (Clontech, USA). The Outer 5′-RACE PCR was performed with primer P2-A (Table 1) and Universal Primer A Mix (UPM, provided in the kit) at 94 °C for 3 min, followed by 25 cycles of amplification (30 s at 94 °C, 30 s at 68 °C and 2 min at 72 °C), and then 10 min extension at 72 °C. The product was used as templates for Inner PCR, which was performed with primer P3-A (Table 1) and Nested Universal Primer A (NUP, supplied in the kit) under 20 cycles of amplification (30 s at 94 °C, 30 s at 68 °C and 2 min at 72 °C). The final product was ligated into pMD19-T vector and cloned in *E. coli* strain DH5α followed by sequencing.

At last, after comparing and aligning the sequences of 5′-RACE and 3′-RACE products, the coding sequence of *EhAACT* was obtained through RT-PCR with primers P4-S and P4-A (Table 1). 2 μL of 5′ RACE-Ready cDNA was used for the PCR in a total volume of 50 μL under the following conditions: 30 cycles of amplification (98 °C for 10 s, 55 °C for 5 s, 72 °C for 2 min). The final product was ligated into pMD19-T vector and cloned in *E. coli* strain DH5α followed by sequencing. Finally, the *EhAACT* sequence had been submitted to NCBI Genebank and the accession number is KP995935.

Bioinformatics analysis

The cDNA sequence of *EhAACT* was compared online in the non-redundant peptide database at the National Center for Biotechnology Information (NCBI) (http://www.ncbi.nlm.nih.gov). A coding sequence was predicted by NCBI ORF Finder (http://www.ncbi.nlm.nih.gov/gorf/orfig.cgi) and compared with other *AACT* by NCBI BLAST (http://blast.ncbi.nlm.nih.gov/Blast.cgi). Subsequently, multiple alignment analysis was performed with DNAMAN 6.0 software. And a phylogenetic tree was constructed using MEGA 6.0 software by applying the neighbor-joining method and was corrected using Poisson correction method.

Expression pattern analysis of *EhAACT* by real-time quantitative PCR

The expression level of *EhAACT* in roots, stems, and leaves were quantified with SYBR® Premix Ex Taq™ kit (Tli RNaseH Plus) (Takara, Japan) in the CFX96™ Real-Time PCR System (Bio-Rad, United States). After an initial denaturation at 95 °C for 10 s, the PCR was carried out with 39 cycles of 95 °C for 10 s, 60 °C for 30 s, and 72 °C for 20 s. The 25 μL reaction mixture included 1 μL of cDNA templates, 12.5 μL of 2 × SBRY Premix ExTaq buffer, 9.5 μL of DEPC-treated water, and 0.4 μmol/L of P5-S and P5-A primers (Table 1). The specificity of PCR products were decided through the melting curve analysis. The relative expression levels were normalized according to the internal standard of *actin* gene using the $2^{-\Delta\Delta Ct}$ method as described by Livak and Schmittgen (Livak and Schmittgen 2001). Experiments were performed in triplicate, and the results were represented as mean values ± standard error (SE).

Prokaryotic expression and antibody preparation

The plasmids were used for overexpressing 6 × His-AACT recombinant protein, which the full open reading frame of *EhAACT* was subcloned into, were transformed into Rosetta cells. The cells were grown at 37 °C for more than 16 h, and then induced by the addition of isopropyl-D-thiogalactopyranoside. Cells were collected by centrifugation and then ultrasonic broken in ice bath until solution becomes clear. The supernatant obtained after centrifugation was collected, filtered and performed affinity chromatography. Recombinant AACT proteins were purified as His fusion proteins using a nickel-nitrilotriacetic acid agarose column according to the manufacturer's instructions. Different concentrations of imidazole (50, 100, 200, 300 and 500 mmol/L) were used to elute, and the best concentration of imidazole elution was determined by SDS-PAGE. Then the purified EhAACT recombinant protein was obtained.

Protein concentrations were determined with SDS-PAGE, and then the gel pieces containing the recombinant proteins were extracted and injected directly into health rabbits. Antibodies from the rabbits were affinity purified using Cyanogen bromide-activated sepharose

(Sigma-Aldrich, USA) conjugated with recombinant proteins and the specificity of antibodies was proved by the ELISA and immunoblot (WB).

Immunoblot analysis

The stems were cut using a scalpel, and white milky sap flowing from the holes of the stems was collected using plastic micropipettes. The sap was treated by TCA-Acetone precipitation methods, and the precipitation was then re-suspended in lysis buffer (7 mol/L urea, 2 mol/L thiourea, 4% CHAPS, 2 mmol/L EDTA, 2 mmol/L Tris, 1 mmol/L PMSF). Total proteins were separated by 12% SDS-PAGE and electrophoretically transferred onto nitrocellulose membrane. The nitrocellulose membrane was incubated with EhAACT antibodies (diluted 1/1000 with TBST) overnight at 4 °C. Following incubation with the secondary antibody (diluted 1/2000 with TBST), horseradish peroxidase conjugated immunodetection was detected using the enhanced chemiluminescence system (Tanon, Shanghai, China).

Electron microscopy analysis

Stem buds were fixed in a solution of 0.25% glutaraldehyde and 4% paraformaldehyde in 0.1 mol/L phosphate buffer (pH 7.2) for 4 h at 4 °C. After rinsing, samples were dehydrated stepwise in an acetone series from 30 to 100% and embedded in LR White resin (Sigma-Aldrich, USA). ULtrathin sections (60–70 nm) collected on nickel grids (1-GN; 150 meshes) were washed with PBS for three times, and blocked in 3% BSA-TBST solution for 1 h at 37 °C. Subsequently, samples were incubated in anti-EhAACT antibodies at a dilution of 1:50 overnight at 4 °C, 37 °C for 1 h, and then washed with PBS three times. The primary antibody binding was detected using anti-rabbit IgG conjugated with Gold particles (10 nm; Bioss, Beijing, China). After washing with PBS and distilled water, the nickel nets were examined and photographed with JEM-1230 (TEM H-7650, Japan).

Results and discussion

Full-length cDNA sequence of *EhAACT* Gene

The full-length cDNA sequence of *EhAACT* was obtained by RT-PCR Using RACE method (GenBank Accession No. KP995935). It was found to be 1847 bp consisted of a 1239 bp coding sequence, a 207 bp 5′ untranslated region, and a 401 bp 3′ untranslated region (Fig. 1). It encoded a peptide of 413 amino acids. By using the software of Computer ProtParam Tool at http://web.expasy.org/protparam, the calculated molecular weight and isoelectric point (pI) of the deduced EhAACT protein were predicted to be 42.61 kDa and 8.7, respectively.

Homologous analysis, multiple sequence alignment and phylogenetic tree construction

Homologous analysis on a database searching with Blast at http://blast.ncbi.nlm.nih.gov/Blast.cgi and the DNAMAN Tool showed that the *EhAACT* shared above 85% identities with *Ricinus communis, Jatropha curcas, Populus euphratica* and *Hevea brasiliensis* respectively. The *AACT* sequences of other 11 species of highly homology from GenBank were taken for multiple sequence alignment analysis (Fig. 2). The result indicated that C-terminal was more conservative than N-terminal.

In order to further determine the EhAACT function, the amino acid sequence of EhAACT was compared with amino acid sequences of other species by phylogenetic tree (Fig. 3). The result showed that AACT was functionally divided into two main branches including MVA cytosolic thiolases and peroxisomal thiolases. Cytosolic thiolases known as thiolase II can catalyze two acetyl-CoA molecules to form an acetoacetyl-CoA in the first step of MVA pathway (Dyer et al. 2009). Whereas, peroxisomal thiolases called thiolases I plays a significant role in β-oxidation process of fatty acid (Hooks 2002). EhAACT were clustered into the same branch with thiolase II of *Populus trichocarpa, Hevea brasiliensis, Medicago sativa* and *Arabidopsis Thaliana* in evolutionary tree, which exhibited highly functional similarity. And it displayed functional difference with thiolases I from *Neurospora crassa* OR74A, *Oryza sativa Japonica* Group, *Arabidopsis thaliana, Rattus Norvegicus* and *Homo sapiens*. The *Arabidopsis thaliana* AACT (BAH19918.1-AT5G48230-acat2) had been reported to locate in the cytosol by subcellular localization of fusion proteins of AACT and GFP, and play an important role in process of terpenoid biosynthesis as a thiolase II (Ahumada et al. 2008). In addition, the protein of AACT1 of *Medicago sativa* (MsAACT1) known as a thiolase II was also found to locate in cytosol and regulated the MVA pathway during abiotic stress adaptation (Soto et al. 2011). In *Hevea brasiliensis*, the gene of acetyl-CoA acetyltransferase has been identified and it was highly expressed in latex, its protein was also functionally classified as a thiolase II and was thought to be a major enzyme in the process of terpenoid biosynthesis (Sando et al. 2008). Therefore, from stated above, it can be verified that the EhAACT belongs to thiolase II on the function and catalyzes terpenoid biosynthesis in MVA pathway.

Organ distribution of *EhAACT*

The mRNA transcripts of *EhAACT* were detected by real-time RT-PCR (Fig. 4). The *EhAACT* expression level was similar in roots and stems, and little higher in leaves, and there was no significant difference ($P < 0.05$). The expression of *AACT* gene which was cloned from *Houttuynia*

```
1       AGCAGTGGTATCAACGCAGAGTACATGGGGAGTAATGATCATATGAAATATGGATATGGC
61      TTGCTGACAAGAGAGACGTATTTACGCATCCCAACGAAGAAAAGCTTAATTCTGATAATT
121     CCATGACGTTATAATTATATATATTATTCTAACAATTCTTCTTTTCATTCCCCTTTCCTT
181     CAATCTCAGCTTGTGTTCTCTCCATCG[ATG]GCTTCTTCTTCTTCTGATCCTATTAAGCCT
1                                   M  A  S  S  S  S  D  P  I  K  P
241     CGAGATGTTTGCATCGTGGGTGTTGCTCGTACGCCGATTGGTTCCTTTCTCGGCTCCCTT
12       R  D  V  C  I  V  G  V  A  R  T  P  I  G  S  F  L  G  S  L
301     TCTTCTATCTCTGCCACAAAGCTTGGCTCTATAGCTATTCAATGTGCTCTTAAAAGGGCA
32       S  S  I  S  A  T  K  L  G  S  I  A  I  Q  C  A  L  K  R  A
361     AATGTTGATCCATCGCTTGTGCAAGAGGTGTTCTTTGGCAATGTTCTTAGTGCTAATTTA
52       N  V  D  P  S  L  V  Q  E  V  F  F  G  N  V  L  S  A  N  L
421     GGACAAGCTCCTGCGAGGCAGGCTGCTTTAGGTGCGGGTATACCTAATTCAGTGATCTGC
72       G  Q  A  P  A  R  Q  A  A  L  G  A  G  I  P  N  S  V  I  C
481     ACCACTATCAATAAAGTTTGTTCATCTGGGATGAAAGCAACAATGCTTGCTGCACAGACT
92       T  T  I  N  K  V  C  S  S  G  M  K  A  T  M  L  A  A  Q  T
541     ATCCAATTAGGTGTCCATGACATTGTTGTCGCTGGTGGGATGGAAAGCATGTCTAATGCA
112      I  Q  L  G  V  H  D  I  V  V  A  G  G  M  E  S  M  S  N  A
601     CCCAAGTATCTCCCAGAAGCAAGAAATGGTTCAAGGCTAGGACATGATACTATCATTGAT
132      P  K  Y  L  P  E  A  R  N  G  S  R  L  G  H  D  T  I  I  D
661     GGCATACTGAAAGATGGCCTCTGGGATGTATATAATGACTTTGGAATGGGAGTTTGTGGA
152      G  I  L  K  D  G  L  W  D  V  Y  N  D  F  G  M  G  V  C  G
721     GAAATATGTGCTGATCAGCACAAAATAACAAGAGAACAGCAGGATGCTTATGCTATCCAG
172      E  I  C  A  D  Q  H  K  I  T  R  E  Q  Q  D  A  Y  A  I  Q
781     AGCTTTGAGCGTGGAATTTCTGCACAAAATGCTGGTCTTTTCTCCTGGGAGATTGCTCCA
192      S  F  E  R  G  I  S  A  Q  N  A  G  L  F  S  W  E  I  A  P
841     GTTGAAGTATCTGCAGGACGAGGAAAATCATCTACTATTGTTGATAAGGATGAAGGTCTA
212      V  E  V  S  A  G  R  G  K  S  S  T  I  V  D  K  D  E  G  L
901     GGAAAGTTTGATGCTGCAAAACTGAGGAAACTTAGACCGAGCTTCAAGAAGAATGGTGGT
232      G  K  F  D  A  A  K  L  R  K  L  R  P  S  F  K  K  N  G  G
961     TCTGTTACTGCTGGAAATGCATCTAGCATAAGTGATGGTGCAGCAGCACTAGTGTTGGTA
252      S  V  T  A  G  N  A  S  S  I  S  D  G  A  A  A  L  V  L  V
1021    AGTGGGGAAAAGGCAATGAAACTTGGTTTGCATGTTATTGCTAGAATACGTGGATATGGT
272      S  G  E  K  A  M  K  L  G  L  H  V  I  A  R  I  R  G  Y  G
1081    GATGCGGCTCAGGCTCCTGAACTATTTACTACTGCACCAGCCCTTGCTATTCCAATAGCT
292      D  A  A  Q  A  P  E  L  F  T  T  A  P  A  L  A  I  P  I  A
1141    ATTTCAAATGCTGGTTTGGATGCTTCAAAGATTGATTATTATGAAATAAATGAAGCATTT
312      I  S  N  A  G  L  D  A  S  K  I  D  Y  Y  E  I  N  E  A  F
1201    GCTGTTGTTGCCCTTGCAAATCAAAAGCTTCTCAATATAAATCCAGAAAGATTAAATGCC
332      A  V  V  A  L  A  N  Q  K  L  L  N  I  N  P  E  R  L  N  A
1261    CATGGCGGAGGTGTATCTTTGGGACATCCAATAGGATGCAGCGGGGCTCGTATATTAGTC
352      H  G  G  G  V  S  L  G  H  P  I  G  C  S  G  A  R  I  L  V
1321    ACATTATTAGGGATACTTAGACATAAGAACGGTAAGTATGGGGTTGCTGGCATCTGCAAT
372      T  L  L  G  I  L  R  H  K  N  G  K  Y  G  V  A  G  I  C  N
1381    GGTGGAGGAGGGGCATCTGCACTTGTTGTGGAGCTCATGTCTGTTCCACTTGGACGTTCA
392      G  G  G  G  A  S  A  L  V  V  E  L  M  S  V  P  L  G  R  S
1441    TCGCTC[TGA]GCAGATTTCTTACTCCAGCATGATATTGTGCAATATAGCCATCCTGTTCTG
412      S  L  *
1501    TCTGTTACACAAATAAACGAGCATGCATAACTCAGTACTCGGGTAAAATCTGATCTATGT
1561    AGATGTTGATATTGTTTGGTCACTATATCAGCTGGCCAGGTATTCAAGATGGGATGCATA
1621    TCTACGAAGAGATCGAATTGCTGATACAGTTACATGCCTTGCGGCTTTCCAAGAAGAGAA
1681    TGTATTTAGTGAAAGAACTATACAGATTTGTATGAATAATTCATAGTTGTACATTATGTA
1741    TCTGCAATCATATGATAGTAGAGCTTTGCCTCAAGCACTTGATTCAAAATGAACGTCTCA
1801    ATTGCTTCAAAAAAAAAAAAAAAAGGATATCACTTTAGTGATCACCT
```

Fig. 1 The full-length cDNA sequence of *EhAACT* and deduced amino acid sequence. "ATG" is initiation codon, "TGA" is termination codon

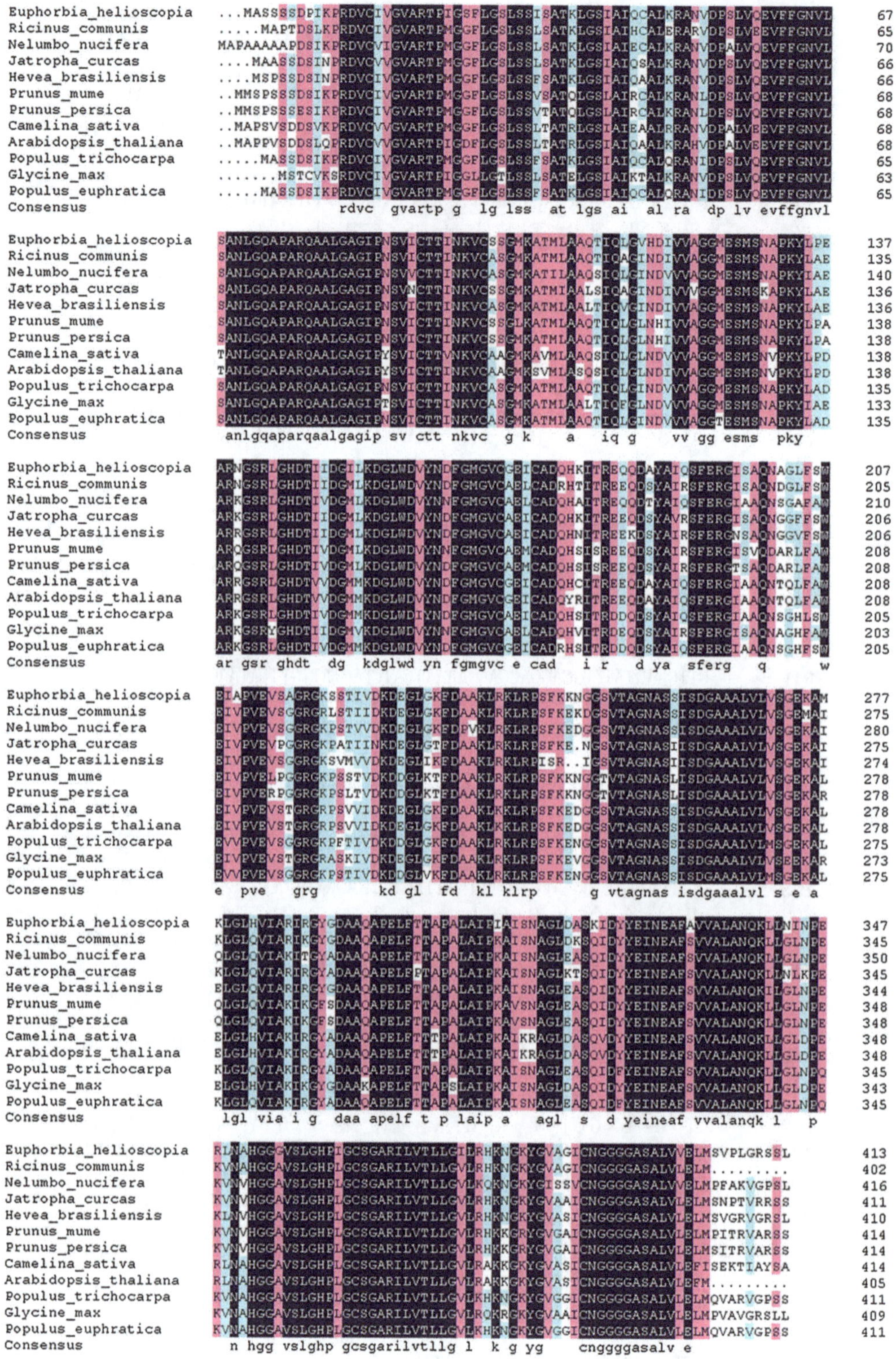

Fig. 2 Multiple sequence alignment of deduced *EhAACT* with those of other species. Blue shading means 100% sequence identity, pink shading means more than 75% sequence identity, light blue means more than 50% sequence identity; underline means specific domain of thiolase

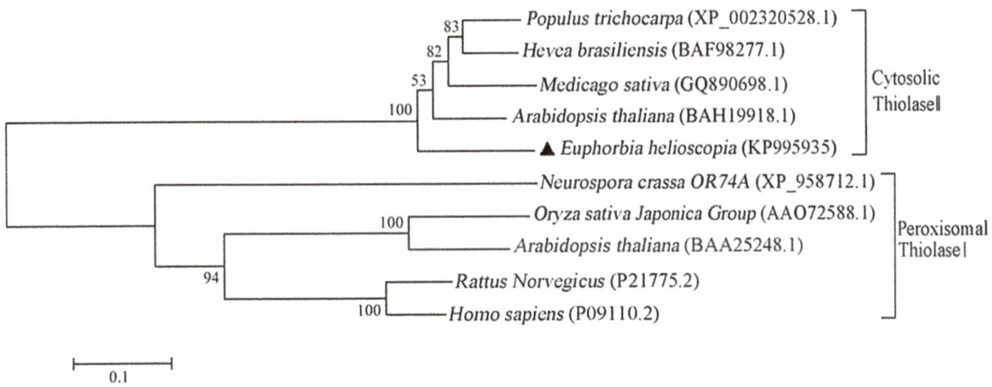

Fig. 3 The phylogenetic tree of acetoacetyl-CoA thiolase (AACT) was constructed with the neighbor-joining method of the program MGEA 6.0 and reflected functional classification of thiolases by comparing with other species. Bootstrap value was calculated with 1000 permutations to identify the homogeneity of the analysis. *Euphorbia helioscopia* AACT exhibited highly functional similar with *Populus trichocarpa, Hevea brasiliensis, Medicago sativa* and *Arabidopsis Thaliana*. And it displayed a more distant relationship with protein of AACT from other species, including *Neurospora crassa* OR74A, *Oryza sativa Japonica* Group, *Arabidopsis thaliana, Rattus Norvegicus* and *Homo sapiens*

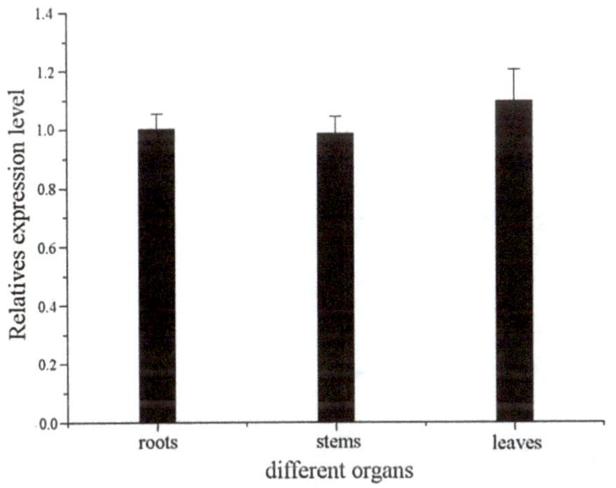

Fig. 4 The relative expression levels of *EhAACT* gene in the roots, stems and leaves of *Euphorbia helioscopia*. There was similar expression level among roots, leaves and stems of *EhAACT* and no significant difference ($P < 0.05$) among them

cordata was highest in stem, followed by the underground stem, the expression in flowers and leaves was relatively low (Yao et al. 2015). The transcription level of *SmAACT* was relatively higher in roots than that in stems and leaves in *Salvia miltiorrhiza* (Cui et al. 2010). Therefore, mRNAs for individual pathway enzymes or isozymes can accumulate to significantly different levels in the different organs or developmental stages (Vranová et al. 2013).

Prokaryotic expression and antibody preparation

It was successful for the prokaryotic expression by using a complete open reading frame of *EhAACT*. All the bacterial solution of prokaryotic expression, the bacterial fluid after induced by IPTG, the supernatant after flow, and the protein solution after affinity chromatography of nickel column was detected by the method of SDS-PAGE (Fig. 5). The molecular weight of Ni column affinity sample containing GST-His-EhAACT recombinant protein product of *E. helioscopia* is about 86 kDa. When vector fragment of 43 kDa is removed, the remaining is about 43 kDa which is the molecule weight of recombinant EhAACT protein and extremely close with the EhAACT molecule weight predicted (42.61 kDa).

The purified AACT protein was used as antigen to immunize rabbits. After seven times of immunization,

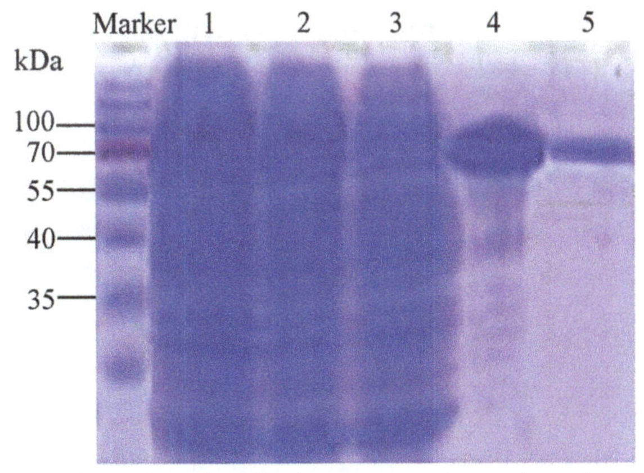

Fig. 5 SDS-PAGE analysis of prokaryotic expression products of EhAACT. Marker: Protein marker; 1: All the bacterial solution of prokaryotic expression; 2: Bacterial solution produced by IPTG induction; 3: The supernatant of breaking bacterial cells; 4: Fusion protein produced by Ni-chelating affinity chromatography; 5: Protein solution diluted tenfolds

the serum was taken to obtain antibodies. The concentration of the protein antigen in the experiment was about 1 μg/mL. The titer of antibody indicates the effective concentration of antibody, which was an important index of antibody quality detection. The antibody titer was detected by indirect ELISA, and the results were shown in Fig. 6. The titer of antibody is about 400 K, which indicates that our EhAACT antibody is qualified to use in the following immunolocalization experiment.

Immunoblotting analysis of EhAACT antibodies

In order to verify if EhAACT functions as endogenous AACT in *E. helioscopia* laticifers, western blotting was proceeded. Immunoblotting analysis found out an obvious band appearing at little above 40 kDa which corresponds to the relative molecular mass 43 kDa of AACT (Fig. 7), and showed that the EhAACT antibodies specifically recognized the endogenous proteins. In addition, this result also proved that the anti-EhAACT antibody is highly specific.

Subcellular localization of AACT in laticifers

Laticifers were polygonal, and their cell wall was thickened obviously, these characteristics make them easier to be identified from the other cells around. Our present studies showed that in the early development of laticifers, cytoplasm is relatively uniform, there are a large number of mitochondria with distinct cristae (Fig. 8a, b), and rich endoplasmic reticulums dispersed in the cytoplasm (Fig. 8b–d). Colloidal gold particles were observed to locate in the cytosol (Fig. 8a, d), endoplasmic reticulum cisterna (Fig. 8b–d), the cytosol around endoplasmic reticulum, and small vacuoles from endoplasmic reticulum (Fig. 8c, d). With the development of laticifers, it could be seen that endoplasmic reticulum cisternae swelled to form lots of vacuoles, and these vacuoles were always separated by cytosol, colloidal gold particles were distributed around endoplasmic reticulum and mainly in these vacuoles (Fig. 8e, f). Some colloidal gold particles were also observed to locate in cytosol and endoplasmic reticulum between vacuoles (Fig. 8g). In the negative control which was not treated with the primary antibody incubation, no colloidal gold particles were found (Fig. 8h), which indicated that the positive reaction in our experiment is reliable. In the past decades of years, many of the mammalian MVA pathway enzymes were mostly reported to locate in peroxisomes (Kovacs et al. 2002, 2007). AACT was also identified as a peroxisomal protein in mammals (Thompson and Krisans 1990; Olivier et al. 2000). However, Hogenboom et al. (2004) found a cytosolic localization of endogenous human MK

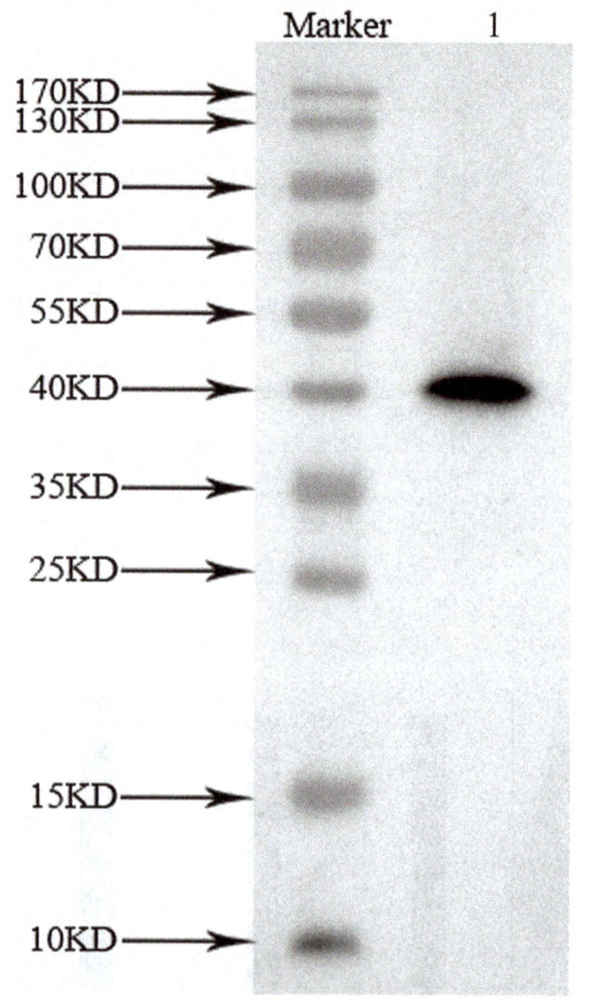

Fig. 7 Immunoblot analysis with EhAACT antibody. Lane 1: marker; Lane 2: the corresponding protein gel blot probed with affinity-purified anti-EhAACT antibody

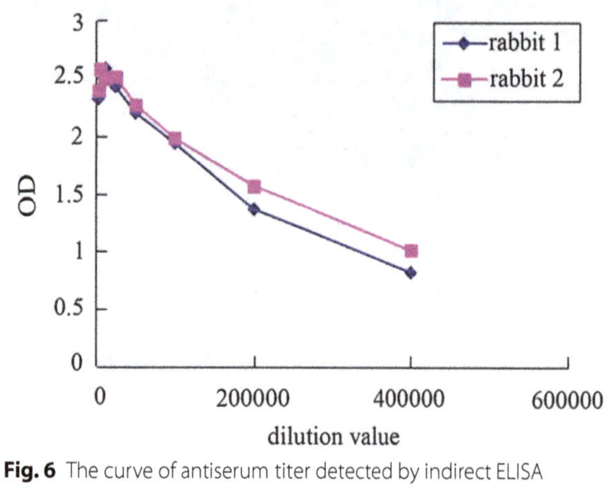

Fig. 6 The curve of antiserum titer detected by indirect ELISA

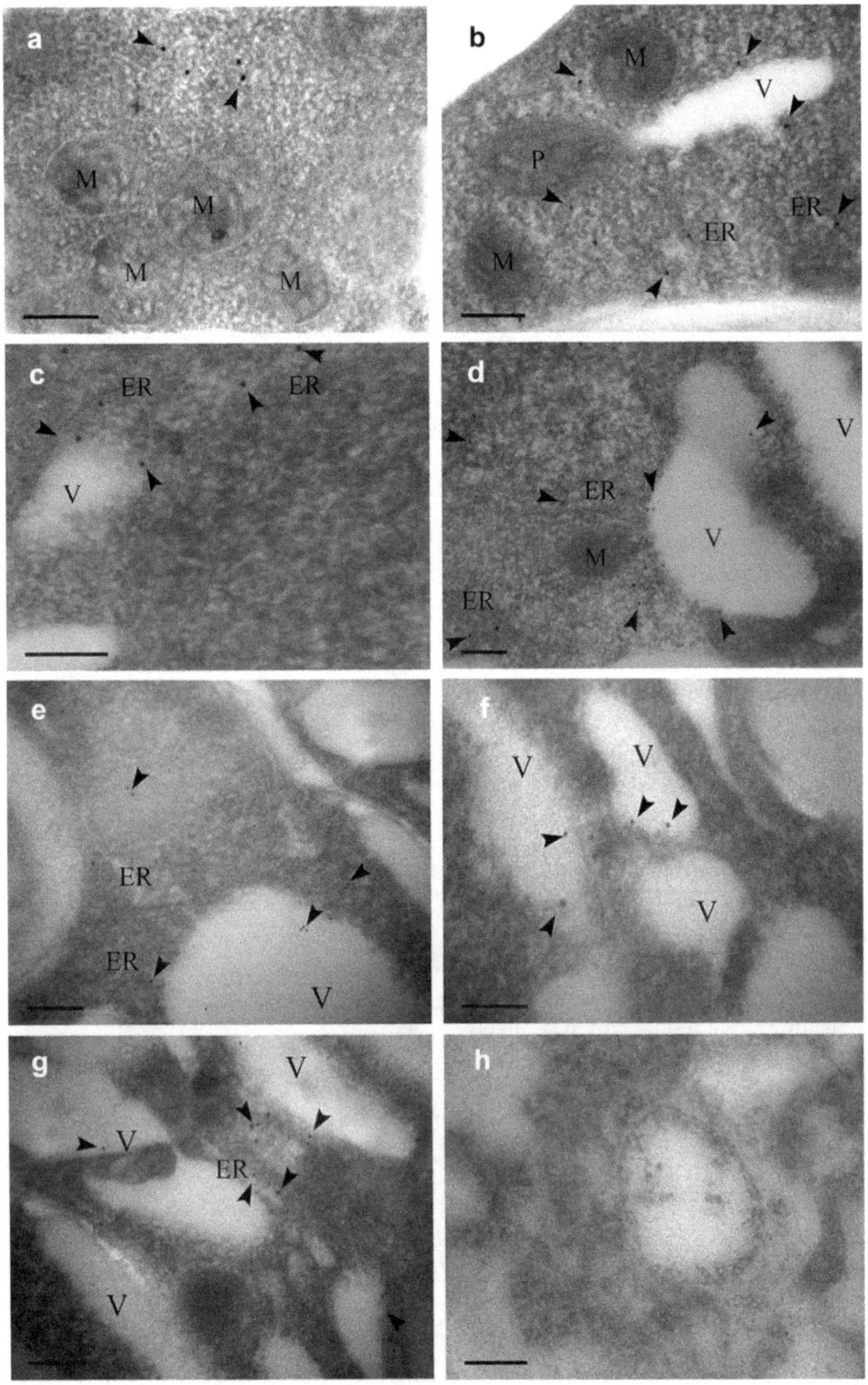

Fig. 8 Immunolocalization of the AACT in *Euphorbia helioscopia* laticifers. **a** AACT signals appeared to locate in the cytosol randomly (arrows). **b** AACT signals appeared to locate mainly in endoplasmic reticulum and cytosol near small vacuoles from endoplasmic reticulum (arrows). **c** AACT signals appeared to locate in the endoplasmic reticulum, cytosol near small vacuoles from endoplasmic reticulum (arrows). **d** Golden particles conjugated AACT antibodies were found to locate in the endoplasmic reticulum, cytosol near small vacuoles from endoplasmic reticulum (arrows). **e** AACT signals locate in small vacuoles from endoplasmic reticulum and cytosol (arrows). **f** AACT signals locate in small vacuoles from endoplasmic reticulum (arrows). **g** Golden particles conjugated AACT antibodies were found to locate in the endoplasmic reticulum, small vacuoles from endoplasmic reticulum and cytosol between small vacuoles (arrows). **h** Negative controls showed no gold labeling in the laticifer cells. Bars = 200 nm. ER, Endoplasmic reticulum; M, Mitochondrion; P, Plastid; V, Vacuole

using immunofluorescence microscopy and immunocytochemistry, the postulated role of peroxisomes in isoprenoid biosynthesis has been challenged (Hogenboom et al. 2004). In plants, Sapir-Mir et al. (2008) pointed out a new model for compartmentalization of the plant MVA pathway in the three cellular compartments: cytosol, ER, and peroxisome. Endogenous *Arabidopsis* HMGR was observed to locate in steady state within ER and predominantly within spherical, vesicular structures that range from 0.2- to 0.6-mm diameter, and locate in the cytoplasm and within the central vacuole in differentiated cotyledon cells (Leivar et al. 2005). Expression of green fluorescent protein (GFP)-tagged versions of squalene synthase (SQS) in onion epidermal cells demonstrated that SQS1 is targeted to the endoplasmic reticulum (ER) membrane and that this location is exclusively dependent on the presence of the SQS1 C-terminal hydrophobic trans-membrane domain (Busquets et al. 2008). AACT was recently identified as a peroxisomal protein by a proteomic approach of *Arabidopsis* (Reumann et al. 2007). In the model of Sapir-Mir et al., AACT were also considered as peroxisomal localization (Sapir-Mir et al. 2008). While, in *Arabidopsis thaliana*, a member of the AACT family called acat2 has been characterized with cytosolic location (Carrie et al. 2007; Ahumada et al. 2008). Ahumada et al. (2008) reported the characterization of two genes from *Arabidopsis thaliana*, ACT1 and ACT2, which encode two closely related AACT isoforms (Ahumada et al. 2008). Transient expression of constructs encoding AACT1 and AACT2 fused to GFP revealed that the two proteins show a different subcellular localization. AACT1 is found in peroxisomes, AACT2 locate in the cytosol and the nucleus. These obtained results are in agreement with the involvement of AACT2 in catalyzing the first step of the MVA pathway. We found EhAACT involving in the MVA pathway was mainly distributed in ER and vacuoles dilated from ER in *E. helioscopia* laticifers by immunogold electron microscopy, which is consistent with reports about HMGR in *Arabidopsis* (Leivar et al. 2005) and SQS1 in onion epidermal cells (Busquets et al. 2008). In addition, we also found diphosphomevalonate decarboxylase (MDC), the last rate-limiting enzyme of generating the IPP of terpenoid in MVA pathway, to locate mainly in the cisternae of endoplasmic reticulum and small vacuoles from endoplasmic reticulum in *E. helioscopia laticifers* (Chai et al. 2017).

Furthermore, Groeneveld purified latex particles from laticifers of a number of *Euphorbias* species by using gel filtration chromatography and showed that the latex particles were composed of triterpenoids (Groeneveld 1976). Besides, latex particles were also found to be triterpenes in *E. lathyris* latex (Skrukrud 1987). Moreover, some reports pointed out that the latex particles were synthesized in vacuoles originated from ER (Fineran 1983; Skrukrud 1987; Cai et al. 2009). Therefore, it can be say that the latex particles are the accumulation site of terpenoid end products in the laticifers, and the mevalonate to terpenoid converting activity is associated with the structure which was identified as a vacuole originated from ER. Based on stated above, we further speculated that as an initial enzyme in MVA pathway for terpenoid backbone biosynthesis, the EhAACT located in cytosol could be transported to small vacuoles originated from ER to synthesize terpenoid backbones. Lastly, the terpenoid was further synthesized into latex particles in these small vacuoles in *E. helioscopia* laticifers.

Abbreviations

IPP: isoprene phosphate; DMAPP: dimethylallyl pyrophosphate; MVA: mevalonate; MEP: 2-C-methyl-D-erythritol-4-phosphate; AACT: acetyl-CoA acetyltransferase; ER: endoplasmic reticulum; RT-PCR: reverse transcription-polymerase chain reaction; MK: mevalonate kinase; HMGR: 3-hydroxy-3-methyl glutaryl CoA reductase; SQS: squalene synthase; GFP: green fluorescent protein; MDC: diphosphomevalonate decarboxylase.

Authors' contributions

All the authors have equally contributed towards this manuscript. All authors read and approved the final manuscript.

Acknowledgements

The authors thank Dr. Danyang Wang for guidance in experiment and critical reading of the manuscript.

Competing interests

The authors declare that they have no competing interests.

Funding

This work was supported by the National Natural Science Foundation of China (31270220) and State Key Laboratory Research Program Funded by Shaanxi Provincial Education Department (12JS084).

References

Ahumada I, Cairó A, Hemmerlin A, González V, Pateraki I, Bach TJ, Rodriguez-concepción M, Campos N, Boronat A (2008) Characterization of the gene family encoding acetoacetyl-coa thiolase in *Arabidopsis*. Funct Plant Biol 35(11):1100–1111

Busquets A, Keim V, Closa M, Del AA, Boronat A, Arró M, Ferrer A (2008) *Arabidopsis thaliana* contains a single gene encoding squalene synthase. Plant Mol Biol 67(1–2):25–36

Cai Y, Wang J, Liang B (1999) Antitumor activity of the root of *Euphorbia helioscopia* in vitro. J Chin Med Mater 22(2):85–87

Cai X, Li W, Yin L (2009) Ultrastructure and cytochemical localization of acid phosphatase of laticifers in *Euphorbia kansui* Liou. Protoplasma 238(1–4):3–10

Carrie C, Murcha MW, Millar AH, Smith SM, Whelan J (2007) Nine 3-ketoacyl-CoA thiolases (KATs) and acetoacetyl-CoA thiolases (ACATs) encoded by five genes in *Arabidopsis thaliana* are targeted either to peroxisomes or cytosol but not to mitochondria. Plant Mol Biol 63(1):97–108

Chai J, Wang D, Peng Y, Zhao XY, Zhang Q, Li P, Fang XA, Wang M, Cai X (2017) Molecular cloning, expression and immunolocalization analysis of diphosphomevalonate decarboxylase involved in terpenoid biosynthesis from *Euphorbia helioscopia* L. Biotechnol Biotec Eq 31(6):1106–1115

Chappell J (1995) The biochemistry and molecular biology of isoprenoid metabolism. Plant Physiol 107(1):1–6

Cui GH, Wang XY, Feng H, Zhao JX, Huang LQ (2010) Molecular cloning and SNP analysis of a acetyl-CoA C-acetyltransferase gene (*SmAACT*) from *Salvia miltiorrhiza*. Acta Pharm Sin 45(6):785–790

Duncan RE, Lau D, Eisohemy A, Archer MC (2004) Gernaiol and β-ionone inhibit proliferation, cell cycle progression, and cyclin-dependent kinase 2 activity in MCF-7 breast cancer cells independent of effects on HMG-CoA reductase activity. Biochem Pharmacol 68(9):1739–1747

Dyer JH, Maina A, Gomez ID, Cadet M, Oeljeklaus S, Schiedel AC (2009) Cloning, expression and purification of an acetoacetyl CoA thiolase from sunflower cotyledon. Int J Biol Sci 5(7):736–744

Eisenreich W, Schwarz M, Cartayrade A, Arigoni D, Zenk MH, Bacher A (1998) The deoxyxylulose phosphate pathway of terpenoid biosynthesis in plants and microorganisms. Chem Biol 5(9):R221–R233

Fineran BA (1983) Differentiation of non-articulated laticifers in *Poinsettia* (*Euphorbia pulcherrima* Willd.). Ann Bot 52(3):279–293

Groeneveld HW (1976) A comparative investigation of the particle fractions from *Hoya*, *Dischidia* and *Euphorbia* latices. Acta Bot Neerl 25(2):167–178

Hogenboom S, Tuyp JJ, Espeel M, Koster J, Wanders RJ, Waterham HR (2004) Phosphomevalonate kinase is a cytosolic protein in humans. J Lipid Res 45(4):697–705

Hooks MA (2002) Molecular biology, enzymology, and physiology of β-oxidation. Plant Peroxisomes. Springer, Netherlands

Hua WF, Cai SH (2006) Basic and clinical studies on antitumor activity of β-elemene. Chin Med Mater 29(1):93–96

Jin H, Song Z, Nikolau BJ (2012) Reverse genetic characterization of two paralogous acetoacetyl-CoA thiolase genes in *Arabidopsis* reveals their importance in plant growth and development. Plant J 70(6):1015–1032

Kovacs WJ, Olivier LM, Krisans SK (2002) Central role of peroxisomes in isoprenoid biosynthesis. Prog Lipid Res 41(5):369–391

Kovacs WJ, Tape KN, Shackelford JE, Duan X, Kasumov T, Kelleher JK, Brunengraber H, Krisans SK (2007) Localization of the pre-squalene segment of the isoprenoid biosynthetic pathway in mammalian peroxisomes. Histochem Cell Biol 127(3):273–290

Lange BM, Wildung MR, McCaskill D, Croteau R (1998) A family of transketolases that directs isoprenoid biosynthesis via amevalonate-independent pathway. Proc Natl Acad Sci USA 95(5):2100–2104

Leivar P, González VM, Castel S, Trelease RN, Lópeziglesias C, Arró M, Boronat A, Campos N, Ferrer A, Fernàndez-Busquets X (2005) Subcellular localization of *Arabidopsis* 3-hydroxy-3-methylglutaryl-coenzyme A reductase. Plant Physiol 137(1):57–69

Livak KJ, Schmittgen TD (2001) Analysis of relative gene expression data using real-time quantitative PCR and the 2(-Delta Delta C(T)) method. Methods 25(4):402–408

Olivier LM, Kovacs W, Masuda K, Keller GA, Krisans SK (2000) Identification of peroxisomal targeting signals in cholesterol biosynthetic enzymes. AA-CoA thiolase, hmg-coa synthase, MPPD, and FPP synthase. J Lipid Res 41(12):1921–1935

Reumann S, Babujee L, Ma C, Wienkoop S, Siemsen T, Antonicelli GE, Rasche N, Luder F, Weckwerth W, Jahn O (2007) Proteome analysis of *Arabidopsis* leaf peroxisomes reveals novel targeting peptides, metabolic pathways, and defense mechanisms. Plant Cell 19(10):3170–3193

Rodríguez-Concepción M, Boronat A (2015) Breaking new ground in the regulation of the early steps of plant isoprenoid biosynthesis. Curr Opin Plant Biol 25:17–22

Sando T, Takaoka C, Mukai Y, Yamashita A, Hattori M, Ogasawara N, Fukusaki E, Kobayashi A (2008) Cloning and characterization of mevalonate pathway genes in a natural rubber producing plant, *Hevea brasiliensis*. Biosci Biotechnol Biochem 72(8):2049–2060

Sapir-Mir M, Mett A, Belausov E, Tal-Meshulam S, Frydman A, Gidoni D, Eyal Y (2008) Peroxisomal localization of *Arabidopsis* isopentenyl diphosphate isomerases suggests that part of the plant isoprenoid mevalonic acid pathway is compartmentalized to peroxisomes. Plant Physiol 148(3):1219–1228

Skrukrud CL (1987) Terpenoid biosynthesis in *Euphorbia lathyris* and *Copaifera* ssp. Biomass Fuels

Soto G, Stritzler M, Lisi C, Alleva K, Pagano ME, Ardila F, Mozzicafreddo M, Cuccioloni M, Angeletti M, Ayub ND (2011) Acetoacetyl-CoA thiolase regulates the mevalonate pathway during abiotic stress adaptation. J Exp Bot 62(15):5699–5711

Thompson SL, Krisans SK (1990) Rat liver peroxisomes catalyze the initial step in cholesterol synthesis. The condensation of acetyl-CoA units into acetoacetyl-CoA. J Biol Chem 265(10):5731–5735

Vranová E, Coman D, Gruissem W (2013) Network analysis of the MVA and MEP pathways for isoprenoid synthesis. Annu Rev Plant Biol 64(1):665–700

Wanders RJ, Waterham HR (2006) Biochemistry of mammalian peroxisomes revisited. Annu Rev Biochem 75(1):295–332

Wang ZY, Liu HP, Zhang YC, Guo LQ, Li ZX, Shi XF (2012) Anticancer potential of *Euphorbia helioscopia* L. extracts against human cancer cells. Anat Rec 295(2):223–233

Yao YZ, Xiao-Ying LI, Wei L, Xian-Jin WU, Liu SG, Tang YL (2015) Cloning, expression, and bioinformatics analysis of acetyl-CoA C-acetyltransferase gene in *Houttuynia cordata*. Chin Tradit Herbal Drugs 46(1):107–111

6

Heterologous expression of rice *9-cis-epoxycarotenoid dioxygenase 4* (*OsNCED4*) in *Arabidopsis* confers sugar oversensitivity and drought tolerance

San-Gwang Hwang[1*], Chia-Yun Lee[2] and Ching-Shan Tseng[3]

Abstract

Background: The *9-cis-epoxycarotenoid dioxygenases OsNCED4* was cloned from rice in conjunction with *OsNCED 1-3* and *5*, of which *3* has been shown to function in ABA biosynthesis and alteration of leaf morphology. In higher plants, NCEDs have been shown to be key enzymes controlling ABA biosynthesis and belong to a differentially expressed gene family. Aside from *OsNCED3*, it remains largely unknown if other *OsNCED* genes are involved in ABA biosynthesis in rice. Thus, transgenic Arabidopsis plants overexpressing *OsNCED4* were generated in the 129B08/*nced3* mutant background to explore *OsNCED4* function in ABA biosynthesis.

Results: Heterologous expression of *OsNCED4* in Arabidopsis increased ABA levels and altered plant size and leaf shape, delayed seed germination, caused sugar oversensitivity in post-germination growth, and enhanced tolerance to drought. The native *OsNCED3* and *OsNCED4* promoters were expressed in an overlapping pattern in rice seeds and young seedlings, suggesting possible functional redundancy between *OsNCED3* and *OsNCED4*. At the one-leaf stage, similar regulation of *OsNCED3* and *OsNCED4* gene expression in roots or leaves in response to moderate salt stress (150 mM NaCl) was observed.

Conclusion: Like *OsNCED3*, *OsNCED4* is functionally active in ABA biosynthesis in rice. *OsNCED3* and *OsNCED4* might play redundant roles in controlling ABA biosynthesis in rice, as suggested by GUS staining assay, but this should be further analyzed through complementation of rice *NCED* knockout mutants.

Keywords: Abscisic acid, 9-*cis*-epoxycarotenoid dioxygenase, Seed germination, Post-germination growth arrest, Drought tolerance

Background

The functions of abscisic acid (ABA) in regulation of seed dormancy, seedling growth and development, stomatal closure, and stress tolerance is well studied (Jiang and Yu 2009; Zhu et al. 2009, 2011; Waterland et al. 2010; Gao et al. 2011; Bauer et al. 2013; Merilo et al. 2015). Connections among ABA, molecular signaling components, and nutrition have been identified. For example, Jiang and Yu (2009) suggested that the transcription factor AtWRKY2 regulates seed germination and post-germination developmental arrest in Arabidopsis via its response to ABA. At the same year, Zhu et al. (2009) reported that the glucose-induced germination delay in rice seeds is due to the prevention of ABA degradation, rather than an increase in ABA biosynthesis. Zhu et al. (2011) further elaborated that downregulation of *CYP707A2* expression, a gene encoding an ABA 8′-hydroxylase involved in ABA catabolism, and the subsequent reduction in ABA degradation, are closely associated with the delay of seed germination and seedling growth in Arabidopsis. More recently, Gao et al. (2011) characterized the protein function of AtCPR5 and revealed that seed germination and early

*Correspondence: sghwang@dragon.nchu.edu.tw
[1] Department of Horticulture, National Chung Hsing University, 145 Xingda Road, South District, Taichung 40227, Taiwan, ROC
Full list of author information is available at the end of the article

seedling growth are independently regulated through the ABA and lipoxygenase (LOX) pathways.

ABA regulates numerous physiological responses in addition to seed germination and seedling growth. Previous reports have shown that ABA synthesis in guard cells is essential and sufficient for stomatal closure in response to declined relative humidity (Bauer et al. 2013; Merilo et al. 2015). Waterland et al. (2010) demonstrated the involvement of ABA in drought tolerance. The roles of ABA and ABA signaling in plant abiotic stress responses, including drought tolerance, have been recently reviewed and discussed (Sah et al. 2016; Vishwakarma et al. 2017).

In higher plants, 9-*cis*-epoxycarotenoid dioxygenases (NCEDs) are thought to be the key enzymes controlling ABA biosynthesis and stress tolerance (Iuchi et al. 2001; Sun et al. 2012; Vishwakarma et al. 2017). To date, five rice *NCED* genes have been reported (Oliver et al. 2007; Welsch et al. 2008; Zhu et al. 2009). However, only *OsNCED3* (GenBank Accession No. AY838899) was characterized as functionally active in ABA biosynthesis (Hwang et al. 2010). The biological functions of the other four rice *NCED* genes, *OsNCED1* (GenBank Accession No. AY838897), *OsNCED2* (GenBank Accession No. AY838898), *OsNCED4* (GenBank Accession No. AY838900) and *OsNCED5* (GenBank Accession No. AY838901) have yet to be deciphered. It remains unknown whether rice *OsNCED4* gene is involved in ABA biosynthesis or ABA-regulated physiological processes.

Previous studies of the *NCED* gene family in Arabidopsis revealed that different *NCED* genes function in different plant tissues to regulate ABA biosynthesis (Iuchi et al. 2001; Tan et al. 2003; Lefebvre et al. 2006; Agustí et al. 2007; Martínez-Andújar et al. 2011). For example, *AtNCED3* is primarily induced at high levels in leaves in response to water stress (Endo et al. 2008), and *AtNCED6* and *AtNCED9* appear to be major players regulating ABA biosynthesis in developing seeds (Lefebvre et al. 2006). More recently, Huo et al. (2013) found that lettuce *LsNCED4* is required for heat-inhibition of seed germination and its expression in leaves is induced by heat but not water stress. In contrast, *LsNCED2* and *LsNCED3* are both induced by water stress but not heat.

A real-time RT-PCR analysis of *OsNCED3*, *OsNCED4*, and *OsNCED5* in roots of rice seedlings subjected to salt or ABA treatment revealed that these three *NCED* genes are salt- and ABA-inducible (Welsch et al. 2008). In transgenic rice plants harboring an *OsNCED3::gfp* transgene, the *OsNCED3* promoter activity is strongly increased in the roots and leaves under drought and high-salt (400 mM NaCl) conditions, but little to no activity can be observed in grains and flowers (Bang et al. 2013). It remains unclear if *OsNCED3* and *OsNCED4* are expressed in an overlapping or non-overlapping pattern in rice seeds and leaves.

In the present study, the rice *OsNCED4* gene was heterologously expressed in the Arabidopsis 129B08/*nced3* mutant to test if *OsNCED4* may complement the 129B08/*nced3* mutant phenotype. We used the Arabidopsis 129B08/*nced3* mutant rather than rice *OsNCED* mutant because no rice *OsNCED* mutant was available at the time when we initiated this study. The 129B08/*nced3* mutant was a T-DNA insertion mutant requested from the Nottingham Arabidopsis Stock Center. The *OsNCED4* overexpression line was used to characterize *OsNCED4* function in ABA biosynthesis, seed germination, post germination growth, and drought tolerance.

Methods

Plant materials and growth conditions

The Arabidopsis wild type and mutant plants used in this study were in the Columbia (Col-0) background. Arabidopsis plants overexpressing the rice *OsNCED4* transgene in the 129B08/*nced3* mutant background were denoted as N4C (for complementation). The growing conditions were described previously (Hwang et al. 2010). Cold-pretreated seeds (4 °C, 4 days) from WT, 129B08/*nced3*, and the transgenic lines N4C-1, and N4C-2 were grown either on water agar medium made up by adding 0.7% Phytoagar (Duchefa Biochemie, Haarlem, The Netherlands) in autoclaved ddH$_2$O, on agar plates supplemented with 0.2 or 4% glucose and half-strength MS, or on soil for the period of time specified below.

Transgenic plant isolation

Rice *OsNCED4* gene cloning, plasmid construction and transformation, and screening of homozygous transgenic lines were performed using the procedures described previously (Hwang et al. 2010), except that *OsNCED4* was overexpressed instead of *OsNCED3* and the binary vector used in this study was modified from pCAMBIA1300 rather than pCAMBIA1281Z. The modified pCAMBIA1300 vector used in this study contained a 35S promoter. Two lines, N4C-1 and N4C-2, were subsequently subjected to phenotypic comparison, germination test, ABA assay, analysis of post-germination growth arrest, and determination of drought tolerance.

To test promoter activity in rice, the *OsNCED3* promoter, encompassing 2407 bp upstream of the ATG start codon of *OsNCED3*, and the *OsNCED4* promoter, a total of 1939 bp upstream of the ATG start codon of *OsNCED4*, were amplified by polymerase chain reaction (PCR) and fused to a β-glucuronidase (GUS) coding region in the pCAMBIA1305.1 binary vector. The binary vector was then used to introduce the *OsNCED3::GUS* or *OsNCED4::GUS* constructs into Japonica rice variety TNG67. Homozygous transgenic rice plants were screened by using 70 mg/L hygromycin B (InvivoGen, USA). Two independent homozygous transgenic lines carrying the *OsNCED3::GUS* construct (denoted N3P-1 and N3P-2) and two independent homozygous transgenic lines carrying *OsNCED4::GUS* (denoted N4P-1 and N4P-2) were randomly chosen for further study.

Reverse-transcription polymerase chain reaction (RT-PCR)

In Arabidopsis, the procedures for total RNA extraction, RNA reverse transcription, and polymerase chain reaction, and the sequences of the *AtUBQ5* and *AtNCED3* gene-specific primers were previously described in Hwang et al. (2010). Other gene-specific primers used in this study were for *OsNCED4*: forward primer 5′-CCGTCCAAGGTGAAGGTGGC-3′, and reverse primer 5′-CTTCTCCGCCGTGCCGCTC-3′.

Germination test

Three independent batches of cold-pretreated seeds from WT, 129B08/*nced3*, N4C-1, and N4C-2 were grown on water agar medium. The total number of germinated seeds was counted daily for a period of 7 days. In another set of experiments, seeds from the above-mentioned plants were grown on agar plates supplemented with 0.2 or 4% glucose (Glc) and half-strength MS for 7 days. Seeds were considered germinated when the radical protruded at least 2 mm from the seed coat.

Phenotypic comparisons

Cold-treated seeds from WT, 129B08/*nced3*, N4C-1, and N4C-2 were grown in soil for 35 days and then subjected to phenotypic comparisons. To compare leaf shape, the fifth leaves were sampled after 35 days for width:length ratio determinations.

ABA assay

Seedlings grown on agar plates supplemented with 0.2 or 4% glucose for 10 days were harvested for ABA analysis. The ABA assay followed the methods described in Lin et al. (2007). Briefly, ABA was first extracted from the plant samples, the extraction supernatants were then dried in a SpeedVac concentrator (miVac Duo concentrator, Genevac Ltd, Ipswich, UK), resuspended, and purified by filtering through a polyvinylpolypyrrolidone column and C18 cartridges (Hsu and Kao 2003). ABA content was determined by using Agdia Phytodetek® ELISA kits according to the manufacturers' instructions.

Expression analysis of ABA-inducible genes by quantitative real-time PCR

To further confirm the involvement of *OsNCED4* in ABA biosynthesis, expression analyses of ABA-regulated genes, such as *AtKIN2* (At5g15970) and *AtRD29A* (At5g52310) (Zimmerli et al. 2008), were performed in WT, 129B08/*nced3*, N4C-1, and N4C-2 by quantitative real-time PCR (qRT-PCR). Seeds from the above-mentioned genotypes were grown on agar plates supplemented with 0.2 or 4% glucose for 10 days. The seedlings were then subjected to total RNA extraction using the RNeasy Plant Mini Kit (Qiagen, Germany). The qRT-PCR assay was followed the procedures described in Chen et al. (2014).

Relative water loss tests

To determine relative water loss, the rosette leaves of 3-week-old soil grown WT, 129B08/*nced3*, N4C-1, N4C-2 plants were excised, put on plastic weigh boats and stored in an electronic dry box (Model-D-60C, EDRY Co., Ltd, Taichung, Taiwan, ROC). The relative humidity inside the electronic dry box was around 60%. The fresh weights of the rosette leaves were recorded every 30 min for a period of 3 h.

Drought treatment

For drought treatment, 9 cm × 7 cm (diameter × height) plastic pots containing 53-day-old, well-watered WT, 129B08/*nced3*, or N4C (N4C-1 and N4C-2) lines were at first watered to field capacity by soaking the pots in a plastic tray containing water until the soil was saturated and removing the excess water in another plastic tray. Water was then withheld for 9 and 13 days to evaluate the drought resistance among the genotypes. The degree of drought resistance was determined by visual symptoms of foliage wilt.

Rice genomic DNA extraction and polymerase chain reaction (PCR)

Seeds derived from TNG67, N3P-1, N3P-2, N4P-1, and N4P-2 rice plants were germinated and grown in half-strength Kimura B solution (Hsu and Kao 2008) for 1 week. Genomic DNA was then extracted from these 1-week-old seedlings using a Plant Genomic DNA Purification Kit (GeneMark, Taiwan). The primer pairs N3P-F (5′-CTGTCAACTTCAAGCTTGGG-3′) and GUS-R (5′-GCACGATACGCTGATCCTTC-3′) or N4P-F (5′-GCAGTGGTATTGTGACAGAC-3′) and GUS-R were used in the PCR reaction to verify the integration of the *OsNCED3::GUS* or *OsNCED4::GUS* constructs in N3P or N4P transgenic rice plants, respectively.

GUS staining of transgenic rice plants

Rice husks harvested at heading stage, dehusked seeds, and dehusked seeds imbibed in water for 24 h at 25 °C from the four transgenic lines (N3P-1, N3P-2, N4P-1, and N4P-2) were subjected to GUS staining according to procedures described by Jefferson et al. (1987). One-week-old seedlings of N3P-1, N3P-2, N4P-1, and N4P-2 grown in half-strength Kimura B solution supplemented with or without 150 mM NaCl were sampled for GUS histochemical staining following the same procedures.

Results

In this study, reverse genetic approaches were used to determine if rice *OsNCED4* is functionally active in ABA biosynthesis. GUS staining of transgenic rice plants harboring an *OsNCED3::GUS* or *OsNCED4::GUS* construct was carried out to determine the expression patterns of these two genes in developing seeds and seedlings under salt stress.

Transgenic Arabidopsis plants overexpressing *OsNCED4*

Comparison of the amino acid sequence between OsNCED3 and OsNCED4 indicated that OsNCED3 and OsNCED4 share approximately 79% sequence homology and 71% sequence identity. Similar to Arabidopsis NCED3, both OsNCED3 and OsNCED4 proteins contain a putative plastid-targeting peptide near the N-terminal end and four conserved histidine residues essential for catalytic acticity (Additional file 1). Our previous study indicated that *OsNCED3* plays a role in ABA biosynthesis (Hwang et al. 2010). To verify whether *OsNCED4* may play a role in ABA biosynthesis, we generated Arabidopsis transgenic plants overexpressing *OsNCED4*.

Reverse transcription (RT)-PCR was used to confirm the presence of the *OsNCED4* transcript in homozygous transgenic Arabidopsis plants. Our results indicated that *OsNCED4* transcripts were present in the two N4C transgenic lines, but not in Col-0 or 129B08/*nced3* mutant plants. *AtNCED3* transcripts were detected only in Col-0 (Fig. 1). These results confirmed the successful transformation and transcription of the *OsNCED4* transgene in the 129B08/*nced3* mutant background in two N4C transgenic lines. The 129B08/*nced3* mutant line maintained the lack of a functional *AtNCED3* gene.

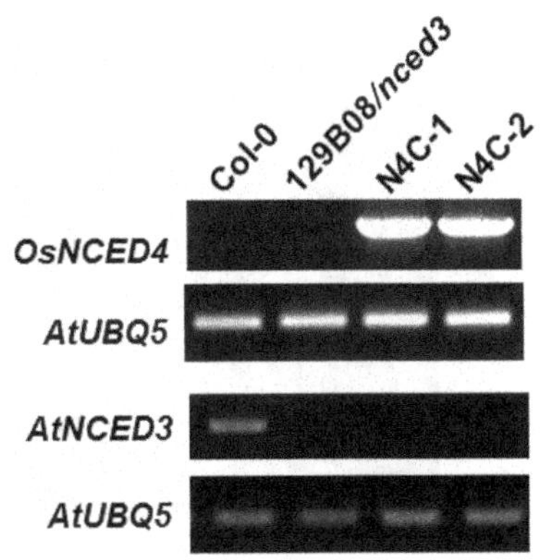

Fig. 1 RT-PCR analysis of native *AtNCED3* and transgenic *OsNCED4* expression in Arabidopsis plants; Col-0, 129B08/*nced3* mutant, and two *OsNCED4*-overexpressing (N4C) transgenic lines

Analysis of leaf morphology and seed germination in *OsNCED4-overexpressing Arabidopsis*

Heterologous expression of *OsNCED4* in the 129B08/*nced3* mutant reduced plant size (Fig. 2a), delayed seed germination (Fig. 2b), caused smaller and rounder rosette leaves (Fig. 2c), and increased the leaf width-to-length ratio to a value significantly closer to one, indicating that the leaf shape was rounder in the two N4C transgenic lines (Fig. 2d).

Overexpression of *OsNCED4* shifted the bulk of the germination by delay 1–2 days compared to wild type and its mutant background. Germination did reach 100% by day 6 which was similar to wild type (Fig. 2b). While the germination rate of the 129B08/*nced3* mutant was similar to that of Col-0 in media either lacking Glc or containing low amounts of Glc (0.2% Glc) (Figs. 2b, 3a), the rate in this mutant was higher than that of Col-0 under high Glc (4.0% Glc) conditions 3 days after sowing (Fig. 3b). On the other hand, the transgenic lines overexpressing *OsNCED4*, N4C-1 and N4C-2, showed a seed germination delay at all levels of Glc in the media, from that lacking Glc (Fig. 2b) to those containing both low and high Glc levels (Fig. 3a, b).

Heterologous expression of *OsNCED4* in the 129B08/*nced3* mutant affected post-germination growth and ABA content under high glucose conditions

After growing on agar medium containing 0.2% Glc for 7 days, the two N4C lines showed slower leaf growth compared to Col-0 and 129B08/*nced3* mutant plants (Fig. 4a). When grown on agar medium supplemented with 4% Glc for 7 days, the N4C transgenic lines showed post-germination developmental arrest, achieving neither cotyledon greening nor leaf formation (Fig. 4b). The percentage of N4C seedlings displaying post-germination developmental arrest was significantly higher than Col-0 and the 129B08/*nced3* mutant under high Glc conditions (4% Glc) (Fig. 4c; $P < 0.001$).

The ABA content was significantly lower in the 129B08/*nced3* mutant but significantly higher in the two N4C lines under both low and high Glc conditions compared to Col-0 (Fig. 5a). The transcripts of two

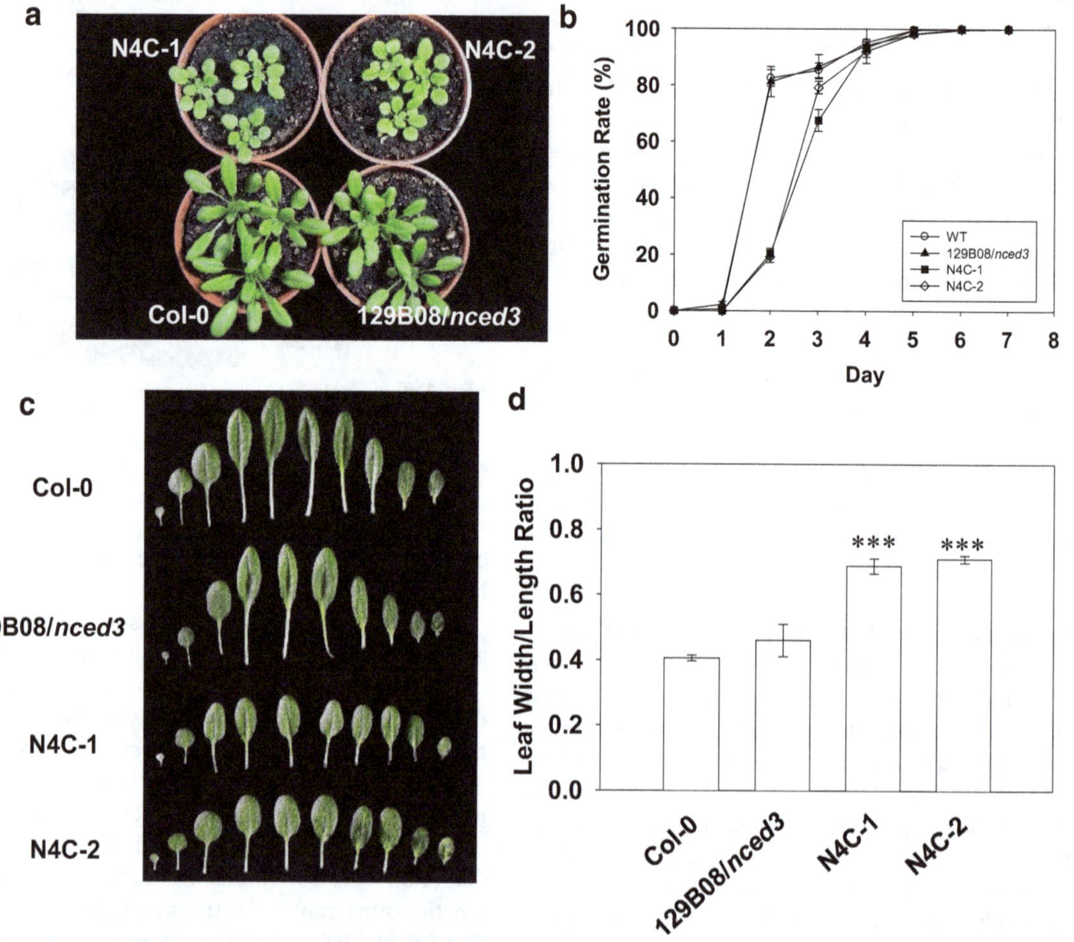

Fig. 2 Phenotypic comparison of 35-day-old, soil-grown Col-0, 129B08/*nced3* mutant, and two *OsNCED4*-overexpressing transgenic lines. **a** Above-ground plant phenotype. **b** Germination rate, calculated from 100 seeds for each line, with mean ± SD of three independent experiments. **c** Leaf shape. **d** Width-to-length ratio of the fifth leaf. In **d**, the values are the mean ± SD of three independent experiments, each with nine plants. ***$P < 0.001$, Student's *t* test

ABA-regulated genes, *AtKIN2* and *AtRD29A*, varied in a pattern similar to the ABA content among the tested genotypes under both low and high Glc conditions (Fig. 5b), except that *AtRD29A* showed only minor signals in all lines under low Glc.

Heterologous expression of *OsNCED4* in the 129B08/*nced3* mutant affected relative water loss and drought tolerance

Relative water loss tests revealed that the 129B08/*nced3* mutant plants displayed a more severe water loss phenotype than the rest genotypes tested (Fig. 6). After 9 days of withholding water, wild type plants began to wilt, 129B08/*nced3* mutant plants showed severe wilt symptoms, and N4C transgenic plants remained healthy (Fig. 7a). It was not until 13 days of water withholding that two N4C lines tested displayed wilt symptoms (Fig. 7b).

Characterization of transgenic rice plants carrying *OsNCED* promoter::*GUS* constructs

Promoter analysis using PlantCare (http://bioinformatics.psb.ugent.be/webtools/plantcare/html/) (Lescot et al. 2002) revealed that *OsNCED3* and *OsNCED4* share many common *cis*-elements such as ABRE, ARE, HSE, and MBS in their promoters within two Kb upstream of ATG start codon (Additional file 2). These data suggest that they might play an overlapping role in response to abiotic stress.

To understand and compare the expression patterns of *OsNCED3* and *OsNCED4* genes under both normal and stress conditions, transgenic rice plants carrying the *OsNCED3::GUS* or *OsNCED4::GUS* construct were generated. The presence of *OsNCED3::GUS* or *OsNCED4::GUS* constructs in the transgenic rice plants were verified by PCR. The forward primers were designed

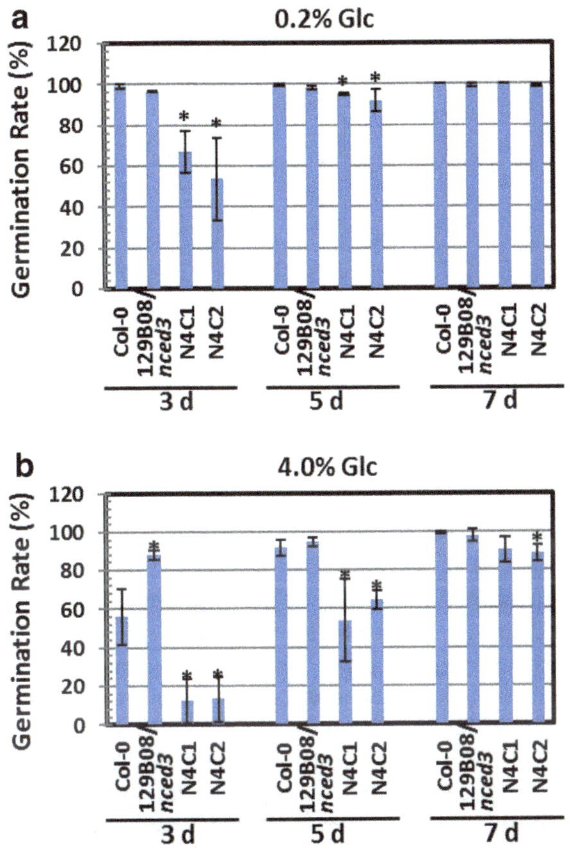

Fig. 3 Heterologous expression of *OsNCED4* delayed seed germination in Arabidopsis. **a** Germination rate of seeds from Col-0, 129B08/*nced3* mutant, and two *OsNCED4*-overexpressing transgenic lines grown on agar medium supplemented with 0.2% Glc for 3, 5, and 7 days. **b** Germination rate of the same transgenic lines grown on agar medium supplemented with 4% Glc for 3, 5, and 7 days. The values are the mean ± SD of three independent experiments, each with 80–100 seeds. *$P < 0.05$, Student's *t* test

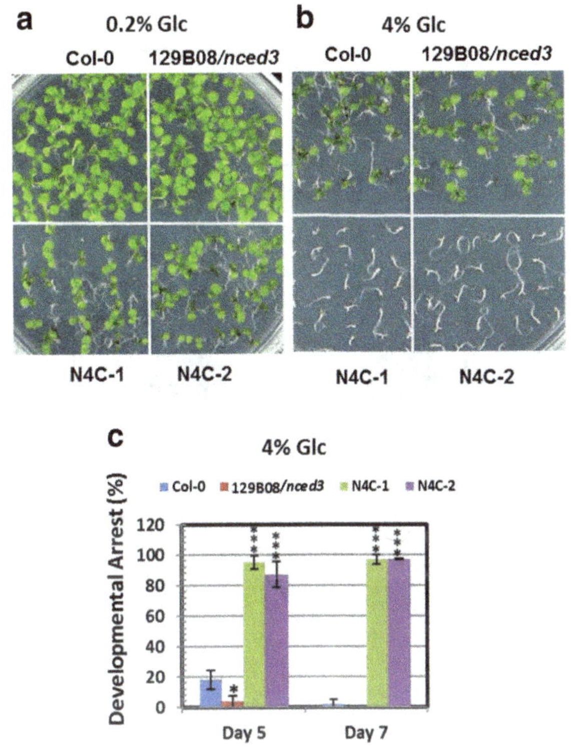

Fig. 4 Heterologous expression of rice *OsNCED4* induced post-germination developmental arrest under high glucose condition. **a** Col-0, 129B08/*nced3* mutant, and N4C-1 and -2 transgenic seedlings grown on agar medium supplemented with 0.2% Glc for 7 days. **b** Col-0, 129B08/*nced3* mutant, and N4C-1 and -2 transgenic seedlings grown on agar medium supplemented with 4% Glc for 7 days. **c** The percentage of seedlings displaying post-germination developmental arrest at 5 and 7 days after plating on media containing 4% Glc. The values are the mean ± SD of three independent experiments, each with 80–100 seeds. *$P < 0.05$; ***$P < 0.001$, Student's *t* test

to bind the *OsNCED3* or *OsNCED4* promoter region and the reverse primer the GUS region. Amplicons of the *OsNCED3::GUS* and *OsNCED4::GUS* constructs were only detectable in homozygous N3P and N4P transgenic rice plants, respectively, but not in the TNG67 rice plant (Fig. 8a), indicating successful integration into the genomes of N3P and N4P transgenic rice plants.

Expression analysis of *OsNCED3* and *OsNCED4* in rice seeds and seedlings

Tissue-specific expression analysis using RiceXPro database (http://ricexpro.dna.affrc.go.jp/) (Sato et al. 2011) indicated that both *OsNCED3* and *OsNCED4* genes are highly expressed during lemma and palea development (Additional file 3). As this prediction is based on bioinformatic analyses and the experimental data are still lacking. Thus, transgenic rice plants expressing *OsNCED3::GUS* and *OsNCED4::GUS* were used to confirm tissue-specific expression of *OsNCED3* and *OsNCED4*, respectively.

Results from GUS staining revealed that both *OsNCED3* and *OsNCED4* were expressed in roots under both no-salt and salt-stress conditions, although the GUS signal appeared to be weaker in the salt-treated roots (Fig. 8b, d). In contrast, GUS signals in leaf blade and leaf sheath were enhanced by salt treatment (Fig. 8c). In N3P and N4P rice, whole grains harvested at heading stage showed strong GUS signals in the veins of the husks (Fig. 9a). In both dry (Fig. 9b) and imbibed (Fig. 9c) N3P and N4P seeds, GUS signals were observed in the aleurone layer. Interestingly, the GUS signals in N4P seeds were stronger than those in dry or imbibed N3P seeds (Fig. 9b, c).

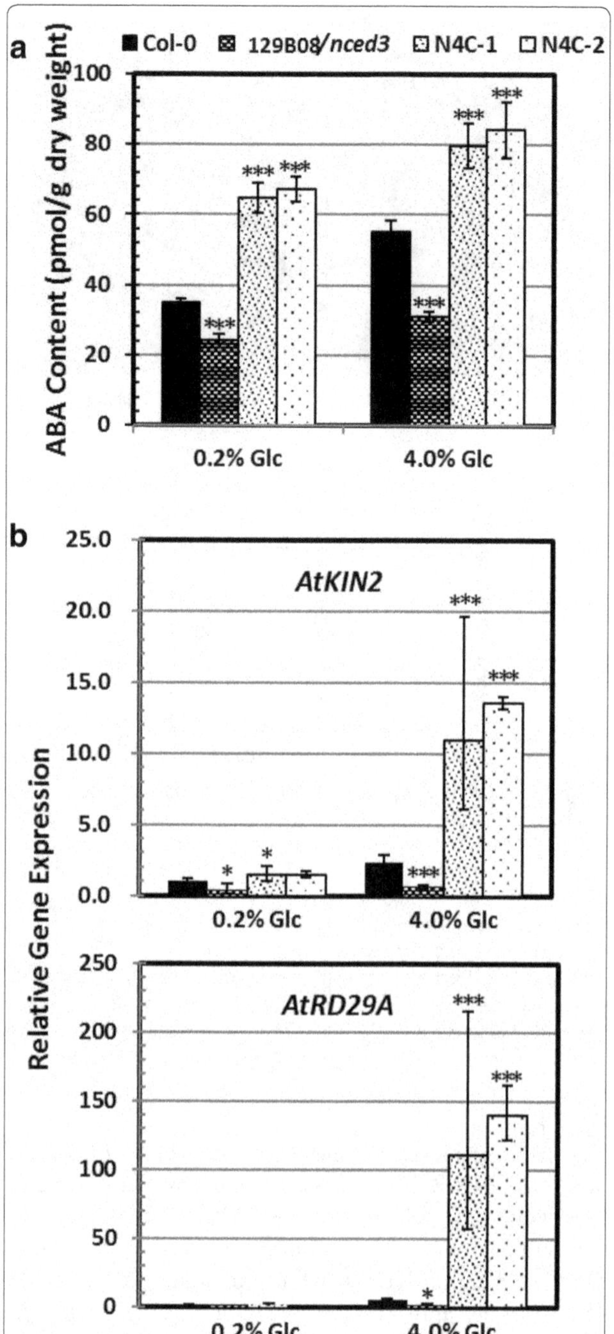

Fig. 5 Transgenic Arabidopsis plants overexpressing *OsNCED4* had increased ABA content and expression of ABA-mediated stress-responsive genes. **a** ABA content in seedlings of Col-0, 129B08/*nced3* mutant, and two N4C transgenic lines grown on agar medium supplemented with 0.2 or 4% Glc for 10 days. **b** Relative transcript levels of *AtKIN2* (upper panel) and *AtRD29A* (lower panel) in seedlings of Col-0, 129B08/*nced3* mutant, and two N4C transgenic lines grown on agar medium supplemented with 0.2 or 4% Glc for 10 days. The values are the mean ± SD of three independent experiments. $^{*}P < 0.05$; $^{***}P < 0.001$, Student's *t* test

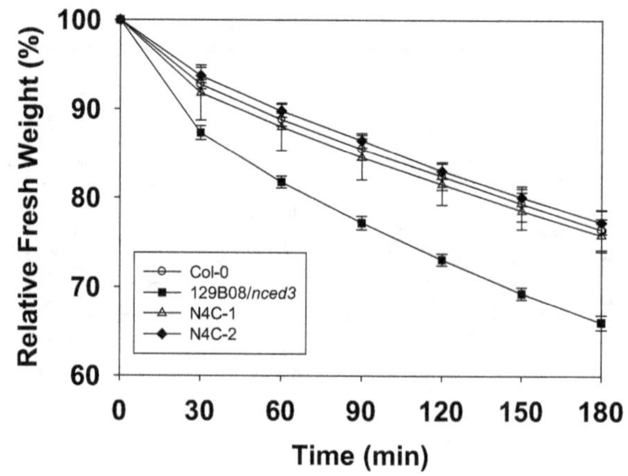

Fig. 6 Relative fresh weight of rosette leaves excised from 3-week-old Col-0, 129B08/*nced3* mutant, and two N4C transgenic lines

Discussion

Based on the NCBI GenBank database, there are more than three *NCED* genes in the rice genome. Welsch et al. (2008) investigated three rice *NCED* genes, namely *OsNCED3, OsNCED4*, and *OsNCED5*, and found that all three *NCED* genes are salt- and ABA-inducible. Previous research indicated that *OsNCED3* is functionally active in ABA biosynthesis, because its heterologous expression in Arabidopsis increases ABA levels (Hwang et al. 2010). The monocot *OsNCED3* gene was shown to alter leaf shape and regulate the development of vascular bundles in Arabidopsis (Hwang et al. 2010). However, it remains largely unknown if other rice *NCED* genes are involved in ABA biosynthesis and leaf morphogenesis.

NCBI blast analysis indicated that amino acid sequences of OsNCED3 and OsNCED4 have high levels of identity (Additional file 1). Additionally, results from promoter analysis of *OsNCED3* and *OsNCED4* within 2407 and 1939 bp upstream of the ATG start codon, respectively using PlantCARE indicated that both gene promoters contain multiple stress-responsive *cis*-acting elements (Additional file 2). Furthermore, tissue-specific expression analysis of *OsNCED3* and *OsNCED4* genes using RiceXPro database revealed that both genes are highly expressed during lemma and palea development (Additional file 3). Taken together, these results led us to investigate the potential function of *OsNCED4* in ABA biosynthesis.

We have previously transformed *35S::OsNCED3* into rice TNG67 and obtained some transgenic rice plants. Unfortunately, the transgene in these transgenic rice plants appeared to be leaky. In other words, the

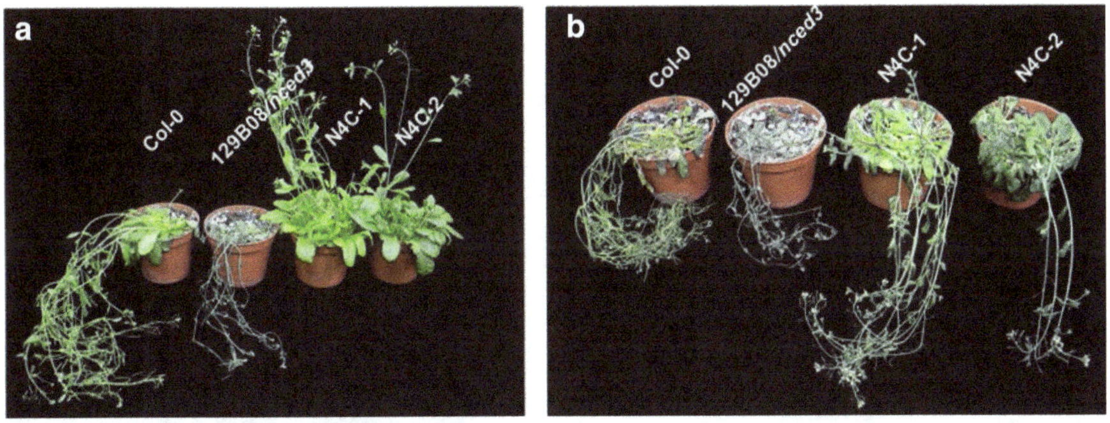

Fig. 7 Overexpression of *OsNCED4* in transgenic Arabidopsis plants conferred drought tolerance. **a** Col-0, 129B08/*nced3* mutant, and two N4C transgenic lines after 9 days of withholding water. **b** Col-0, 129B08/*nced3* mutant, and two N4C transgenic lines after 13 days of withholding water

transgenic rice plants overexpressing *35S::OsNCED3* had transgene integration (based on antibiotic selection), but without transgene expression. It is most likely that the ABA balance is required during somatic embryogenesis of rice (a monocot crop) and disturbance of ABA balance will block somatic embryogenesis during rice transformation. To ensure accuracy of *35S::OsNCED3* molecular construct, *35S::OsNCED3* was heterologously transformed into Arabidopsis and the overexpression of *OsNCED3* was able to complement the ABA-deficient phenotypes in the 129B08/*nced3* mutant and increase ABA content in wild type (Hwang et al. 2010). Following the same reasoning, *35S::OsNCED4* transgene was overexpressed in Arabidopsis in this study. On the other hand, functional transgenic rice plants harboring *OsNCED3::GUS* or *OsNCED4::GUS* were obtained because not *NCED* coding sequence but *NCED* promoter sequence was present in the transgenic region and the ABA content and signaling pathway presumably were not changed in these transgenic rice plants.

Heterologous expression of *OsNCED4* in the 129B08/*nced3* mutant delays seed germination and plays a role in shaping leaf morphology

To investigate the potential role of *OsNCED4* in ABA biosynthesis, a complementation test was performed using transgenic Arabidopsis plants carrying the *35S::OsNCED4* transgene in the 129B08/*nced3* mutant background. In the two complementation lines (N4C-1 and N4C-2), delays in seed germination were noticed under no, low- and high-Glc conditions (Figs. 2b, 3a, b). These results strongly suggested the possible involvement of *OsNCED4* in ABA biosynthesis, because ABA delays seed germination (Lin et al. 2007; Liu et al. 2016). A positive feedback regulation of ABA biosynthesis may cause the higher germination rate observed in the 129B08/*nced3* mutant relative to Col-0 under high Glc condition, a stress coping mechanism suggested by Xiong and Zhu (2003). Additionally, results from this research suggested that *OsNCED4* might share a redundant role with *OsNCED3* in regulating leaf morphology and ABA biosynthesis (Hwang et al. 2010), since heterologous expression of these two genes in Arabidopsis exerted similar effects on plant size, leaf shape and seed germination.

Heterologous expression of *OsNCED4* in the 129B08/*nced3* mutant confers sugar oversensitivity

Post-germination growth was slightly reduced when the two *OsNCED4*-overexpressing lines were grown on medium containing low amounts of Glc, but it was severely inhibited under high Glc condition (Fig. 4a, b). These results were consistent with previous research indicating that Glc-induced physiological responses are different between low and high concentrations (Price et al. 2003). Interestingly, the percentage of seedlings that displayed post-germination developmental arrest was in accordance with seedling ABA content among all genotypes tested under high Glc conditions (Figs. 4c, 5a). In lines with this observation, the transcription levels of two ABA-regulated genes, *AtKIN2* and *AtRD29A*, varied with ABA content in the seedlings of the different genotypes exposed to high Glc (Fig. 5b). These results were in agreement with a previous finding suggesting that sugar-induced seedling growth arrest involves ABA biosynthesis and ABA signaling (Cheng et al. 2002; Dekkers et al. 2008). The sugar-oversensitive phenotype and the higher ABA content in the two *OsNCED4*-overexpressing lines together suggested a role for *OsNCED4* in ABA biosynthesis.

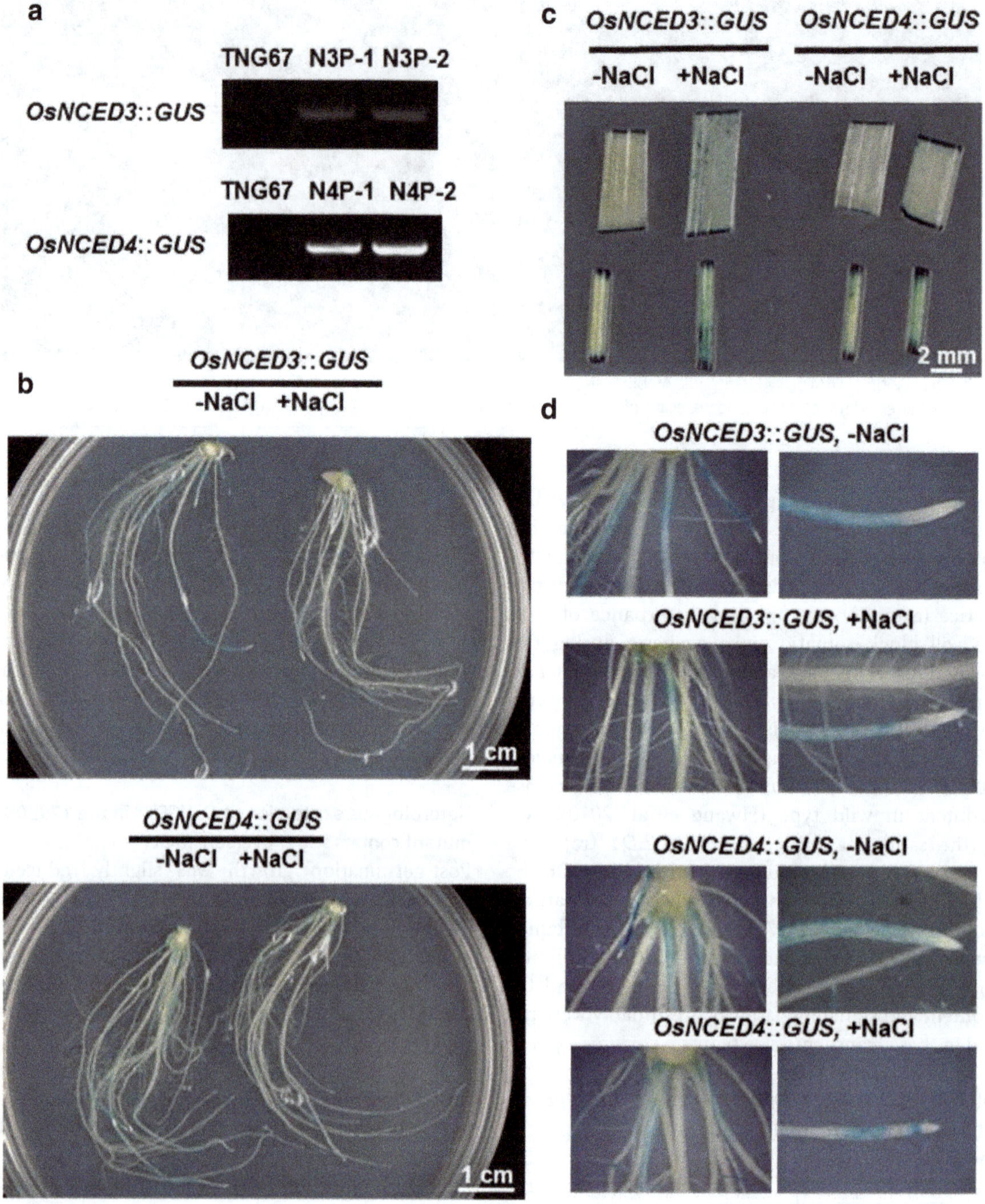

Fig. 8 Comparison of tissue-specific expression of *OsNCED3* and *OsNCED4* promoters in transgenic rice plants under salt stress. **a** PCR analysis of *OsNCED3::GUS* or *OsNCED4::GUS* constructs in N3P or N4P transgenic rice plants, respectively. **b** GUS staining in roots of 7-day-old, solution-grown transgenic rice seedlings treated with (+ NaCl) or without (− NaCl) 150 mM NaCl. **c** GUS staining in leaf blade and leaf sheath of 7-day-old, transgenic rice seedlings grown in solution with or without 150 mM NaCl. **d** GUS staining in root segments of 7-day-old transgenic rice seedlings grown in solution with or without 150 mM NaCl

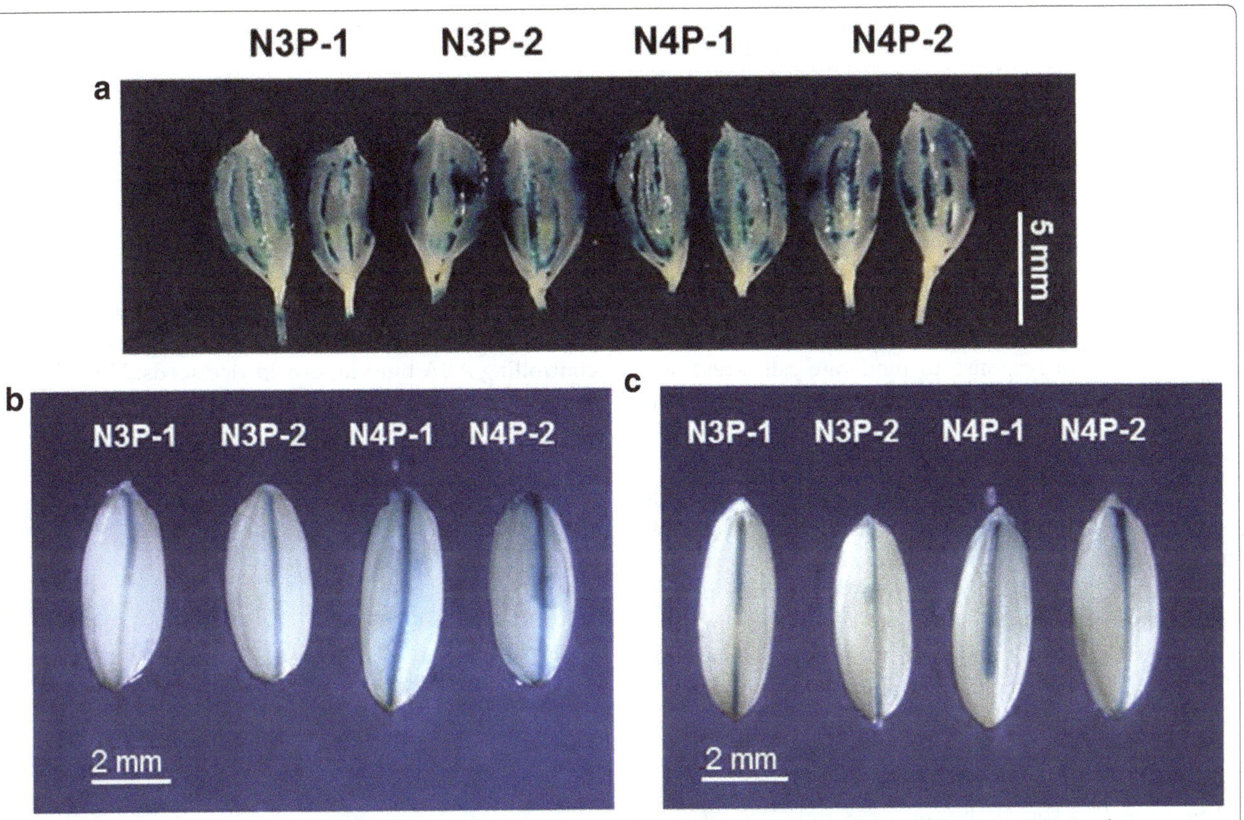

Fig. 9 Tissue-specific expression of *OsNCED3* and *OsNCED4* in the husks and seeds of transgenic rice plants. **a** GUS staining in husks of *OsNCED3::GUS* (N3P) and *OsNCED4::GUS* (N4P) transgenic rice plants. **b** GUS staining in dry seeds of N3P and N4P transgenic rice plants. **c** GUS staining in imbibed seeds of N3P and N4P transgenic rice plants. In **a**, the husks were harvested at heading stage. In **c**, the seeds were imbibed in water at 25 °C for 24 h

Heterologous expression of *OsNCED4* complements the relative water loss and increases drought tolerance in the 129B08/*nced3* mutant

The roles of ABA in inhibiting stomatal opening and stimulating stomatal closure to achieve reduction in water loss are well documented in the literature (e.g., Kim et al. 2010; Sah et al. 2016). Results from relative water loss tests clearly showed that overexpression of *OsNCED4* in the 129B08/*nced3* mutant background successfully complements the severe water loss phenotype of the mutant to a level similar to that of Col-0 (Fig. 6). These results were consistent with the idea that *OsNCED4* is functionally active in ABA biosynthesis.

To determine the effects on drought tolerance of heterologous expression of *OsNCED4* in the 129B08/*nced3* mutant, 53-day-old plants (Col-0, 129B08/*nced3*, *OsNCED4-OE*) were left unwatered. After 9 or 13 days of withholding water, rosette leaves in the two *OsNCED4* complementation lines were less wilted than those of Col-0 or 129B08/*nced3* (Fig. 7a, b). These results indicated that heterologous expression of *OsNCED4* in 129B08/*nced3* not only complements the severe wilting phenotype of the mutant but also increases its drought tolerance to a level higher than that of Col-0. Results from this experiment further supported that *OsNCED4* is involved in ABA biosynthesis.

Overlapping expression patterns of *OsNCED3::GUS* and *OsNCED4::GUS* reporter genes in rice seeds and seedlings

In this study, the GUS reporter gene was used to detect promoter activity of *OsNCED3* and *OsNCED4* under normal and salt-stress conditions. In 1-week-old transgenic rice seedlings at the one-leaf stage, overlapping GUS expression patterns were noticed. Interestingly, control roots (− NaCl) showed stronger GUS signal compared to those treated with 150 mM NaCl. This result was different from previous result documented by Welsch et al. (2008), where both *OsNCED3* and *OsNCED4* genes were found to be upregulated in 3-week-old seedling roots under salt stress (250 mM NaCl). This discrepancy may be explained by various seedling age (1- vs 3-weeks old), different levels of salt stress (150 mM vs 250 mM NaCl), and/or different lengths of salt treatment (7 days vs

1–3 h). Furthermore, stronger GUS signals were identified in newly developed adventitious roots, whereas older adventitious roots with root hairs showed much weaker signals (Fig. 8b, d). Opposite responses to salt were seen in the leaf sheath and blade, in which stronger GUS signals were detected in response to salt stress (Fig. 8c). Taken together, these results suggested that these two rice *OsNCED* genes are similarly regulated in roots or leaves at the one-leaf stage under long-term, moderate salt stress (150 mM NaCl). It is not clear why *OsNCED* genes are down-regulated in 1-week-old, solution-grown seedling roots in response to moderate salt stress, and the physiological meaning behind this requires further exploration.

Using *OsNCED3::gfp* transgenic plants, Bang et al. (2013) reported that *OsNCED3* promoter activity was not detected in grains and only barely detectable in flowers under normal growth conditions. Our GUS staining data indicated that promoter activity was clearly present in the husk at the heading stage and in dry and imbibed transgenic rice seeds under normal growth conditions (Fig. 9a–c). This discrepancy may be due to the more sensitive nature of GUS in detecting weak promoter activity compared to green fluorescent protein (Mantis and Tague 2000).

The revelation that both *OsNCED3* and *OsNCED4* are expressed in rice seeds in an overlapping pattern means that further assays are needed to understand how these genes work together to control ABA biosynthesis in rice seeds. Future studies on expression of *OsNCED3* and *OsNCED4* in dry and imbibed seeds will require using quantitative real-time PCR, seed ABA assays, and germination tests on wild-type and *OsNCED3* and *OsNCED4* knockout mutant seeds. It is interesting to note that *OsNCED1* and *OsNCED3*, but not *OsNCED4*, were suggested to be major players in controlling ABA biosynthesis in rice roots (Shi et al. 2015).

Conclusions

Heterologous expression of rice *OsNCED4* in the 129B08/*nced3* Arabidopsis mutant background resulted in smaller and rounder leaves, delayed seed germination, post-germination sugar oversensitivity, increased ABA content, and enhanced tolerance to drought. These results indicated that in addition to *OsNCED3*, *OsNCED4* is also functionally active in ABA biosynthesis at least in Arabidopsis. Results from GUS staining with transgenic rice plants further revealed that *OsNCED3* and *OsNCED4* are expressed in seeds and one-leafed seedlings in an overlapping pattern and that both genes are downregulated in roots and upregulated in leaves in response to long-term, moderate salt stress. Our results demonstrated the possible functional redundancy between the *OsNCED3* and *OsNCED4* genes, although results from GUS staining alone are not sufficient to reach this conclusion. In light of different results in previous reports indicating upregulation of *NCED* genes in response to salt stress, the physiological importance of downregulation of *OsNCED* genes in roots of young seedlings under salt stress remains elusive and awaits further investigation. The fact that stronger GUS signals were detected in N4P seeds compared to N3P seeds suggested that *OsNCED4* may play a more prominent role in controlling ABA biosynthesis in rice seeds. Nonetheless, more research is required to confirm this hypothesis.

Abbreviations

ABA: abscisic acid; LOX: lipoxygenase; NCED: 9-*cis*-epoxycarotenoid dioxygenase; PCR: polymerase chain reaction; GUS: β-glucuronidase; RT-PCR: reverse-transcription polymerase chain reaction; Glc: glucose; qRT-PCR: quantitative real-time polymerase chain reaction; GFP: green fluorescent protein.

Authors' contributions

SGH conceived and designed the experiments; CYL, CST and SGH performed the experiments and analyzed the data; SGH wrote the manuscript. All authors read and approved the final manuscript.

Author details

[1] Department of Horticulture, National Chung Hsing University, 145 Xingda Road, South District, Taichung 40227, Taiwan, ROC. [2] Institute of Plant and Microbial Biology, Academia Sinica, Taipei 11529, Taiwan, ROC. [3] Taiwan Agricultural Research Institute, Council of Agriculture, Executive Yuan, No.189, Zhongzheng Road, Wufeng District, Taichung 41362, Taiwan, ROC.

Acknowledgements

We would like to thank Dr. Wan-Hsing Cheng from Institute of Plant and Microbial Biology, Academia Sinica for his critical reading of the manuscript and Miss Yi-Chun Chen from National Chung Hsing University for her valuable assistance in the laboratory.

Competing interests

The authors declare that they have no competing interests.

Funding

This study was supported by the Ministry of Science and Technology through Grant No. 102-2313-B-005-035 to SGH, Taiwan, Republic of China.

References

Agustí J, Zapater M, Iglesias DJ, Cercós M, Tadeo FR, Talón M (2007) Differential expression of putative 9-*cis*-epoxycarotenoid dioxygenases and abscisic acid accumulation in water stressed vegetative and reproductive tissues of citrus. Plant Sci 172:85–94

Bang SW, Park SH, Jeong JS, Kim YS, Jung H, Ha SH, Kim JK (2013) Characterization of the stress-inducible *OsNCED3* promoter in different transgenic rice organs and over three homozygous generations. Planta 237:211–224

Bauer H, Ache P, Lautner S, Fromm J, Hartung W, Al-Rasheid KA, Sonnewald S, Sonnewald U, Kneitz S, Lachmann N, Mendel RR, Bittner F, Hetherington AM, Hedrich R (2013) The stomatal response to reduced relative humidity requires guard cell-autonomous ABA synthesis. Curr Biol 23:53–57

Chen YH, Shen HL, Hsu PJ, Hwang SG, Cheng WH (2014) *N*-acetylglucosamine-1-P uridylyltransferase 1 and 2 are required for gametogenesis and embryo development in *Arabidopsis thaliana*. Plant Cell Physiol 55(11):1977–1993

Cheng WH, Endo A, Zhou L, Penney J, Chen HC, Arroyo A, Leon P, Nambara E, Asami T, Seo M, Koshiba T, Sheen J (2002) A unique short-chain dehydrogenase/reductase in Arabidopsis glucose signaling and abscisic acid biosynthesis and functions. Plant Cell 14:2723–2743

Dekkers BJ, Schuurmans JA, Smeekens SC (2008) Interaction between sugar and abscisic acid signalling during early seedling development in Arabidopsis. Plant Mol Biol 67:151–167

Endo A, Sawada Y, Takahashi H, Okamoto M, Ikegami K, Koiwai H, Seo M, Toyomasu T, Mitsuhashi W, Shinozaki K, Nakazono M, Kamiya Y, Koshiba T, Nambara E (2008) Drought induction of Arabidopsis 9-*cis*-epoxycarotenoid dioxygenase occurs in vascular parenchyma cells. Plant Physiol 147:1984–1993

Gao G, Zhang S, Wang C, Yang X, Wang Y, Su X, Du J, Yang C (2011) Arabidopsis *CPR5* independently regulates seed germination and postgermination arrest of development through LOX pathway and ABA signaling. PLoS ONE 6(4):e19406

Hsu YT, Kao CH (2003) Role of abscisic acid in cadmium tolerance of rice (*Oryza sativa* L.) seedlings. Plant Cell Environ 26(6):867–874

Hsu YT, Kao CH (2008) Distinct roles of abscisic acid in rice seedlings during cadmium stress at high temperature. Bot Stud 49:335–342

Huo H, Dahal P, Kunusoth K, McCallum CM, Bradford KJ (2013) Expression of *9-cis-epoxycarotenoid dioxygenase4* is essential for thermoinhibition of lettuce seed germination but not for seed development or stress tolerance. Plant Cell 25:884–900

Hwang SG, Chen HC, Huang WY, Chu YC, Shii CT, Cheng WH (2010) Ectopic expression of rice *OsNCED3* in Arabidopsis increases ABA level and alters leaf morphology. Plant Sci 178(1):12–22

Iuchi S, Kobayashi M, Taji T, Naramoto M, Seki M, Kato T, Tabata S, Kakubari Y, Yamaguchi-Shinozaki K, Shinozaki K (2001) Regulation of drought tolerance by gene manipulation of 9-*cis*-epoxycarotenoid dioxygenase, a key enzyme in abscisic acid biosynthesis in Arabidopsis. Plant J 27:325–333

Jefferson RA, Kavanagh TA, Bevan MW (1987) GUS fusions: β-glucuronidase as a sensitive and versatile gene fusion marker in higher plants. EMBO J 20:3901–3907

Jiang W, Yu D (2009) *Arabidopsis WRKY2* transcription factor mediates seed germination and post germination arrest of development by abscisic acid. BMC Plant Biol 9:96

Kim T, Böhmer M, Hu H, Nishimura N, Schroeder JI (2010) Guard cell signal transduction network: advances in understanding abscisic acid, CO_2, and Ca^{2+} signaling. Annu Rev Plant Biol 61:561–591

Lefebvre V, North H, Frey A, Sotta B, Seo M, Okamoto M, Nambara E, Marion-Poll A (2006) Functional analysis of Arabidopsis *NCED6* and *NCED9* genes indicates that ABA synthesized in the endosperm is involved in the induction of seed dormancy. Plant J 45:309–319

Lescot M, Déhais P, Thijs G, Marchal K, Moreau Y, Van de Peer Y, Rouzé P, Rombauts S (2002) PlantCARE, a database of plant *cis*-acting regulatory elements and a portal to tools for in silico analysis of promoter sequences. Nucleic Acids Res 30:325–327

Lin PC, Hwang SG, Endo A, Okamoto M, Koshiba T, Cheng WH (2007) Ectopic expression of *ABSCISIC ACID 2/GLUCOSE INSENSITIVE 1* in Arabidopsis promotes seed dormancy and stress tolerance. Plant Physiol 143:745–758

Liu X, Hu PW, Huang MK, Tang Y, Li Y, Li L, Hou XL (2016) The NF-YC-RGL2 module integrates GA and ABA signalling to regulate seed germination in *Arabidopsis*. Nat Commun 7:12768

Mantis J, Tague BW (2000) Comparing the utility of β-glucuronidase and green fluorescent protein for detection of weak promoter activity in *Arabidopsis thaliana*. Plant Mol Biol Rep 18:319–330

Martínez-Andújar C, Ordiz MI, Huang Z, Nonogaki M, Beachy RN, Nonogaki H (2011) Induction of 9-*cis*-epoxycarotenoid dioxygenase in *Arabidopsis thaliana* seeds enhances seed dormancy. Proc Natl Acad Sci USA 108:17225–17229

Merilo E, Jalakas P, Kollist H, Brosche M (2015) The role of ABA recycling and transporter proteins in rapid stomatal responses to reduced air humidity, elevated CO , and exogenous ABA. Mol Plant 8:657–659

Oliver SN, Dennis ES, Dolferus R (2007) ABA regulates apoplastic sugar transport and is a potential signal for cold-induced pollen sterility in rice. Plant Cell Physiol 48:1319–1330

Price J, Li TC, Kang SG, Na JK, Jang JC (2003) Mechanisms of glucose signaling during germination of Arabidopsis. Plant Physiol 132:1424–1438

Sah SK, Reddy KR, Li J (2016) Abscisic acid and abiotic stress tolerance in crop plants. Front Plant Sci 7:571

Sato Y, Antonio BA, Namiki N, Takehisa H, Minami H, Kamatsuki K, Sugimoto K, Shimizu Y, Hirochika H, Nagamura Y (2011) RiceXPro: a platform for monitoring gene expression in japonica rice grown under natural field conditions. Nucleic Acids Res 39:D1141–D1148

Shi L, Guo M, Ye N, Liu Y, Liu R, Xia Y, Cui S, Zhang J (2015) Reduced ABA accumulation in the root system is caused by ABA exudation in upland rice (*Oryza sativa* L. var. Gaoshan1) and this enhanced drought adaptation. Plant Cell Physiol 56(5):951–964

Sun L, Sun Y, Zhang M, Wang L, Ren J, Cui M, Wang Y, Ji K, Li P, Li Q, Chen P, Dai S, Duan C, Wu Y, Leng P (2012) Suppression of 9-*cis*-epoxycarotenoid dioxygenase, which encodes a key enzyme in abscisic acid biosynthesis, alters fruit texture in transgenic tomato. Plant Physiol 158:283–298

Tan BC, Joseph LM, Deng WT, Liu L, Li QB, Cline K, McCarty DR (2003) Molecular characterization of the *Arabidopsis* 9-*cis* epoxycarotenoid dioxygenase gene family. Plant J 35:44–56

Vishwakarma K, Upadhyay N, Kumar N, Yadav G, Singh J, Mishra RK, Kumar V, Verma R, Upadhyay RG, Pandey M, Sharma S (2017) Abscisic acid signaling and abiotic stress tolerance in plants: a review on current knowledge and future prospects. Front Plant Sci 8:161

Waterland N, Campbell CA, Finer JJ, Jones ML (2010) Abscisic acid application enhances drought stress tolerance in bedding plants. Hortsci 45:409–413

Welsch R, Wüst F, Bär C, Al-Babili S, Beyer P (2008) A third phytoene synthase is devoted to abiotic stress-induced abscisic acid formation in rice and defines functional diversification of phytoene synthase genes. Plant Physiol 147:367–380

Xiong L, Zhu JK (2003) Regulation of abscisic acid biosynthesis. Plant Physiol 133:29–36

Zhu G, Ye N, Zhang J (2009) Glucose induced delay of seed germination in rice is mediated by the suppression of ABA catabolism rather than an enhancement of ABA biosynthesis. Plant Cell Physiol 50:644–651

Zhu G, Liu Y, Ye N, Liu R, Zhang J (2011) Involvement of the abscisic acid catabolic gene *CYP707A2* in the glucose-induced delay in seed germination and post-germination growth of Arabidopsis. Physiol Plant 143:375–384

Zimmerli L, Hou BH, Tsai CH, Jakab G, Mauch-Mani B, Somerville S (2008) The xenobiotic β-aminobutyric acid enhances Arabidopsis thermotolerance. Plant J 53:144–156

7

Ethylene signaling modulates contents of catechin and ability of antioxidant in *Camellia sinensis*

Shun-Wun Ke[1], Guan-Heng Chen[2], Chung-Tse Chen[2], Jason T. C. Tzen[2] and Chin-Ying Yang[1*]

Abstract

Background: Tea is one of the most popular beverages in the world. There are many secondary metabolites can be found in tea such as anthocyanins, proanthocyanidins, flavonols and catechins. These secondary metabolites in plants are proved to act protective components for human health effect. Plant hormone ethylene is considered to have an important role in regulation of plant development and signal transduction. This study evaluated the effect of ethylene signaling regulation in phenolic compounds in tea plants. The ethylene precursor 1-aminocyclopropane-1-carboxylic acid (ACC) enhanced contents of total catechin in treated oolong tea seedlings.

Results: The degree of epigallocatechin and epicatechin galloylation was increased after ACC treatment in oolong tea seedlings by high performance liquid chromatography determination. The contents of anthocyanins, flavonoids, and total polyphenol were higher after ACC treatment in comparison with control. Antioxidant enzyme such as catalase, superoxide dismutase, and total peroxidase decreased their antioxidant activities after ACC treatment, yet the activity of ascorbate peroxidase is increased. The ability of oxygen radical absorption and 2,2-diphenyl-1-picrylhydrazyl was used to evaluate the antioxidant activity, which was enhanced by ACC treatment.

Conclusions: Taken together the results of this study demonstrate that the ethylene signaling is involved in modulation of secondary metabolites accumulation and antioxidant ability that to enhance the benefit of human health in tea products.

Keywords: Ethylene, Antioxidant, Catechins, Polyphenol, Tea plant

Background

Tea (*Camellia sinensis* L.) is an evergreen shrub and prevalent commercial crop grown in more than 50 countries. Tea plants are rich in secondary metabolites, which have potential health benefits for humans (Zhou et al. 2016). Secondary metabolic products in tea include flavonoids, polyphenols, theanine, and alkaloids. Flavonoids are a large group of phenolic secondary metabolites including anthocyanins, proanthocyanidins, phenolic acids, flavonols and catechins (Abeynayake et al. 2012; Sun et al. 2016). The predominant flavonoids in tea are catechins (flavan-3-ols), which are generally classified into the following groups such as epicatechin (EC), epicatechin-gallate (ECG), epigallocatechin (EGC) and epigallocatechin-gallate (EGCG) (Perva-Uzunalic et al. 2006).

The production of secondary metabolites in higher plants is often affected by hormone regulation. Ethylene is a plant hormone that affects the plants growth and development and is sometimes required for processes including ripening, senescence, and abscission (Schaller 2012). In tea plants, floral buds are more sensitive to ethylene than are leaves (Woolf et al. 1999). Ethylene is critical for signal transduction and its effect on the accumulation of secondary metabolites. Ethylene regulates flavonol biosynthesis through the MYB12 transcription factor, which affects the expression of chalcone synthase,

*Correspondence: emiyang@dragon.nchu.edu.tw
[1] Department of Agronomy, National Chung Hsing University, Taichung 40227, Taiwan
Full list of author information is available at the end of the article

chalcone isomerase, and flavonol synthase (Lewis et al. 2011). Ethylene signaling is involved in the defense resistance of plants by enhancing polyphenol oxidase activity during stress (Bosch et al. 2014; Thipyapong and Steffens 1997). Studies on rice coleoptiles have shown that ethylene induces the accumulation of polyamine (Lee and Chu 1992).

Previous studies have shown that ethylene signaling contributes to the accumulation of secondary metabolites to facilitate defense resistance in plants. However, little is known about ethylene signaling regulation in the phenolic compounds in tea plants. 1-aminocyclopropane-1-carboxylic acid (ACC) is known to be a direct precursor of ethylene and is transported throughout the plant over short and long distances (Boller et al. 1979), it is believed that the responses observed in this study is caused by ethylene. Our results demonstrated that after treatment with the ethylene precursor ACC, the total catechin content in oolong tea seedlings increased. High-performance liquid chromatography (HPLC) indicated that the degrees of EGC and EC galloylation were increased by ACC treatment in seedlings of oolong tea. The accumulation of phenolic compounds was enhanced after ACC treatment such as anthocyanins, flavonoids, and total polyphenol. The activities of catalase (CAT), superoxide dismutase (SOD), and total peroxidase (POX) decreased and ascorbate peroxidase (APX) increased after ACC treatment by antioxidant enzyme activity assay. Ability of antioxidant was enhanced after ACC treatment via evaluated according to the oxygen radical absorption capacity (ORAC), and 2,2-diphenyl-1-picrylhydrazyl (DPPH). Our results demonstrate that the contents of secondary metabolites and capacity of antioxidant can be increased through modulating ethylene signaling pathways in tea.

Methods

Plant materials and treatments

Tea [*Camellia sinensis* (L.) Kuntze] cultivar Chin-Shin Oolong was used in the experiments in this study. All seedlings of the tea plant were grown in the same area in Nantou, Central Taiwan, over a 2-year period. The tea plant seedlings were approximately 50 cm high. For ACC treatment, 2-year-old tea seedlings were irrigated with 20 mL of 100 μM ACC per day for 5 days; the control (CK) samples were irrigated with 20 mL of water instead. Fresh samples each with one tip and two leaves were collected from tea seedlings, frozen immediately in liquid nitrogen, and stored at − 80 °C until analysis.

Preparation of extracts from young tea seedling leaves

The samples were weighed, and lyophilized samples were homogenized to powder by using a disposable pestle. The extraction of tea was prepared with 1.0 mL of 75% ethanol and shaking the mixture vigorously with an Intelli-Mixer (MyLabTM RM-2 M) for 2 h at room temperature. The mixture extraction of tea was filtered by 0.45 μm polyvinylidene difluoride membrane filter (Pall Corporation, Glen Cove, NY, USA) before HPLC analysis.

HPLC analysis

Tea extracts were analyzed on a liquid chromatography system coupled with a Model 600E photodiode array detector (Waters Corporation, Milford, MA, USA). The chromatographic separation was performed on a Mightysil RP-18 GP column (250 × 4.6 mm i.d., 5 μm; Kanto Chemical Co., Tokyo, Japan). The mobile phase consisted of water containing 0.5% acetic acid (solvent A) and acetonitrile (solvent B).

The samples were eluted using the following gradient program, 0 min, 95% A and 5% B; 100 min, 75% A and 25% B. The column was kept at room temperature, the volume of each injection was 10 μL, and the flow rate was 1 mL/min. The PDA detector was set at 280 nm. The four major catechins (EC, EGC, ECG, and EGCG) shown in the HPLC profiles were identified as described in a previous study (Chen et al. 2014).

Determination of anthocyanins, flavonoids, and total phenol

The tea seedling samples (0.1 g) were excised and immediately used for anthocyanin, flavonoid, and total phenol content assays. For the anthocyanin and flavonoid content assays, the sample extracts in 2 mL of potassium phosphate buffer (100 mM, pH 7.8) were ground into powder with liquid nitrogen. The mixture of tea extraction was centrifuged at 16,000*g* and 4 °C for 30 min, and the resultant supernatant was measured at 600 and 320 nm by using a spectrophotometer (Metertec SP8001). One absorbance unit was defined as the quantity of anthocyanins and flavonoids that exhibits an absorbance of 1 at 600 and 320 nm. The total phenolic content assay was performed using a modified form of the procedure described by Singleton and Rossi (1965). The sample extracts in 1.0 mL of 75% ethanol were ground into powder with liquid nitrogen. The homogenate was centrifuged at 10,000*g* and 4 °C for 10 min, and 50 μL of the resultant supernatant was diluted with distilled water and added to 250 μL of the Folin–Ciocalteu reagent (0.2 mol/L). The reaction mixture reacted for 5 min before 50 μL of 20% Na_2CO_3 was added. The samples were then incubated at room temperature in darkness for 60 min, and the absorbance was measured at 760 nm. A calibration curve was obtained using 0–4000 mg of gallic acid per milliliter and used to calculate the total phenolic content of the tea seedling leaves.

Determination of antioxidant enzyme activity

The fresh leaves (0.2 g) of tea seedling were used immediately for enzyme extraction after excised, and subsequently homogenized with liquid nitrogen. Sodium phosphate buffer (50 mM; pH 6.8) was used as extraction buffer. The mixture of tea leaves extraction was then centrifuged at 15,000*g* for 30 min, and the resultant supernatant was used in the following enzyme activity assays. CAT activity was analyzed as described in previous studies (Chao et al. 2012; Kato and Shimizu 1985). Decreasing in absorbance at 240 nm was observed using a spectrophotometer (Metertec SP8001) which implied the decrease of H_2O_2. Activity of CAT was calculated on the basis of the extinction coefficient (40 mM^{-1} cm^{-1} at 240 nm) of H_2O_2. One unit of CAT was defined as the enzyme amount that degraded 1 μmol of H_2O_2 per minute. SOD activity assay was conducted based on Chao et al. (2012). The reaction buffer was Triethanolamine–diethanolamine buffer (100 mM; pH 7.4) containing ethylenediaminetetraacetic acid/$MnCl_2$ (100 mM/50 mM, pH 7.4), 7.5 mM β-nicotinamide adenine dinucleotide (β-NADH), and 10 mM 2-mercaptoethanol and mixed with enzyme extract. The enzyme reaction was initiated by the addition of β-NADH, and detected the absorbance at 340 nm for 10 min. One unit of SOD was defined as the enzyme amount that inhibited the rate of β-NADH oxidation by 50%. For APX activity assay, the decrease in the ascorbic acid (AsA) concentration was determined according to the decline in absorbance at 290 nm, and the activity was calculated based on the extinction coefficient (2.8 mM^{-1} cm^{-1} at 290 nm) for AsA (Chao et al. 2012; Nakano and Asada 1981). Both soluble and ionically bound peroxidase (Total POX) activity assay was modified from previous studies (Lin and Kao 1999; MacAdam et al. 1992; Wu and Yang 2016). Samples were extracted by homogenized in liquid nitrogen with 50 mM potassium phosphate buffer (pH 5.8) containing 0.8 M KCl buffer. The 50 mM potassium phosphate buffer (pH 5.8), 21.6 mM guaiacol, and 39 mM H_2O_2 were added into enzyme extract for reaction. The enzyme reaction was initiated by adding H_2O_2, the absorbance was measured at 470 nm for 3 min. Total POX activity was calculated on the basis of the extinction coefficient of 26.6 mM^{-1} cm^{-1} at 470 nm for tetraguaiacol. One unit of POX was defined as the enzyme amount that caused the formation of 1 μmol of tetraguaiacol per minute.

Determination of DPPH radical scavenging activity

Potential antioxidant activity was determined using DPPH (2,2-diphenyl-1-picrylhydrazyl) according to Tadolini et al. (2000) with some modifications. The sample (0.1 g) extract was added to 1.0 mL of 75% ethanol. The mixture was shaken for 120 min through vortexing and left to centrifuge at 10,000*g* and room temperature in darkness for 10 min. The absorbance for the sample was measured using a SpectraMax M2 spectrophotometer at 521 nm against an ethanol blank. A control sample (ΔA_{DPPH}) was extracted after adding 0.19 mM DPPH solution to 0.2 mL of the respective extraction solvent. Every sample was extracted in triplicate, and the results were calculated based on four biologically independent experiments. The percentage of DPPH free radicals scavenged in the sample was calculated using the following equation:

$$\text{DPPH free radical scavenging ratio (\%)} = \left[\left(\Delta A_{DPPH} - \Delta A_{sample+DPPH}\right) / \left(\Delta A_{DPPH}\right)\right] \times 100$$

Oxygen radical absorbance capacity (ORAC) measurement

The ORAC assay was by using a modified form of the method described by Ou et al. (2001). The ORAC assay was measured the capacity of antioxidative compounds in test materials to inhibit the decrease in fluorescence induced by the peroxyl radical 2,2′-azobis (2-amidinopropane) dihydrochloride (AAPH). The samples (0.1 g) were homogenized with liquid nitrogen and added to 1 mL of double distilled H_2O for incubation 10 min at 100 °C. The 20 μL extract was added to 120 μL 120 nM disodium fluorescein solution (72 nM, final concentration) and preincubated for 15 min at 37 °C. The mixture was rapidly added to 60 μL of 40 mM AAPH in the well of the microplate and the microplate was immediately placed in the reader, and the fluorescence was recorded at an 485 nm (excitation wavelength) and 535 nm (emission wavelength) every 1 min for 60 min. The blank (FL + AAPH) with phosphate buffer instead of the antioxidant solution and Trolox was used as the standard. All the reaction mixtures were detected in duplicate, and at least three independent assays were performed for each sample. Fluorescence was measured and recorded every 1 min (*f*0, *f*1, *f*2, *f*3,..., *f*60) until the fluorescence began to decline. The curves of antioxidant were initially normalized to the blank curve corresponding to the same assay by the factor $\text{fluorescence}_{blank,t=0}/\text{fluorescence}_{sample,t=0}$. Based on the normalized curves, the area under the fluorescence decay curve (AUC) was calculated as follows:

$$\text{AUC} = 1 + \sum_{i=1}^{i=60} fi/f0$$

$$\text{ORAC}_{FL} = \frac{\text{AUC}_{sample} - \text{AUC}_{blank}}{\text{AUC}_{trolox} - \text{AUC}_{blank}} \times \text{Trolox molarity} \times \text{sample dilution.}$$

Each sample was calculated by subtracting the AUC corresponding to the blank. Equations for regression between the net AUC and antioxidant concentration were calculated for all samples. The concentrations are expressed as micrograms of Trolox equivalents per 100 mg of fresh weight.

Results

Content and degree of catechin galloylation were higher in response to ethylene signaling

The level of catechin galloylation was determined through HPLC after 5 days of treatment with the ethylene precursor ACC, and the content of total catechin and values of ECG/(EC + ECG) and EGCG/(EGC + EGCG) were calculated. The HPLC profiles revealed the corresponding peaks to caffeine and the four major catechins including EC, ECG, EGC, and EGCG (Fig. 1a). The total catechin content was higher in the tea seedlings treated with ACC than in the control sample (Fig. 1b). The degrees of EGC and EC galloylation were higher after ACC treatment in comparison with the control sample (Fig. 1c).

Content of anthocyanins, flavonoids, and total polyphenol was enhanced after the ACC treatment

To investigate the effects of ethylene signaling in the secondary metabolic pathway, the contents of anthocyanins, flavonoids, and total polyphenol were determined after ACC treatment in tea seedlings. The contents of anthocyanins (Fig. 2a), flavonoids (Fig. 2b), and total polyphenols (Fig. 2c) were higher in the tea seedlings that underwent ACC treatment than in the control sample. These findings show that ethylene signaling was involved in the

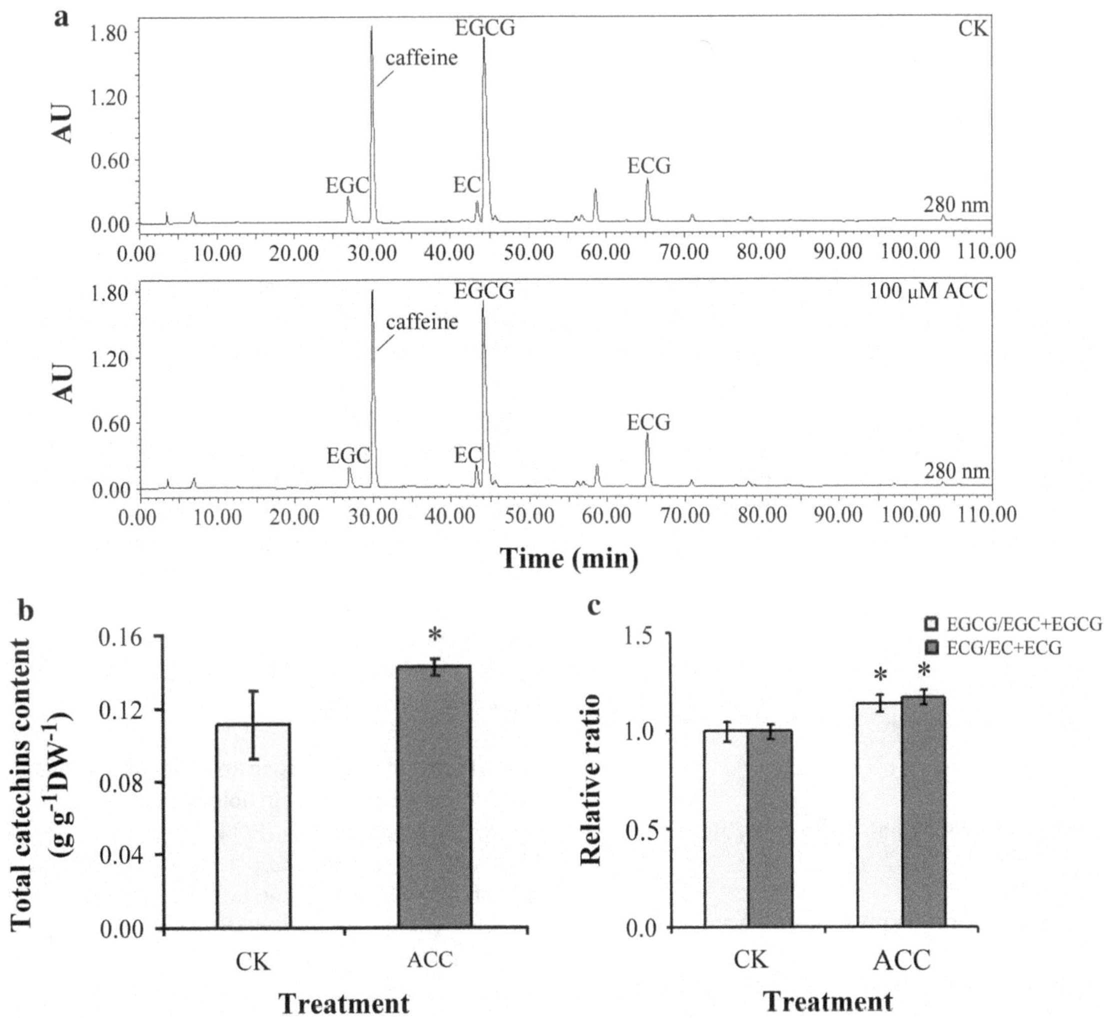

Fig. 1 Effects of ethylene signaling on the content and degree of catechin galloylation in tea seedlings. **a** Caffeine and the four major catechins, EC, EGC, ECG, and EGCG, were detected. **b** Total catechin content. **c** The relative ratio was calculated as follows: ECG/(EC + ECG) and EGCG/(EGC + EGCG). Data points are mean ± SD of five biologically independent experiments. Asterisks indicate significant differences from the control sample ($P < 0.05$; Student's *t* test)

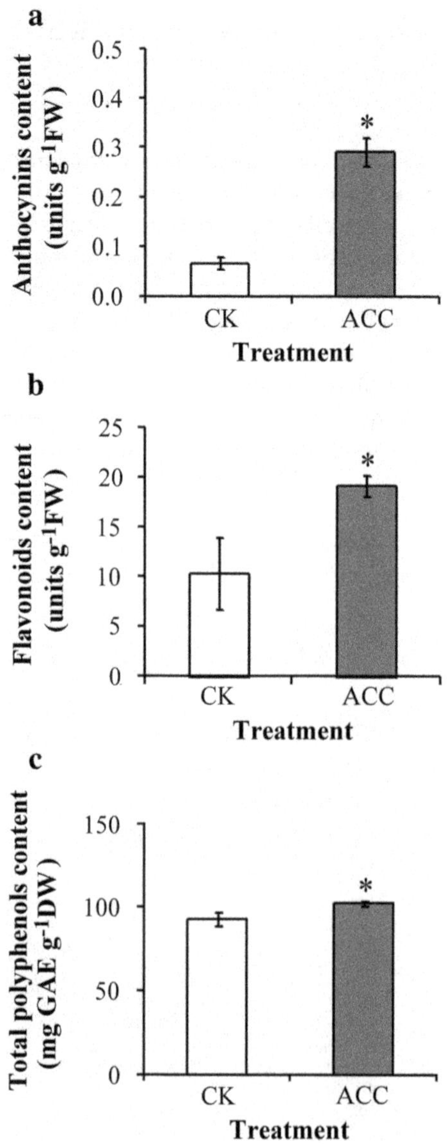

Fig. 2 Effects of ethylene signaling on the content of secondary metabolites in tea seedlings. The content of **a** anthocyanins, **b** flavonoids, and **c** total phenol was determined. Data points are mean ± SD of three biologically independent experiments. Asterisks indicate significant differences from the control sample ($P < 0.05$; Student's *t* test)

modulation of the metabolic pathway in the oolong tea seedlings.

Effects of ethylene signaling on antioxidant enzyme activity

To evaluate the cellular free radical scavenging capacity of the antioxidant enzymes in oolong tea seedlings after ACC treatment, the activity of antioxidant enzymes such as SOD, CAT, APX, and total POX was evaluated. Our results show that the activity of SOD (Fig. 3a), CAT (Fig. 3b), and POX (Fig. 3d) decreased under ACC treatment for 5 days. By contrast, APX activity significantly increased under ACC treatment for 5 days (Fig. 3c).

Antioxidant capacity was increased on modulation of ethylene signaling

To evaluate the effects of ethylene signaling on the antioxidant capacity of tea seedlings, assays of DPPH and ORAC were conducted. After ACC treatment for 5 days, 64% of the DPPH radicals were scavenged, whereas only 57% were scavenged in the control sample (Fig. 4). The ORAC-FL assay indicated that the ORAC was enhanced by ACC treatment (Fig. 5).

Discussion

Ethylene signaling involved in the flavonoid synthesis pathway in tea seedlings

Ethylene is a vital phytohormone in plants. Some previous studies have demonstrated that ethylene signaling is involved in metabolic pathways; for example, ethylene treatment increased total polyphenol and anthocyanin after grape harvesting (Bellincontro et al. 2006). Studies have indicated that exogenous application of α-aminoisobutyric acid as an analogue of ACC maintained higher catechin content in longan fruit pericarp during storage (Wang et al. 2015). Catechins are the most vital secondary metabolites in tea plants which already demonstrated galloylated catechins presented stronger positive health benefit potential than that of nongalloylated catechins (Chen et al. 2014). The results of this study show increases in not only the total catechin content but also the level of EGC and EC galloylation after ACC treatment in seedlings of oolong tea (Fig. 1). Moreover, the content of anthocyanins, flavonoids, and total polyphenol was enhanced by ACC treatment in oolong tea seedlings (Fig. 2). Our results indicate that ethylene signaling was involved in the flavonoid biosynthesis pathway and modulated the catechin galloylation level in oolong tea seedlings.

Role of ethylene signaling in antioxidant enzyme activity and antioxidant capacity in oolong tea seedlings

Reactive oxygen species (ROSs) are produced in chloroplasts, mitochondria, and peroxisomes as a typical product of plant cellular metabolism. Several studies have reported that plants exhibited excessive ROS production when they were subjected to biotic or abiotic stress, and the consequent oxidant–antioxidant imbalance activated factors responsible for cell injury (Liu et al. 2013; Wang et al. 2012; Wu and Yang 2016). Antioxidants are crucial because their main responsibility is the scavenging

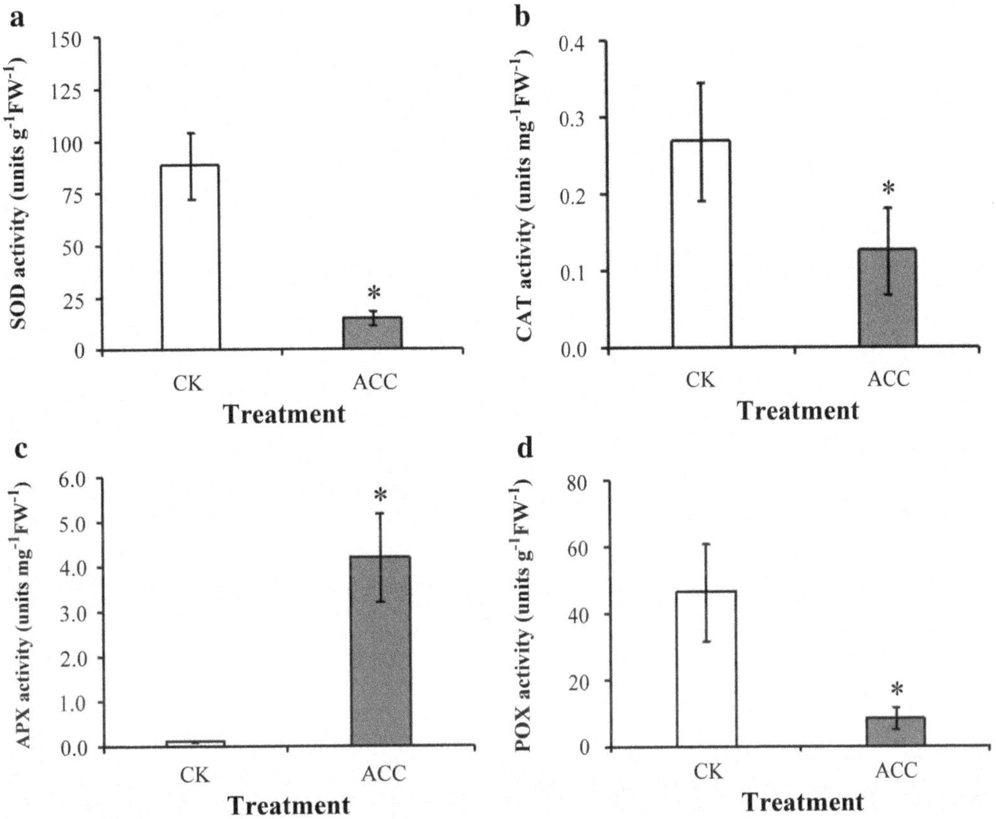

Fig. 3 CAT, APX, SOD, and POX activity in tea seedlings on modulation of ethylene signaling. Enzyme activity was detected from the detached tea seedling samples with one tip and two leaves after treatment with the control (CK) or 100 μM ACC for 5 days.CAT, APX, SOD, and total POX activity was determined. Data points are mean ± SD of five biologically independent experiments. Asterisks indicate significant differences from the control sample ($P < 0.05$; Student's *t* test)

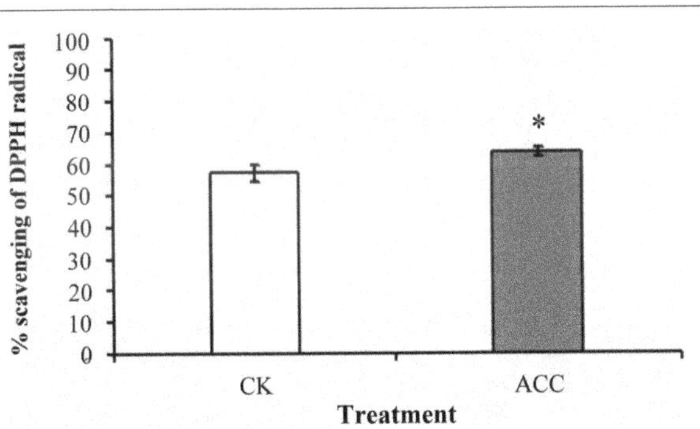

Fig. 4 DPPH radical scavenging activity in tea seedlings on modulation of ethylene signaling. The scavenging ratios of DPPH radicals were determined from the detached tea seedling samples after treatment with the control (CK) or 100 μM ACC for 5 days. Data points are mean ± SD of three biologically independent experiments. Asterisks indicate significant differences from the control sample ($P < 0.05$; Student's *t* test)

of ROSs, which is indispensable in protecting cell constituents from damage. Studies on phenolic compounds have shown that they are involved in plant antioxidant defense systems (Posmyk et al. 2009). Ethylene is known involved in the abscission of vegetative and reproductive organs and senescence under stress conditions. Ethylene reduced chilling damage during fruit storage and increased the activity of APX enzymes (Lafuente et al. 2004). Our data showed that using the ethylene precursor ACC to treat tea seedlings promoted the induction of phenolic compounds (Fig. 2). Therefore, ethylene signaling presumably affects the tea plant antioxidant system balance. The antioxidant enzyme activity assay revealed that the SOD, CAT, and POX activity in tea seedlings (Fig. 3a, b, d, respectively) was decreased after ACC treatment. By contrast, the APX activity was significantly increased after ACC treatment (Fig. 3c). These findings suggest that ethylene signaling promoted higher phenolic compound accumulation, which has implications for antioxidant enzyme activity in tea seedlings. The antioxidant capacity was determined by evaluating ORAC and DPPH free radical scavenging (Figs. 4 and 5). Our results indicate that higher antioxidant capacity in oolong tea seedlings can be achieved through ethylene signaling.

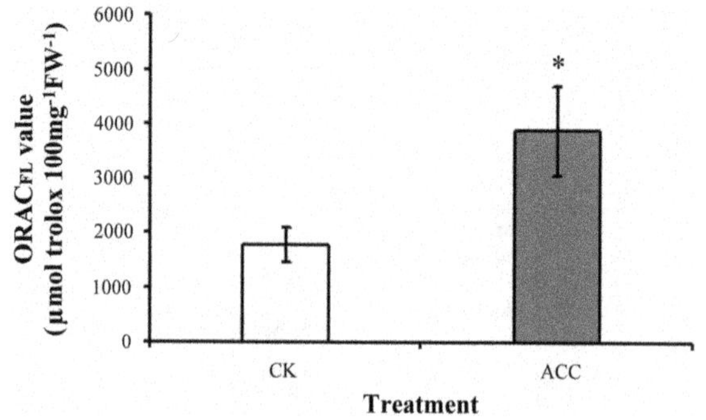

Fig. 5 ORAC in tea seedlings on modulation of ethylene signaling. ORAC-FL values were determined from the detached tea seedling samples after treatment with the control (CK) or 100 μM ACC for 5 days. Data points are mean ± SD of three biologically independent experiments. Asterisks indicate significant differences from the control sample ($P < 0.05$; Student's *t* test)

Conclusions

According to our results suggest that the ethylene signaling involved in the secondary metabolic pathway induced the accumulation of phenolic compounds and catechin galloylation and reduced antioxidant enzyme activity in the tea seedlings. The antioxidant capacity was positively correlated with phenolic compound accumulation during ethylene signaling. Whether or not the ACC signal is possible involved to regulate plant secondary metabolic pathway requires further research.

Abbreviations

ACC: 1-aminocyclopropane-1-carboxylic acid; EC: epicatechin; ECG: epicatechin-gallate; EGC: epigallocatechin; EGCG: epigallocatechin-gallate; HPLC: high-performance liquid chromatography; CAT: catalase; SOD: superoxide dismutase; POX: total peroxidase; APX: ascorbate peroxidase; ORAC: oxygen radical absorption capacity; DPPH: 2,2-diphenyl-1-picrylhydrazyl; ROS: reactive oxygen species.

Authors' contributions

The work presented here was carried out in collaboration between all authors. SWK, GHC and CTC carried out the experiments and analyzed the data. Dr. JTCT helps the analysis of HPLC data. Dr. CYY defined the research theme, designed experiments and wrote the manuscript. All authors read and approved the final manuscript.

Author details

[1] Department of Agronomy, National Chung Hsing University, Taichung 40227, Taiwan. [2] Graduate Institute of Biotechnology, National Chung Hsing University, Taichung 40227, Taiwan.

Acknowledgements

Not applicable.

Competing interests

The authors declare that they have no competing interests.

Funding

This study was funded by the National Science Council, Taiwan, R.O.C. (NSC 104-2311-B-005-006).

References

Abeynayake SW, Panter S, Chapman R, Webster T, Rochfort S, Mouradov A, Spangenberg G (2012) Biosynthesis of proanthocyanidins in white clover flowers: cross talk within the flavonoid pathway. Plant Physiol 158(2):666–678

Bellincontro A, Fardelli A, De Santis D, Botondi R, Mencarelli F (2006) Postharvest ethylene and 1-MCP treatments both affect phenols, anthocyanins, and aromatic quality of Aleatico grapes and wine. Aust J Grape Wine R 12(2):141–149

Boller T, Herner RC, Kende H (1979) Assay for and enzymatic formation of an ethylene precursor, 1-aminocyclopropane-1-carboxylic acid. Planta 145(3):293–303

Bosch M, Berger S, Schaller A, Stintzi A (2014) Jasmonate-dependent induction of polyphenol oxidase activity in tomato foliage is important for defense against *Spodoptera exigua* but not against *Manduca sexta*. BMC Plant Biol 14:257

Chao YY, Chou TS, Kao CH (2012) Involvement of abscisic acid and hydrogen peroxide in regulating the activities of antioxidant enzymes in leaves of rice seedlings under magnesium deficiency. Plant Growth Regul 66:1–8

Chen GH, Yang CY, Lee SJ, Wu CC, Tzen JTC (2014) Catechin content and the degree of its galloylation in oolong tea are inversely correlated with cultivation altitude. J Food Drug Anal 22(3):303–309

Kato M, Shimizu S (1985) Chlorophyll metabolism in higher plants VI. involvement of peroxidase in chlorophyll degradation. Plant Cell Physiol 26(7):1291–1301

Lafuente MT, Sala JM, Zacarias L (2004) Active oxygen detoxifying enzymes and phenylalanine ammonia-lyase in the ethylene-induced chilling tolerance in citrus fruit. J Agric Food Chem 52(11):3606–3611

Lee TM, Chu C (1992) Ethylene-induced polyamine accumulation in rice (*Oryza sativa* L.) coleoptiles. Plant Physiol 100(1):238–245

Lewis DR, Ramirez MV, Miller ND, Vallabhaneni P, Ray WK, Helm RF, Winkel BSJ, Muday GK (2011) Auxin and ethylene induce flavonol accumulation through distinct transcriptional networks. Plant Physiol 156(1):144–164

Lin CC, Kao CH (1999) NaCl induced changes in ionically bound peroxidase activity in roots of rice seedlings. Plant Soil 216:147–153

Liu J, Wu YC, Kan J, Wang Y, Jin CH (2013) Changes in reactive oxygen species production and antioxidant enzyme activity of *Agaricus bisporus* harvested at different stages of maturity. J Sci Food Agric 93(9):2201–2206

MacAdam JW, Nelson CJ, Sharp RE (1992) Peroxidase activity in the leaf elongation zone of tall fescue. Plant Physiol 99(3):872–878

Nakano Y, Asada K (1981) Hydrogen peroxide is scavenged by ascorbate-specific peroxidase in spinach chloroplasts. Plant Cell Physiol 22(5):867–880

Ou BX, Hampsch-Woodill M, Prior RL (2001) Development and validation of an improved oxygen radical absorbance capacity assay using fluorescein as the fluorescent probe. J Agric Food Chem 49(10):4619–4626

Perva-Uzunalic A, Skerget M, Knez Z, Weinreich B, Otto F, Gruner S (2006) Extraction of active ingredients from green tea (*Camellia sinensis*): extraction efficiency of major catechins and caffeine. Food Chem 96(4):597–605

Posmyk MM, Kontek R, Janas KM (2009) Antioxidant enzymes activity and phenolic compounds content in red cabbage seedlings exposed to copper stress. Ecotoxicol Environ Saf 72(2):596–602

Schaller GE (2012) Ethylene and the regulation of plant development. BMC Biol 10:9

Singleton VL, Rossi JA (1965) Colorimetry of total phenolics with phosphomolybdic–phosphotungstic acid reagents. Am J Enol Vitic 16:144–158

Sun BM, Zhu ZS, Cao PR, Chen H, Chen CM, Zhou X, Mao YH, Lei JJ, Jiang YP, Meng W, Wang YX, Liu SQ (2016) Purple foliage coloration in tea (*Camellia sinensis* L.) arises from activation of the R2R3-MYB transcription factor CsAN1. Sci Rep 6:32534. https://doi.org/10.1038/srep32534

Tadolini B, Juliano C, Piu L, Franconi F, Cabrini L (2000) Resveratrol inhibition of lipid peroxidation. Free Radic Res 33(1):105–114

Thipyapong P, Steffens JC (1997) Tomato polyphenol oxidase—differential response of the polyphenol oxidase F promoter to injuries and wound signals. Plant Physiol 115(2):409–418

Wang HB, Zhang ZQ, Xu LY, Huang XM, Pang XQ (2012) The effect of delay between heat treatment and cold storage on alleviation of chilling injury in banana fruit. J Sci Food Agric 92(13):2624–2629

Wang H, Zhi W, Qu HX, Lin HT, Jiang YM (2015) Application of alpha-aminoisobutyric acid and beta-aminoisobutyric acid inhibits pericarp browning of harvested longan fruit. Chem Cent J. https://doi.org/10.1186/s13065-015-0124-1

Woolf AB, Plummer JA, Clemens J (1999) Maturity and temperature influence ethylene-promoted organ abscission in *Camellia*. N Z J Crop Hortic 27(1):33–41

Wu YS, Yang CY (2016) Physiological responses and expression profile of NADPH oxidase in rice (*Oryza Sativa*) seedlings under different levels of submergence. Rice 9(1):2

Zhou QQ, Sun WJ, Lai ZX (2016) Differential expression of genes in purple-shoot tea tender leaves and mature leaves during leaf growth. J Sci Food Agric 96(6):1982–1989

Expression of ovate family protein 8 affects epicuticular waxes accumulation in *Arabidopsis thaliana*

Yao Tang[1], Wei Zhang[2], Yan-Li Yin[1], Peng Feng[1], Hong-ling Li[3] and Ying Chang[1*]

Abstract

Background: Transcription factors could regulate multiple aspects of plants growth and development, which is significant to plants. Ovate family proteins (OFPs) that are named due to contain OVATE domain, a 70-AA C-terminal conserved domain from the protein *OVATE* gene encodes, are plant-specific transcription factors family. Some members of OFPs have been shown to function as transcription factors to regulate plant growth and development, but little is known about the function of AtOFP8.

Results: Here, we found AtOFP8 maybe involve in transcriptional regulation of the epicuticular waxes in *Arabidopsis thaliana*. First, we observed that the distribution of epicuticular waxes of wild type plants was more than that of *Atofp8-1* deletion mutants, but less than that of *35S:HA-AtOFP8* transgenic overexpression lines not only on the leaves but also on the stems utilizing scanning electron microscopes. Second, we extracted waxes from leaves and stems of three types of plants respectively to measure the waxes content and composition by gas chromatography–mass spectrometer (GC–MS), and the results of the total content of waxes were consistent with the results of scanning electron microscopes. Finally, we found that the expression of 12 genes related to the synthesis and metabolism of waxes was changed in the *Atofp8-1* mutants and *35S:HA-AtOFP8* transgenic lines compared with wild type plants.

Conclusions: These findings suggest that AtOFP8 could change the accumulation of epicuticular waxes.

Keywords: AtOFP8, Epicuticular waxes, Transcription factor

Background

Ovate family proteins (OFPs) that are named due to contain OVATE domain, a 70-AA C-terminal conserved domain from the protein *OVATE* gene encodes, are plant-specific family of regulatory proteins (Hackbusch et al. 2005; Liu et al. 2002). All the time, we think that there are 18 genes that are predicted to encode proteins containing OVATE domain in *Arabidopsis thaliana* (Wang et al. 2007). It was not until 2014 that we knew that there are 19 OFPs genes in *A. thaliana* rather than 18 (Liu et al. 2014). *A. thaliana* ovate family proteins (AtOFPs) have been proved to be a novel transcriptional factor family, which could control multiple aspects of plant growth and development (Wang et al. 2011). So far, only a few members of AtOFPs have been shown to function as transcription factors to regulate plant growth and development (Li et al. 2011; Pagnussat et al. 2007; Wang et al. 2007). AtOFP1 was shown to regulate cell elongation by suppressing the expression of *Gibberellin 20 oxidase 1* (*GA20ox1*), a gene encoding the key enzyme in gibberellin (GA) biosynthesis (Wang et al. 2007). AtOFP5 was reported to negatively regulate the activity of a *BELL-like* homeodomain protein 1-*knotted1-like* homeodomain protein 3 (BLH1-KNAT3) complex during early embryo sac development (Pagnussat et al. 2007). AtOFP4 was demonstrated to participate in the regulation of secondary cell wall formation by interacting with KNAT7 (Li et al. 2011). However, little is known about the function of AtOFP8.

*Correspondence: 2427472516@qq.com
[1] Northeast Agricultural University, Harbin 150030, China
Full list of author information is available at the end of the article

As we all know, epicuticular waxes are complex mixtures of very long chain fatty acids (VLCFAs) and their derivatives, which are significant to plants (Shepherd and Griffiths 2006). They could form an important protective layer to protect the internal structure of the plant body (Koch and Ensikat 2008). Except control the loss of water by stomatal transpiration, plants also reduce the water loss by epicuticular waxes, which plays a crucial role in resisting water stress (Sieber et al. 2000). Waxes layer on the surface of the leaves has reflection effect to ultraviolet rays and light, which can protect the leaves from the dangers of radiation (Holmes and Keiller 2002). Until now, lots of genes related to the biosynthesis and metabolism of epicuticular waxes have been reported. *LTP4*, a gene of lipid transfer proteins gene family, could bind to fatty acids and transfer phospholipids between the membranes (Arondel et al. 2000). *CER2* gene expressed in an organ- and tissue-specific manner affects epicuticular waxes accumulation and the stems of *cer2* mutants are bright green (Xia et al. 1996). *CER4* encoded an alcohol-forming fatty acyl-coenzyme A reductase (FAR) is responsible for the synthesis of primary alcohols, which involves in epicuticular waxes production (Rowland et al. 2006). *FAR4* could affect the formation of primary alcohols and the encoded enzyme could increase the length of the fatty acid chain (Domergue et al. 2010). Researchers generally believe that if stems or leaves have a glossy green phenotype, it is indicative of waxes deficiency, and conversely, if plants have whitish green stems, it is most likely displayed more waxes (Go et al. 2014).

During our cultivation of Arabidopsis, we observed that *35S:HA-AtOFP8* transgenic lines had whitish green leaves and *Atofp8-1* mutants were lighter green than wild-type plants. According to this phenomenon, we guessed the function of AtOFP8 maybe associate with the epicuticular waxes. To testify whether transcription factor AtOFP8 could affect epicuticular waxes accumulation and how to change epicuticular waxes content and composition, we had verified from multiple perspectives. In our study, we reported the difference of the shape and the content of epicuticular waxes by comparing *Atofp8-1* mutants, wild-type plants and *35S:HA-AtOFP8* transgenic lines. Moreover, the expression of 12 genes related to the biosynthesis and metabolism of waxes was significant difference among them. These results suggest transcription factor AtOFP8 could change epicuticular waxes content and composition, perhaps by changing genes expression.

Materials and methods

Plant materials and growth conditions

Arabidopsis thaliana (Columbia-0 ecotype) was used as wild-type plants in our experiment. *35S:HA-AtOFP8* transgenic lines and *Atofp8-1* mutants were previously reported (Wang et al. 2011). All seeds were planted in the 1/2 Murashige and Skoog (MS) medium containing 0.8% agar and 3% sucrose and transferred into the soil a week later. Plants were grown under long-day conditions (14 h light/10 h dark cycles) at 23 °C.

Scanning electron microscopy

The stems and the rosette leaves of 5-week-old plants were chosen to dry in the oven at 50 °C after fixing on the foam board. Then they were plated by gold membrane on the surface of the samples glued to the sample table using ion sputtering coating machine and epicuticular waxes were observed using a scanning electron microscope.

Transmission electron microscopy

The stems and the rosette leaves of 5-week-old plants were chosen to fix in a solution containing 2.5% glutaraldehyde in 0.1 mol L^{-1} phosphate buffer, pH 6.8, at 4 °C for 4 h. The samples were then rinsed in 0.1 mol L^{-1} phosphate buffer, pH 6.8, three times and further fixed in 1% osmium tetroxide for 3 h at 4 °C. After rinsing with 0.1 mol L^{-1} phosphate buffer at normal temperature, the samples were dehydrated and embedded in Spurr's resin. After processing embedded block, thin sections (60 nm thickness) were prepared with an ultramicrotome. After staining, the sections were observed under a transmission electron microscope.

GC–MS analysis

150 mg of the stems and the rosette leaves of 5-week-old plants were weight to put into small beakers respectively and 5 mL of chloroform were added into small beakers for 1 min at room temperature. After adding 5 μg of C24 into the extracted chloroform solvent as internal reference, the solvent was transferred into GC bottles and evaporated under a gentle stream of nitrogen. Then 10 μL bis(trimethylsilyl)trifluoroacetamide (BSTFA) and 10 μL pyridine were added into the GC bottles to redissolve waxes samples and the waxes mixtures were heated for 1 h at 70 °C. Samples were measured by injecting 1 μL of sample and helium (2 mL min^{-1}) was used as a carrier gas during which the oven temperature was increased from 80 to 290 °C at a rate of 4 °C/min and finally 30 min at 290 °C.

Gene chip analysis

When the plants grew up to 3 weeks old, leaves were cut off and sent to Shanghai Bohao Biotechnology Corporation to make agilent chip to find differentially expressed genes. All chip results were analyzed by SBC analysis system (SAS), an expression microarray data analysis software of this corporation.

Table 1 qRT-PCR primer sequence

Gene	Primer name	Primer sequence (5′–3′)
LTP4	LTP4-F	CACCTCCGTGCTGTGCAG
	LTP4-R	TGGCGCAGTTGGTGCTC
At2g18370	At2g18370-F	GCATCAAGTCAGTGGCTAATAGTGT
	At2g18370-R	CATTCGCCTACATATATTATACGGC
CER4	CER4-F	TTGATGCTGTTTCTGATGTGATGC
	CER4-R	ACGACCTTCCCTTCTTTGTTGG
KCS2	KCS2-F	GGTTGTAGTGCTGGTCTTATCTCC
	KCS2-R	GGACGGTGTGGATGAGTTGG
ADH1	ADH1-F	ATGCCTTCAAGACTCATCCGA
	ADH1-R	TTGATTTCCGAGAATGGCACT
KCS10	KCS10-F	CTCGACGTCAGTTTAATGCATCT
	KCS10-R	GGGTTCTTGGTAGCGTATTCGT
FAR4	FAR4-F	CACAATGGCTCAGTTCAGTTTCTAC
	FAR4-R	CGCAGTTTCTCAGTATTCGTATCG
GL2	GL2-F	TAGAGATGAAGCTCGTCGGCAT
	GL2-R	GTTGCCTCTGTCTTGTCCCTT
DFR	DFR-F	GGTTACTTTGTTCGTGCCACC
	DFR-R	CCTCAGATAAATCAGCCTTCC
At2g42990	At2g42990-F	GCCTCGAAAAGATGTGAAGAG
	At2g42990-R	TGAATTCTTAACACGCCAGAGG
At1g06360	At1g06360-F	GACCCACATAGCCCTATCGAA
	At1g06360-R	TGCTTCAAGTCCATCACGTT
CER2	CER2-F	CGAGCCGTCTACTTCTTCAAGG
	CER2-R	GTTGCAGCGAATGTAAGGTATGG
ACT2	ACT2-F	CCAGAAGGATGCATATGTTGGTGA
	ACT2-R	GAGGAGCCTCGGTAAGAAGA

Table 2 The transcription level of *AtOFP8* in *35S:HA-AtOFP8* overexpression plants and *Atofp8-1* mutant

Locus	Symbol	Accession	FC in *35S:HA-AtOFP8*	FC in *Atofp8-1*
AT5G19650	OFP8	NC_003076	323.1614	−13.7830

Quantitative real-time RT-PCR analysis

RNA of 3-week-old plants leaves was extracted using RNA extraction kit according to the manufacturer's instructions. Reverse transcription (RT) was performed according to reverse transcription kit instructions to synthesize first-strand cDNA. Quantitative RT-PCR reactions were performed using SYBR green fluorescent quantitation PCR kit. The PCR primers used were listed in Table 1. The values for each set of primers were normalized relative to the *ACTIN 2* gene. qRT-PCR reactions were performed in biological triplicates using RNA samples extracted from three independent replicate samples. The comparative $\Delta\Delta C_T$ method was used for the calculation of the results.

Chlorophyll leaching assays

When the plants grew up to 5 weeks old, 1 g of rosette leaves was weight and put into erlenmeyer flasks wrapped with aluminum foil. Then 30 mL of 80% ethanol were poured into erlenmeyer flasks. Remove 3 mL solution from erlenmeyer flasks to measure absorbance at 647 and 664 nm using a spectrophotometer as the amount of extracted chlorophylls every 10 min. The solution using for measuring was back into erlenmeyer flasks every time. The absorbance measured after 24 h was as the total amount of chlorophylls. The calculation of chlorophyll content was based on the method previously (Lolle et al. 1998).

Water loss assays

The rosette leaves of 5-week-old plants were cut out and put into water for 1 h in the dark to make water contents be same. After drying the water on the surface of leaves, the leaves were weight using a microbalance every 20 min. The water loss rate (v) was calculated by the formula:

$$V = (W_0 - W_1)/W_0 \times 100\%$$

Where W_0, weight of leaves before dehydration; W_1, weight of leaves after dehydration.

Results

Identification of plant materials

35S:HA-AtOFP8 transgenic plants and *Atofp8-1* mutant were previously reported (Wang et al. 2011). Gene chip results showed that the transcription level of *AtOFP8* in *35S:HA-AtOFP8* overexpression plants increased more than 300-fold, while decreased by nearly 14-fold in the *Atofp8-1* mutant (Table 2).

Leaves are whitish green in *35S:HA-AtOFP8*

After transplanting plants in the soil, we gradually found there were significant differences in the color and phenotype of leaves among them. Although there was no significant difference in the phenotype of leaves between *Atofp8-1* and Col-0, leaves of *35S:HA-AtOFP8* were thicker and flatter than that of Col-0. Moreover, leaves of *35S:HA-AtOFP8* were more whitish green than that of Col-0, whereas leaves in *Atofp8-1* were lighter green than that of Col-0 (Fig. 1). We guessed this phenomenon might be related to epicuticular waxes.

The distribution of epicuticular waxes is thin in *Atofp8-1*, whereas dense in *35S:HA-AtOFP8*

Depending on the difference in the color and phenotype of leaves, we guessed this was most likely caused by the different distribution of epicuticular waxes. In order to verify whether this was related to the distribution of

Fig. 1 Image of seedlings 25 days old. **a** Col-0, **b** *Atofp8-1*, **c** *35S:HA-AtOFP8*. Bar = 1 cm

epicuticular waxes, surface of stems and leaves were observed respectively using the scanning electron microscope. On the surface of stems, the morphology of epicuticular waxes in *Atofp8-1*, Col-0 and *35S:HA-AtOFP8* are clearly different each other. Moreover, epicuticular waxes in Col-0 seem to lie more densely than that in *Atofp8-1*, but not so densely as the *35S:HA-AtOFP8* (Fig. 2a). Changes in content were the same in the leaves (Fig. 2b). Then we did transmission electron microscopy analysis to leaves and stems of *Atofp8-1*, Col-0 and *35S:HA-AtOFP8*. The results showed that there were no significant differences in the thickness of the cuticle in neither leaves nor stems (Fig. 3a, b). These observations imply that AtOFP8 conld only affect the distribution of epicuticular waxes without changing the cuticle waxes.

The content of waxes is reduced in *Atofp8-1*, while increased in *35S:HA-AtOFP8*

To verify further whether waxes content was changed, GC-MS was used to measure the content in the stems and leaves respectively. Although the content of waxes was not changed exponentially, there was still some difference among the totals of waxes. Results showed that the totals of waxes were 387.56, 356.05 and 412.23 $\mu g\ g^{-1}$ respectively in the stems of Col-0, *Atofp8-1* and *35S:HA-AtOFP8*. Compared with the wild type plants, the total content of waxes of *Atofp8-1* was slightly lower, while the total content of waxes in *35S:HA-AtOFP8* transgenic plants was slightly higher (Fig. 4a). The different components of the stems waxes also changed. The content of primary alcohols and ketone in *35S:HA-AtOFP8* transgenic plants was significantly higher than that in wild type, while *Atofp8-1* did not change significantly with wild type (Fig. 4a). However, the totals of waxes in the leaves were almost approximate (Fig. 4b), but there were distinct variations between the specific components. In the leaves of the *35S:HA-AtOFP8*, alkanes were significantly increased compared with the wild type plants, while primary alcohols were significantly reduced (Fig. 4c, d). These results indicate that AtOFP8 could change the content and composition of waxes.

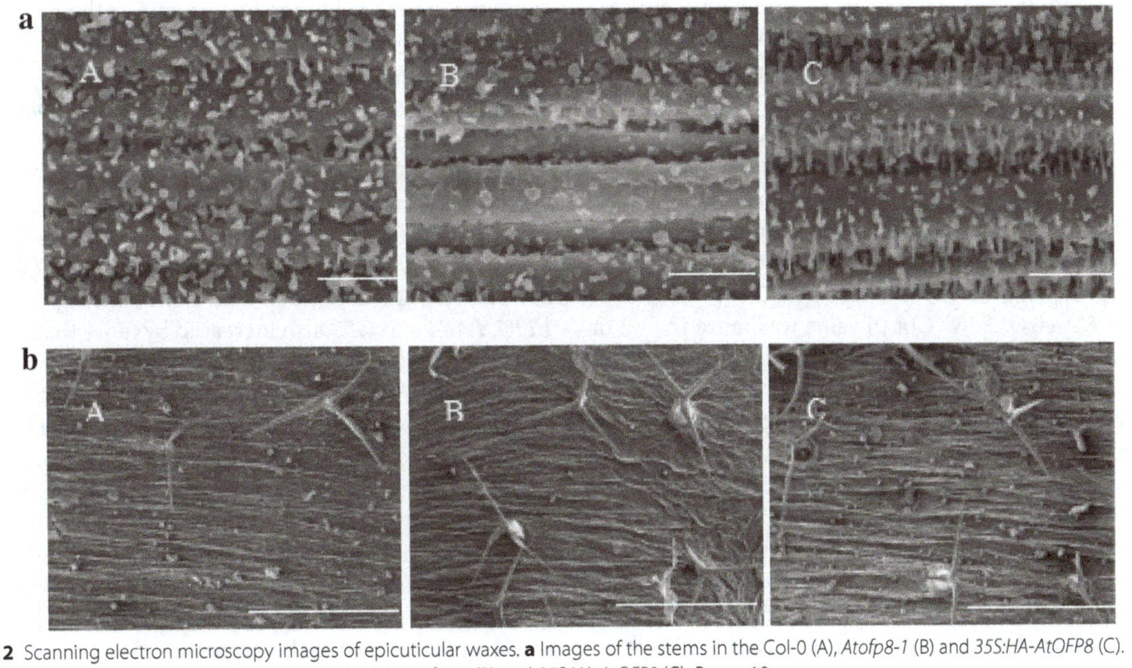

Fig. 2 Scanning electron microscopy images of epicuticular waxes. **a** Images of the stems in the Col-0 (A), *Atofp8-1* (B) and *35S:HA-AtOFP8* (C). Bars = 10 μm. **b** Images of the leaves in the Col-0 (A), *Atofp8-1* (B) and *35S:HA-AtOFP8* (C). Bars = 10 μm

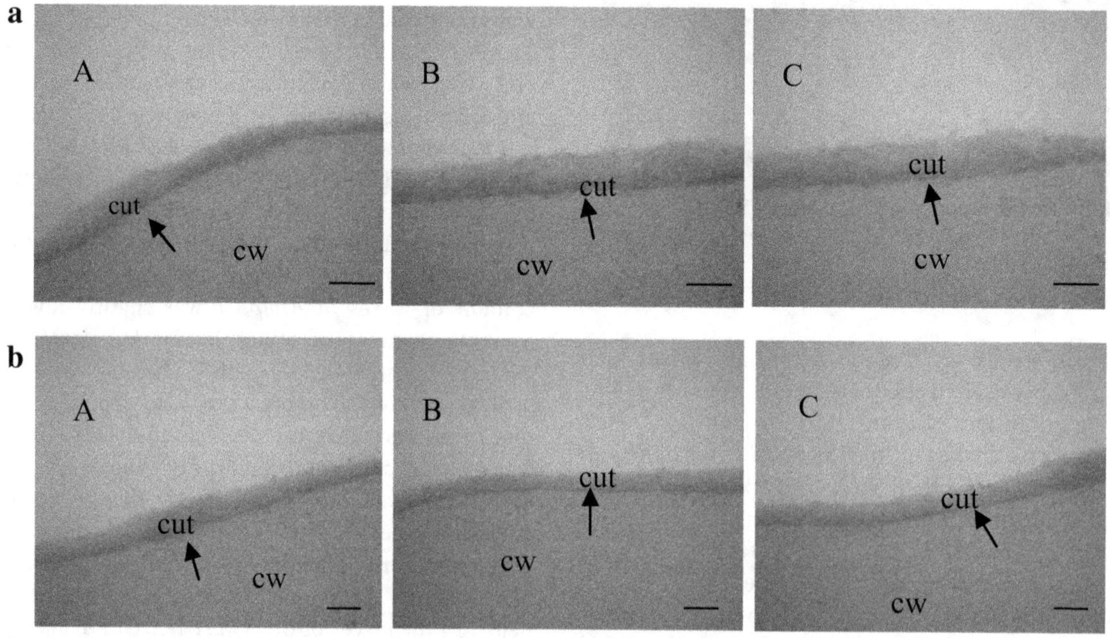

Fig. 3 Transmission electron microscopy images of epicuticular waxes. *cut* cuticle, *cw* cell wall. **a** Images of the stems in the Col-0 (A), *Atofp8-1* (B) and *35S:HA-AtOFP8* (C). Bars = 200 nm. **b** Images of the leaves in the Col-0 (A), *Atofp8-1* (B) and *35S:HA-AtOFP8* (C). Bars = 200 nm

Gene chip results show that AtOFP8 affects many aspects of Arabidopsis growth and development

To find out which genes caused by the change of epicuticular waxes, we made gene chip analysis among the *Atofp8-1*, *35S:HA-AtOFP8* and Col-0. Results showed that the expression of abundant genes was changed in the *Atofp8-1* and *35S:HA-AtOFP8* compared with Col-0. FC (*35S:HA-AtOFP8* vs Col-0) represented the ratio of gene expression level in *35S:HA-AtOFP8* transgenic plants to Col-0 in the data analysis. FC (*Atofp8-1* vs Col-0) represented the gene in *Atofp8-1* mutant and Col-0 expression levels in the ratio. It was generally considered that there was no significant difference in gene expression in the range of 0.5–2.0 for FC values, while genes with FC values greater than 2 or less than 0.5 were considered significant expression differences. There were 300 genes whose FC (*35S:HA-AtOFP8* vs Col-0) value was more than 2 or less than 0.5, among which 192 genes were up-regulated, which were mainly related to photosynthesis, glucose metabolism, amino acid metabolism, fatty acid metabolism, alkanes and flavonoids biosynthesis (Fig. 5a). 108 genes were down-regulated, and these genes were associated with plant hormone signal transduction, tryptophan metabolism, nitrogen metabolism, endoplasmic reticulum protein processing and zeatin biosynthesis (Fig. 5b). There were 434 genes with FC (*Atofp8-1* vs Col-0) value more than 2 or less than 0.5, which 268 genes were up-regulated, mainly related to DNA replication, plant hormone signal transduction, glucose metabolism, photosynthesis, alkanes, pyrimidines and organisms alkaloid biosynthesis processes (Fig. 5c). 166 down-regulated genes were mainly associated with processes such as glyceride metabolism, endoplasmic reticulum protein processing, plant pathogen interactions, flavonoid biosynthesis, amino acid metabolism, and anthocyanin synthesis (Fig. 5d).

The expression of genes related to epicuticular waxes varies greatly in *35S:HA-AtOFP8*

We chose some genes involved in epicuticular waxes synthesis and metabolism from gene chip results to test by qRT-PCR (Table 3). Results showed that the expression of *KCS10* was up-regulated by more than tenfold in *35S:HA-AtOFP8* compared with Col-0. The expression of *LTP4*, *FAR4* and *GL2* also increased by more than 5 times in *35S:HA-AtOFP8* (Fig. 6a). In addition, the expression of *CER4*, *ADH1* and *At2g18370* was raised about twice in *35S:HA-AtOFP8*, while significantly down-regulated in *Atofp8-1* compared with Col-0 (Fig. 6b). Contrary to these, the expression of *CER2*, *At2g42990* and *At1g06360* in *35S:HA-AtOFP8* was less than half of Col-0, while up-regulated in *Atofp8-1* compared with Col-0 (Fig. 6c). These results show that AtOFP8 could change the expression of genes involving in epicuticular waxes accumulation, largely promoting the expression.

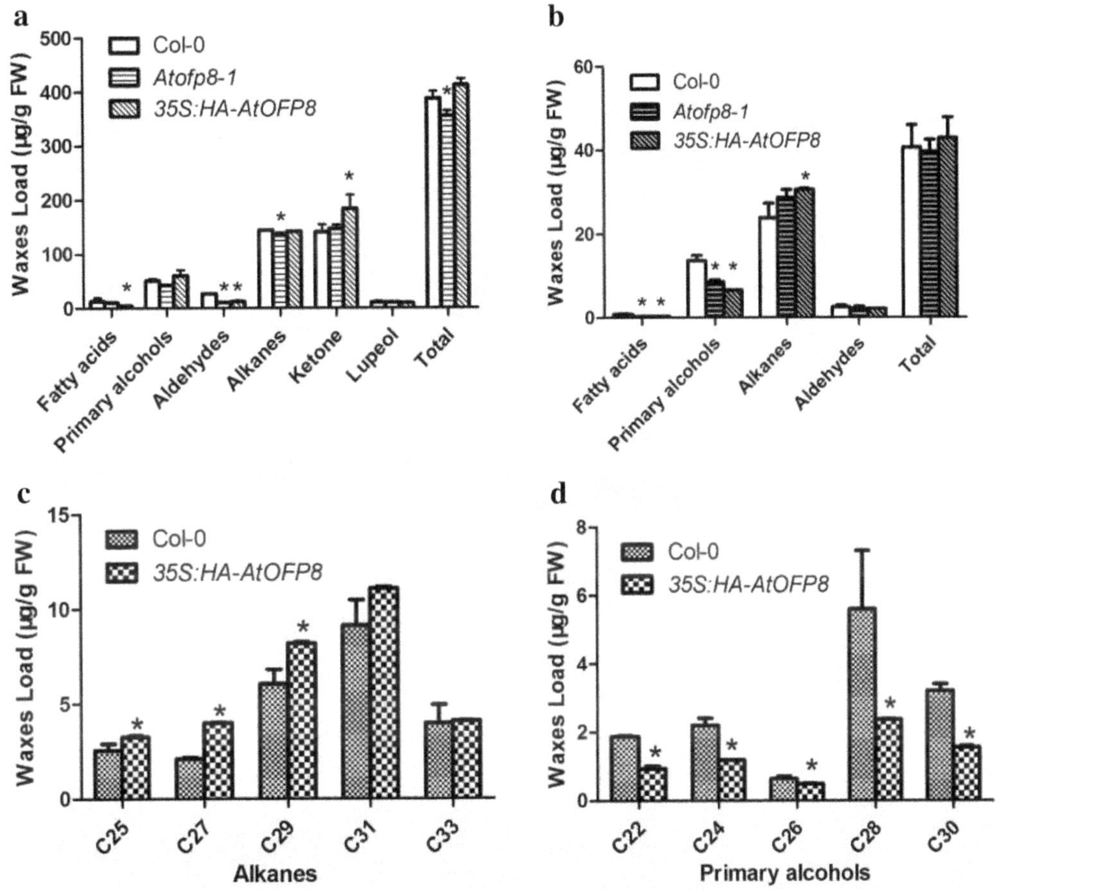

Fig. 4 Epicuticular waxes accumulation in the stems (**a**) and leaves (**b–d**). 5 weeks old stems and leaves were used. Biological triplicates, each with technical duplicates, were averaged. Statistical significance of the measurements was by comparing Col-0 (*$P < 0.05$). Bars indicate SE of the mean. *FW* fresh weight

Cuticle transpiration slows in *35S:HA-AtOFP8*

After observing so many changes related to epicuticular waxes, we made water loss assays to test whether stratum corneum changed. Results showed that water loss in Col-0 was slower than that in *Atofp8-1*, but faster than that in *35S:HA-AtOFP8* (Fig. 7a). Further chlorophyll leaching experiment showed that the percentage is the lowest in *35S:HA-AtOFP8* after 70 min (Fig. 7b). These results indirectly prove that there are more epicuticular waxes distributed on the leaves surface of *35S:HA-AtOFP8*.

Discussion

To explore the function of AtOFP8, we designed this experiment using *Atofp8-1*, Col-0 and *35S:HA-AtOFP8*. We analyzed microscopic distribution of epicuticular waxes, all kinds of component content, gene expression difference, which all results could demonstrate that AtOFP8 was able to regulate epicuticular waxes accumulation in some way. This laboratory finding provides a new direction for research AtOFP8 even AtOFPs.

Although the total amount of waxes in leaves was little difference, changes in specific components were obvious. Maybe AtOFP8 only changed the conversion of some certain components in leaves, so the total amount of waxes in leaves changed little. For example, in the *35S:HA-AtOFP8* leaves, alkanes were significantly increased compared with the wild type plants, while primary alcohols were significantly reduced (Fig. 4c, d). Another reason was the presence of functional redundancy. Loss of AtOFP8 might induce expression of other genes that had the same function, so the content of epicuticular waxes in the *Atofp8-1* did not change significantly.

Due to diverse component of epicuticular waxes and its complex metabolic process, we do not determine what a role AtOFP8 plays in the metabolic process of waxes. In addition, functional redundancy exists in the transcription factors AtOFPs all the time, so it is difficult to find the its specific function invisibly. This is another

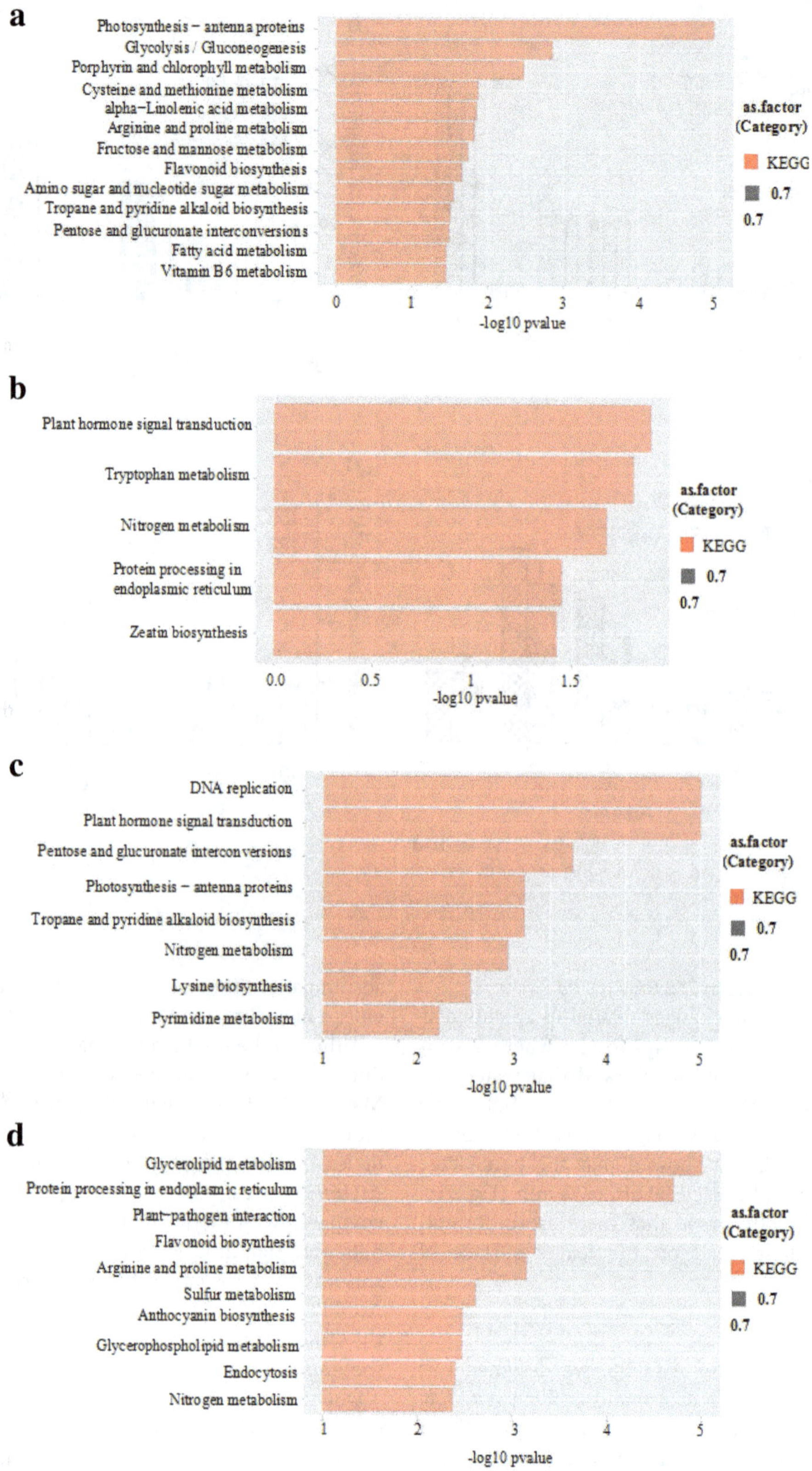

Fig. 5 Pathway enrichment affected by AtOFP8 gene. **a** Pathway enrichment of *35S:HA-AtOFP8* vs Col-0 up-regulated genes. **b** Pathway enrichment of *35S:HA-AtOFP8* vs Col-0 down-regulated genes. **c** Pathway enrichment of *Atofp8-1* vs Col-0 up-regulated genes. **d** Pathway enrichment of *Atofp8-1* vs Col-0 down-regulated genes

Table 3 Genes about epicuticular waxes accumulation

Gene	Symbol	FC in *35S:HA-AtOFP8*	FC in *Atofp8-1*	Description
At5g59310	LTP4	4.6633	1.1343	Lipid transfer protein 4
At2g18370		2.2816	−1.7933	Lipid transfer protein
At4g33790	CER4	1.9469	−1.2743	Fatty acyl-CoA reductase
At1g04220	KCS2	1.9174	−1.0575	3-Ketoacyl-CoA synthase 2
At1g77120	ADH1	1.9057	−2.2203	Alcohol dehydrogenase 1
At2g26250	KCS10	1.8824	−1.2770	3-Ketoacyl-CoA synthase 10
At3g44540	FAR4	1.8416	−1.2542	Fatty acyl-CoA reductase 4
At1g79840	GL2	1.6498	−1.3211	Lipid binding function
At5g42800	DFR	1.4015	−2.5641	Dihydroflavonol 4-reductase
At2g42990		−1.0988	2.9812	Lipid catabolic process
At1g06360		−1.2762	1.7454	Fatty acid desaturase family protein
At4g24510	CER2	−1.1792	1.4884	Fatty acid elongation

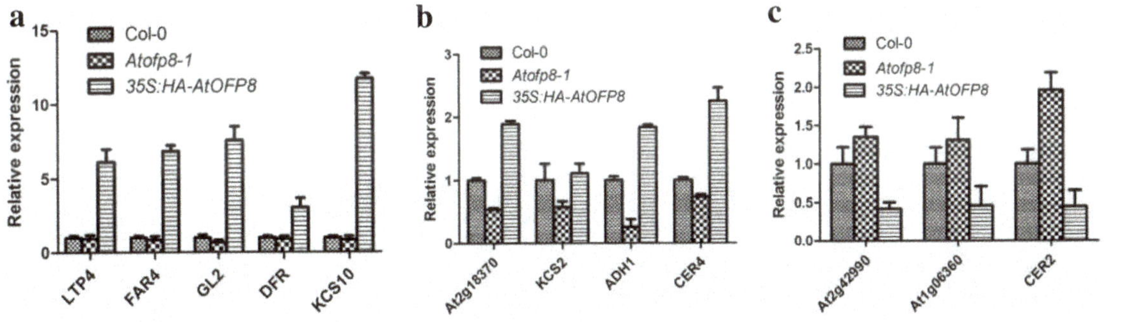

Fig. 6 Expression of genes that relate to waxes in Arabidopsis (**a**, **b**, **c**). The leaves of 3 weeks old plants were used. Bars indicate SE of the mean

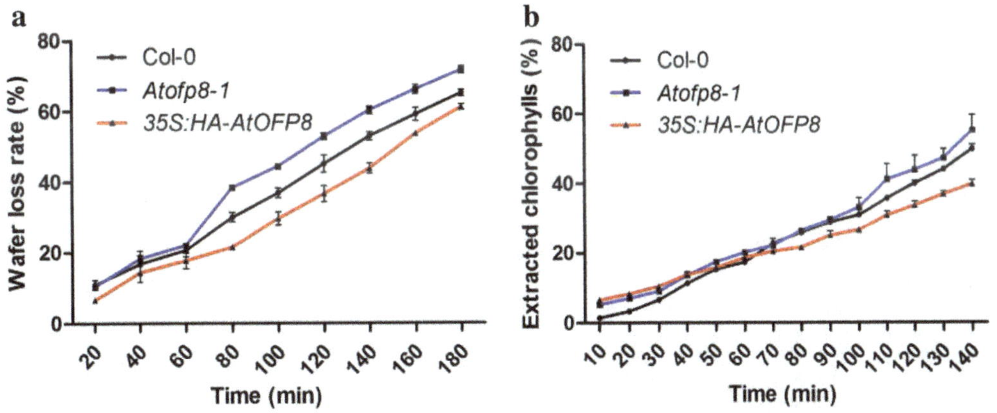

Fig. 7 Cuticular transpiration in Col-0, *Atofp8-1* and *35S:HA-AtOFP8* leaves. 5 weeks old leaves were used. Bars indicate SE of the mean. **a** Water loss assays. **b** Chlorophyll leaching assays

reason that we cannot understand which part AtOFP8 affects epicuticular waxes accumulation. In spite of these, various aspects of changes of epicuticular waxes are obvious enough to prove that it plays an important role that AtOFP8 could change genes expression to affect epicuticular waxes accumulation.

Abbreviations

AtOFP: *Arabidopsis thaliana* ovate family proteins; BSTFA: bis(trimethylsilyl) trifluoroacetamide; GC–MS: gas chromatography–mass spectrometer; MS: Murashige and Skoog; OFPs: ovate family proteins; SAS: SBC analysis system; VLCFAs: very long chain fatty acids.

Authors' contributions

YC designed the experiments. YT, WZ, YY, PF, HL and YC performed the experiments and analyzed the data. YT drafted the initial manuscript. All the authors agreed on the contents of the paper. All authors read and approved the final manuscript.

Author details

[1] Northeast Agricultural University, Harbin 150030, China. [2] Keshan Branch of Heilongjiang Academy of Agricultural Sciences, Keshan 160000, Heilongjiang Province, China. [3] College of Pharmacy, Qiqihar Medical University, Qiqihar 161006, China.

Acknowledgements

This work was supported by the National Natural Science Foundation of China (Project No. 30870144).

Competing interests

The authors declare that they have no competing interests.

Funding

This work was supported by the National Natural Science Foundation of China (Project No. 30870144).

References

Arondel V, Vergnolle C, Cantrel C, Kader JC (2000) Lipid transfer proteins are encoded by a small multigene family in *Arabidopsis thaliana*. Plant Sci 157:1–12

Domergue F, Vishwanath SJ, Joubes J, Ono J, Lee JA, Bourdon M, Alhattab R, Lowe C, Pascal S, Lessire R, Rowland O (2010) Three Arabidopsis fatty acyl-coenzyme A reductases, FAR1, FAR4, and FAR5, generate primary fatty alcohols associated with suberin deposition. Plant Physiol 153:1539–1554

Go YS, Kim H, Kim HJ, Suh MC (2014) Arabidopsis cuticular wax biosynthesis is negatively regulated by the *DEWAX* gene encoding an AP2/ERF-type transcription factor. Plant Cell 26:1666–1680

Hackbusch J, Richter K, Muller J, Salamini F, Uhrig JF (2005) A central role of *Arabidopsis thaliana* ovate family proteins in networking and subcellular localization of 3-aa loop extension homeodomain proteins. Proc Natl Acad Sci USA 102:4908–4912

Holmes MG, Keiller DR (2002) Effects of pubescence and waxes on the reflectance of leaves in the ultraviolet and photosynthetic wavebands: a comparison of a range of species. Plant Cell Environ 25:85–93

Koch K, Ensikat HJ (2008) The hydrophobic coatings of plant surfaces: epicuticular wax crystals and their morphologies, crystallinity and molecular self-assembly. Micron 39:759–772

Li EY, Wang SC, Liu YY, Chen JG, Douglas CJ (2011) Ovate family protein 4 (OFP4) interaction with KNAT7 regulates secondary cell wall formation in *Arabidopsis thaliana*. Plant J 67:328–341

Liu JP, Eck VJ, Cong B, Tanksley SD (2002) A new class of regulatory genes underlying the cause of pear-shaped tomato fruit. Proc Natl Acad Sci USA 99:13302–13306

Liu D, Sun W, Yuan YW, Zhang N, Hayward A, Liu YL, Wang Y (2014) Phylogenetic analyses provide the first insights into the evolution of OVATE family proteins in land plants. Ann Bot 113:1219–1233

Lolle SJ, Hsu W, Pruitt RE (1998) Genetic analysis of organ fusion in *Arabidopsis thaliana*. Genetics 149:607–619

Pagnussat GC, Yu HJ, Sundaresan V (2007) Cell-fate switch of synergid to egg cell in Arabidopsis eostre mutant embryo sacs arises from misexpression of the BEL1-like homeodomain gene *BLH1*. Plant Cell 19:3578–3592

Rowland O, Zheng HQ, Hepworth SR, Lam P, Jetter R, Kunst L (2006) *CER4* encodes an alcohol-forming fatty acylcoenzyme A reductase involved in cuticular wax production in Arabidopsis. Plant Physiol 142:866–877

Shepherd T, Griffiths DW (2006) The effects of stress on plant cuticular waxes. New Phytol 171:469–499

Sieber P, Schorderer M, Ryser U, Buchala A, Kolattukudy P, Metraux JP, Nawrath C (2000) Transgenic Arabidopsis plants expressing a fungal cutinase show alterations in the structure and properties of the cuticle and postgenital organ fusions. Plant Cell 12:721–738

Wang SC, Chang Y, Guo JJ, Chen JG (2007) Arabidopsis ovate family protein 1 is a transcriptional repressor that suppresses cell elongation. Plant J 50:858–872

Wang SC, Chang Y, Guo JJ, Zeng QN, Ellis BE, Chen JG (2011) Arabidopsis Ovate Family Proteins, a novel transcriptional repressor family, control multiple aspects of plant growth and development. PLoS ONE 6:e23896

Xia Y, Nikolau BJ, Schnable PS (1996) Cloning and characterization of *CER2*, an Arabidopsis gene that affects cuticular wax accumulation. Plant Cell 8:1291–1304

9

Cryopreservation without vitrification suitable for large scale cryopreservation of orchid seeds

Emily Schofield, Edward P. Jones and Viswambharan Sarasan*

Abstract

Background: Orchids are under threat from human activities and climate change, with populations limited to small geographic hotspots. This makes them ideal candidates for ex situ conservation. Orchid seeds are desiccation tolerant, but often have poor longevity in seed banks and cryopreservation of orchid protocorms is complex and expensive. Therefore, simple methods for large-scale storage programs are essential to store orchid seeds of different life forms. Seeds of five species representing epiphytic, lithophytic and terrestrial orchids from the Central Highlands of Madagascar were studied to find a simple and effective system of cryopreservation. The use of a vitrification solution prior to cryopreservation to improve survival was investigated, as well as the use of symbiotic and asymbiotic germination media to maximise germination after cryopreservation. Using the filter paper packet method, dried seeds were stored in vapour phase above liquid nitrogen and recovered after thawing with both symbiotic and asymbiotic media.

Results: The study revealed that cryoprotection is not essential for the species in this study, which represented a range of lifeforms. Vitrification generally led to a decrease in germination post cryopreservation. The use of a symbiotic germination medium post cryopreservation was found to be successful in the species in which it was tested. However, the use of an asymbiotic medium was successful for all the species in this study.

Conclusions: Vitrification was not essential for the species in this study as the orchid seeds were already ultralow temperature and desiccation tolerant. However, further studies using more species are required to validate this approach. This may be an ecophysiological or genetic trait of these species. Therefore, this form of dry seed cryopreservation could form part of ex situ orchid seed conservation using a standard method. The methods developed here will store greater genetic diversity compared to what can be achieved with protocorms and are suitable for both asymbiotic and symbiotic recovery after cryopreservation. This will help reduce the time and cost of ex situ conservation, and help develop universal protocols for large genera, compared to custom protocols required for protocorm cryopreservation.

Keywords: Epiphyte, Ex situ conservation, Endemic, Madagascar, Lithophyte, Orchid seed, Terrestrial

Background

The Orchidaceae is probably the largest family of flowering plants with more than 27,000 species (Zotz 2013) and is the most diverse plant group (Swarts and Dixon 2009), with high levels of ecological specialisation for pollinators and mycorrhizal fungi (Bailarote et al. 2012). The Itremo region in the Central Highlands of Madagascar is a micro-hotspot for orchids (Yokoya et al. 2015). The taxa we studied were from granite and quartz outcrops, and montane grassland. Seed banking is one of the cornerstones of ex situ biodiversity conservation. The majority of epiphytic orchids are desiccation tolerant, and can be stored at −20 °C in seed banks (Seaton et al. 2010). According to Merritt et al. (2014) seed life span depends on seed traits, along with the biotic and abiotic components of their environment. Desiccation tolerance is common in most orchids, but longevity in storage using

*Correspondence: v.sarasan@kew.org
Natural Capital and Plant Health, Royal Botanic Gardens, Kew, Richmond TW9 3AE, UK

conventional seed banking is poor (Li and Pritchard 2009). Therefore, cryopreservation offers a viable long-term seed storage option (Popova et al. 2016).

Seed and protocorm cryopreservation utilises a range of different available methods including: direct freezing, encapsulation-vitrification and droplet-freezing, each tested in several taxa (Sakai and Engelmann 2007; Sopalun et al. 2010; Jitsopakul et al. 2012; Hu et al. 2013; Galdiano et al. 2014; Teixeira da Silva et al. 2014; Popova et al. 2016). However, protocorms, a widely used propagule for cryopreservation, are prone to cryopreservation injury due to their cellular structure and high-water content. Our preliminary studies using protocorms of selected species yielded no successful recovery (unpublished data). Use of seeds reduces the intrinsic risk of intracellular damage due to ice crystal formation (Kulus and Zalewska 2014) owing to low water content of mature seeds (Jitsopakul et al. 2008). Thousands of seeds can be stored in a relatively small space with low maintenance requirements (Popova et al. 2003). There is also the added benefit of storing a wider genetic diversity compared to what can be achieved using protocorms. This can be achieved either by storage in liquid nitrogen at − 196 °C or in its vapour phase around − 150 °C (Vendrame et al. 2014). Therefore, we are exploring the feasibility of dry seed cryopreservation without the use of a cryoprotectant and comparing symbiotic and asymbiotic media for post-cryopreservation recovery.

Methods

The current study examines the viability of dry seed as propagules for the cryopreservation. To test this, three *Angraecum* species (*A. protensum*, *A. rutenbergianum* and *A. sesquipedale*) were cryopreserved with and without cryoprotectant. Vitrification using plant vitrification solution 2 (PVS2) was used in this study, due to its ability to remove and replace a large proportion of intracellular water from plant tissue (Kulus and Zalewska 2014). In addition, to test the use of symbiotic recovery after cryopreservation, *Angraecum magdalenae* and *Benthamia cinnabarina* were recovered using both asymbiotic and symbiotic methods. Popova et al. (2016) reviewed published seed cryopreservation to date and cryopreservation of both orchid and its fungal symbiont has been reported before as an experimental model (Batty et al. 2001; Teixeira da Silva et al. 2014). Merritt et al. (2014) proposed more advanced cryopreservation methods than currently available and simultaneous storage of seeds and fungal symbionts in a single system as potential area for future development. Our study takes the first step towards this by exploring possible simplification of the method of dry orchid seed cryopreservation. Therefore, our primary question concerns the development of a cost-effective and simple model to cryopreserve seeds from different life forms of orchids using dry seeds. Results will be discussed in the context of large scale ex situ programmes for threatened orchids from habitats where the threat to biodiversity is greater than anywhere else in the world.

Plant species

Five taxa were chosen for this study representing epiphytic, lithophytic and terrestrial life forms (Table 1). Lithophytic and terrestrial groups require niche habitats and are less abundant in Madagascar compared to epiphytic taxa. The genus *Angraecum* consists of 210 species, of which 128 are endemic to Madagascar (Cribb and Hermans 2009). The research permits from the Madagascar Forestry Authority and CITES (Convention of International Trade in Endangered Species of Wild Flora and

Table 1 Orchid taxa and orchid mycorrhizal fungi studied, their life form and habitat in the wild

Orchid	Orchid life form	Habitat
Angraecum magdalenae	Lithophyte	On or near quartz boulders
Angraecum protensum	Lithophyte	Granite and quartz outcrops
Angraecum sesquipedale	Epiphyte	Sloping tree trunks in humid forest (collected from glasshouse at Royal Botanic Gardens Kew)
Angraecum rutenbergianum	Epiphyte	Tree trunks of evergreen humid forests
Benthamia cinnabarina	Terrestrial	Montane grassland in sandy soil at the borders of granite and quartz boulders
OTU (operational taxonomic unit) name of the mycorrhizal fungus (OMF)	**Orchid life form of OMF partner**	**Habitat**
Tul1	*Angraecum magdalenae* (Lithophyte)	Montane grassland in a crack in granite
Tul2	*Benthamia cinnabarina* (Terrestrial)	Montane grassland in sandy soil at the borders of granite and quartz boulders

Fauna) permit for collecting the seeds are included as Additional file 1. We selected two lithophytic *Angraecum* species; *A. protensum* and *A. magdalenae* from quartz and granite outcrops, both of which grow on boulders or large rocks. Two epiphytic *Angraecum* species were collected, *A. rutenbergianum* from the Central Highlands of Madagascar (CHM) gallery forest and *A. sesquipedale* from a glasshouse collection in the Royal Botanic Gardens, Kew to investigate whether the observed desiccation tolerance traits were genetic or ecophysiological. The terrestrial species tested was *Benthamia cinnabarina,* a montane grassland species found in sandy soil at the borders of granite and quartz boulders (Table 1). Due to restrictions on collecting, a maximum of three capsules were collected from the wild, however, in many cases only one or two capsules were available for collecting at each site. The seeds used for each species were a mixture of seeds from at least 3 capsules. They were mixed before being stored at 4 °C.

Orchid mycorrhizal fungi (OMF)

The different OMF (orchid mycorrhizal fungi) isolates (Tul denotes *Tulasnella*) Tul1 and Tul2 (Table 1) were collected from the Central Highlands of Madagascar. The isolates used in this study were collected, isolated and cultured as described by Yokoya et al. (2015) and isolates were selected from in vitro cultures stored at 18 °C.

Seed sowing

Freshly collected mature seeds were cleaned and dried over LiCl for 2 weeks and stored at 4 °C until used for the study (approximately 1 year). Lithium chloride-mediated drying is a commonly used method (Galdiano et al. 2014). A pinch of seeds (ranging from 300 to 2000 seeds depending on the species) were transferred to hardened filter paper (Whatman, UK) and stapled. Due to the rare nature of the orchids and therefore limited seeds available, only one replicate per treatment was possible. Seed packets destined for direct sowing were surface sterilised in Magenta™ jars containing 50 ml of 0.5% sodium dichloroisocyanurate for 1 h. The other seeds were treated with or without vitrification solution. The vitrified seeds were then recovered and sterilised as described above. Once sterilised all seeds were washed in sterile water for 4 min and sown onto recovery medium 1 (RM1), Phytamax (Sigma, UK) with 3% sucrose and 0.15% activated charcoal, and recovery medium 2 (RM2), Oat Meal Agar medium (Clements 1988). RM2 was used for both asymbiotic and symbiotic recovery. The symbiotic RM2 consists of either OMF isolates Tul1 and Tul2 and (Rafter et al. 2016) (Table 1). Tul1 (originated from a lithophyte) and Tul2 (originated from a terrestrial) were specifically selected as these fungi were isolated from the mature roots of *Angraecum magdalenae* and *Benthamia cinnabarina* respectively, two of the taxa studied here (Table 1). Yokoya et al. (2015) were unsuccessful to isolate OMF from roots of both *A. protensum* and *A. sesquipedale* roots, therefore, only asymbiotic (RM1) medium was used for these species. Packets were cut open and seeds spread onto both asymbiotic (RM1) and symbiotic media (RM2). Non-cryopreserved controls were treated with PVS2 and then sown as above.

Cryopreservation

Effect of vitrification on recovery of seeds after cryopreservation

Seeds, in seed packets as described above but without the sterilisation step, were treated directly in PVS2 for different times (0, 30, 45 and 60 min), and then transferred into Nunc™ cryovials (Sigma-Aldrich, Gillingham, UK) and covered with PVS2. Seeds with 0 min treatment were not treated with PVS2. Lids were put on both sets of cryovials, with and without PVS2 and dipped into liquid nitrogen on multi-vial trays, then put directly in the vapour phase above Liquid Nitrogen (LN) (-170 ± 5 °C) in a D4000 Cryo Dewar (International Cryogenics, Indiana, USA) and stored for 1 week.

Recovery

Cryovials containing seeds were removed from the Dewar and put in a water bath at 38 °C for 2 min. PVS2 solution was removed from the vial using Pasteur pipette and replaced with 1 ml of 1.2 M sucrose solution to completely cover the packet and then repeated twice. The 0 min PVS2 control vial was transferred directly into water bath for 2 min for thawing without sucrose solution. Seeds were sterilised after thawing, before sowing onto media as described above.

PVS toxicity

After the initial cryopreservation experiments *Angraecum protensum* seeds treated with plant vitrification solution 2 (PVS2) achieved very little germination. Therefore, the effect of vitrification was tested using Plant Vitrification Solutions PVS2 (30% (v/v) glycerol (Merck, Hertfordshire, UK), 15% (v/v) ethylene glycol (Sigma-Aldrich, Gillingham, UK), 15% (v/v) Dimethyl sulfoxide (DMSO) (Sigma-Aldrich, Gillingham, UK) and 0.4 M sucrose ½ Murashige and Skoog solution (½ MS) (Duchefa Biochemie, Haarlem, the Netherlands)) (Sakai et al. 1990) and PVS3 (50% glycerol, 50% sucrose (Sigma-Aldrich, Gillingham, UK)) (Nishizawa et al. 1993). *A. protensum* seeds were cryopreserved with either 30 or 60 min PVS2, PVS3 or no PVS.

Data analysis

Germination was assessed 12 weeks after thawing as described by Yamazaki and Miyoshi (2006). The process of seed germination was divided into six categories according to the developmental stage of the embryos, adapted from Miyoshi and Mii (1995), as follows; Stage 0—cultured seed; Stage 1—imbibed seed; Stage 2—testa broken by emerging protocorm Stage 3—protocorm has emerged from the testa; Stage 4—protocorms with rhizoids; Stage 5—shoot primordium visible. Germination was determined to have occurred at stage 4 and 5 (protocorm and seedling stage). Assessments of germination were done under a stereo-microscope (Leica, Germany). Cryopreservation data was analysed using either an ANOVA or Two Sample T test in Microsoft Excel (Microsoft, Reading, UK). The effect of PVS2 and the recovery media used was investigated in cryopreserved and non-cryopreserved seeds.

Results

Effect of vitrification solutions on recovery of seeds after cryopreservation

In order to test the effect of vitrification in *Angraecum*, *A. protensum* was tested for PVS2 toxicity. Germination (protocorm and seedling stage combined) after PVS2 and PVS3 treatments was recorded to assess seed viability after vitrification and cryopreservation. The highest germination percentage after cryopreservation occurred when the seeds were not vitrified before cryopreservation (Fig. 1).

Effect of vitrification on non-cryopreserved seeds

PVS2 reduced the average seed germination of all the tested species apart from *Angraecum magdalenae* when recovered with recovery medium 1 (RM1) without undergoing cryopreservation (Fig. 2). However, when the mean germination of all the tested species combined was analysed using an ANOVA test there was no significant difference between the average germination of seeds treated with or without PVS2 ($F_2=0.004$, $P=0.99$).

Effect of vitrification on cryopreservation of seeds

A similar pattern as that of the non-cryopreserved seed was found when seeds were treated with PVS2 and cryopreserved. The only species to have increased germination in response to cryopreservation following PVS2 treatment was *A. magdalenae.* A decrease in average germination was found for *A. protensum*, *A. magdalenae* and *A. rutenbergianum* (Fig. 2). There was also no significant difference in the mean germination across the four species when analysed using an ANOVA test ($F_{10}=0.02$, $P=0.97$) (Fig. 3).

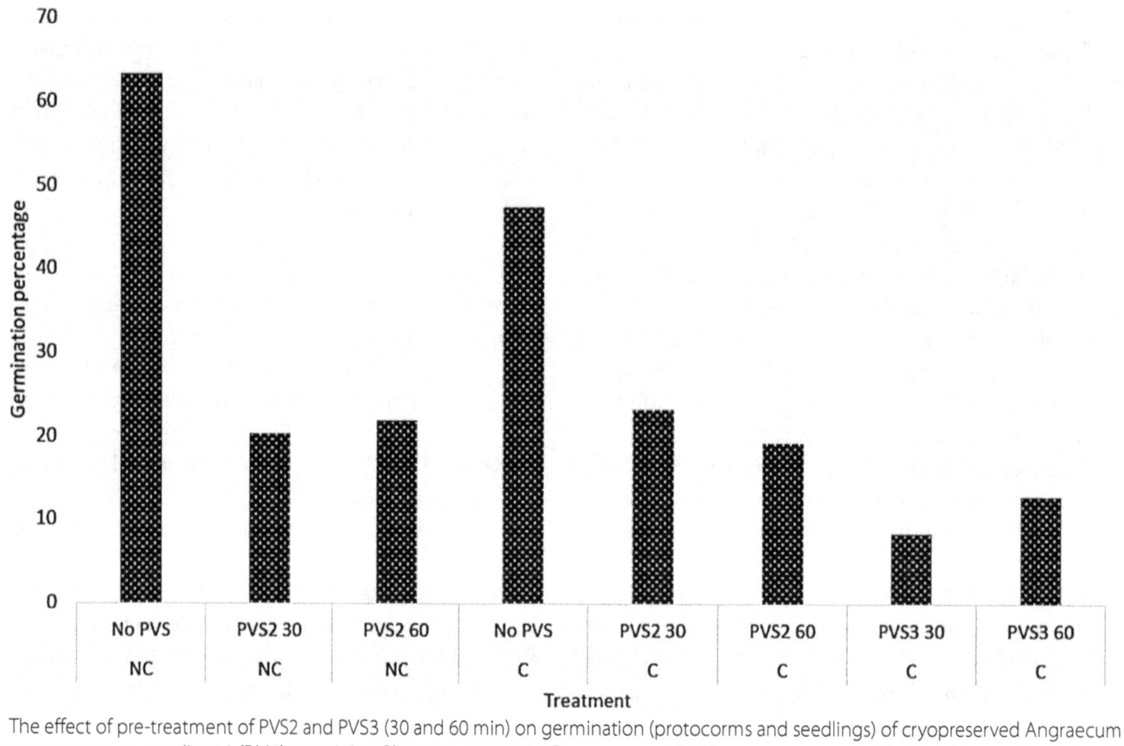

Fig. 1 The effect of pre-treatment of PVS2 and PVS3 (30 and 60 min) on germination (protocorms and seedlings) of cryopreserved Angraecum protensum on recovery medium 1 (RM1) containing Phytamax nutrients. *C* cryopreserved, *NC* non-cryopreserved. Total viable seed tested = 10,535

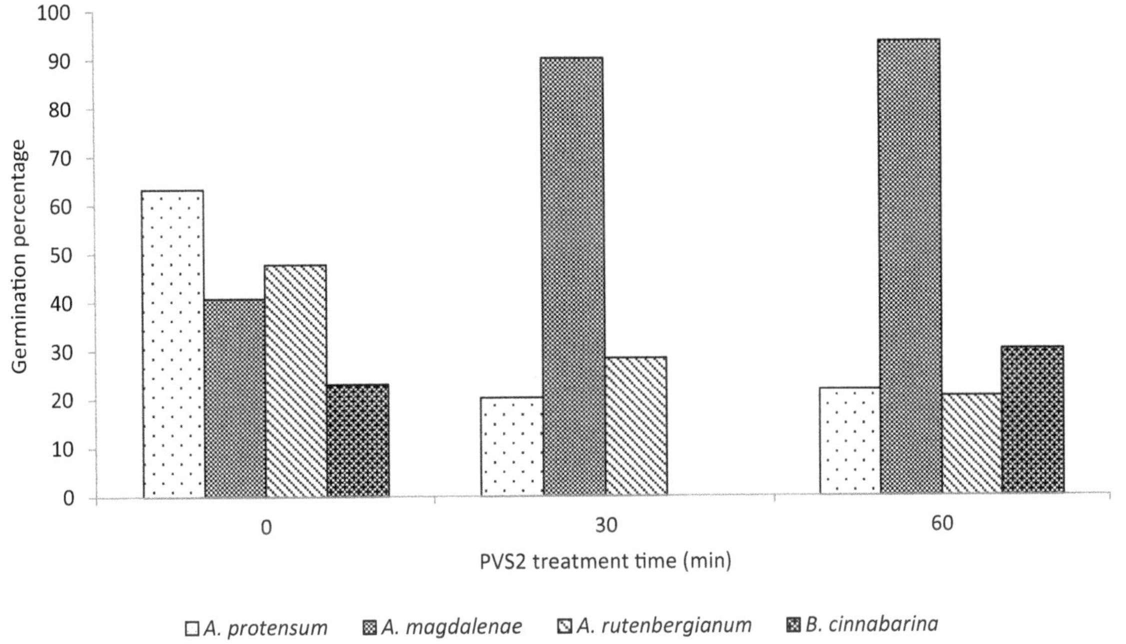

Fig. 2 Percentage of germination (protocorms and seedlings) of non-cryopreserved seeds of four species of orchids (*Angraecum magdalenae*, *A. protensum*, *Benthamia cinnabarina* and *A. rutenbergianum*) treated with PVS2 and recovered on recovery medium 1 (RM1) containing Phytamax nutrients. Total viable seed tested = 3084

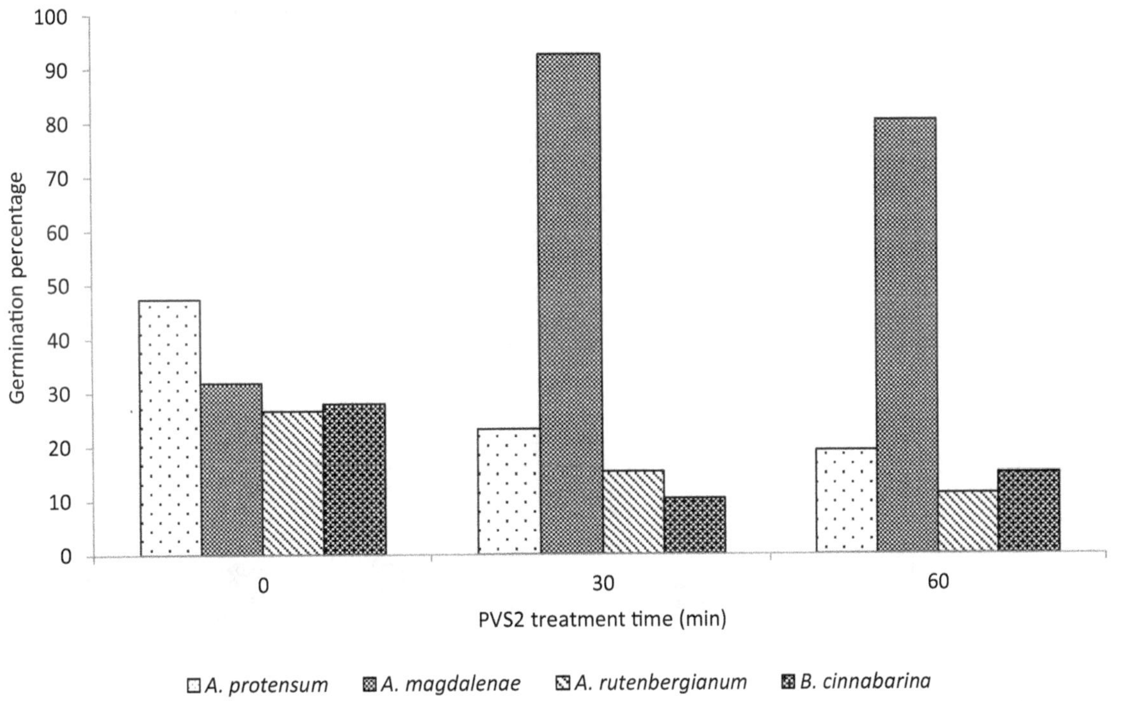

Fig. 3 Germination (protocorms and seedlings) percentage of cryopreserved seeds of four species of orchids (Angraecum magdalenae, *A. protensum*, *A. rutenbergianum* and *Benthamia cinnabarina*) treated with PVS2 and recovered on recovery medium 1 containing Phytamax nutrients. Total viable seed = 3082

Traits of orchids from the same region

Angraecum magdalenae and *B. cinnabarina* occur in the same area of the Central Highlands of Madagascar. Therefore, these species were compared to test whether a requirement for PVS2 vitrification before cryopreservation is an ecophysiological or species-specific effect.

A. magdalenae

The lithophytic *A. magdalenae* behaved in a similar way to the terrestrial *B. cinnabarina*. Cryopreservation did not significantly reduce germination, compared to non-cryopreserved controls with 30 min PVS2 ($\chi_1=0.0002$, $P=0.98$). However, unlike *B. cinnabarina*, *A. magdalenae* benefitted from PVS2 treatment (Fig. 4a). Significantly more recovery was found after cryopreservation with 30 min PVS2 compared to no PVS2 treatment ($\chi_1=192.23$, $P=<0.01$).

B. cinnabarina

Cryopreservation without vitrification was not harmful to seed recovery, with no significant difference in germination between the non-cryopreserved and cryopreserved treatments without PVS2 ($\chi_1=0.34$, $P=0.55$), (Fig. 4b). Unlike *A. magdalenae,* PVS2 had no significant benefit to the recovery of cryopreserved *B. cinnabarina* seed, with no significant difference in recovery between 0 and 60 min PVS2 treatment ($\chi_1=0.48$, $P=0.48$).

Symbiotic vs. asymbiotic recovery

Angraecum magdalenae and *B. cinnabarina* were recovered symbiotically and asymbiotically to understand the relative benefits of symbiotic recovery in two species with different lifeforms (Table 2).

Table 2 Germination (protocorms and seedlings) of *Angraecum magdalenae* and *Benthamia cinnabarina* after cryopreservation with the optimum PVS2 treatment for each species

Medium	*A. magdalenae*	*B. cinnabarina*
Symbiotic (RM2 + fungus)	79.6	15.0
Asymbiotic (RM1)	92.5	23.2

RM2 tul1 was the symbiotic medium for *A. magdalenae* and RM2 tul2 for *B. cinnabarina*

Benthamia cinnabarina

The symbiotic medium RM2 tul2 produced the maximum germination after 60 min PVS2 and cryopreservation, although this was not significantly more than the asymbiotic RM1 medium ($\chi_1=1.04$, $P=0.31$). Both asymbiotic RM1 and symbiotic RM2 tul2 were found to facilitate germination to seedling stage.

Angraecum magdalenae

No significant difference was found between recovery of cryopreserved seeds on the asymbiotic medium RM1 compared to the symbiotic medium RM2 tul1 ($\chi_1=0.38$, $P=0.54$). Pre-treatment with 30 min PVS2 with recovery medium RM2 tul1 led to 65% germination, but only 13% progression to seedling stage. This contrasts with asymbiotic germination using RM1, which facilitated 66% of the germinated seeds to progress to seedling stage. Using a symbiotic recovery medium led to a high percentage of seeds becoming stuck at protocorm stage which could not progress to full seedling stage. This did not occur when the seeds were recovered on RM1.

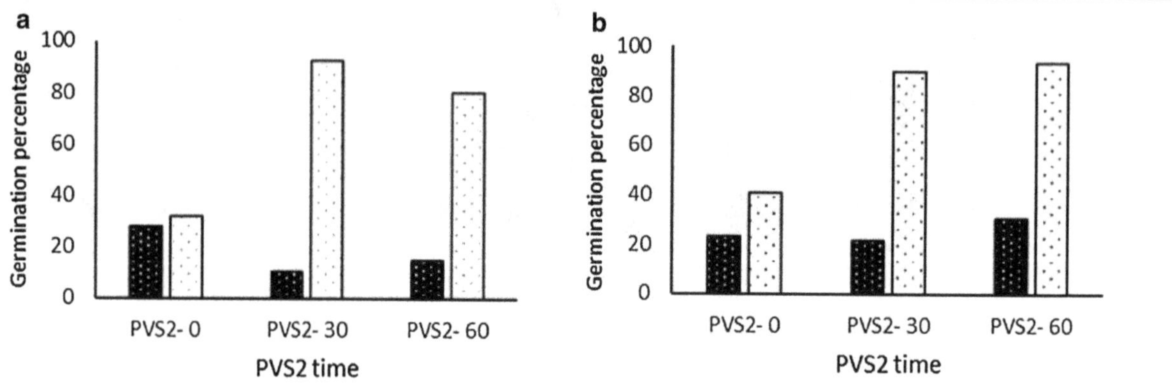

Fig. 4 **a** Germination (protocorms and seedlings) percentage of *B. cinnabarina* and *A. magdalenae* cryopreserved with and without PVS2 treatment. Total seeds = 1969. **b** The germination percentage (protocorm and seedling) of non-cryopreserved *B. cinnabarina* and *A. magdalenae*. All seeds were recovered on recovery medium 1 (RM1) containing Phytamax nutrients. Total viable seeds tested = 1908

Discussion

Need for vitrification before cryopreservation

This study clearly demonstrates that for the tested orchid seeds of *Angraecum* spp. and *B. cinnabarina* vitrification is not critical. Although vitrification is considered essential for the cryopreservation of protocorm and shoot tissue, direct cryopreservation of seeds without vitrification has been reported in other taxa (Popova et al. 2016). As more than one lifeform were tested, the lack of requirement for vitrification is unlikely to be due to a species or lifeform specific factor. This is in contrast to a number of previous studies of other tropical and subtropical species (Nikishina et al. 2001; Yamazaki and Miyoshi 2006), that are not traditionally considered desiccation and ultra-low temperature-tolerant (Popova et al. 2016). Orchid seeds are generally classified as orthodox with very low water content. However, further studies are required to determine the molecular mechanisms that allow cryopreservation without vitrification.

Angraecum magdalenae was the only species tested that significantly benefitted from vitrification before cryopreservation. Pre-treatment with 30 min PVS2 promoted germination to 92% compared to only 40% without PVS2 treatment. Despite the impressive germination after cryopreservation with PVS2, this is an exception to the general trend observed in this study. Previous studies have also successfully recovered *A. magdalenae* without vitrification (Nikishina et al. 2001). Therefore, we argue that the percentage of seeds successfully recovered without the use of PVS2 would be satisfactory for the long term cryogenic storage of *A. magdalenae*. If a less than optimum germination percentage could be accepted, then the need for expensive and time consuming extensive pre-screening of each individual orchid species before cryopreservation would be avoided.

Benthamia cinnabarina is from the granite and quart outcrops in the Central Highlands of Madagascar, like *A. protensum* and *A. magdalenae* studied here. As a terrestrial species we expected that *B. cinnabarina* would be more likely to require cryoprotection compared to the lithophytic species studied. However, *B. cinnabarina* did not require vitrification, whereas the lithophytic *A. magdalenae* benefitted from 30 min PVS2 pre-treatment. The mechanism behind the need for vitrification might be due to ecophysiological factors, as the terrestrial *B. cinnabarina* is adapted to a low moisture environment. However, *A. sesquipedale* seeds collected from the glasshouse at Kew also demonstrated desiccation and ultralow temperature tolerance, being cryopreserved most successfully without vitrification. This suggests that this response may also have a genetic basis. By conducting larger trials in different species of *Benthamia* and *Angraecum* from a range of habitat types will shed light on the traits that are linked to seed tolerance of ultralow temperature.

A universal cryopreservation method has yet to be found that is suitable for the seeds of a range of orchid species (Popova et al. 2016). Therefore, the observation that vitrification is not critical for cryopreservation in the studied *Angraecum* and *Benthamia* species from Madagascar has significance to the conservation of other related species from the genera and beyond. This finding could contribute to further studies in 128 species of *Angraecum* and more than 35 species of *Benthamia* for quick screening and development of generalised cryopreservation protocols. Habitat destruction continues at an unprecedented pace in the Central Highlands of Madagascar, leading to habitat fragmentation (Ramiadantsoa et al. 2015), and driving many endemic orchids to the verge of extinction.

Symbiotic recovery of seeds

Asymbiotic recovery media is the current standard approach to recover cryopreserved orchid seeds. However, symbiotic germination of orchid seeds is more successful when reintroducing orchids into the wild (Johnson et al. 2007). Most mature orchids produce multiple seed pods each season, each containing thousands of seeds (Swarts and Dixon 2009). Therefore, 30–50% recovery after cryopreservation would still be able to potentially produce hundreds or even thousands of plants for re-establishment of orchid populations in the wild. Cryopreservation without vitrification will help develop low-cost cryopreservation methods, as described by Li and Pritchard (2009).

Maximum *A. magdalenae* germination was achieved with asymbiotic recovery. Despite the tul2 isolate being able to promote germination after cryopreservation, it was unable to promote germination to seedling stage. Phytamax was effective at germinating all the species in this study. It contains nitrogen in a form which is accessible to the plant, organic compounds and sucrose, all components required for the germination of many different orchid species (Dutra et al. 2008). The selected fungus used for symbiotic recovery in this study may have been unable to provide the appropriate nutrients or sugars to the germinating orchid seed. This highlights the vital importance of selecting the appropriate mycorrhizal fungus to promote germination to seedling stage, and the need for large scale screening of mycorrhizal fungal isolates to identify these for each cryopreserved orchid species.

Previous studies have either focussed on encapsulation of the seed and mycorrhizal fungus (Wood et al. 2000) or separate storage and recovery of the orchid and its symbiont (Sommerville et al. 2008; Ercole et al. 2013). These require initial investment into finding an appropriate mycorrhizal partner and optimisation of storage conditions for both the fungus and orchid. This is likely to be a secondary step and less suitable for orchid species that are under immediate threat, due to the time required for optimisation.

Conclusions

This study across many orchid species has found that PVS2 is not a requirement for ex situ storage of orchid seeds from the Central Highlands of Madagascar. Despite *A. magdaleane* benefitting from PVS2 pre-treatment before cryopreservation, sufficient germination after cryopreservation without vitrification was achieved for potential reintroduction to take place. This approach could be the basis of a large scale ex situ conservation strategy for orchids under immediate threat of extinction. Aiming for sufficient instead of optimum germination is a good initial step to protect critically endangered orchids. This can then be followed up with a more specific symbiotic storage method using a compatible mycorrhizal fungus to aid the production of seedlings suitable for long term reintroduction to the wild.

Abbreviations

PVS2: plant vitrification solution 2; PVS3: plant vitrification solution 3; CITES: Convention on International Trade of Endangered Species of Wild Flora and Fauna; CHM: Central Highlands of Madagascar; OTU: operational taxonomic unit; OMF: orchid mycorrhizal fungi; RM1: recovery medium 1; RM2: recovery medium 2; Tul1: tulasnella isolate 1; Tul2: tulasnella isolate 2.

Authors' contributions

ES and EPJ carried out the cryopreservation experiments, ES analysed the data. ES and VS wrote the article; VS conceived the idea and designed the project. All authors read and approved the final manuscript.

Acknowledgements

We kindly acknowledge Jonathan Kendon and Kazutomo Yokoya for collecting and storing the orchid seed, and culturing the orchid mycorrhizal isolates used in this study. We also thank Rui Fang for assisting data analysis.

Competing interests

The authors declare that they have no competing interests.

Confirmation that all authors have approved the manuscript for submission

All authors have approved the manuscript for submission to Botanical Studies.

Funding

We acknowledge the financial support received from the Sainsbury Orchid Endowment.

References

Bailarote BC, Lievens B, Jacquemyn H (2012) Does mycorrhizal specificity affect orchid decline and rarity? Am J Bot 99:1655–1665. https://doi.org/10.3732/ajb.1200117

Batty AL, Dixon KW, Brundrett M, Sivasithamparam K (2001) Constraints to symbiotic germination of terrestrial orchid seed in a mediterranean bushland. New Phytol 152:511–520. https://doi.org/10.1046/j.0028-646X.2001.00277

Clements M (1988) Orchid mycorrhizal associations. Lindleyana 3:73–86

Cribb P, Hermans J (2009) Field guide to the orchids of Madagascar orchids. R Bot Gard, Kew

Dutra D, Johnson TR, Kauth PJ et al (2008) Asymbiotic seed germination, in vitro seedling development, and greenhouse acclimatization of the threatened terrestrial orchid *Bletia purpurea*. Plant Cell Tissue Organ Cult 94:11–21. https://doi.org/10.1007/s11240-008-9382-0

Ercole E, Rodda M, Molinatti M et al (2013) Cryopreservation of orchid mycorrhizal fungi: a tool for the conservation of endangered species. J Microbiol Methods 93:134–137. https://doi.org/10.1016/j.mimet.2013.03.003

Galdiano RF, de Macedo Lemos EG, de Faria RT, Vendrame WA (2014) Seedling Development and evaluation of genetic stability of cryopreserved *Dendrobium* hybrid mature seeds. Appl Biochem Biotechnol 172:2521–2529. https://doi.org/10.1007/s12010-013-0699-8

Hu WH, Yang YH, Liaw SI, Chang C (2013) Cryopreservation the seeds of a Taiwanese terrestrial orchid, *Bletilla formosana* (Hayata) Schltr. by vitrification. Bot Stud. 54:33. https://doi.org/10.1186/1999-3110-54-33

Jitsopakul N, Thammasiri K, Ishikawa K (2008) Cryopreservation of *Bletilla striata* mature seeds, 3-day germinating seeds and protocorms by droplet-vitrification. CryoLetters. 29:517–526

Jitsopakul N, Thammasiri K, Yukawa T, Ishikawa K (2012) Effect of cryopreservation on seed germination and protocorm development of *Vanda tricolor*. ScienceAsia 38:244. https://doi.org/10.2306/scienceasia1513-1874.2012.38.244

Johnson TR, Stewart SL, Dutra D et al (2007) Asymbiotic and symbiotic seed germination of *Eulophia alta* (Orchidaceae)—preliminary evidence for the symbiotic culture advantage. Plant Cell Tissue Organ Cult 90:313–323. https://doi.org/10.1007/s11240-007-9270-z

Kulus D, Zalewska M (2014) Cryopreservation as a tool used in long-term storage of ornamental species—a review. Sci Hort 168:88–107

Li DZ, Pritchard HW (2009) The science and economics of ex situ plant conservation. Trends Plant Sci 14:614–621

Merritt DJ, Hay FR, Swarts ND et al (2014) Ex situ conservation and cryopreservation of orchid germplasm. Int J Plant Sci 175:46–58. https://doi.org/10.1086/673370

Miyoshi K, Mii M (1995) Phytohormone pre-treatment for the enhancement of seed germination and protocorm formation by the terrestrial orchid, *Calanthe discolor* (Orchidaceae), in asymbiotic culture. Sci Hortic (Amsterdam) 63:263–267. https://doi.org/10.1016/0304-4238(95)00813-9

Nikishina TV, Popov AS, Kolomeitseva GL, Golovkin BN (2001) Cryopreservation of seeds of some tropical orchids. Dokl Biochem Biophys 378:231–233. https://doi.org/10.1023/A:1011585801668

Nishizawa S, Sakai A, Amano Y, Matsuzawa T (1993) Cryopreservation of asparagus (*Asparagus officinalis* L.) embryogenic suspension cells and subsequent plant regeneration by vitrification. Plant Sci 91:67–73

Popova EV, Nikishina TV, Kolomeitseva GL, Popov AS (2003) The effect of seed cryopreservation on the development of protocorms by the hybrid orchid Bratonia. Russ J Plant Physiol 50:672–677. https://doi.org/10.1023/A:1025600526588

Popova E, Kim HH, Saxena PK et al (2016) Frozen beauty: the cryobiotechnology of orchid diversity. Biotechnol Adv 34:380–403. https://doi.org/10.1016/j.biotechadv.2016.01.001

Rafter M, Yokoya K, Schofield EJ et al (2016) Non-specific symbiotic germination of *Cynorkis purpurea* (Thouars) Kraezl., a habitat-specific terrestrial orchid from the Central Highlands of Madagascar. Mycorrhiza 26:541–552. https://doi.org/10.1007/s00572-016-0691-6

Ramiadantsoa T, Ovaskainen O, Rybicki J, Hanski I (2015) Large-scale habitat corridors for biodiversity conservation: a forest corridor in Madagascar. PLoS ONE 10:e0132126. https://doi.org/10.1371/journal.pone.0132126

Sakai A, Engelmann F (2007) Vitrification, encapsulation-vitrification and droplet-vitrification: a review. Cryo-Letters 28:151–172

Sakai A, Kobayashi S, Oiyama I (1990) Cryopreservation of nucellar cells of navel orange (*Citrus sinensis* Osb. var. *brasiliensis* Tanaka) by vitrification. Plant Cell Rep 9:30–33. https://doi.org/10.1007/BF00232130

Seaton PT, Hu H, Perner H, Pritchard HW (2010) Ex situ conservation of orchids in a warming world. Bot Rev 76:193–203. https://doi.org/10.1007/s12229-010-9048-6

Sommerville KD, Siemon JP, Wood CB, Offord CA (2008) Simultaneous encapsulation of seed and mycorrhizal fungi for long-term storage and propagation of terrestrial orchids. Aust J Bot 56:609. https://doi.org/10.1071/BT08008

Sopalun K, Thammasiri K, Ishikawa K (2010) Vitrification-based cryopreservation of *Grammatophyllum speciosum* protocorms. Cryo-Letters 31:347–357

Swarts ND, Dixon KW (2009) Terrestrial orchid conservation in the age of extinction. Ann Bot 104:543–556. https://doi.org/10.1093/aob/mcp025

Teixeira da Silva JA, Zeng S, Galdiano RF et al (2014) In vitro conservation of *Dendrobium* germplasm. Plant Cell Rep 33:1413–1423

Vendrame W, Faria RT, Sorace M, Sahyun SA (2014) Orchid cryopreservation. Ciência e Agrotecnologia 38:213–229. https://doi.org/10.1590/S1413-70542014000300001

Wood CB, Pritchard HW, Miller AP (2000) Simultaneous preservation of orchid seed and its fungal symbiont using encapsulation-dehydration is dependent on moisture content and storage temperature. Cryo-Letters 21:125–136

Yamazaki J, Miyoshi K (2006) In vitro asymbiotic germination of immature seed and formation of protocorm by *Cephalanthera falcata* (Orchidaceae). Ann Bot 98:1197–1206. https://doi.org/10.1093/aob/mcl223

Yokoya K, Zettler LW, Kendon JP et al (2015) Preliminary findings on identification of mycorrhizal fungi from diverse orchids in the Central Highlands of Madagascar. Mycorrhiza 25:611–625. https://doi.org/10.1007/s00572-015-0635-6

Zotz G (2013) The systematic distribution of vascular epiphytes—a critical update. Bot J Linn Soc 171:453–481. https://doi.org/10.1111/boj.12010

10

Influence of harvest stage on the pharmacological effect of *Angelica dahurica*

Wei-Hong Liang[1], Tung-Wu Chang[2] and Yuh-Chyang Charng[1*]

Abstract

Background: Baizhi (*Angelica dahurica*) has been widely used as a traditional Chinese herbal medicine, functional food and cosmetic product ingredient, mostly because of the high furanocoumarin compounds in roots. The cropping system of Baizhi with its unique summer dormancy feature, is easily affected by the transition of its growth stages. The aim of this study was to analyze the quantity (size, form and dry weight [DW]) and quality (antioxidant and furanocoumarin content) of taproot and lateral root from three growth stages of Baizhi; vegetative (V-stage), summer dormancy (S-stage) and bolting stage (B-stage).

Results: Root length and diameter were lower at V-stage than the other two stages, and S-stage had higher lateral root to total root ratio. However, the highest root DW was observed at S-stage. Antioxidant activity was revealed by 2,2-diphenyl-ʟ-picrylhydrazyl and Fe^{2+} chelating assay, and the content of six furanocoumarin compounds, including xanthotoxin, bergapten, oxypeucedanin, imperatorin, phellopterin and isoimperatorin, was analyzed by liquid chromatography. Although the antioxidant activity was less at S-stage than the other stages, furanocoumarin contents showed little variation.

Conclusion: Considering the high DW and stable furanocoumarin composition, S-stage is the best harvest stage than the other stages because of its richer total pharmacological content.

Keywords: *Angelica dahurica*, Summer dormancy, Harvest time, Furanocoumarin, Antioxidant

Background

Baizhi (*Angelica dahurica* BENTH. et HOOK.) is a perennial Apiaceae plant, originated in Taiwan (Wang et al. 2001) and found abundantly in Korea, China, Japan and Russia (Sarker and Nahar 2004; Li et al. 2014a, b). The root of Baizhi is a traditional Chinese medicine known for its antipyretic and analgesic properties since thousands of years ago. Today, researchers are analyzing the multiple biological functions and investigating the chemical constituents of Baizhi.

The multiple pharmacological effects of Baizhi root extracts include antioxidation (Piao et al. 2004), anti-inflammatory (Pervin et al. 2014), antiproliferative (Yousif et al. 1999), skin whitening (Cho et al. 2006), antitumor (Kim et al. 2007), antimicrobial (Kwon et al. 1997) and anti-Alzheimer effects (Marumoto and Miyazawa 2010). These effects are believed to be attributed to the plant's rich furanocoumarin compounds such as imperatorin and isoimperatorin. Imperatorin has anti-inflammatory (Garcia et al. 2000), anticonvulsant (Luszczki et al. 2007), hepatoprotective (Oh et al. 2002), myorelaxant (Chiou et al. 2001), vasodilator (Wang et al. 2016) and anti-cancer effects (Li et al. 2014a, b). Isoimperatorin has anti-inflammatory (Abad et al. 2001), antiallergic (Ryu et al. 2001) and antimicrobial effects (Suleimenov 2009). Beside the pharmacological effects of Baizhi furanocoumarins, root extracts contain a variety of phenolic compounds that are connected to its strong antioxidant activity (Wang et al. 2017). Recently, foods with antioxidant effects become valued for their ability to remove reactive oxygen species, which oxidize lipids, DNA, membranes and proteins, and are involved in atherosclerosis, cancers and other diseases (Leopold and Loscalzo 2009).

*Correspondence: bocharng@ntu.edu.tw
[1] Department of Agronomy, National Taiwan University, Taipei, Taiwan, Republic of China
Full list of author information is available at the end of the article

Growth of Baizhi has three stages: V-stage, vegetative stage; B-stage, bolting stage; and S-stage, summer dormancy stage. Summer dormancy in herbaceous perennials is characterized by: (1) cessation or reduction of leaf meristem growth; (2) senescence of most or all above-ground herbage; (3) possible dehydration of the bases of the youngest leaves at the base of vegetative tillers; and (4) possible preceding formation of resting organs in the form of swollen basal internodes (corm/tuber) or swollen leaf bases (such as bulbs) (Volaire and Norton 2006). Although a report indicated NtFTs are target genes of temperature signal pathway to regulate the dormancy process in *Narcissus tazetta* var *Chinensis* (Feng et al. 2015), the relationship between biochemical regular and summer dormancy still lacked conclusive researches (Gillespie and Volaire 2017). For environmental drivers, vegetative organs develop under increasing day-length and temperature at the end of spring will induce summer dormancy (Volaire et al. 2009). In addition, water deficit was shown to be an induction factor for *P. bulbosa* L. (Ofir and Kigel 2007) and *D. glomerata* (Volaire 2002). Accordingly, after the V-stage, instead of transitioning to the B-stage, Baizhi enters summer dormancy (S-stage) to survive hot and dry environmental conditions. Its above-ground portions die during summer, and its root becomes the storage organ and therefore becomes heavier, a mechanism that implies more harvest yield than at other growth stages. Thus, it is expected that Baizhi roots should be harvested at S-stage, although the S-stage is not necessary to complete the regular life cycle of Baizhi.

A previous study arranged a series of sowing dates, from September to November, to analyze the B-/S-stage ratio in Baizhi (Xingfu et al. 1999). However, few analyses have revealed the content of pharmacological compounds in Baizhi roots extracted from different stages. Accordingly, the quantities in root may change during the growth stages. For example, with *Angelica sinensis* (known as Danggui in China), a well-known functional food for its roots, the growth stage is the most important factor affecting functional compound composition (Lü et al. 2009). Another study indicated that the content of six phthalides and four aromatic acids in *A. sinensis* varied with harvest timing (Qian et al. 2013).

In this study, we aimed to determine the functional activities of Baizhi roots at different growth stages, specifically the S-stage, which rarely occurs in other root-based medicinal plant species. To this end, we determined the activity of six furanocoumarin compounds, xanthotoxin, bergapten, oxypeucedanin, imperatorin, phellopterin and isoimperatorin, and antioxidants in Baizhi primary taproots and lateral roots at V-, S- and B-stages. The results will provide experimental proof for the harvest timing of Baizhi, especially for functional quality.

Materials and methods

Plant materials

The Baizhi field experiment of our study was grown in year 2014–2015 on a sandy loam soil at the Hualien District Agricultural Research and Extension Station (121°33E, 23°58N), Council of Agriculture, Executive Yuan. The 30 main plots $10 \times 1.2 \times 0.3$ m (L × W × H) were sown in October and November 2014. Seed was initially sown to dry seedbeds and transplanted as one plant per hill to the puddled field plots at 3–4 leaves stage. The site soils had pretreated (2 weeks before transplantation) with organic matter ($N:P_2O_5:K_2O = 4.9:2.1:1.9$; 8000 kg ha^{-1}). The experiment was conducted one plant per hill to the puddled field with regular water application. Hills were spaced at a distance of 120 cm with 60 cm between planted rows (hill spacing = 120 cm × 60 cm). During the culture period, the total rainfall is about 900 mm, the mean air temperatures range from about 17 to 29 °C, the mean monthly sunshine hours range from 70 to 230 h. It was harvested at the three harvest stages, vegetative (V-stage), bolting (B-stage) and summer dormancy (S-stage), during the following year in May, July and August based on plant appearance (Fig. 1). Ten plantlets were harvested at each stage. The taproots and lateral roots (Fig. 1d) were separated, and the length, diameter and fresh weight were measured. The roots were then sliced, and part of the fresh root slices were freeze-dried (LYPH. LOCK 18, Labconco, USA) as HPLC–DAD samples; residues were completely dried in a 40 °C oven. The total weight of the two parts of dry roots were the dry weight.

Reagents and chemicals

All solvents were purchased from Merck (Darmstadt, Germany) and were of analytical grade or high-performance liquid chromatography (HPLC) grade. 1,1-Diphenyl-2-picrylhydrazyl (DPPH) and (3-(2-pyridyl)-5,6,diphenyl-1,2,4-triazine-p, p′-disulfonic acid monosodium salt hydrate were from Merck, and Fe^{2+} chloride tetrahydrate ferrozine, and furanocoumarin compound standards, including xanthotoxin, bergapten, oxypeucedanin, imperatorin, phellopterin and isoimperatorin were from Merck. The standards for HPLC with diode-array detection (HPLC–DAD) were dissolved in methanol to a concentration of 100 μg/ml and were stored at − 20 °C.

Root extracts and sample processing

Freeze-dried root slices were crushed to a power by using a high-speed disintegrator (Solar Energy Co, Taiwan). Extraction for HPLC–DAD was performed as described (Shiau et al. 2011) with some modifications. Methanol (20 ml) was added to 0.2 g ground powder, vortexed and

Fig. 1 The appearance of Baizhi at three growth stages: V-stage, vegetative (23 weeks after sowing, **a**); B-stage, bolting (34 weeks after sowing, **b**); S-stage, summer dormancy (38 weeks after sowing, **c**); and roots harvested at V-stage (**d**)

let stand for 30 min. The powder with methanol was then ultrasonic shocked (DC600H, Delta, Taiwan) for 30 min, then centrifuged at 3320 rcf for 3 min. The supernatant was collected and the residue was re-extracted under same conditions for complete extraction. Finally, the supernatants were mixed and filtered through a 0.2-μm Acrodisc Syringe Filter (Pall, USA) before being collected in the Automatic Sampler-approved certified vials (Agilent, USA). Antioxidant assay extraction was performed as described (Shimada et al. 1992) with some modifications. Methanol (20 ml) was added to 1.0 g ground powder; vortexed and shaken overnight (50 rpm), then centrifuged at 3320 rfc for 3 min. The supernatant was collected as the sample.

HPLC–DAD analysis

To identify the furanocoumarin compounds in root extracts, 25 μl samples were separated by using an analytical column (Syncronis C18, 250 × 4.6 mm, 5 μm particle; ThermoFisher, USA) with flow rate 1.0 ml/min at 30 °C and absorbance 310 nm monitored by the ThermoFisher LC system (ThermoFisher, USA) equipped with a ThermoFinnigan UV detector and ThermoFisher Automatic Sampler. The column was protected by a C18 Guard column (HI-5C18-10C5, Hichrom, UK). A binary gradient was used for H_2O (Solvent A) and methanol (Solvent B). Optimized pigment separation was achieved with a linear gradient of 40–20% Solvent A for 0–5 min, 20–10% Solvent A for 5–7 min, and 10–0% Solvent A for 7–9 min.

DPPH assay

DPPH radical scavenging activity was as described (Shimada et al. 1992) with some modifications. The DPPH solution in methanol (0.4 mM) was prepared daily, and antioxidant samples were diluted or concentrated to at least seven concentrations. An amount of 50 μl sample solutions was loaded on a 96-well plate. Each concentration was loaded in six wells. Three were mixed thoroughly with 150 μl fresh DPPH solution (A1) and the others were mixed thoroughly with 150 μl methanol (A2). Pure methanol was used as the control (C1, C2). The loaded 96-well plates were incubated for 30 min in the dark, then the decrease in absorbance of light at 517 nm was measured. The DPPH radical scavenging effects were calculated as $[1-(A1_{517}-A2_{517})/(C1_{517}-C2_{517})]$.

Fe^{2+} chelating assay

Fe^{2+} chelating activity was measured as described (Dinis et al. 1994) with some modifications. Antioxidant samples were diluted or concentrated to at least seven concentrations. The Fe^{2+} chloride tetrahydrate solution in methanol (12 mM) was prepared, and 10.5 μl of this solution was mixed thoroughly with 1400 μl sample solution. Ferrozine solution in methanol (5 mM) was prepared. An amount of 200 μl sample solution was loaded on 96-well plates, with each concentration loaded in six wells. Three were thoroughly mixed with 20 μl ferrozine solution (A1) and the others were thoroughly mixed with 150 μl methanol (A2). Pure methanol was used as the control (C1, C2). The loaded 96-well plates were left to stand for 5 min, then the decrease in absorbance of light at 562 nm was measured. The Fe^{2+} chelating effects were calculated as $[1-(A1_{562}-A2_{562})/(C1_{562}-C2_{562})]$.

Statistical analysis

Data were analysed by using ANOVA. Source of variation was harvest stage. Significant differences by harvest stage were calculated by the Fisher least significant difference (LSD) tests ($p<0.05$). Data are expressed as mean ± SD. The IC50 value of DPPH radical scavenging activity and Fe^{2+} chelating activity was predicted by Loess regression. All analyses involved use of R (version 3.2.2) software. $p<0.05$ was considered statistically significant.

Results

Root weight, length, diameter and moisture content

The transition of Baizhi from vegetative to bolting or summer dormancy, and subsequent root mass is determined by sowing dates (Xingfu et al. 1999), fertilizing management (Derong et al. 1999) and environmental factors such as latitude. Root length and diameter were lower at V-stage than the other two stages, with no significant ($p<0.05$) difference between B- and S-stage. In Chinese medicine market, lateral roots of ginseng were commercialized independently, which termed as ginseng "fibers" or "beards". To this, root samples were divided into 'Taproot' and 'Lateral root' as shown in Table 1. The root lengths were 23.91 ± 1.97, 29.60 ± 5.46 and

Table 1 Weight, length, diameter and moisture content of Baizhi roots at three growth stages

	V-stage	B-stage	S-stage
Taproot weight (g)	$171.45 \pm 99.53^{b*}$	153.27 ± 37.89^{b}	406.96 ± 188.43^{a}
Lateral root weight (g)	61.12 ± 35.42^{b}	19.66 ± 13.08^{c}	133.19 ± 85.02^{a}
Total root weight (g)	232.57 ± 128.05^{b}	172.93 ± 38.95^{c}	540.16 ± 258.82^{a}
Lateral root/total root (%)	26	11	25
Root length (cm)	23.91 ± 1.97^{b}	29.60 ± 5.46^{a}	30.40 ± 5.83^{a}
Root diameter (mm)	76.45 ± 17.92^{b}	106.93 ± 20.08^{a}	106.87 ± 38.41^{a}

Data are mean ± SD of 10 independent plants

V-stage, vegetative stage; B-stage, bolting stage; S-stage, summer dormancy stage

* Different letters indicate statistical significance ($p<0.05$) by LSD test

30.40 ± 5.83 cm at V-, B- and S-stage, respectively, and the root diameters were 76.45 ± 17.92, 106.93 ± 20.08 and 106.87 ± 38.41 cm, respectively. Root dry weight was highest at S-stage and lowest at B-stage. For taproots, the DW was 2.4 times greater at S- than V-stage and was about 2.7 times greater than at B-stage. For lateral roots, the DW was about 2.2 times greater at S- than V-stage and about 6.8 times greater than at B-stage. V-stage had the highest lateral root to total root ratio, and the proportions of lateral roots were about 26, 11 and 25% at V-, B- and S-stage, respectively. The findings could be explained by a portion of the biomass contributing to the growth of bolting, which may lead to about 2 m of flower moss. In contrast, the above-ground portions die during S-stage. Accordingly, harvesting Baizhi at S-stage rather than at V- or B-stage would allow for better yield. The variation in shoot may affect the harvest weight of root.

Antioxidant activity

Extracts of Baizhi root have anti-oxidant properties (Pervin et al. 2014; Wang et al. 2017). Both reports have successfully used BHT as a standard to determine the antioxidant activity. Accordingly, in our study, we focused on the content of functional compounds in different harvest stages. We determined DPPH scavenging activity and Fe^{2+} chelating assay by abundant samples with statistical results without the usage of BHT. First, we determined the antioxidant activity of the three stages by DPPH assay and observed a dose-dependent relation in DPPH scavenging activity of Baizhi root extracts (Fig. 2a, b). For taproots, the IC50 value of the extracts were 5.70 ± 4.54, 6.46 ± 3.06 and 10.92 ± 3.82 mg/ml at V-, B- and S-stage, respectively (Table 2). The activity was about 2 times less at S-stage than the other two stages. However, from Table 1, for taproot, the DW was about 2.5 times greater for S-stage than the other stages. The low activity for S-stage may be due to its high DW, which mostly consists of storage compounds such as starch. For extracts of lateral root, the three stages did not differ in activities. The IC50 value of the lateral root extracts were 5.07 ± 5.20, 6.76 ± 5.37 and 6.12 ± 5.12 mg/ml at V-, B- and S-stage, respectively (Table 2). Furthermore, the DW of lateral roots was about 12 times greater at S-stage than the other stages, which indicates that 12 times total activity could be obtained from S-stage than the other stages.

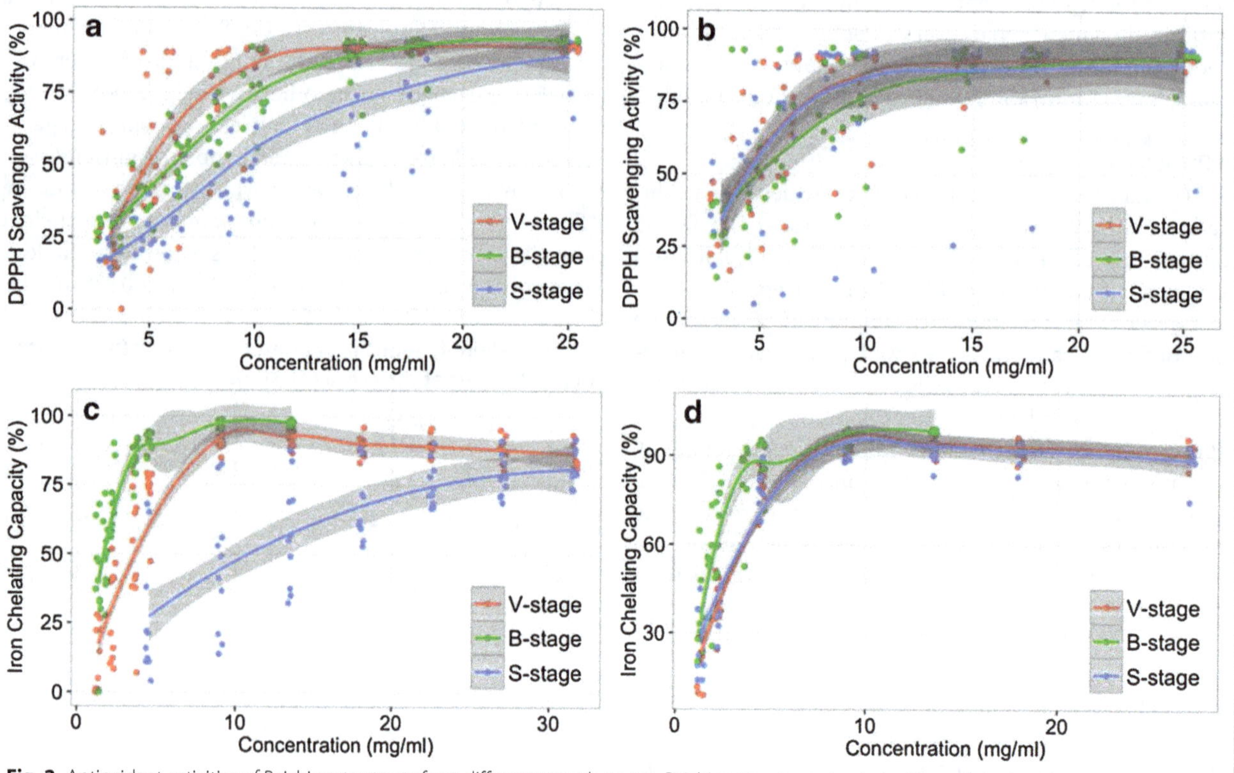

Fig. 2 Antioxidant activities of Baizhi root extracts from different growth stages. Baizhi roots were extracted with methanol, and extracts were diluted or concentrated to at least seven concentrations, and subjected to DPPH assay of taproot (**a**) and lateral root (**b**) and Fe^{2+} chelating assay of taproot (**c**) and lateral root (**d**). The gray-line zones indicate 95% confidence intervals for the loess regression line derived from ten plant samples for each concentration

Table 2 Antioxidant activity of Baizhi taproot and lateral root at three growth stages

Harvest stage	Taproot	Lateral root
DPPH radical scavenging capacity		
V-stage	5.70±4.54[a]*	5.07±5.20[a]
B-stage	6.46±3.06[a]	6.76±5.37[a]
S-stage	10.92±3.82[b]	6.12±5.12[a]
Fe^{2+} chelating capacity		
V-stage	2.52±7.13[a]	2.55±6.13[a]
B-stage	1.66±1.70[a]	1.90±1.74[a]
S-stage	11.94±6.09[b]	2.65±6.33[a]

Roots were extracted with methanol and subjected to DPPH and Fe^{2+} chelating assay. Activity of extracts is shown as mean ± SD IC50 (mg/ml) predicted by Loess regression

* Different letters indicate statistical significance ($p < 0.05$) by *t* test

Taken together, the highest DPPH radical scavenging activity was obtained from roots harvested at S-stage.

We also determined the antioxidant activity of roots by Fe^{2+} chelating assay and found a dose-dependent relation in Fe^{2+} chelating assay of Baizhi root extracts (Fig. 2c, d). For taproots, the IC50 value of the extracts were 2.52±7.13, 1.66±1.70 and 11.94±6.09 mg/ml for V-, B- and S-stage, respectively (Table 2). The activity was about 5 times less at S-stage than the other two stages (Table 2). Activity in lateral roots did not differ among the stages. The IC50 value for lateral root extracts were 2.55±6.13, 1.90±1.74 and 2.65±6.33 mg/m for V-, B- and S-stage, respectively (Table 2). However, the DW was about 2.5 times greater at S-stage than the other stages. Hence, the active compounds for DPPH and Fe^{2+} chelating mechanisms differed. Baizhi root harvested at S-stage provided antioxidant activity primarily via a DPPH radical scavenging instead of Fe^{2+} chelating mechanism.

Characterization of six furanocoumarin compounds in Baizhi root

In traditional Chinese medicine, crude Baizhi root materials are used as medicine or functional food for antipyretic and analgesic activities because of abundant furanocoumarin compounds (Li et al. 2014a, b). For example, isoimperatorin, imperatorin and oxypeucedanin may be useful for treating Alzheimer disease (Marumoto and Miyazawa 2010). By analyzing the inhibitory activity on LPS-induced PGE2 production, it has been demonstrated that Isoimperatorin, imperatorin and phellopterin have potential as anti-inflammatory drugs (Ban et al. 2003; Deng et al. 2015). For other furanocoumarin compounds, xanthotoxin, isoimperatorin, imperatorin and oxypeucedanin have anti-cancer abilities (Kim et al. 2007).

The chemicals described above involve major furanocoumarin compounds, which we extracted from Baizhi root for HPLC–DAD chromatogram assay; seven clear

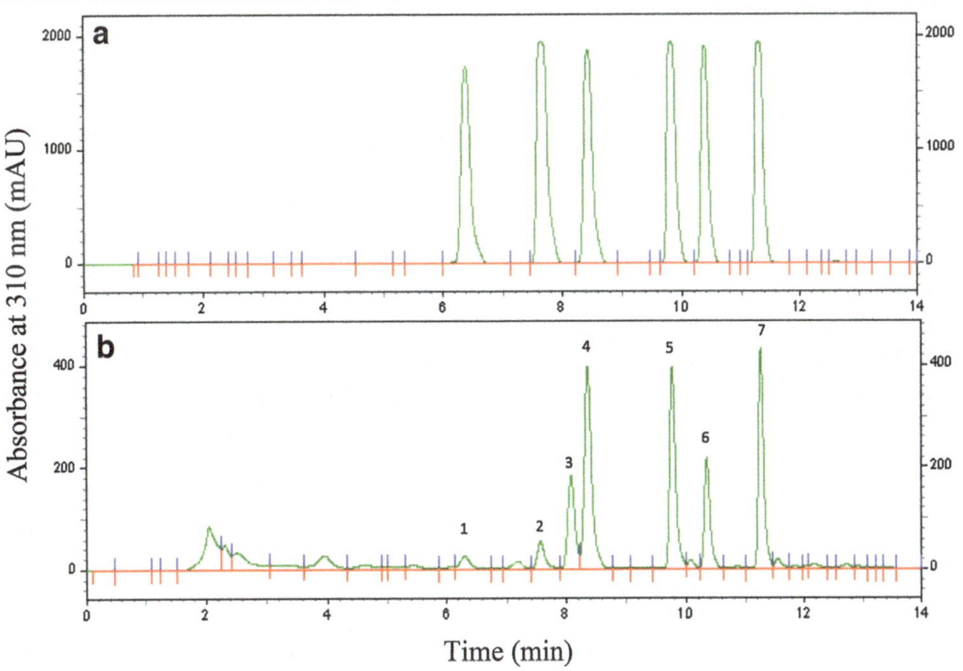

Fig. 3 HPLC-DAD chromatogram of standard (**a**) compared with Baizhi root tissue (**b**). Peaks 1, 2, 4, 5, 6 and 7 were identified as xanthotoxin, bergapten, oxypeucedanin, imperatorin, phellopterin and isoimperatorin, respectively

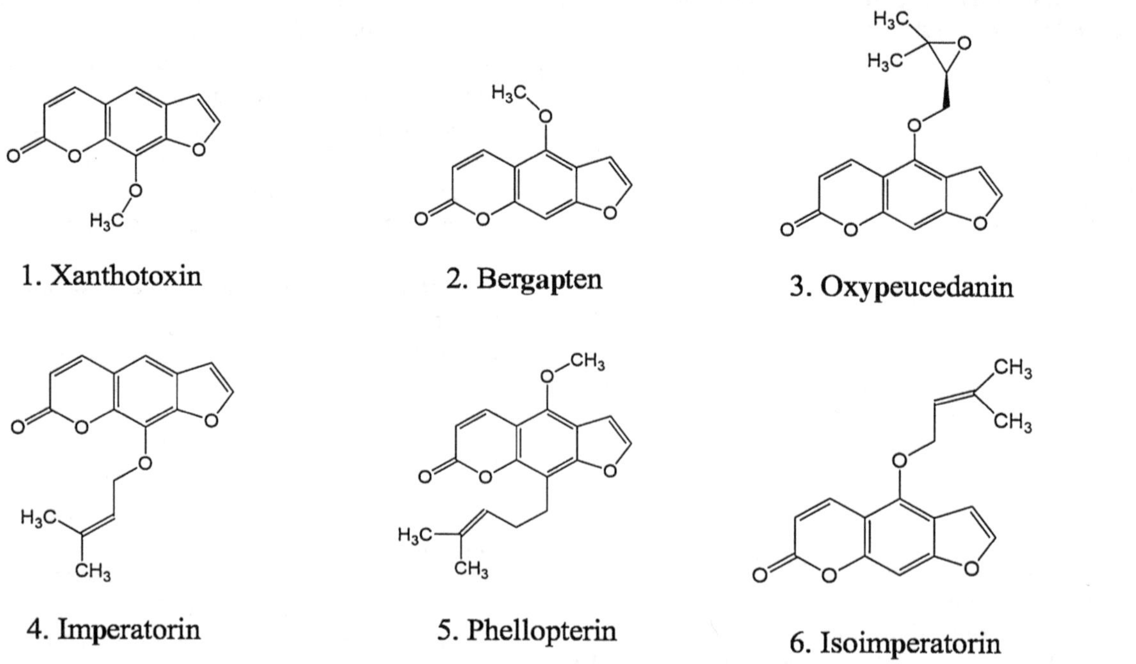

Fig. 4 Chemical structure of furanocoumarins identified from root of Baizhi

peaks were obtained. By using reference compounds, we identified peaks 1, 2, 4, 5, 6 and 7 as xanthotoxin, bergapten, oxypeucedanin, imperatorin, phellopterin and isoimperatorin, respectively (Fig. 3) and the chemical structures are in Fig. 4. Then, we analyzed these six furanocoumarins in Baizhi root and found different patterns of furanocoumarin compounds from each harvest stage. The furanocoumarin content in taproot (or lateral root) at each harvest stage showed different patterns (Fig. 5). Oxypeucedanin had the highest yield of the six furanocoumarins (1.5–3.0 mg/g) and the other five yielded from 0.15 mg/g (bergapten) to 1.6 mg/g (phellopterin). The yield of each compound varied less than 50% at different harvest stages except for isoimperatorin from taproot. Similarly, the yield of taproot and lateral root varied less than 50% except for imperatorin and isoimperatorin at B-stage. Additionally, we have also analyzed the furanocoumarin content in shoot (above ground part) of V stage, only imperatorin was detected as about 0.21 mg/g.

To analyze the total amount of the six furanocoumarins in Baizhi roots, the results of harvest yield and six furanocoumarin contents were integrated (Fig. 6). Total furanocoumarin content was highest at S-stage. Also, furanocoumarin content was generally higher at S-stage than the other two stages; especially, the yield of xanthotoxin, bergapten and phellopterin was 2 times higher at S-stage than the other 2 stages. This result strongly supports that S-stage is the best harvest stage than the other stages because of its richer total furanocoumarin content.

Discussion

The root of Baizhi has been used as a functional food and traditional medicine. In this study, we analyzed the antioxidant activity and total furanocoumarin content of Baizhi root and concluded that the summer dormancy stage is the best stage for harvesting.

Furanocoumarins are believed to function as phytoalexin (Beier and Oertli 1983) or allelochemical compound (Baskin et al. 1967). These compounds are relevant for being potential substitutes for pesticides in crop protection (Eljarrat and Barceló 2001). In wild parsnip (*Pastinaca sativa,* also an Apiaceae plant), the quantity of furanocoumarins allocated to floral units at different stages of development is consistent with the value of those floral units (Nitao and Zangerl 1987). Additionally, studies of field-grown plants showed that FCs storage is closely related to organs and phenological stages. Leaves, stems, roots and fruits do not have equal concentrations

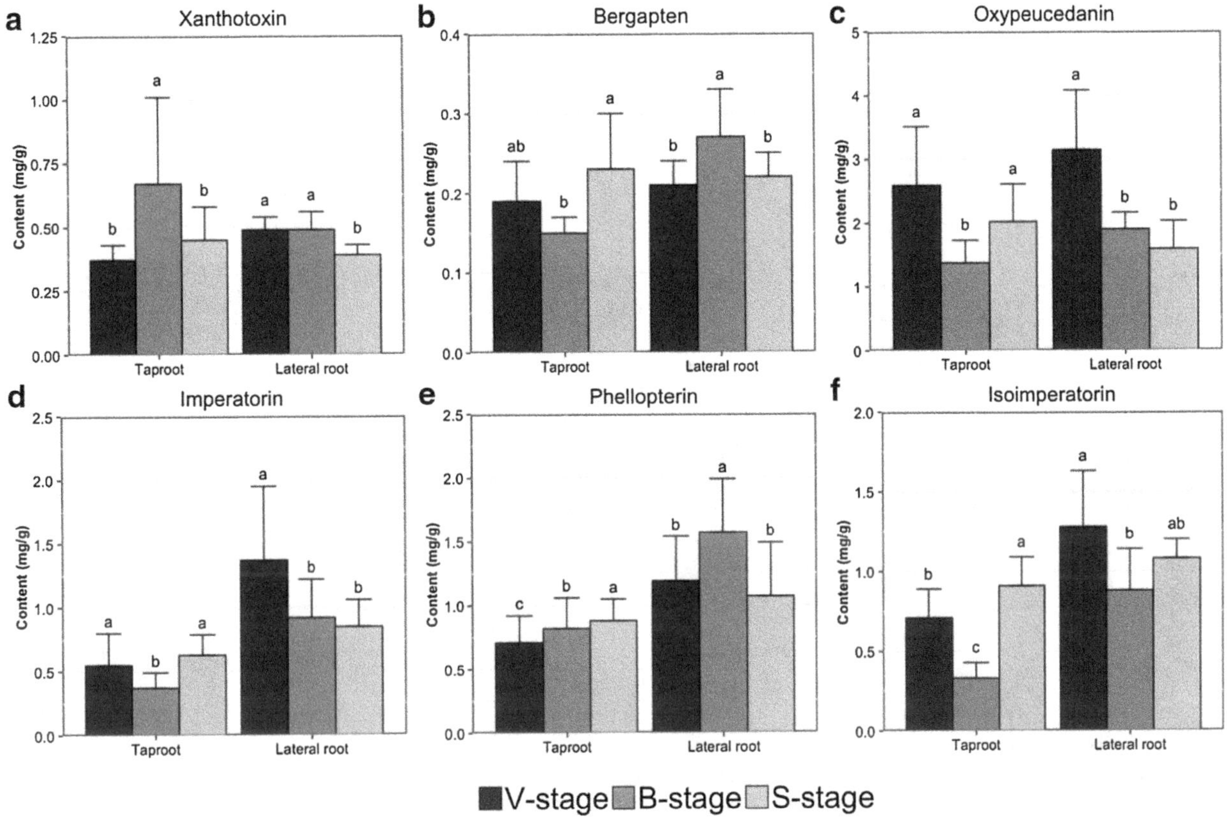

Fig. 5 Xanthotoxin (**a**), bergapten (**b**), oxypeucedanin (**c**), imperatorin (**d**), phelllopterin (**e**) and isoimperatorin (**f**) content in Baizhi taproot and lateral root harvested at three stages. Roots were extracted with methanol and subjected to HPLC-DAD analysis ($n = 10$). Data are mean ± SD furanocoumarin content (mg/g sample). Different letters indicate statistical significance ($p < 0.05$) by LSD test

of FCs even if psoralen, xanthotoxin, bergapten and isopimpinellin were detected in all tissues (Milesi et al. 2001). Therefore, it is believed that FCs act, specifically in storage organs, as protection chemicals against biostress. As for the case of Baizhi entering the S-stage, the above ground parts occur yellowing, compounds redistribution. FCs are then yielded and accumulated in storage organ (root), specifically in cortex. We suggest that there may be two strategies for Baizhi to survive the stress of summer season. The first strategy is the avoidance trait, in which Baizhi will bolt early in May and adapt the summer season with seed dormancy. The second strategy is the resistance trait, the above-ground parts of Baizhi will senescence and the root will become the resting organs which contain higher dry weigh.

Conclusions

The results indicate that Baizhi roots should be harvested at S-stage. However, it is possible that the growth of Baizhi switch from vegetative stage to bolting stage directly, without a summer dormancy stage. Our preliminary studies indicated that sowing earlier by 1 month (in November) may result in early bolting in May. Many studies have reported that stage-transition of Baizhi is determined by, for example, sowing date, fertilizing management and other environmental factors. In the near future, global warming is expected to affect phenological events such as flowering and fruiting in plants and will hasten development time in plant species that respond to cues such as degree days (Hughes 2000). Under the threat of global warming, analyzing the stage-transition physiology of Baizhi will facilitate the modification of the cropping system to achieve a stable and uniform S-stage of Baizhi.

Abbreviations

V-stage: vegetative stage; S-stage: summer dormancy stage; B-stage: bolting stage; DW: dry weight; HPLC–DAD: high-performance liquid chromatography with diode-array detection; DPPH: 2,2-diphenyl-L-picrylhydrazyl.

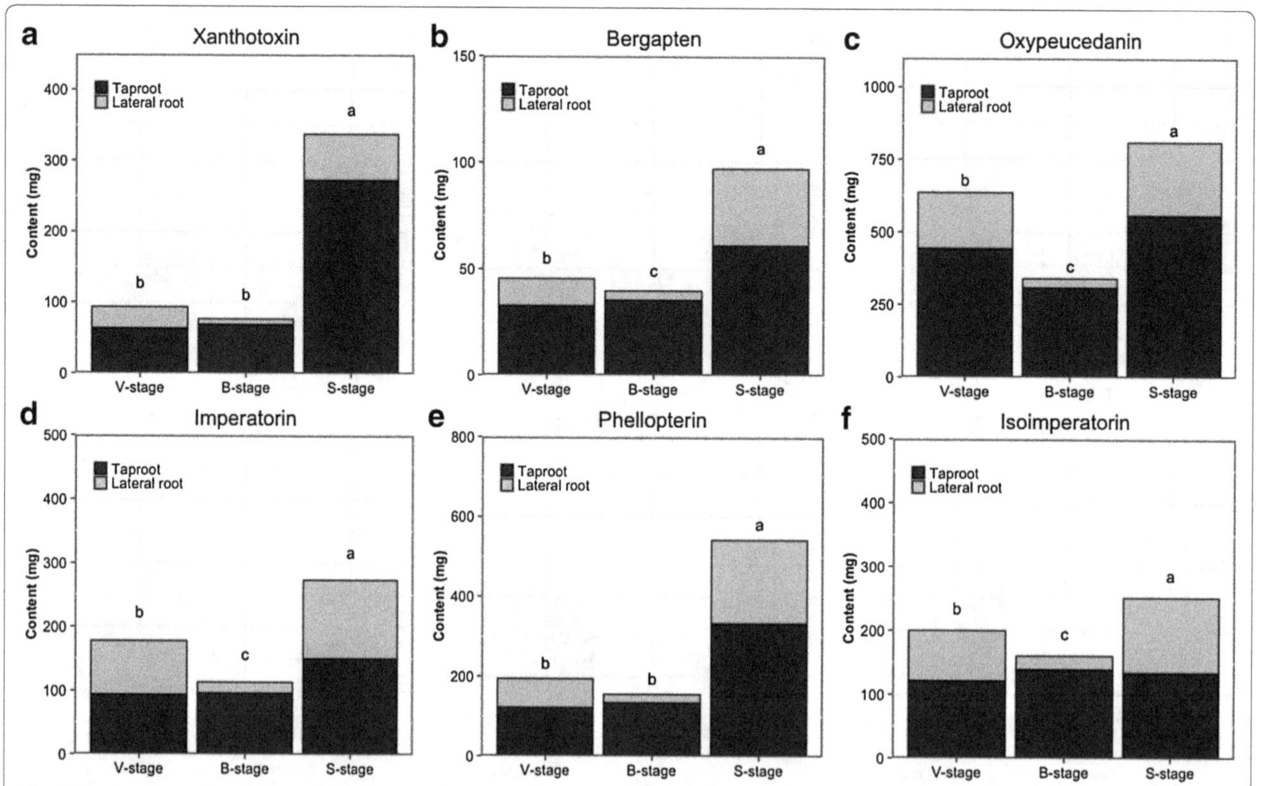

Fig. 6 Total content of xanthotoxin (**a**), bergapten (**b**), oxypeucedanin (**c**), imperatorin (**d**), phellopterin (**e**), phellopterin (**f**) in Baizhi taproots and lateral root harvest at three stages. Data are mean of the DW shown at Table 1 multiplied the content of each furanocoumarin at each stage's taproot (black bars) and lateral root (gray bars). Different letters indicate statistical significance ($p < 0.05$) by LSD test

Authors' contributions

YCC and WHL conceived the experiment, analyzed the data and drafted the manuscript. WHL, TWC performed the experiment. All authors read and approved the final manuscript.

Author details

[1] Department of Agronomy, National Taiwan University, Taipei, Taiwan, Republic of China. [2] Hualien District Agricultural Research and Extension Station, Council of Agriculture, Executive Yuan, Hualien, Taiwan, Republic of China.

Acknowledgements

We thank Professor Li-yu D. Liu and Chen-An Tsai for description of statistic results.

Competing interests

The authors declare that they have no competing interests.

Funding

This project was supported by the Ministry of Science and Technology and Council of Agriculture, Executive Yuan of Taiwan.

References

Abad MJ, Heras BDL, Silvan AM, Pascual R, Bermejo P, Rodriguez B, Villar AM (2001) Effects of furocoumarins from *Cachrys trifida* on some macrophage functions. J Pharm Pharmacol 53:1163–1168

Ban HS, Lim SS, Suzuki K, Jung SH, Lee S, Lee YS, Shin KH, Ohuchi K (2003) Inhibitory effects of furocoumarins isolated from the roots of *Angelica dahuricae* on prostaglandin E2 production. Planta Med 69:408–412

Baskin JM, Ludlow CJ, Harris TM, Wolf FT (1967) Psoralen, an inhibitor in the seeds of *Psoralea subacaulis* (Leguminosae). Phytochemistry 6:1209–1213

Beier RC, Oertli EH (1983) Psoralen and other linear furocoumarins as phytoalexins in celery. Phytochemistry 22:2595–2597

Chiou WF, Huang YL, Chen CF, Chen CC (2001) Vasorelaxing effect of coumarins from *Cnidium monnieri* on rabbit corpus cavernosum. Planta Med 67:282–284

Cho YH, Kim JH, Park SM, Lee BC, Pyo HB, Park HD (2006) New cosmetic agents for skin whitening from *Angelica dahurica*. J Cosmet Sci 57:11–21

Deng GG, Wei W, Yang XW, Zhang YB, Xu W, Gong NB, Lü Y, Wang FF (2015) New coumarins from the roots of *Angelica dahurica var. formosana cv. Chuanbaizhi* and their inhibition on NO production in LPS-activated RAW264.7 cells. Fitoterapia 101:194–200

Derong D, Jin LU, Xingfu C, Mingliang S, Deming XIE, Huafen LI (1999) Studies on different methods of fertilizer application to curtail earlier bolting for a better yield of taiwan angelica (*Angelica dahurica var. formosasa*). Chin Tradit Herbal Drugs 2:135–136

Dinis TC, Madeira VM, Almeida LM (1994) Action of phenolic derivatives (acetaminophen, salicylate, and 5-aminosalicylate) as inhibitors of membrane lipid peroxidation and as peroxyl radical scavengers. Arch Biochem Biophys 315:161–169

Eljarrat E, Barceló D (2001) Sample handling and analysis of allelochemical compounds in plants. Trends Anal Chem 20:584–590

Feng Y, Zhu L, Pan T, Guo Z, Zhong X, Ding A, Pan D (2015) Characterization of summer dormancy in *Narcissus tazetta var. Chinensis* and the role of *NtFTs* in summer dormancy and flower differentiation. Sci Hortic 183:109–117

Garcia A, Apan AN, Delgado TOR, Velázquez HP, Martínez-Vázquez M (2000) Anti-inflammatory activity of coumarins from *Decatropis bicolor* on TPA ear mice model. Planta Med 66:279–281

Gillespie LM, Volaire FA (2017) Are winter and summer dormancy symmetrical seasonal adaptive strategies? The case of temperate herbaceous perennials. Ann Bot 119:311–323

Hughes L (2000) Biological consequences of global warming: is the signal already apparent? Trends Ecol Evol 15:56–61

Kim YK, Kim YS, Ryu SY (2007) Antiproliferative effect of furanocoumarins from the root of *Angelica dahurica* on cultured human tumor cell lines. Phytother Res 21:288–290

Kwon YS, Kobayashi A, Kajiyama SI, Kawazu K, Kanzaki H, Kim CM (1997) Antimicrobial constituents of *Angelica dahurica* roots. Phytochemistry 44:887–889

Leopold JA, Loscalzo J (2009) Oxidative risk for atherothrombotic cardiovascular disease. Free Radical Biol Med 47:1673–1706

Li B, Zhang X, Wang J, Zhang L, Gao B, Shi S, Tu P (2014a) Simultaneous characterisation of fifty coumarins from the roots of *Angelica dahurica* by off-line two-dimensional high-performance liquid chromatography coupled with electrospray ionisation tandem mass spectrometry. Phytochem Anal 25:229–240

Li X, Zeng X, Sun J, Li H, Wu P, Fung KP, Liu F (2014b) Imperatorin induces Mcl-1 degradation to cooperatively trigger Bax translocation and Bak activation to suppress drug-resistant human hepatoma. Cancer Lett 348:146–155

Lü JL, Zhao J, Duan JA, Yan H, Tang YP, Zhang LB (2009) Quality evaluation of *Angelica sinensis* by simultaneous determination of ten compounds using LC-PDA. Chromatographia 70:455–465

Luszczki JJ, Glowniak K, Czuczwar SJ (2007) Imperatorin enhances the protective activity of conventional antiepileptic drugs against maximal electroshock-induced seizures in mice. Eur J Pharmacol 574:133–139

Marumoto S, Miyazawa M (2010) β-Secretase inhibitory effects of furanocoumarins from the root of *Angelica dahurica*. Phytother Res 24:510–513

Milesi S, Massot B, Gontier E, Bourgaud F, Guckert A (2001) *Ruta graveolens* L.: a promising species for the production of furanocoumarins. Plant Sci 161:189–199

Nitao JK, Zangerl AR (1987) Floral development and chemical defense allocation in wild parsnip (*Pastinaca sativa*). Ecology 68:521–529

Ofir M, Kigel J (2007) Regulation of summer dormancy by water deficit and ABA in *Poa bulbosa* ecotypes. Ann Bot 99:293–299

Oh H, Lee HS, Kim T, Chai KY, Chung HT, Kwon TO, Jun JY, Kim YC, Yun YG (2002) Furocoumarins from *Angelica dahurica* with hepatoprotective activity on tacrine-induced cytotoxicity in Hep G2 cells. Planta Med 68:463–464

Pervin M, Hasnat MD, Debnath T, Park SR, Kim DH, Lim BO (2014) Antioxidant, anti-inflammatory and antiproliferative activity of *Angelica dahurica* root extracts. J Food Biochem 38:281–292

Piao XL, Park IH, Baek SH, Kim HY, Park MK, Park JH (2004) Antioxidative activity of furanocoumarins isolated from *Angelicae dahuricae*. J Ethnopharmacol 93:243–246

Qian Y, Wang Y, Sa R, Yan H, Pan X, Yang Y, Sun Y (2013) Metabolic fingerprinting of *Angelica sinensis* during growth using UPLC-TOFMS and chemometrics data analysis. Chem Cent J 7:42

Ryu SY, Kou NY, Choi HS, Ryu H, Kim TS, Kim KM (2001) Cnidicin, a coumarin, from the root of *Angelica koreana*, inhibits the degranulation of mast cell and the NO generation in RAW 264.7 cells. Planta Med 67:172–174

Sarker SD, Nahar L (2004) Natural medicine: the genus *Angelica*. Curr Med Chem 11:1479–1500

Shiau YJ, Jeng TL, Kao JL, Lai JS, Huang CH (2011) Extraction of imperatorin from root slices and explants of *Angelica dahurica*. J Taiwan Agric Res 60:101–114

Shimada K, Fujikawa K, Yahara K, Nakamura T (1992) Antioxidative properties of xanthan on the autoxidation of soybean oil in cyclodextrin emulsion. J Agric Food Chem 40:945–948

Suleimenov EM (2009) Components of *Peusedanum morisonii* and their antimicrobial and cytotoxic activity. Chem Nat Compd 45:710

Volaire F (2002) Drought survival, summer dormancy and dehydrin accumulation in contrasting cultivars of *Dactylis glomerata*. Physiol Plant 116:42–51

Volaire F, Norton M (2006) Summer dormancy in perennial temperate grasses. Ann Bot 98:927–933

Volaire F, Seddaiu G, Ledda L, Lelievre F (2009) Water deficit and induction of summer dormancy in perennial Mediterranean grasses. Ann Bot 103:1337–1346

Wang NH, Huang LQ, Yang B, Kimie B, Masahiko T, Yuan CQ, Qin HZ, Shu P (2001) Studies on original plant of traditional Chinese drug"Bai Zhi" (*Radix Angelicae dahuricae*) and its closely related wild plants discussion on original plant and cultivation history of traditional chinese drug "Bai Zhi" and evolution of its closely related wild plants. China J Chin Mater Med 26:733–736

Wang C, Wang T, Huang L, Lu W, Zhang J, He H (2016) Synthesis and fluorescent study of 5-phenyl furocoumarin derivatives as vasodilatory agents. Bioorg Med Chem Lett 26:640–644

Wang GH, Chen CY, Tsai TH, Chen CK, Cheng CY, Huang YH, Hsieh MC, Chung YC (2017) Evaluation of tyrosinase inhibitory and antioxidant activities of *Angelica dahurica* root extracts for four different probiotic bacteria fermentations. J Biosci Bioeng 123:679–684

Xingfu C, Jin LU, Derong D (1999) Bo zhong qi dui bai zuo zao qi chou tai ying xiang de yan jiu. China J Chin Mater Med 24:211

Yousif AN, Scaman CH, Durance TD, Girard B (1999) Flavor volatiles and physical properties of vacuum-microwave-and air-dried sweet basil (*Ocimum basilicum* L.). J Agric Food Chem 47:4777–4781

11

The roles of *Arabidopsis HSFA2, HSFA4a*, and *HSFA7a* in the heat shock response and cytosolic protein response

Kuan-Fu Lin†, Meng-Yu Tsai†, Chung-An Lu, Shaw-Jye Wu and Ching-Hui Yeh*

Abstract

Previously, we found that *Arabidopsis* plants transformed with a construct containing the promoter of *Oshsp17.3* from rice fused to the β-glucuronidase gene (*GUS*), *Oshsp17.3Pro::GUS* (*Oshsp17.3p*), showed a *GUS* signal after heat shock (HS) or azetidine-2-carboxylic acid (AZC) treatment. HS and AZC trigger the heat shock response (HSR) and cytosolic protein response (CPR), respectively, in the cytosol by modulating specific heat shock factor (HSF) activity. Here we further identified that *AtHSFA2* (*At2g26150*), *AtHSFA7a* (*At3g51910*), *AtHSFB2a* (*At5g62020*), and *AtHSFB2b* (*At4g11660*) are HS- and AZC-inducible; *AtHSFA4a* (*At4g18880*) is AZC-inducible; and *AtHSFA5* (*At4g13980*) is less AZC- and HS-inducible. To investigate the roles of these 6 *AtHSFs* in the HSR or CPR, we crossed two independent *Oshsp17.3p* transgenic *Arabidopsis* plants with the *AtHSF*-knockout mutants *athsfa2* (SALK_008978), *athsfa4a* (GABI_181H12), *athsfa5* (SALK_004385), *athsfa7a* (SALK_080138), *athsfb2a* (SALK_137766), and *athsfb2b* (SALK_047291), respectively. As compared with the wild type, loss-of-function mutation of *AtHSFA2*, *AtHSFA4a*, and *AtHSFA7a* decreased HS and AZC responsiveness, so these 3 *AtHSFs* are essential for the HSR and CPR. In addition, loss-of-function results indicated that *AthsfB2b* is involved in regulating the HSR in *Arabidopsis*. Furthermore, analysis of the relative GUS activity of two double knockout mutants, *athsfA2/athsfA4a* and *athsfA2/athsfA7a*, revealed that *AtHSFA2*, *AtHSFA4a*, and *AtHSFA7a* function differentially in the HSR and CPR. Transcription profiling in *athsf* mutants revealed positive or negative transcriptional regulation among the 6 *AtHSFs* in *Arabidopsis* plants under HS and AZC conditions. Tunicamycin treatment demonstrated that these 6 *AtHSFs* are not involved in the unfolded protein response.

Keywords: Azetidine-2-carboxylic acid, Cytosolic protein response, Heat shock factor, Heat shock protein, Heat shock response, Unfolded protein response

Background

Protein homeostasis is crucial for maintaining normal cellular function. Plants, being sessile organisms, cannot escape from their growing environments. Extremes in environmental factors can result in stressful conditions that inevitably damage proteins directly or cause cells to synthesize misfolded proteins, which can lead to perturbed cell function and stress-induced cell death. Plants have evolved an extensive network of chaperone systems to restore protein folding or to remove irreversibly unfolded proteins (Mehdy 1994; Shinozaki and Yamaguchi-Shinozaki 1996; Bukau et al. 2006; Cramer et al. 2011; Redondo-Gómez 2013).

Accumulation of unfolded proteins within cells, eliciting compartment-specific chaperones and pathways, is termed the unfolded protein response (UPR). The UPR initiates the dissociation of the endoplasmic reticulum (ER) chaperone, immunoglobulin binding protein, and ER master sensors, such as inositol-requiring 1 and protein kinase R-like ER kinase, to activate downstream effectors to restore protein homeostasis in the lumen of the ER. A cytosolic process, the cytoplasmic protein response (CPR), increases the synthesis of molecular chaperones such as heat shock proteins (HSPs). In

*Correspondence: yehch63@cc.ncu.edu.tw
†Kuan-Fu Lin and Meng-Yu Tsai contributed equally to this work
Department of Life Sciences, National Central University, Taoyuan 32001, Taiwan

contrast to the better-understood UPR of the ER, the regulatory molecules in the CPR are not well elucidated.

The heat-shock response (HSR), predominantly a response to maintain protein-folding homeostasis in the cytosol, causes transcriptional activation of HSPs under thermal stress (Aparicio et al. 2005; Jungkunz et al. 2011). The expression of *HSP* genes is mainly under the control of heat shock transcription factors (HSFs) (Schöffl et al. 1998; Nover et al. 2001). The number of HSFs is characteristically higher in plants than in other organisms. For example, *Arabidopsis* and rice have 21 and 25 HSFs, respectively, but *Drosophila, C. elegans* and yeast have only one HSF (Nover et al. 2001; Guo et al. 2008; Scharf et al. 2012). The multiplicity of members of the HSF family in plants may contribute to their fitness to face varied environmental challenges such as extreme temperatures, drought, and salinity (Busch et al. 2005).

Plant HSFs are classified into three classes (A, B, and C) on the basis of structural characteristics and phylogenetic comparison. Class A HSFs contain a DNA binding domain, an oligomerization domain, nuclear localization domains, and transcriptional activation domains. Classes B and C lack a transcriptional activation domain (Nover et al. 2001). Recent studies of tomato *HSFA1a* mutants and an *Arabidopsis HSFA1a/1b/1d/1e* quadruple mutant revealed that members of *HSFA1* genes can function as master regulators for the HSR and play important roles in cross-regulation for abiotic stress responses (Mishra et al. 2002; Liu et al. 2011). Increasing evidence shows functional diversification among different HSF members.

In addition to heat shock (HS), a proline analog, azetidine-2-carboxylic acid (AZC), can induce accumulation of abnormal-misfolded proteins in the cytosol to trigger the CPR by modulating HSFA2 activity (Yeh et al. 2007; Sugio et al. 2009; Nishizawa-Yokoi et al. 2011). In the current study, we fused the promoter of AZC-inducible small *HSP* (*sHSP*), *Oshsp17.3*, with the *β-glucuronidase* gene (*GUS*) (*Oshsp17.3Pro::GUS*) and transformed into *Arabidopsis AtHSF* mutants, and detected GUS activity in response to AZC and HS (Guan et al. 2010). Our results allowed us to characterize the roles of *Arabidopsis HSFs* in the HSR and CPR.

Methods

Plant materials

The *Arabidopsis thaliana* ecotype *Col-0* was used in this study as the wild type (WT). Seeds were surface-sterilized in commercial bleach that contained 5% (v/v) sodium hypochlorite and 0.1% (v/v) Triton X-100 solution for 10 min, rinsed in sterilized water, and stratified at 4 °C for 2 days in the dark. Seeds were germinated on growth agar plates [1/2 Murashige and Skoog medium (MS; Duchefa), 1% sucrose (w/v), 0.8% agar (w/v)]. The T-DNA insertion lines SALK_008978 (*athsfa2*), GABI_181H12 (*athsfa4a*), SALK_004385 (*athsfa5*), SALK_080138 (*athsfa7a*), SALK_137766 (*athsfb2a*), and SALK_047291 (*athsfb2b*) mutants were obtained from the Arabidopsis Biological Resources Center (ABRC, Columbus, OH, USA) (Liu et al. 2011; Kleinboelting et al. 2012). The *athsfa2/athsfa4a* and the *athsfa2/athsfa7a* double mutants were generated by crossing *athsfa2* with *athsfa4a* and *athsfa7a* mutants. Mutant seeds were germinated and selected on selection agar plates [1/2 MS, 1% sucrose (w/v), 25 μg/ml hygromycin, 0.8% agar (w/v)]. All seedlings were grown at 23 °C in a 16-h light/8-h dark cycle in a growth chamber with 60% relative humidity.

RNA isolation and RT-PCR

Total RNA was extracted from 10-day-old *Arabidopsis* seedlings as described (Guan et al. 2010). The first-strand cDNA was synthesized with 1 μg total RNA by using the SuperScript III First-Stand Synthesis System (Invitrogen). PCR amplification corresponding to different *AtHSFs* shown in Fig. 3 were 30 s at 94 °C, 30 s at 52 °C, and 30 s at 72 °C, then 5 min at 72 °C. Primers used for analysis of gene expression were designed by use of NCBI Primer-BLAST (https://www.ncbi.nlm.nih.gov/tools/primer-blast/) and are in Table 1. DNA from 15 μl of each PCR reaction was fractionated by electrophoresis on 1.2% (w/v) agarose gel with 0.01% (w/v) ethidium bromide in 1× Tris–Acetate EDTA buffer. The gel was

Table 1 Oligonucleotides used in RT-PCR

Gene	Primer name	Sequence
AtHSFA2	AtHSFA2-Fw	5′-CCATGGAAGAACTGAAAGTGGAAATGGAGG-3′
	AtHSFA2-Rv	5′-GCGGCCGCAGGTTCCGAACCAAG-3′
AtHSFA4a	AtHSFA4a-Fw	5′-CATCAAGTGGAACAGTTAGA-3′
	AtHSFA4a-Rv	5′-ACTCCGGCTTTATCTTTATC-3′
AtHSFA5	AtHSFA5-Fw	5′-AGCAAGAGTGAATGATGTAT-3′
	AtHSFA5-Rv	5′-CTACTTACGCTTTTTCAGTC-3′
AtHSFA7a	AtHSFA7a-Fw	5′-ATCAAAGCTATGGAACAGAG-3′
	AtHSFA7a-Rv	5′-AACTCTCATCACTAAGCAAC-3′
AtHSFB2a	AtHSFB2a-Fw	5′-TTGAGACATTATAATCGAAC-3′
	AtHSFB2a-Rv	5′-TCTAAAAATGTACTTGTGAT-3′
AtHSFB2b	AtHSFB2b Fw	5′-GAGGAGAATAACTCCGGTAA-3′
	AtHSFB2b Rv	5′-ATGCAATGGGGATCAGTAAC-3′
AtTubulin	AtTubulin Fw	5′-GCCAATCCGGTGCTGGTAACA-3′
	AtTubulin Rv	5′-CATACCAGATCCAGTTCCTCCTCCC-3′
AtbZIP60	AtbZIP60-Fw	5′-AGGACGTATGCTTGAGTGCTTCGT-3′
	AtbZIP60-Rv	5′-TTCTGGACGTAGGAGGCAACACT-3′
GUS	GUS-Fw	5′-GGCCTGTGGGCATTCAGTCT-3′
	GUS-Rv	5′-AGTTCAGTTCGTTGTTCACACAA-3′

digitally photographed and the corresponding DNA signal was quantified by using ImageJ (http://rsbweb.nih.gov/ij/) (Schneider et al. 2012) and normalized to *18S* rRNA expression.

Preparation of DNA constructs and transformation

Oshsp17.3Pro::GUS (*Oshsp17.3p*) and *Oshsp17.3Pro-ΔAZRE::GUS* (*Oshsp17.3pΔAZRE*) were constructed and transformed *Arabidopsis* plants as described (Guan et al. 2010). Transgenic plants #5 and #11 of *Oshsp17.3Pro::GUS* (*Oshsp17.3p5* and *Oshsp17.3p11*, respectively), which showed *GUS* expression induced by HS and AZC (Guan et al. 2010), were selected to cross with *AtHSF* mutants *athsfA2, athsfA4a, athsfA5, athsfA7a, athsfB2a, athsfB2b, athsfA2/athsfA4a*, and *athsfA2/athsfA7a* mutants, respectively. F2 lines *Oshsp17.3p5/athsfA2, Oshsp17.3p5/athsfA4a, Oshsp17.3p5/athsfA5, Oshsp17.3p5/athsfA7a, Oshsp17.3p5/athsfB2a, Oshsp17.3p5/athsfB2b, Oshsp17.3p5/athsfA2/athsfA4a, Oshsp17.3p5/athsfA2/athsfA7a, Oshsp17.3p11/athsfA2, Oshsp17.3p11/athsfA4a, Oshsp17.3p11/athsfA5, Oshsp17.3p11/athsfA7a, Oshsp17.3p11/athsfB2a, Oshsp17.3p11/athsfB2b, Oshsp17.3p11/athsfA2/athsfA4a*, and *Oshsp17.3p11/athsfA2/athsfA7a* were obtained and then self-pollinated to produce the F3 generation, which was used for analysis of HS and AZC responsiveness in this study. In addition, transgenic plants #2 and #7 of *Oshsp17.3Pro-ΔAZRE::GUS* (*Oshsp17.3pΔAZRE*), which showed weak *GUS* expression with HS and AZC treatment (Guan et al. 2010), were used as the negative control.

Stress treatment of transgenic *Arabidopsis* mutants

For HS treatment, 10-day-old F3-generation *Arabidopsis* seedlings were incubated in shaking buffer [1% sucrose (w/v), 5 mM potassium phosphate buffer, pH 6.8] at 39 °C for 1 h, then 23 °C for 20 h of recovery. For AZC treatment, 10-day-old F3-generation *Arabidopsis* seedlings were incubated in shaking buffer with or without 5 mM AZC at 23 °C for 4 h, rinsed in sterilized water, then incubated in shaking buffer at 23 °C for 15 h of recovery. For tunicamycin (Tm) treatment, 10-day-old F3-generation *Arabidopsis* seedlings were incubated in shaking buffer with or without 5 μg/ml Tm at 23 °C for 4 h, rinsed in sterilized water, then incubated in shaking buffer at 23 °C for 15 h of recovery. All samples were frozen by liquid nitrogen and stored at − 80 °C.

GUS staining

GUS staining was described previously (Guan et al. 2010). In brief, 10-day-old seedlings were treated and incubated in the fixation solution (0.3% formaldehyde, 0.1% Triton X-100, 0.1% β-mercaptoethanol, 100 mM sodium phosphate buffer, pH 7.0) for 60 min. Then the fixation solution was replaced with washing solution (100 mM sodium phosphate buffer, 1 mM EDTA, pH 7.0) twice for 15 min. Washed seedlings were vacuum-infiltrated for 5 min in GUS staining buffer (1 mM X-Gluc, 0.5 mM ferricyanide, 0.5 mM ferrocyanide, 0.1% Triton X-100, 10 mM EDTA, 100 mM sodium phosphate buffer, pH 7.0), then incubated at 37 °C for 24 h. The staining reaction was stopped by adding distilled water, the color of chlorophyll was removed with 70% ethanol (v/v) several times, and seedlings were soaked in 95% ethanol (v/v) for 1 h. Plants were photographed to record deposition of the GUS.

Analysis of GUS activity

Seedlings after HS or AZC treatment were powdered in liquid nitrogen and extracted with GUS extraction buffer (50 mM sodium phosphate buffer, 10 mM EDTA, 0.1% SDS, 0.1% triton X-100, 0.1% β-mercaptoethanol, 1 mM PMSF, pH 7.0). After centrifugation, 10-μl protein extract was mixed with 990-μl GUS assay solution [2.5 mM MUG, 50 mM NaPO4, 10 mM EDTA, 10 mM DTT, 2% Leupeptin (w/v), 20% methanol (v/v), pH 7.0], which was preheated in 37 °C for 5 min, and extract was incubated in 37 °C for 1 h. For GUS activity assay, the fluorescence was measured in a Fluoroskan Ascent FL fluorometer (Labsystems, Helsinki, Finland).

Statistical analysis

Data are shown as mean ± SE from three independent experiments. Statistical differences were analyzed by Student *t* test or Duncan multiple range test. $P < 0.05$ was considered statistically significant.

Results

Transcript levels of *AtHSFs* under heat and AZC stress

HSFA2, HSFA7a, HSFB1, HSFB2a, and *HSFB2b* were previously found as AZC- and HS-inducible *HSFs* in *Arabidopsis* seedlings (Sugio et al. 2009). To further confirm the responsiveness of *Arabidopsis HSFs* to AZC and HS under our test conditions, we analyzed transcript levels of *Arabidopsis HSFs* under AZC and heat treatments (data not shown). We selected highly AZC- and HS-inducible *AtHSFA2* (*At2g26150*; 41.8–21.8-fold and 31.3–5-fold induction, respectively), *AtHSFA7a* (*At3g51910*; 4.1–2.9-fold and 8.8–2.2-fold, respectively), *AtHSFB2a* (*At5g62020*; 26.8–18.6-fold and 8.7–6.7-fold, respectively), and *AtHSFB2b* (*At4g11660*; 4.5–2.9-fold and 8.3–3.5-fold, respectively) as candidate *HSFs* for further study (Fig. 1). In addition, *AtHSFA4a* (*At4g18880*), which showed AZC responsiveness (3.9–3.2-fold), and *AtHSFA5* (*At4g13980*), which showed less AZC and HS responsiveness, were included in the test.

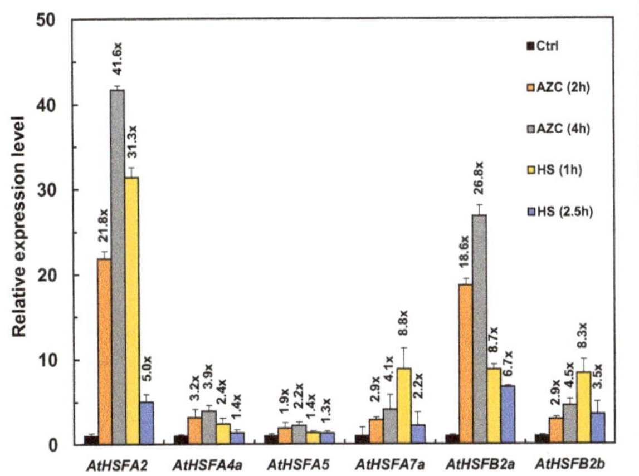

Fig. 1 Expression profiles of *AtHSFs* in *Arabidopsis* plants under AZC and HS treatment. RT-PCR analysis of transcript levels in wild-type (WT) seedlings treated with AZC and HS for the indicated times. Data are mean ± SE expression relative to that of non-stressed control (Ctrl) from three independent experiments and the fold expression is indicated

AtHSFA2, *AtHSFA4a*, and *AtHSFA7a* genes function differentially in response to heat and AZC stress

To investigate whether the *AtHSFs* examined are involved in the HS or AZC responsiveness, Two independent *Oshsp17.3Pro::GUS* transgenic plants, *Oshsp17.3p5* and *Oshsp17.3p11*, were separately crossed with *athsfA2*, *athsfA4a*, *athsfA5*, *athsfA7a*, *athsfB2a*, and *athsfB2b* mutants. Lines *Oshsp17.3p5/athsfA2*, *Oshsp17.3p5/athsfA4a*, *Oshsp17.3p5/athsfA5*, *Oshsp17.3p5/athsfA7a*, *Oshsp17.3p5/athsfB2a*, *Oshsp17.3p5/athsfB2b*, *Oshsp17.3p5/athsfA2/athsfA4a*, *Oshsp17.3p5/athsfA2/athsfA7a*, *Oshsp17.3p11/athsfA2*, *Oshsp17.3p11/athsfA4a*, *Oshsp17.3p11/athsfA5*, *Oshsp17.3p11/athsfA7a*, *Oshsp17.3p11/athsfB2a*, *Oshsp17.3p11/athsfB2b*, *Oshsp17.3p11/athsfA2/athsfA4a*, and *Oshsp17.3p11/athsfA2/athsfA7a* were obtained for analyzing HS and AZC responsiveness.

Under the HS condition (39 °C for 1 h), *Oshsp17.3p5* plants showed GUS staining; *Oshsp17.3p5/athsfA2*, *Oshsp17.3p5/athsfA4a*, *Oshsp17.3p5/athsfA7a*, and *Oshsp17.3p5/athsfB2b* plants showed reduced GUS expression; and GUS staining was similar in *Oshsp17.3p5/athsfA5* and *Oshsp17.3p5/athsfB2a* plants (Fig. 2a). Under AZC treatment (5 mM AZC for 4 h), both cotyledons and true leaves of *Oshsp17.3p5/athsfA2* and *Oshsp17.3p5/athsfA7a* plants did not show any GUS signal (Fig. 2a), and true leaves of *Oshsp17.3p5/athsfA4a*, *Oshsp17.3p5/athsfB2a*, and *Oshsp17.3p5/athsfB2b* plants showed little or no GUS signal; the profile of GUS staining was similar in *Oshsp17.3p5* and *Oshsp17.3p5/athsfA5* plants. Similar HS- and AZC-induced profile of GUS staining was found in *Oshsp17.3p11*, *Oshsp17.3p11/athsfA2*, *Oshsp17.3p11/athsfA4a*, *Oshsp17.3p11/athsfA5*, *Oshsp17.3p11/athsfA7a*, *Oshsp17.3p11/athsfB2a*, *Oshsp17.3p11/athsfB2b*, *Oshsp17.3p11/athsfA2/athsfA4a*, and *Oshsp17.3p11/athsfA2/athsfA7a* (data not shown).

The reduction in HS and AZC responsiveness measured by GUS activity was further confirmed quantitatively. With HS treatment, GUS activity was about 55% lower for *Oshsp17.3p5/athsfA2* than *Oshsp17.3p5* plants (Fig. 2b). Also, GUS activity was lower for *Oshsp17.3p5/athsfA4a*, *Oshsp17.3p5/athsfA7a*, and *Oshsp17.3p5/athsfB2b* than *Oshsp17.3p5* plants (36, 24–34, and 36% reduction, respectively). Similar reduction of GUS activity was further confirmed in *Oshsp17.3p11/athsfA2*, *Oshsp17.3p11/athsfA4a*, *Oshsp17.3p11/athsfA7a*, and *Oshsp17.3p11/athsfB2b* compared with *Oshsp17.3p11* plants (Fig. 2c). We did not find a significant difference in GUS activity among *Oshsp17.3p5*, *Oshsp17.3p11*, *Oshsp17.3p5/athsfA5*, *Oshsp17.3p11/athsfA5*, *Oshsp17.3p5/athsfB2a* and *Oshsp17.3p11/athsfB2a* plants. These loss-of-function results indicate that mutation of *AthsfA2*, *AthsfA4a*, *AthsfA7a*, and *AthsfB2b* may alter HS responsiveness in *Arabidopsis* plants.

We then compared the effect of *AtHSF* mutation on AZC responsiveness. With AZC treatment, relative GUS activity was lower for *Oshsp17.3p5/athsfA2*, *Oshsp17.3p5/athsfA4a*, and *Oshsp17.3p5/athsfA7a* than *Oshsp17.3p5* plants (65–67, 46–48, and 40–42% reduction, respectively) (Fig. 2d) but did not significantly differ among *Oshsp17.3p5*, *Oshsp17.3p5/athsfA5*, *Oshsp17.3p5/athsfB2a*, and *Oshsp17.3p5/athsfB2b* plants. Similar reduction of GUS activity was further confirmed in *Oshsp17.3p11/athsfA2*, *Oshsp17.3p11/athsfA4a*, and *Oshsp17.3p11/athsfA7a* compared with *Oshsp17.3p11* plants (Fig. 2e). Thus, on GUS activity analysis of HS- and AZC-treated seedlings, *AtHSFA2*, *AtHSFA4a*, and *AtHSFA7a* were important for the HSR and AZC response in *Arabidopsis*.

Furthermore, we crossed *athsfA2*/*athsfA4a* and *athsfA2*/*athsfA7a* plants with *OsHsp17.3p5* and *OsHsp17.3p11* transgenic *Arabidopsis*, respectively and obtained *Oshsp17.3p5/athsfA2*/*athsfA4a*, *Oshsp17.3p11/athsfA2*/*athsfA4a*, *Oshsp17.3p5/athsfA2*/*athsfA7a*, and *Oshsp17.3p11/athsfA2*/*athsfA7a* *Arabidopsis* plants for testing HS and AZC responsiveness. With HS treatment, GUS signal was absent in true leaves of *Oshsp17.3p5/athsfA2*/*athsfA4a* and cotyledons of *Oshsp17.3p5/athsfA2*/*athsfA7a*, and AZC-induced GUS signal was not significant in *Oshsp17.3p5/athsfA2*/*athsfA4a* or *Oshsp17.3p5/athsfA2*/*athsfA7a* *Arabidopsis* plants, which was similar to *Oshsp17.3p5/athsfA2* and *Oshsp17.3p*Δ*AZRE* plants (Fig. 2a). Similar HS- and AZC-induced profile of GUS staining was found in *Oshsp17.3p11/athsfA2*/*athsfA4a* and *Oshsp17.3p11/*

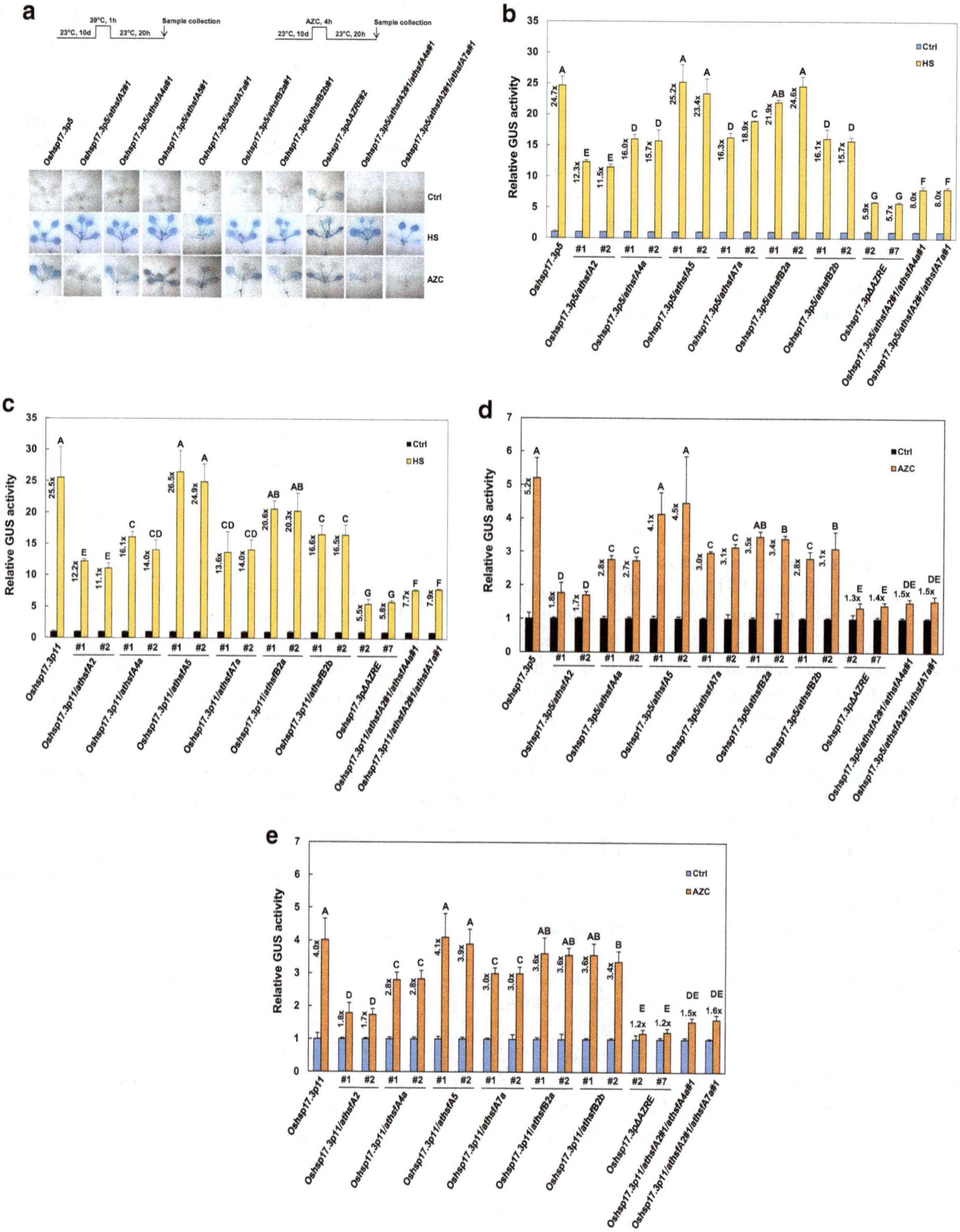

Fig. 2 Analysis of HS and AZC responsiveness of *AtHSF* mutants by GUS staining. *AtHSF* mutant plants were transformed with a chimeric *Oshsp17.3Pro::GUS* gene as described. **a** Seedlings from independent transgenic lines underwent HS or AZC treatment as indicated and GUS histochemical staining. Non-stress control condition (Ctrl). Relative GUS activity of seedlings treated with **b**, **c** HS and **d**, **e** AZC. Data are mean ± SE GUS activity relative to that of the Ctrl from three independent experiments and the fold expression is indicated. Bars with the same letter are not significantly different at $P < 0.05$

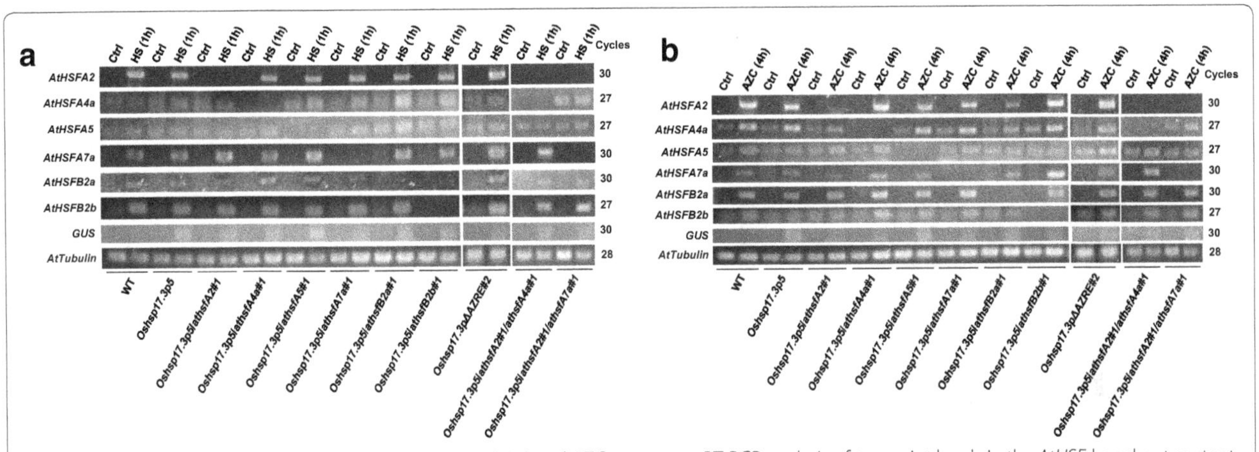

Fig. 3 Expression profiles of *AtHSFs* in *AtHSF* mutants with HS and AZC treatment. RT-PCR analysis of transcript levels in the *AtHSF*-knockout mutant seedlings treated with **a** HS and **b** AZC for the indicated times. *AtTubulin* level was an internal control. Two biological repeats were performed, and similar results were obtained. Non-stress control condition (Ctrl)

athsfA2/athsfA7a Arabidopsis plants (data not shown). On quantitative analysis, with HS treatment, relative GUS activity was significantly lower for *Oshsp17.3p5/athsfA2/athsfA4a* and *Oshsp17.3p5/athsfA2/athsfA7a* than *Oshsp17.3p5/athsfA2*, *Oshsp17.3p5/athsfA4a*, and *Oshsp17.3p5/athsfA7a* plants (Fig. 2b). Also, GUS activity was lower for *Oshsp17.3p11/athsfA2/athsfA4a* and *Oshsp17.3p11/athsfA2/athsfA7a* than *Oshsp17.3p11/athsfA2*, *Oshsp17.3p11/athsfA4a*, and *Oshsp17.3p11/athsfA7a* plants (Fig. 2c). With AZC treatment, the GUS activity for *Oshsp17.3p5/athsfA2/athsfA4a* and *Oshsp17.3p5/athsfA2/athsfA7a* plants dropped to a level (38 and 40% of GUS activity, respectively, versus *Oshsp17.3p5* plants) comparable to that for *Oshsp17.3p5/athsfA2* and *Oshsp17.3pΔAZRE* plants (Fig. 2d). In addition, GUS activity did not significantly differ among *Oshsp17.3p11/athsfA2/athsfA4a*, *Oshsp17.3p11/athsfA2/athsfA7a*, and *Oshsp17.3p11/athsfA2* plants (Fig. 2e). These results suggest that *AtHSFA2*, *AtHSFA4a*, and *AtHSFA7a* genes function independently in the HSR of *Arabidopsis* plants.

Positive and negative regulation among the *AtHSFs*

Data in Fig. 1 revealed that *AtHSFA2*, *AtHSFA7a*, *AtHSFB2a*, and *AtHSFB2b* were HS- and AZC-inducible and *AtHSF4a* was AZC-inducible. We examined the expression profiles of the 6 *AtHSFs* in the mutants under stress. After 1 h of heat treatment, compared with the WT and *Oshsp17.3p5* plants, *AtHSFA4a* transcript level was significantly elevated in *Oshsp17.3p5/athsfA5* and *Oshsp17.3p5/athsfB2a* plants and *AtHSFA7a* level was increased in *Oshsp17.3p5/athsfA2*, *Oshsp17.3p5/athsfA5*, *Oshsp17.3p5/athsfB2a*, and *Oshsp17.3p5/athsfA2/athsfA4a* plants, with no significant change in *AtHSFA2*, *AtHSFA5*, *AtHSFB2a*, and *AtHSFB2b* levels in the mutant plants tested (Fig. 3a). With AZC treatment, *AtHSFA2* and *AthsfA4a* levels were reduced in *Oshsp17.3p5/athsfB2a* and *Oshsp17.3p5/athsfA2* plants, respectively, and *AtHSFA7a* level was increased in *Oshsp17.3p5/athsfA4a* and *Oshsp17.3p/athsfB2b* plants (Fig. 3b). Similar expression profiles of the 6 *AtHSFs* were also found in *Oshsp17.3p11*, *Oshsp17.3p11/athsfA2*, *Oshsp17.3p11/athsfA4a*, *Oshsp17.3p11/athsfA5*, *Oshsp17.3p11/athsfA7a*, *Oshsp17.3p11/athsfB2a*, *Oshsp17.3p11/athsfB2b*, *Oshsp17.3p11/athsfA2/athsfA4a*, and *Oshsp17.3p11/athsfA2/athsfA7a* plants under HS and AZC conditions (data not shown). These results suggest a finely tuned activation and repression of the expression of *HSFs* under HS and AZC stress.

AtHSFA2, *AtHSFA4a*, and *AtHSFA7a* are not responsive to Tm

AZC typically induces the UPR and CPR. The data in Fig. 2 indicated that *AtHSFA2*, *AtHSFA4a*, and *AtHSFA7a* are essential for the HSR and AZC response in *Arabidopsis*. Studies have shown *AtHSFA2* as a crucial regulatory component of the CPR (Sugio et al. 2009). To understand whether these *AtHSFs* are involved in the UPR, we examined the effect of Tm treatment (UPR induction) in the *AtHSF* mutants tested. Tm did not activate the expression of the 6 *AtHSF* genes (Fig. 4a). On GUS analysis, no Tm responsiveness was detected in the mutant plants tested (Fig. 4b, c). These results confirmed that *AtHSFA2*, *AtHSFA4a*, and *AtHSFA7a* function in the CPR.

Discussion

To adapt to biotic and abiotic stresses, plants have evolved a complex set of molecular responses, which often exhibit features sharing substantial overlap

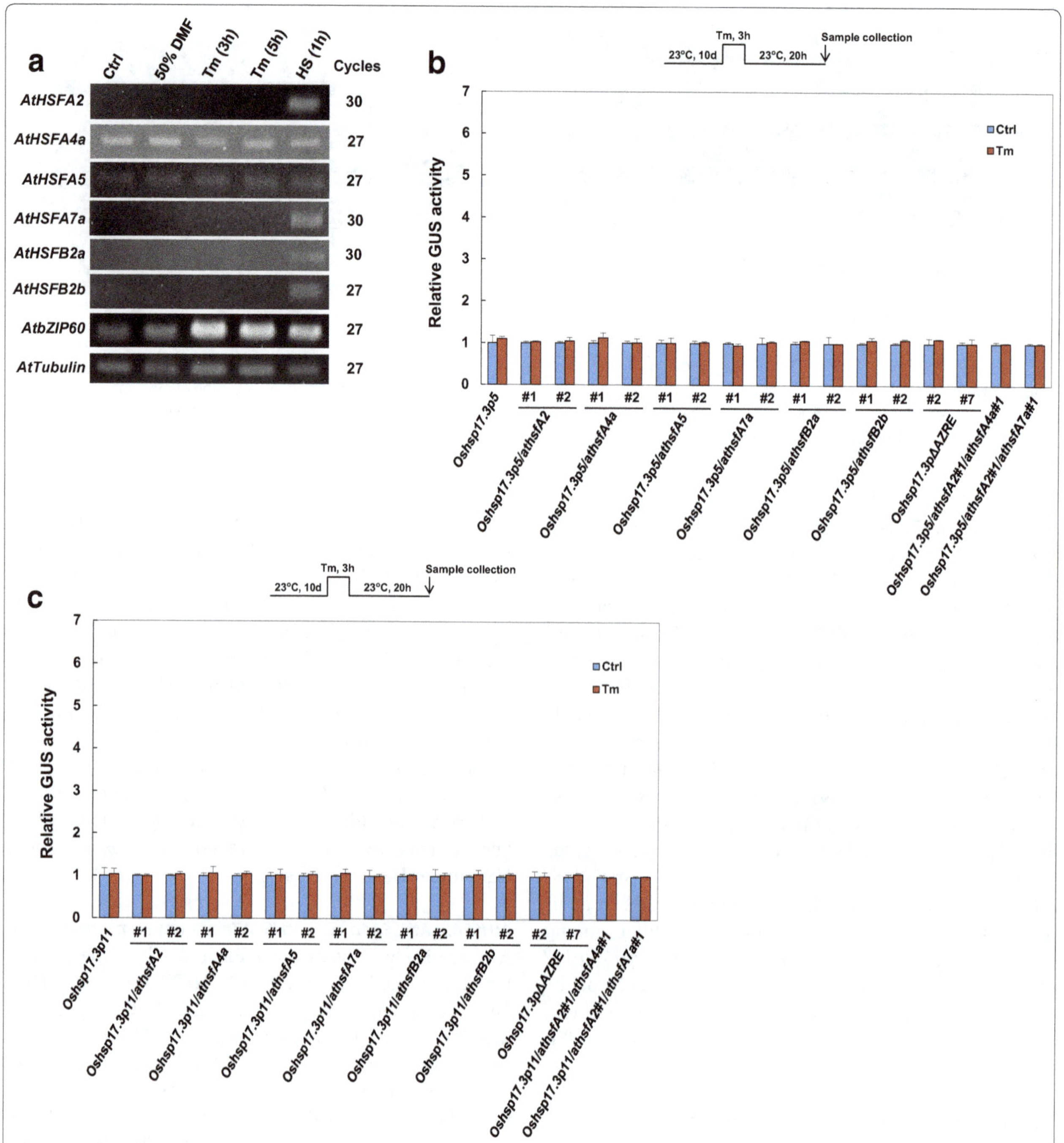

Fig. 4 Analysis of tunicamycin (Tm) responsiveness of *AtHSF* mutants. **a** RT-PCR analysis of transcript levels in WT seedlings treated with Tm or solvent (dimethylformamide; DMF) and HS for the indicated times. mRNA expression of *AtbZIP60* was a positive control for HS and Tm treatment. *AtTubulin* level was an internal control. Two biological repeats were performed, and similar results were obtained. **b**, **c** Relative GUS activity of seedlings treated with Tm. Data are mean ± SE GUS activity relative to that of non-stress control condition (Ctrl) from three independent experiments

pathways and components. *HSF/HSP* responses are recognized as central chaperone components against unfolded protein accumulation, a signal for triggering HSR, UPR, or CPR based on distinct subcellular localization (Aparicio et al. 2005; Swindell et al. 2007; Yeh et al. 2007). Many reports have shown that HSFs are important for resistance to heat and other environmental stresses (Mishra et al. 2002; Charng et al. 2007; Banti et al. 2010; Liu et al. 2011). Using an HS- and AZC-sensitive promoter-*GUS* fusion system (Guan et al. 2010) together with knockout plants, we aimed to identify the contribution of *AtHSFA2, AtHSFA4a, AtHSFA5, AtHSFA7a, AtHSFB2a*, and *AtHSFB2b* to the responses induced by HS, AZC, and Tm.

Plant *HSFs* are regulated by HS and AZC, including up- and downregulation. We found the expression of *AtHSFA2, AtHSFA4a, AtHSFA7a, AtHSFB2a*, and *AtHSFB2b* induced > twofold with 1-h HS treatment and then reduced after prolonged heat incubation (Fig. 1). As well, AZC upregulated *AtHSFA2, AtHSFA4a, AtHSFA7a, AtHSFB2a*, and *AtHSFB2b* expression > 2.9-fold during treatment. However, Tm did not affect the expression of the 6 *AtHSFs* (Fig. 4a). Despite a slight difference in plant material and treatment time, the results are similar to published microarray data (Busch et al. 2005; Schramm et al. 2008; Sugio et al. 2009), finding that *AtHSFA2, AtHSFA4a, AtHSFA7a, AtHSFB2a*, and *AtHSFB2b* are important for stress response networks.

Studies have shown that *AtHSFA2* and *AtHSFA7a* knockout mutants lose acquired thermotolerance, and *AtHSFA2* mutants also show reduced tolerance to AZC (Charng et al. 2007; Siddique et al. 2008; Sugio et al. 2009). In this study, loss-of-function mutation of *AtHSFA2* significantly repressed relative GUS activity under HS and AZC treatment (Fig. 2b–e). By contrast, null mutation of *AtHSFA4a* and *AtHSFA7a* only slightly repressed relative GUS activity under HS and AZC stress. These results agree with others showing that *AtHSFA2* is closely related to the regulation of HSR as well as CPR (Busch et al. 2005; Nishizawa et al. 2006; Ogawa et al. 2007; Sugio et al. 2009; Jung et al. 2010), whereas *AtHSFA4a* and *AtHSFA7a* have a lesser effect on HSR and CPR. Furthermore, as compared with *AtHSFA2* knockout alone, double knockout with *AtHSFA2* and *AtHSFA4a* or *AtHSFA7a* showed more significant repression of HS-induced GUS activity (Fig. 2b–e). Thus, *AtHSFA2, AtHSFA4a*, and *AtHSFA7a* may be linked to activation of different target genes/pathways in the HSR. However, *AtHSFA2* appears to be a functionally redundant factor to *AtHSFA4a* and *AtHSFA7a* for AZC-induced CPR because the GUS activity of *AtHSFA2*-knockout plants was similar to that with double knockout of *AtHSFA2* and *AtHSFA4a* or *AtHSFA7a* under AZC treatment (Fig. 2b–e).

Ikeda et al. (2011) reported that *AtHsfB1* and *AtHsfB2b*, sharing functional redundancy in repressive activities, were able to suppress the accumulation of *AtHSFA2* and *AtHSFA7a* transcripts and were indispensable for acquired thermotolerance. As compared with *AtHSFA2* knockout, *AtHsfB2b* knockout slightly repressed GUS activity in response to HS treatment (Fig. 2b, c). We also revealed no significant change in HS-induced *AtHSFA2* and *AtHSFA7a* transcript levels with *AtHsfB2b* knockout (Fig. 3a). These results suggest that *AtHSFB2b* may mediate the HSR but not CPR. Of note, *AtHSFB2a* is highly AZC- and HS-inducible, but we did not find a significant reduction in GUS activity with *AtHsfB2a* knockout during AZC treatment. However, we cannot absolutely exclude the role of *AtHsfB2a* in AZC-induced CPR because of its high expression under AZC and HS treatment.

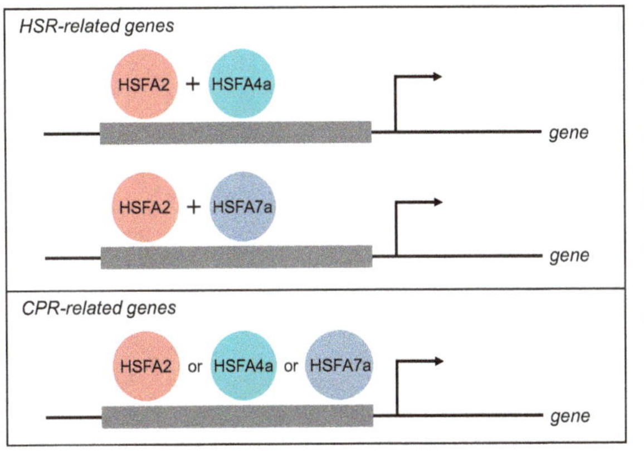

Fig. 5 Working model of the AtHSFs involved in HSR and CPR. Gray bar indicates the promoter region of HSR- and CPR-related genes. Arrow indicates transcription starting site

In conclusion, we confirmed and characterized the roles of *AtHSFA2, AtHSFA4a, AtHSFA5, AtHSFA7a, AtHSFB2a*, and *AtHSFB2b* in the HSR and CPR. For simplifying our result, we propose a working model to show the roles of following AtHSFs in CPR and HSR (Fig. 5). *AtHSFA2, AtHSFA4a*, and *AtHSFA7a* function independently in the HSR, but *AtHSFA2* may function redundantly with *AtHSFA4a* and *AtHSFA7a* in the CPR. *AtHSFB2b* has some role in mediating the HSR, and *AtHSFA5 and AtHSFB2a* cannot mediate the HSR and CPR. These 6 *AtHSFs* are not involved in the UPR.

Abbreviations

AZC: azetidine-2-carboxylic acid; AZRE: azetidine-2-carboxylic acid response element; CPR: cytosolic protein response; HSF: heat shock factor; HSP: heat shock protein; HSR: heat shock response; UPR: unfolded protein response; Tm: tunicamycin.

Authors' contributions

C-AL, S-JW, and C-HY conceived the concept and designed the experiment. K-FL and M-YT have equally contributed towards this manuscript. K-FL and M-YT performed the experiments and analyzed the data. C-HY wrote the manuscript. All authors read and approved the final manuscript.

Acknowledgements

We are grateful to Dr. Yee-Yung Charng for technical assistance. We thank ABRC for providing seeds.

Competing interests

The authors declare that they have no competing interests. And there have neither financial competing interests nor other competing interests.

Funding

This work was supported by the National Science Council, Taiwan, ROC (NSC101-2313-B-008-001) to C-HY.

References

Aparicio F, Aparicio F, Thomas CL, Lederer C, Niu Y, Wang D, Maule AJ (2005) Virus induction of heat shock protein 70 reflects a general response to protein accumulation in the plant cytosol. Plant Physiol 138:529–536. https://doi.org/10.1104/pp.104.058958.1

Banti V, Mafessoni F, Loreti E, Alpi A, Perata P (2010) The heat-inducible transcription factor *HsfA2* enhances anoxia tolerance in *Arabidopsis*. Plant Physiol 152:1471–1483. https://doi.org/10.1104/pp.109.149815

Bukau B, Weissman J, Horwich A (2006) Molecular chaperones and protein quality control. Cell 125:443–451. https://doi.org/10.1016/j.cell.2006.04.014

Busch W, Wunderlich M, Schöffl F (2005) Identification of novel heat shock factor-dependent genes and biochemical pathways in *Arabidopsis thaliana*. Plant J 41:1–14. https://doi.org/10.1111/j.1365-313X.2004.02272.x

Charng Y-Y, Liu H-C, Liu N-Y, Chi W-T, Wang C-N, Chang S-H, Wang T-T (2007) A heat-inducible transcription factor, HsfA2, is required for extension of acquired thermotolerance in *Arabidopsis*. Plant Physiol 143:251–262. https://doi.org/10.1104/pp.106.091322

Cramer GR, Urano K, Delrot S, Pezzotti M, Shinozaki K (2011) Effects of abiotic stress on plants: a systems biology perspective. BMC Plant Biol 11:163. https://doi.org/10.1186/1471-2229-11-163

Guan JC, Yeh CH, Lin YP, Ke YT, Chen MT, You JW, Liu YH, Lu CA, Wu SJ, Lin CY (2010) A 9 bp cis-element in the promoters of class I small heat shock protein genes on chromosome 3 in rice mediates L-azetidine-2-carboxylic acid and heat shock responses. J Exp Bot 61:4249–4261. https://doi.org/10.1093/jxb/erq230

Guo J, Wu J, Ji Q, Wang C, Luo L, Wang Y, Wang J (2008) Genome-wide analysis of heat shock transcription factor families in rice and *Arabidopsis*. J Genet Genomics 35:105–118. https://doi.org/10.1016/S1673-8527(08)60016-8

Ikeda M, Mitsuda N, Ohme-Takagi M (2011) Arabidopsis HsfB1 and HsfB2b act as repressors of the expression of heat-inducible Hsfs but positively regulate the acquired thermotolerance. Plant Physiol 157(3):1243–1254

Jung K-H, Cao P, Seo Y-S, Dardick C, Ronald PC (2010) The Rice Kinase Phylogenomics Database: a guide for systematic analysis of the rice kinase super-family. Trends Plant Sci 15:595–599. https://doi.org/10.1016/j.tplants.2010.08.004

Jungkunz I, Link K, Vogel F, Voll LM, Sonnewald S, Sonnewald U (2011) AtHsp70-15-deficient *Arabidopsis* plants are characterized by reduced growth, a constitutive cytosolic protein response and enhanced resistance to TuMV. Plant J 66:983–995. https://doi.org/10.1111/j.1365-313X.2011.04558.x

Kleinboelting N, Huep G, Kloetgen A, Viehoever P, Weisshaar B (2012) GABI-Kat SimpleSearch: new features of the *Arabidopsis thaliana T-DNA* mutant database. Nucleic Acids Res 40:1211–1215. https://doi.org/10.1093/nar/gkr1047

Liu HC, Liao HT, Charng YY (2011) The role of class A1 heat shock factors (*HSFA1s*) in response to heat and other stresses in *Arabidopsis*. Plant Cell Environ 34:738–751. https://doi.org/10.1111/j.1365-3040.2011.02278.x

Mehdy MC (1994) Active oxygen species in plant defense against pathogens. Plant Physiol 105:467–472

Mishra SK, Tripp J, Winkelhaus S, Tschiersch B, Theres K, Nover L, Scharf K-D (2002) In the complex family of heat stress transcription factors, *HsfA1* has a unique role as master regulator of thermotolerance in tomato. Genes Dev 16:1555–1567. https://doi.org/10.1101/gad.228802

Nishizawa A, Yabuta Y, Yoshida E, Maruta T, Yoshimura K, Shigeoka S (2006) *Arabidopsis* heat shock transcription factor *A2* as a key regulator in response to several types of environmental stress. Plant J 48:535–547. https://doi.org/10.1111/j.1365-313X.2006.02889.x

Nishizawa-Yokoi A, Nosaka R, Hayashi H, Tainaka H, Maruta T (2011) *HsfA1d* and *HsfA1e* involved in the transcriptional regulation of *HsfA2* function as key regulators for the *HSF* signaling network in response to environmental stress. Plant Cell Physiol 52:933–945. https://doi.org/10.1093/pcp/pcr045

Nover L, Bharti K, Döring P, Mishra SK, Ganguli A, Scharf KD (2001) *Arabidopsis* and the heat stress transcription factor world: how many heat stress transcription factors do we need? Cell Stress Chaperones 6:177–89. 10.1379/1466-1268(2001)006<0177:aathst>2.0.co;2

Ogawa D, Yamaguchi K, Nishiuchi T (2007) High-level overexpression of the *Arabidopsis HsfA2* gene confers not only increased thermotolerance but also salt/osmotic stress tolerance and enhanced callus growth. J Exp Bot 58:3373–3383. https://doi.org/10.1093/jxb/erm184

Redondo-Gómez S (2013) Abiotic and biotic stress tolerance in plants. In: Rout GR, Das AB (eds) Molecular stress physiology of plants. Springer, New Delhi, pp 1–20

Scharf K-D, Berberich T, Ebersberger I, Nover L (2012) The plant heat stress transcription factor (Hsf) family: structure, function and evolution. Biochim Biophys Acta 1819:104–119. https://doi.org/10.1016/j.bbagrm.2011.10.002

Schneider CA, Rasband WS, Eliceiri KW (2012) NIH Image to ImageJ: 25 years of image analysis. Nat Methods 9:671–675. https://doi.org/10.1038/nmeth.2089

Schöffl F, Prandl R, Reindl A (1998) Update on signal transduction regulation of the heat-shock response. Plant Physiol 117:1135–1141. https://doi.org/10.1104/pp.117.4.1135

Schramm F, Larkindale J, Kiehlmann E, Ganguli A, Englich G, Vierling E, von Koskull-Döring P (2008) A cascade of transcription factor *DREB2A* and heat stress transcription factor *HsfA3* regulates the heat stress response of *Arabidopsis*. Plant J 53:264–274. https://doi.org/10.1111/j.1365-313X.2007.03334.x

Shinozaki K, Yamaguchi-Shinozaki K (1996) Molecular responses to drought and cold stress. Curr Opin Biotechnol 7:161–167. https://doi.org/10.1016/S0958-1669(96)80007-3

Siddique M, Gernhard S, von Koskull-Döring P, Vierling E, Scharf K-D (2008) The plant sHSP superfamily: five new members in *Arabidopsis thaliana* with unexpected properties. Cell Stress Chaperones 13:183–197. https://doi.org/10.1007/s12192-008-0032-6

Sugio A, Dreos R, Aparicio F, Maule AJ (2009) The cytosolic protein response as a subcomponent of the wider heat shock response in *Arabidopsis*. Plant Cell 21:642–654. https://doi.org/10.1105/tpc.108.062596

Swindell WR, Huebner M, Weber AP (2007) Transcriptional profiling of *Arabidopsis* heat shock proteins and transcription factors reveals extensive overlap between heat and non-heat stress response pathways. BMC Genomics 8:125. https://doi.org/10.1186/1471-2164-8-125

Yeh C-H, Wu S-J, Tsai Y-F, Chen H-Y, Lin C-Y (2007) Physiological effects of azetidine on cellular leakage in soybean seedlings. Plant Sci 172:1124–1130. https://doi.org/10.1016/j.plantsci.2007.02.021

12

Arabidopsis and rice showed a distinct pattern in ZIPs genes expression profile in response to Cd stress

Xin Zheng[1], Liang Chen[1] and Xiaofang Li[1,2*]

Abstract

Background: Plant ZIP genes represent an important transporter family involved in metal transport. Evidence has implied that some ZIPs may contribute to plant Cd uptake, but a genome-wide examination of ZIPs' role in Cd tolerance and uptake has rarely been reported. In this study, a genome-wide bioinformatic screening of candidate ZIP genes in *Arabidopsis* and rice was performed, followed by a systematic determination of their expression profile in response to Cd stress. Typical up-regulated ZIPs genes were then expressed in yeast cells to examine their effect on hosts' Cd uptake.

Results: A total of 27 ZIP genes in *Arabidopsis* and rice were screened out based on sequence similarity. In *Arabidopsis*, Cd exposure strongly impacted the expression of most ZIPs, among which *AtIRT1*, *AtIRT2*, *AtIRT4 AtZIP9*, *AtZIP10* and *AtZIP12* were sharply up-regulated and *AtIRT3*, *AtIRT5* were significantly down-regulated in root. In rice, all tested genes in shoot except for *OsIRT1* and *OsIRT12* were sharply up-regulated, while *OsIRT1* and *OsZIP1* in root were significantly down-regulated. Interestingly, some genes like *AtIRT3*, *AtZIP5*, *AtZIP12*, *OsIRT1* and *OsZIP1* showed converse expression regulation when subject to the tested Cd stress. When expressed in yeast cells, three ZIPs, *AtIRT1*, *OsZIP1* and *OsZIP3*, caused a substantial increase in Cd sensitivity and Cd accumulation of the host cells.

Conclusions: In conclusion, this study revealed a distinct pattern in ZIPs family genes expression between *Arabidopsis* and rice in response to Cd stress. *Arabidopsis* mainly up-regulated root ZIPs genes, while rice mainly up-regulated shoot ZIPs genes. Three genes, *AtIRT1*, *OsZIP1* and *OsZIP3*, conferred an increased Cd accumulation and sensitivity to Cd stress when expressed in yeast cells, further implying their roles in Cd uptake in plants.

Keywords: ZIP family, Cd, Metal cation transporter, Gene expression, Cd uptake

Background

The zinc(Zn)-regulated/iron(Fe)-regulated transporter-like family proteins (ZIPs) are membrane-located proteins for cations transport (Eng et al. 1998; Guerinot 2000). They have been found to exist broadly in prokaryotic cells, fungi, plants and mammalians. In plants, ZIPs have been identified in both dicots and monocots, such as *Arabidopsis* (Grotz et al. 1998; Milner et al. 2013), rice (Chen et al. 2008), maize (Li et al. 2013), medicago (Lopez-Millan et al. 2004; Stephens et al. 2011) and barely (Tiong et al. 2015). Grotz et al. identified five ZIP genes (*IRT1*, *ZIP1-4*) in *Arabidopsis* (Grotz et al. 1998), and later up to 11 ZIP genes from *Arabidopsis* were detected bioinformatically (Guerinot 2000). Roles of *ZIP1-12* from *Arabidopsis* in Zn transport were explored experimentally (Milner et al. 2013). More recently, 18 ZIPs from *Arabidopsis* and 16 ZIPs from rice were annotated (Ivanov and Bauer 2017).

In *Arabidopsis* and rice, only a small number of ZIPs have been examined for biological functions in plant till now. *Arabidopsis IRT1* is a well-studied ZIP gene first identified as a crucial transporter for plant Fe uptake

*Correspondence: xfli@sjziam.ac.cn; x.li10@uq.edu.au
[1] Key Laboratory for Agricultural Water Resources, Center for Agricultural Resources Research, Institute of Genetics and Developmental Biology, Chinese Academy of Sciences, Shijiazhuang 050021, Hebei, People's Republic of China
Full list of author information is available at the end of the article

(Varotto et al. 2002; Vert et al. 2002). *Arabidopsis IRT1* can be induced by iron deficiency (Korshunova et al. 1999; Connolly et al. 2002), and may play a role in Mn/Zn transport as well (Korshunova et al. 1999; Rogers et al. 2000; Connolly et al. 2002). Biological functions in Zn/Fe transport of *AtIRT2* (Vert et al. 2001, 2009), *AtZIP1/2* (Grotz et al. 1998; Wintz et al. 2003; Milner et al. 2013), *OsIRT1* (Nakanishi et al. 2006; Lee and An 2009; Ishimaru et al. 2006; Bughio et al. 2002) and *OsZIP4/5/8* (Ishimaru et al. 2005; Chen et al. 2008; Lee et al. 2010a, b; Yang et al. 2009) have also been examined in the past decade.

A few studies have also implied that ZIPs may be involved in Cd transport. Yeast cells expressing *AtIRT1* showed increased Cd sensitivity (Rogers et al. 2000; Vert et al. 2001), and IRT1-dependent Fe/Mn/Zn uptake was inhibited by excess Cd (Eide et al. 1996; Korshunova et al. 1999). The *Arabidopsis IRT1* knock-out mutant *irt1-1* exhibited reduced Cd sensitivity and Cd accumulation (Vert et al. 2002; Fan et al. 2014), while overexpression of *AtIRT1* increased Cd sensitivity in *Arabidopsis* (Connolly et al. 2002). *AtIRT2*, phylogenetically similar to *AtIRT1*, increased Cd uptake when overexpressed in *Arabidopsis* (Vert et al. 2009), though the yeast cells expressing *AtIRT2* exhibited no altered Cd sensitivity (Vert et al. 2001). In rice, expression of *OsIRT1* and *OsIRT2* made the cells more sensitive to Cd and increased Cd accumulation (Nakanishi et al. 2006; Lee and An 2009). Nonetheless, we still know little about the roles of most of the ZIPs genes in Cd stress response in *Arabidopsis* and rice.

In this study, genome-wide ZIPs identification in *Arabidopsis* and rice was performed with rigorous evolutional analysis. A comparative examination of genome-wide expression profile of ZIPs in *Arabidopsis* and rice in response to Cd stress were carried out. Their role in Cd uptake of typical ZIPs responding to Cd stress was further tested by expressing them in yeast. As expected, most identified ZIPs gene expression responded remarkably to Cd stress, while unexpectedly it was found that *Arabidopsis* and rice showed a distinct pattern in ZIPs genes expression profile. These results may help to elucidate the plants' genetic basis for Cd translocation via a ZIPs-dependent pathway.

Materials and methods

Bioinformatics

Genomic query of *Arabidopsis* and rice ZIP family genes was performed online using the PLAZA database (http://bioinformatics.psb.ugent.be/plaza/). The sequences of 27 ZIP genes were retrieved manually from the TAIR database (http://www.arabidopsis.org/index.jsp) and the TIGR database (http://rice.plantbiology.msu.edu/index.shtml).

TM regions and other domains of the identified ZIPs gens were predicted through the TMHMM Server (http://www.cbs.dtu.dk/services/TMHMM-2.0/) and UniProtKB database (http://www.uniprot.org/), following a routine procedure.

Experimental design

Arabidopsis thaliana ecotype Col-0 and *Oryza sativa ssp. japonica* (*cv.* Taichung65) were subject to Cd inhibition test. For *Arabidopsis*, plants were germinated on Murashige and Skoog (MS; pH 5.7) solid medium containing 1% (w/v) sucrose. A total of 60 1-week-old seedlings were transferred to MS (control) or MS with 300 μM $CdCl_2$ (Cd stress treatment) solid medium, and grown for 3 days in a controlled chamber environment under a 16/8 h photoperiod at 22 °C. For rice, seedlings were germinated hydroponically in distilled water. A total of six 10-day-old seedlings were then subject to a hydroponic culture in distilled water (control) or 300 μM $CdCl_2$ solution (Cd stress treatment) for 3 days under 16/8 h photoperiod at 25 °C. The Cd concentration used in this study was selected based on our pilot experiment.

After Cd stress treatment, the shoot and root tissues were harvested and frozen immediately in liquid nitrogen. Total RNA was isolated from the tissues using Trizol reagent (Invitrogen, Corp., Carlsbad, CA, USA) and treated with DNase I (Promega, Madison, WI, USA). A total of 5 μg RNA was used for reverse transcription with PrimeScript™ RT reagent Kit (Takara Biotechnology Co. Ltd., Dalian, China) following the manufacturer's protocol.

Quantitative Real-Time PCR (qPCR) was performed in a Bio-Rad CFX Connect™ Real-Time PCR Detection System (Hercules, CA, USA) using a SYBR Green Premix Ex Taq (Takara). The PCR parameters were set as: 95 °C for 5 min, followed by 40 cycles of 95 °C for 10 s and 60 °C for 30 s. *Arabidopsis ACTIN* gene (GenBank accession number NM_179953) and rice *ACTIN* gene (GenBank accession number XM_015774830) were used as internal references. Relative gene expression levels were detected using the $2^{-\Delta\Delta CT}$ method (Livak and Schmittgen 2001). Gene expression level was normalized using shoot expression level of each gene in the controls as a calibrator. All primer sequences are listed in the Additional file 1: Table S1.

Cd sensitivity analysis was performed using drop assay. Full-length coding sequence (CDS) was obtained via PCR amplification (see primers in Additional file 1: Table S2), and ligated into pCEV-G1-Km vector under the *PGK1* promoter. The recombinant plasmids were then introduced into *Saccharomyces cerevisiae* (strain AH109) using a lithium acetate-based method. Transformed cells were cultured in Yeast Extract Peptone Dextrose (YPD) media with 300 μg/mL geneticin (G418), harvested by centrifugation, and resuspended in water ($OD_{600}=1.0$), followed by a serial dilution. A total of 5 μL of each dilution was inoculated onto the YPD plates containing 300 μg/mL G418 and 50 μM $CdCl_2$. Cells harbouring empty pCEV-G1-Km were used as a negative control. The plates were incubated at 28 °C for 5 days and the growth of the colonies was subsequently observed.

For the determination of Cd concentration in transformed yeast cells, cells expressing *ZIPs* were harvested after 12 h with 50 μM $CdCl_2$ treatment. Cd was determined using a flame atomic absorption spectrometry (F-AAS) quantitative method. In Brief, cells in the liquid culture were harvested by centrifugation at 4000×*g* and washed three times with 3% NaCl solution. The cells were then oven-dried, weighed and digested using 4 mL 65% HNO_3. The digested mixture was dissolved in 3 mL Millipore® water and subject to Cd determination using a Zeenit 700 P Atomic Absorption Spectrometer (Analytik Jena, Germany) equipped with a flame atomizer. CRM Laver (GWB10023, certified by IGGE) was used as a standard reference material for Cd determination.

Data analysis

Phylogenetic analysis was performed using MEGA 7 (Kumar et al. 2016). The model of ZIP gene structure was constructed using Gene Structure Display Server (http://gsds.cbi.pku.edu.cn/).

Statistical analysis was performed using SPSS 21.0 (IBM, New York, USA). Unpaired two-tailed t test was performed for comparison between the controls and the Cd stress group.

Results and discussion

In this study, 15 candidate ZIP genes from *Arabidopsis* and 12 from rice were screened out based on sequence similarity. The number of ZIPs identified here was similar to previous studies (Ivanov and Bauer 2017; Guerinot 2000). Evolutionary analysis further indicated that all of these ZIP genes contain 1–3 introns (Additional file 1: Figures S1 and S2), whose protein precursors comprise eight TM regions (~20 aa length), one variable region with a conserved HG repeat and a typical signal peptide (SP) located on the N-terminal (Fig. 1). AtZIP13 and OsZIP13, which were previously annotated as putative Zn transporter (Ivanov and Bauer 2017), contain more TM domains. *AtZTP29, AtIAR1, OsIAR1, OsZIP11* and *OsZIP12* contain more than 10 exons. These ZIP-like genes seem to be phylogenetically distant from *SpZRT1* and *AtIRT1* and were not tested in this study. Phylogenetic clustering of the tested 27 ZIPs identified three subgroups, which is similar to previous study (Ivanov and Bauer 2017), namely the seed plant-specific group, the mixed plant group, and the mixed group1/2 (Fig. 1).

Most previous studies on ZIPs' biological functions focused on Zn/Fe/Mn/Cu uptake in yeast cells (Table 1), and expression profile of most ZIPs (except for *AtIRT1/2* and *OsIRT1/2*) in response to Cd remained unknown. In this study, the genome-wide expression profile of ZIP genes in response to Cd stress in *Arabidopsis* and rice were quantified using qPCR. To induce a substantial stress response, 300 μM Cd in culture medium was applied based on our pilot experiment. The 3 days' treatment obviously inhibited seedling growth and root elongation of both *Arabidopsis* and rice, and rice seedling height was also reduced (Fig. 2a). It was reported that even moderate Cd exposure can cause toxic symptoms and increased Cd accumulation in *Arabidopsis* (Fan et al. 2014) and rice (Rafiq et al. 2014). The Cd level used here significantly reduced the root length and seedling dry weight (Additional file 1: Figure S3), and was thus supposed to induce rapid expressional changes in the tested plants.

In *Arabidopsis*, Cd exposure impacted the expression of all ZIPs significantly. Strikingly, *AtIRT1* was induced with a 525-fold increase in shoot and a 22-fold increase in root (Fig. 2b). As abovementioned, some evidence already pointed to the Cd transport role of *AtIRT1* in yeast cells (Korshunova et al. 1999; Rogers et al. 2000; Vert et al. 2001; Eide et al. 1996) and in *Arabidopsis* (Fan et al. 2014; Connolly et al. 2002; Vert et al. 2002). Considering that *AtIRT1* is mainly expressed in root (Vert et al. 2002), AtIRT1 may function as a pump absorbing Cd from soil into root under sever Cd stress. A sharp increase of *AtIRT1* expression in shoot was also observed, indicating its potential role in Cd transport in shoot. Indeed, overexpression of *AtIRT1* in yeast increased the hosts' sensitivity substantially (Fig. 3). Cd accumulation of yeast cells expressing *AtIRT1* was also increased by 40.1%, compared with the control (Additional file 1: Figure S4). Taken together, the results here further confirmed the role of *AtIRT1* in plant Cd uptake implied in previous studies (Rogers et al. 2000).

Like *AtIRT1*, *AtIRT2* was induced with a 1452-fold increase in shoot and a fourfold increase in root (Fig. 2b). Previous studies showed that *AtIRT2* overexpression increased Cd uptake of transgenic *Arabidopsis*, probably through the induction of *AtIRT1* expression (Vert et al. 2001, 2009). In this study, while both *AtIRT2* and *AtIRT1* were coincidently sharply induced when subject to Cd stress, overexpression of *AtIRT2* caused no significant changes in neither Cd sensitivity nor Cd accumulation (Fig. 3 and Additional file 1: Figure S4). It is thus very likely that *AtIRT2* worked indirectly and synergistically with *AtIRT1* in response to the Cd stress.

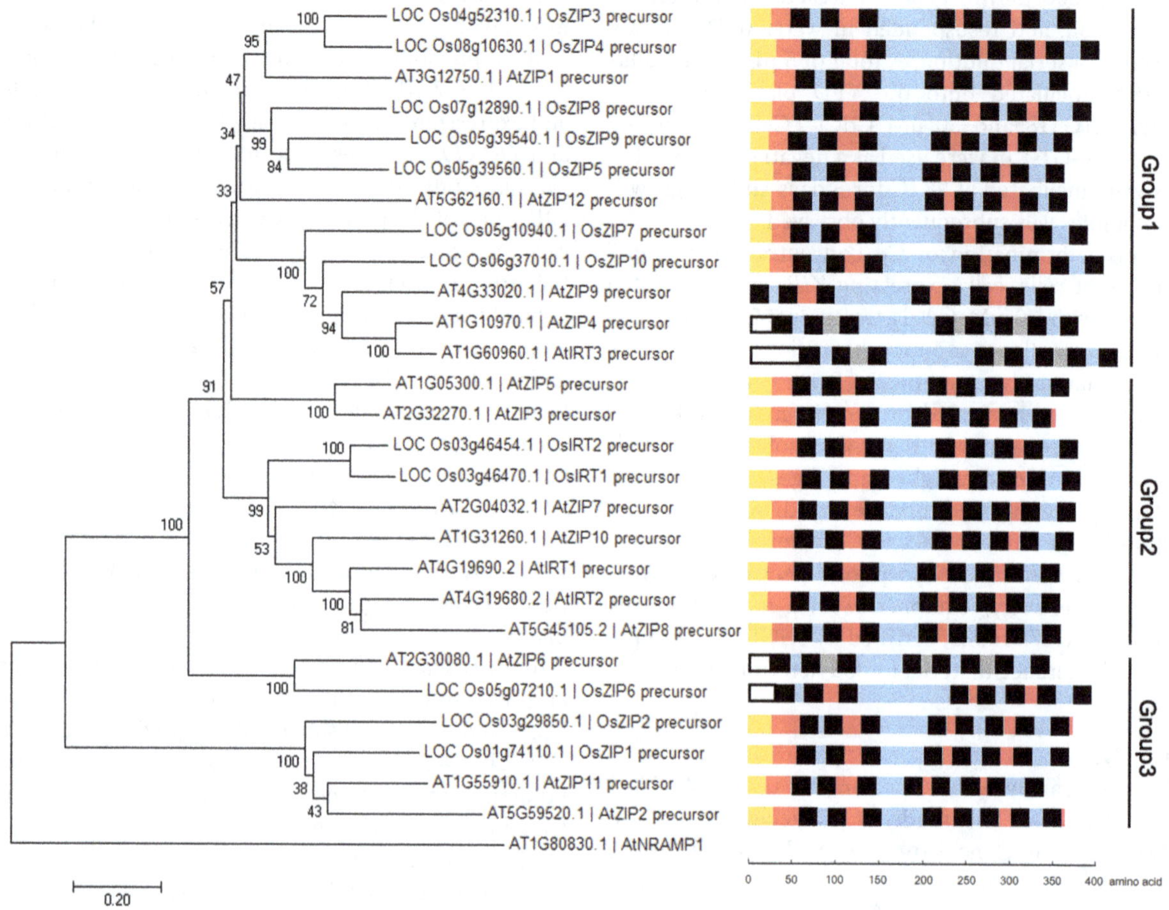

Fig. 1 Phylogenetic relationship of identified ZIPs in *Arabidopsis* and rice. The Neighbor-Joining tree was generated using MEGA7 with 1000 bootstrap replicates, and rooted to the AtNRAMP1. Topological structure was predicted using UniProtKB and TMHMM. Black boxes indicate TM regions. Pink boxes indicate extracellular regions. Blue boxes represent cytoplasmic regions. Gray boxes represent luminal regions. Yellow boxes represent SP region. Uncharacteristic SPs are represented by the white boxes

It was also highlighted that the expression of *AtZIP9* was significantly increased by ninefolds in shoot and 57-folds in root after Cd stress (Fig. 2b). Till now no evidence showed any role of *AtZIP9* in Cd uptake. The strong induction by Cd stress may imply its role in Cd transport, and its overexpression in yeast cells moderately increased hosts' sensitivity to Cd. Conversely, expression of *AtZIP9* did not increase the Cd accumulation of host cells (Additional file 1: Figure S4). As a hypothetic transmembrane ion transporter, *AtZIP9* might affect the growth of host cells by a Cd-independent way. In addition, *AtIRT3, AtZIP4, AtZIP5, AtZIP11* and *AtZIP12* showed converse expression regulation when subject to the tested Cd stress, and *AtZIP7* was reduced in shoot and was under the detection limit in root (Fig. 2b). Their potential roles in Cd transport merit a further investigation.

In rice, homologous ZIPs responded differently from *Arabidopsis* to the Cd stress. Unlike in *Arabidopsis*, Cd stress increased the expression of most rice ZIPs in shoot but not root. These results imply that all these Cd-induced ZIPs involve in plant response to Cd. Except for *OsIRT2*, all ZIPs were significantly induced in rice shoot (Fig. 2b). Like *AtIRT3*, expression changes of *OsIRT1* and *OsZIP1* were converse in shoot and root (Fig. 2b). The positive role of *OsIRT1* and *OsZIP1* was demonstrated in the response of yeast and/or plant to Cd stress (Nakanishi et al. 2006; Lee and An 2009; Ramesh et al. 2003). Rice over-expressing *OsIRT1* showed reduced plant height and increased Cd accumulation under 300 μM Cd stress (Lee and An 2009), and the growth of *OsZIP1*-expressing yeast cells was inhibited by 10 μM Cd stress. In this study, the expression regulation of *OsIRT1* and *OsZIP1* in response to Cd stress was contrary between root and shoot.

Table 1 Locations and known functions of ZIP proteins in *Arabidopsis* and rice

Gene name	Locus	Complementation of yeast metal uptake mutants (y/n)	Validated location in plant	Experimental evidence for potential function in Cd uptake	References
AtIRT1	At4g19690	Δ*fet3*Δ*fet4* (Y); Δ*zrt1*Δ*zrt2* (Y); Δ*ctr1* (N); Δ*smf1* (Y)	Early endosome, vacuole, trans-Golgi network and cell membrane; root epidermis, flower	Increased Cd sensitivity of overexpression plant/yeast Reduced Cd sensitivity of *irt1* Inhibited IRT1-dependent Fe/Mn/Zn uptake by Cd in yeast Reduced Cd uptake of *irt1*	Eide et al. (1996), Vert et al. (2002, 2001), Korshunova et al. (1999), Rogers et al. (2000), Connolly et al. (2002), Varotto et al. (2002), Henriques et al. (2002), Nishida et al. (2011), Shin et al. (2013), Potocki et al. (2013), Fan et al. (2014), Barberon et al. (2014), Blum et al. (2014)
AtIRT2	At4g19680	Δ*fet3*Δ*fet4* (Y); Δ*zrt1*Δ*zrt2* (Y); Δ*smf1* (N)	Intracellular vesicles; root epidermis	No altered Cd sensitivity of overexpression yeast Increased Cd uptake and *IRT1* expression of overexpression plant	Vert et al. (2001, 2009), Wintz et al. (2003), Varotto et al. (2002)
AtIRT3	At1g60960	Δ*Spzrt1* (Y); Δ*zrt1*Δ*zrt2* (Y); Δ*fet3*Δ*fet4* (Y); Δ*smf1* (N)	Cell membrane; broadly expressed	No altered Cd sensitivity of overexpression yeast	Lin et al. (2009), Talke et al. (2006), Shanmugam et al. (2011), Hammes et al. (2005)
AtZIP1	At3g12750	Δ*zrt1*Δ*zrt2* (Y); Δ*fet3*Δ*fet4* (N); Δ*ctr1*Δ*ctr3* (N); Δ*smf1* (Y)	Vacuolar; predominantly root stele and leaf vasculature	Inhibited ZIP1-dependent Zn uptake by Cd in yeast, to a less extent	Grotz et al. (1998), Milner et al. (2013)
AtZIP2	At5g59520	Δ*zrt1*Δ*zrt2* (Y); Δ*fet3*Δ*fet4* (N); Δ*ctr1*Δ*ctr3* (N); Δ*smf1* (Y); Δ*ctr1* (Y)	Cell membrane; predominantly mature root stele	Inhibited ZIP2-dependent Zn uptake by Cd in yeast	Grotz et al. (1998), Milner et al. (2013), Wintz et al. (2003)
AtZIP3	At2g32270	Δ*zrt1*Δ*zrt2* (Y); Δ*fet3*Δ*fet4* (N); Δ*ctr1*Δ*ctr3* (N); Δ*smf1* (N)	Predominantly root	Inhibited ZIP3-dependent Zn uptake by Cd in yeast, to a less extent	Grotz et al. (1998), Talke et al. (2006), Milner et al. (2013)
AtZIP4	At1g10970	Δ*zrt1*Δ*zrt2* (N); Δ*fet3*Δ*fet4* (N); Δ*ctr1* (Y)	Root and leaf	N/A	Grotz et al. (1998), Talke et al. (2006), Wintz et al. (2003)
AtZIP5	At1g05300	Δ*zrt1*Δ*zrt2* (N); Δ*fet3*Δ*fet4* (N); Δ*ctr1*Δ*ctr3* (N); Δ*smf1* (Y)	Root	N/A	Milner et al. (2013), Wintz et al. (2003)
AtZIP6	At2g30080	Δ*zrt1*Δ*zrt2* (N); Δ*fet3*Δ*fet4* (N); Δ*ctr1*Δ*ctr3* (N); Δ*smf1* (Y)	Root	N/A	Milner et al. (2013), Hammes et al. (2005)
AtZIP7	At2g04032	Δ*zrt1*Δ*zrt2* (Y); Δ*fet3*Δ*fet4* (Y); Δ*ctr1*Δ*ctr3* (N); Δ*smf1* (Y)	N/A	N/A	Milner et al. (2013)
AtZIP8	At5g45105	Δ*zrt1*Δ*zrt2* (N); Δ*fet3*Δ*fet4* (N); Δ*ctr1*Δ*ctr3* (N); Δ*smf1* (N)	N/A	N/A	Milner et al. (2013)
AtZIP9	At4g33020	Δ*zrt1*Δ*zrt2* (N); Δ*fet3*Δ*fet4* (N); Δ*ctr1*Δ*ctr3* (N); Δ*smf1* (Y)	Root and shoot	N/A	Talke et al. (2006), Milner et al. (2013), Wintz et al. (2003), Inaba et al. (2015)
AtZIP10	At1g31260	Δ*zrt1*Δ*zrt2* (Y); Δ*fet3*Δ*fet4* (N); Δ*ctr1*Δ*ctr3* (N); Δ*smf1* (N)	N/A	N/A	Milner et al. (2013)
AtZIP11	At1g55910	Δ*zrt1*Δ*zrt2* (Y); Δ*fet3*Δ*fet4* (N); Δ*ctr1*Δ*ctr3* (N); Δ*smf1* (N)	N/A	N/A	Milner et al. (2013)
AtZIP12	At5g62160	Δ*zrt1*Δ*zrt2* (Y); Δ*fet3*Δ*fet4* (N); Δ*ctr1*Δ*ctr3* (N); Δ*smf1* (N)	Root	N/A	Milner et al. (2013), Inaba et al. (2015)

Table 1 (continued)

Gene name	Locus	Complementation of yeast metal uptake mutants (y/n)	Validated location in plant	Experimental evidence for potential function in Cd uptake	References
OsIRT1	LOC_Os03g46470	Δ*fet3*Δ*fet4* (Y); Δ*ctr1* (N); Δ*zrt1*Δ*zrt2* (N); Δ*smf1* (N); Δ*frt1*Δ*fet4*Δ*fre1*(Y); Δ*frt1*Δ*fet1*Δ*fre3*(Y)	Cell membrane; mainly in root epidermis (the inner layer of the cortex, and the stele) and stems (companion cells)	Increased Cd sensitivity of overexpression plant Increased Cd sensitivity and Cd uptake of overexpression yeast	Ishimaru et al. (2006), Bughio et al. (2002), Lee and An (2009), Nakanishi et al. (2006), Ishimaru et al. (2005)
OsIRT2	LOC_Os03g46454	Δ*fet3*Δ*fet4* (Y); Δ*ctr1* (N); Δ*zrt1*Δ*zrt2* (N); Δ*smf1* (N)	Cell membrane	Increased Cd sensitivity and Cd uptake of overexpression yeast	Ishimaru et al. (2006), Nakanishi et al. (2006)
OsZIP1	LOC_Os01g74110	Δ*zrt1*Δ*zrt2* (Y); Δ*smf1* (Y); Δ*fet3*Δ*fet4* (N)	Broadly expressed	Increased Cd sensitivity of overexpression yeast Inhibited ZIP1-dependent Zn uptake by Cd in yeast	Ramesh et al. (2003, Ishimaru et al. (2005), Chen et al. (2008)
OsZIP2	LOC_Os03g29850	Δ*zrt1*Δ*zrt2* (N);	N/A	No altered Cd sensitivity of overexpression yeast	Ramesh et al. (2003)
OsZIP3	LOC_Os04g52310	Δ*zrt1*Δ*zrt2* (Y); Δ*smf1* (Y); Δ*fet3*Δ*fet4* (N)	Mainly induced by zinc deficiency to higher levels in roots	No altered Cd sensitivity of overexpression yeast Mildly increased ZIP3-dependent Zn uptake by Cd in yeast	Ramesh et al. (2003, Ishimaru et al. (2005), Chen et al. (2008)
OsZIP4	LOC_Os08g10630	Δ*zrt1*Δ*zrt2* (Y); Δ*frt1*Δ*fet1*Δ*fre3*(N)	Cell membrane; phloem cells of leaves, roots and meristem	N/A	Ishimaru et al. (2005), Chen et al. (2008)
OsZIP5	LOC_Os05g39560	Δ*zrt1*Δ*zrt2* (Y)	Cell membrane; mainly panicle	N/A	Chen et al. (2008), Lee et al. (2010a)
OsZIP6	LOC_Os05g07210	N/A	Root, shoot and panicle	Increased Cd uptake of overexpression cells	Chen et al. (2008)
OsZIP7	LOC_Os05g10940	N/A	Root, shoot and panicle	N/A	Chen et al. (2008), Yang et al. (2009)
OsZIP8	LOC_Os07g12890	N/A	Cell membrane; mainly root and panicle	N/A	Chen et al. (2008), Lee et al. (2010b), Yang et al. (2009)
OsZIP9	LOC_Os05g39540	N/A	Root, shoot and panicle	N/A	Chen et al. (2008)
OsZIP10	LOC_Os06g37010	N/A	N/A	N/A	N/A

N/A represents not available

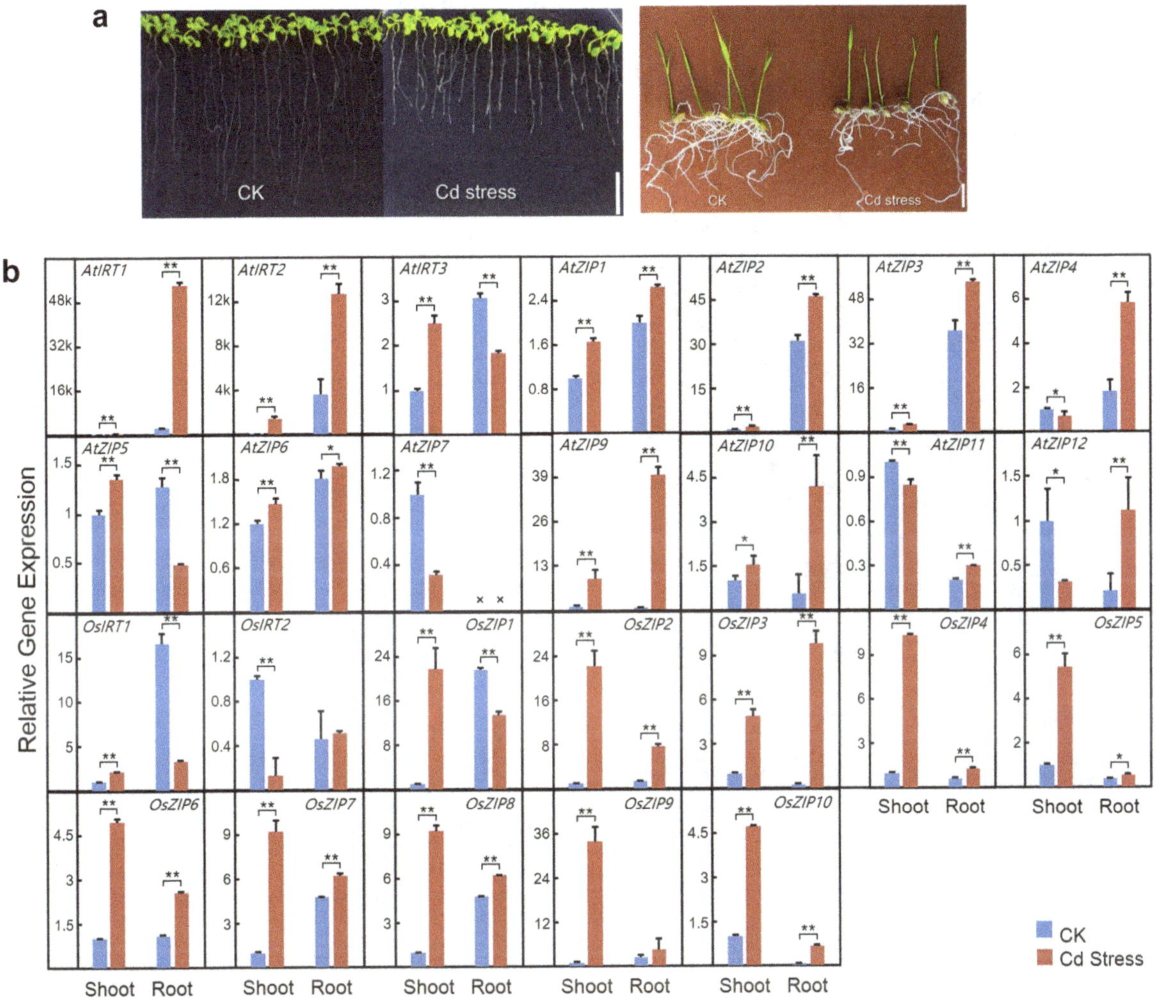

Fig. 2 Expression profiles of ZIP genes of *Arabidopsis* and rice in response to Cd stress. **a** 1-week-old and 10-day-old seedlings of *Arabidopsis* and rice were treated with 300 μM $CdCl_2$ for 3 days. Scale bars indicate 1 cm. **b** Changes in the expression of the 26 ZIP genes in response to Cd exposure Gene expression level was normalized using shoot expression level of each gene in the controls as the calibrator. (n = 3, Student *t* test, *indicates P < 0.05, **indicates P < 0.01)

Rice might have a feedback regulation of *OsIRT1* and *OsZIP1* in root to prevent increasing Cd uptake from soil.

Os*ZIP1-10* were subject to Cd sensitivity and Cd accumulation tests. The expression of *OsZIP1* and *OsZIP3* in yeast caused an increased Cd sensitivity and Cd accumulation (Fig. 3 and Additional file 1: Figure S4), suggesting their potential roles in Cd uptake. This result is different from those by Ramesh et al. (2003), where yeast ZHY3 strains were used and different culture medium was applied. It was also noticed that *OsZIP6* did not caused an obvious increasing in Cd sensitivity (Fig. 3). This is not consistence with previous report, in which *Xenopus laevis* oocytes was used to test the Cd sensitivity (Kavitha et al. 2015). Different host and micro-environment may cause the altered conformation and activity of tested proteins. Expression of *OsZIP5-10* failed to alter Cd sensitivity and Cd accumulation of host cells obviously, implying that these ZIPs probably did not uptake Cd individually. Considering that *AtIRT2* involves in indirect Cd uptake in *Arabidopsis*, these Cd-induced ZIPs may also play roles in Cd uptake or transport indirectly. Their potential roles under Cd stress need further investigation using transgenic plants.

Indeed, this study showed that many ZIPs were significantly induced by Cd stress even the growth of seedling was inhibited obviously, and some of them increased hosts' Cd sensitivity or Cd accumulation. These results will help to elucidate the genetic basis

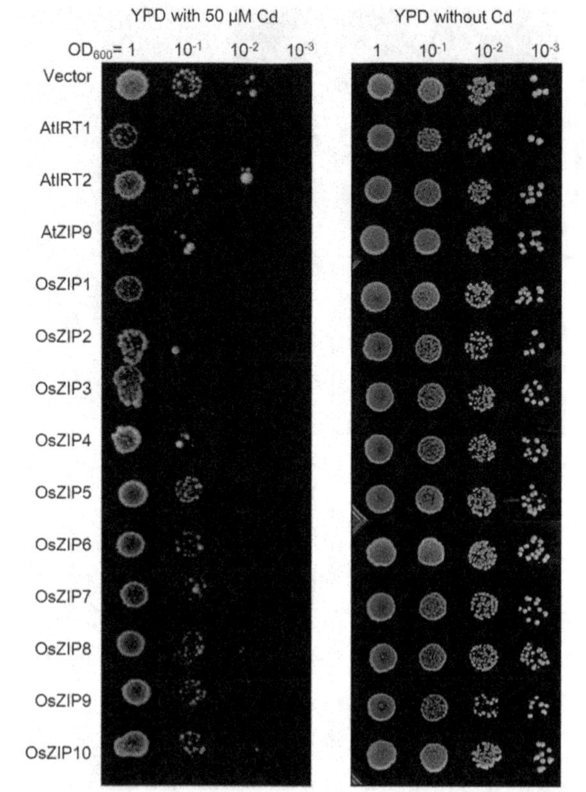

Fig. 3 Drop assay for Cd sensitivity of yeast cells (*S. cerevisiae* AH109) expressing representative *ZIPs* tested in this study. The transformed cells expression *ZIPs* were subjected to a serial dilution (0–10^{-4}) drop assay on YPD plates. 300 μg/mL G418 was added to maintain the vectors. Plates containing 50 μM $CdCl_2$ were incubated at 28 °C for 5 days and growth state was subsequently observed. This experiment was performed three times

for Cd accumulation via a ZIP-dependent pathway in plants. Further analysis using transgenic plants will clarify the biological function of these ZIPs in plant Cd uptake and transport.

Conclusions

In conclusion, this study revealed a distinct pattern in ZIPs genes expression regulation in response to Cd stress between *Arabidopsis* and rice. *Arabidopsis* mainly up-regulated root ZIPs genes, while rice mainly up-regulated shoot ZIPs genes. Interestingly, some genes like *AtIRT3, AtZIP5, AtZIP12, OsIRT1* and *OsZIP1* showed contrary expression regulation when subject to the tested Cd stress. Three genes, *AtIRT1, OsZIP1* and *OsZIP3,* conferred an increased sensitivity to Cd stress and more Cd accumulation when expressed in yeast cells, implying a role in direct Cd uptake in plants.

Additional file

Additional file 1: Table S1. The qPCR primers used in this study. **Table S2.** Primers used in plasmid construction. **Figure S1.** Genome locations of 27 ZIP genes in *Arabidopsis* (A) and rice (B). Information were acquired in the PLAZA database and plotted using Photoshop CS6. **Figure S2.** Evolutionary relationships of ZIP family genes and their structures. The Neighbor-Joining tree was produced using MEGA7 with 1,000 bootstrap replicates, and the gene structures was predicted using Gene Structure Display Server. Dark blue boxes indicate exons; black lines indicate introns; light blue boxes indicate untranslated regions. **Figure S3.** Effect of Cd stress on root length (A and B) and dry weight (C and D) of *Arabidopsis* and rice. (for root length, n=20; for dry weight, n=3. Student *t* test, * indicates P<0.05). **Figure S4.** Effect of *ZIPs* on Cd accumulation. Cells expressing *ZIPs* were incubated using liquid YPD medium plus 300 μg/mL G418 and 50 μM Cd for 12 h, after which the Cd concentration of each strain was measured by an atomic absorption spectrometer method. Cells harboring empty pCEV-G1-Km (Vector) was used as a negative control. (n=3, student *t* test, * P < 0.05).

Abbreviations

ZIPs: zinc(Zn)-regulated/iron(Fe)-regulated transporter-like family proteins; Cd: cadmium; TM: transmembrane; Zn: zinc; Fe: iron; Mn: manganese; Cu: copper.

Authors' contributions

XZ and XL initiated the project. XZ designed the experiment. XZ and LC carried out the experiments and analyzed the data. All authors wrote and revised the manuscript. All authors read and approved the final mansucript.

Author details

[1] Key Laboratory for Agricultural Water Resources, Center for Agricultural Resources Research, Institute of Genetics and Developmental Biology, Chinese Academy of Sciences, Shijiazhuang 050021, Hebei, People's Republic of China. [2] CMLR, Sustainable Minerals Institute, The University of Queensland, Brisbane, QLD 4072, Australia.

Acknowledgements

Not applicable.

Competing interests

The authors declare that they have no competing interests.

Funding

This work was supported by the Pioneer "Hundred Talents Program" of the Chinese Academy of Sciences (Y726012203), the National Key Research and Development Plan (2018YFD0800306) and the Hebei Science Fund for Distinguished Young Scholars (D2018503005).

References

Barberon M, Dubeaux G, Kolb C, Isono E, Zelazny E, Vert G (2014) Polarization of iron-regulated transporter 1 (IRT1) to the plant-soil interface plays crucial role in metal homeostasis. Proc Natl Acad Sci USA 111(22):8293–8298. https://doi.org/10.1073/pnas.1402262111

Blum A, Brumbarova T, Bauer P, Ivanov R (2014) Hormone influence on the spatial regulation of IRT1 expression in iron-deficient *Arabidopsis thaliana* roots. Plant Signal Behav 9(4):e28787

Bughio N, Yamaguchi H, Nishizawa NK, Nakanishi H, Mori S (2002) Cloning an iron-regulated metal transporter from rice. J Exp Bot

53(374):1677–1682

Chen WR, Feng Y, Chao YE (2008) Genomic analysis and expression pattern of OsZIP1, OsZIP3, and OsZIP4 in two rice (*Oryza sativa* L.) genotypes with different zinc efficiency. Russ J Plant Physl+ 55(3):400–409. https://doi.org/10.1134/s1021443708030175

Connolly EL, Fett JP, Guerinot ML (2002) Expression of the IRT1 metal transporter is controlled by metals at the levels of transcript and protein accumulation. Plant Cell 14(6):1347–1357

Eide D, Broderius M, Fett J, Guerinot ML (1996) A novel iron-regulated metal transporter from plants identified by functional expression in yeast. Proc Natl Acad Sci USA 93(11):5624–5628

Eng BH, Guerinot ML, Eide D, Saier MH Jr (1998) Sequence analyses and phylogenetic characterization of the ZIP family of metal ion transport proteins. J Membr Biol 166(1):1–7

Fan SK, Fang XZ, Guan MY, Ye YQ, Lin XY, Du ST, Jin CW (2014) Exogenous abscisic acid application decreases cadmium accumulation in *Arabidopsis* plants, which is associated with the inhibition of IRT1-mediated cadmium uptake. Front Plant Sci 5:721. https://doi.org/10.3389/fpls.2014.00721

Grotz N, Fox T, Connolly E, Park W, Guerinot ML, Eide D (1998) Identification of a family of zinc transporter genes from Arabidopsis that respond to zinc deficiency. Proc Natl Acad Sci USA 95(12):7220–7224

Guerinot ML (2000) The ZIP family of metal transporters. Biochim Biophys Acta 1465(1–2):190–198

Hammes UZ, Schachtman DP, Berg RH, Nielsen E, Koch W, McIntyre LM, Taylor CG (2005) Nematode-induced changes of transporter gene expression in *Arabidopsis* roots. Mol Plant Microbe Interact 18(12):1247–1257. https://doi.org/10.1094/MPMI-18-1247

Henriques R, Jasik J, Klein M, Martinoia E, Feller U, Schell J, Pais MS, Koncz C (2002) Knock-out of Arabidopsis metal transporter gene IRT1 results in iron deficiency accompanied by cell differentiation defects. Plant Mol Biol 50(4–5):587–597

Inaba S, Kurata R, Kobayashi M, Yamagishi Y, Mori I, Ogata Y, Fukao Y (2015) Identification of putative target genes of bZIP19, a transcription factor essential for *Arabidopsis* adaptation to Zn deficiency in roots. Plant J 84(2):323–334. https://doi.org/10.1111/tpj.12996

Ishimaru Y, Suzuki M, Kobayashi T, Takahashi M, Nakanishi H, Mori S, Nishizawa NK (2005) OsZIP4, a novel zinc-regulated zinc transporter in rice. J Exp Bot 56(422):3207–3214. https://doi.org/10.1093/jxb/eri317

Ishimaru Y, Suzuki M, Tsukamoto T, Suzuki K, Nakazono M, Kobayashi T, Wada Y, Watanabe S, Matsuhashi S, Takahashi M, Nakanishi H, Mori S, Nishizawa NK (2006) Rice plants take up iron as an Fe^{3+}-phytosiderophore and as Fe^{2+}. Plant J 45(3):335–346. https://doi.org/10.1111/j.1365-313X.2005.02624.x

Ivanov R, Bauer P (2017) Sequence and coexpression analysis of iron-regulated ZIP transporter genes reveals crossing points between iron acquisition strategies in green algae and land plants. Plant Soil 418(1–2):61–73. https://doi.org/10.1007/s11104-016-3128-2

Kavitha PG, Kuruvilla S, Mathew MK (2015) Functional characterization of a transition metal ion transporter, OsZIP6 from rice (*Oryza sativa* L.). Plant Physiol Biochem 97:165–174. https://doi.org/10.1016/j.plaphy.2015.10.005

Korshunova YO, Eide D, Clark WG, Guerinot ML, Pakrasi HB (1999) The IRT1 protein from *Arabidopsis thaliana* is a metal transporter with a broad substrate range. Plant Mol Biol 40(1):37–44

Kumar S, Stecher G, Tamura K (2016) MEGA7: molecular evolutionary genetics analysis version 7.0 for bigger datasets. Mol Biol Evol 33(7):1870–1874. https://doi.org/10.1093/molbev/msw054

Lee S, An G (2009) Over-expression of OsIRT1 leads to increased iron and zinc accumulations in rice. Plant Cell Environ 32(4):408–416. https://doi.org/10.1111/j.1365-3040.2009.01935.x

Lee S, Jeong HJ, Kim SA, Lee J, Guerinot ML, An G (2010a) OsZIP5 is a plasma membrane zinc transporter in rice. Plant Mol Biol 73(4–5):507–517. https://doi.org/10.1007/s11103-010-9637-0

Lee S, Kim SA, Lee J, Guerinot ML, An G (2010b) Zinc deficiency-inducible OsZIP8 encodes a plasma membrane-localized zinc transporter in rice. Mol Cells 29(6):551–558. https://doi.org/10.1007/s10059-010-0069-0

Li S, Zhou X, Huang Y, Zhu L, Zhang S, Zhao Y, Guo J, Chen J, Chen R (2013) Identification and characterization of the zinc-regulated transporters, iron-regulated transporter-like protein (ZIP) gene family in maize. BMC Plant Biol 13:114. https://doi.org/10.1186/1471-2229-13-114

Lin YF, Liang HM, Yang SY, Boch A, Clemens S, Chen CC, Wu JF, Huang JL, Yeh KC (2009) *Arabidopsis* IRT3 is a zinc-regulated and plasma membrane localized zinc/iron transporter. New Phytol 182(2):392–404. https://doi.org/10.1111/j.1469-8137.2009.02766.x

Livak KJ, Schmittgen TD (2001) Analysis of relative gene expression data using real-time quantitative PCR and the 2(-Delta Delta C(T)) Method. Methods 25(4):402–408. https://doi.org/10.1006/meth.2001.1262

Lopez-Millan AF, Ellis DR, Grusak MA (2004) Identification and characterization of several new members of the ZIP family of metal ion transporters in *Medicago truncatula*. Plant Mol Biol 54(4):583–596. https://doi.org/10.1023/B:PLAN.0000038271.96019.aa

Milner MJ, Seamon J, Craft E, Kochian LV (2013) Transport properties of members of the ZIP family in plants and their role in Zn and Mn homeostasis. J Exp Bot 64(1):369–381. https://doi.org/10.1093/jxb/ers315

Nakanishi H, Ogawa I, Ishimaru Y, Mori S, Nishizawa NK (2006) Iron deficiency enhances cadmium uptake and translocation mediated by the Fe^{2+} transporters OsIRT1 and OsIRT2 in rice. Soil Sci Plant Nutr 52(4):464–469. https://doi.org/10.1111/j.1747-0765.2006.00055.x

Nishida S, Tsuzuki C, Kato A, Aisu A, Yoshida J, Mizuno T (2011) AtIRT1, the primary iron uptake transporter in the root, mediates excess nickel accumulation in *Arabidopsis thaliana*. Plant Cell Physiol 52(8):1433–1442. https://doi.org/10.1093/pcp/pcr089

Potocki S, Valensin D, Camponeschi F, Kozlowski H (2013) The extracellular loop of IRT1 ZIP protein–the chosen one for zinc? J Inorg Biochem 127:246–252. https://doi.org/10.1016/j.jinorgbio.2013.05.003

Rafiq MT, Aziz R, Yang XE, Xiao WD, Rafiq MK, Ali B, Li TQ (2014) Cadmium phytoavailability to rice (*Oryza sativa* L.) grown in representative Chinese soils. A model to improve soil environmental quality guidelines for food safety. Ecotox Environ Safe 103:101–107. https://doi.org/10.1016/j.ecoenv.2013.10.016

Ramesh SA, Shin R, Eide DJ, Schachtman DP (2003) Differential metal selectivity and gene expression of two zinc transporters from rice. Plant Physiol 133(1):126–134

Rogers EE, Eide DJ, Guerinot ML (2000) Altered selectivity in an *Arabidopsis* metal transporter. Proc Natl Acad Sci USA 97(22):12356–12360. https://doi.org/10.1073/pnas.210214197

Shanmugam V, Lo JC, Wu CL, Wang SL, Lai CC, Connolly EL, Huang JL, Yeh KC (2011) Differential expression and regulation of iron-regulated metal transporters in *Arabidopsis halleri* and *Arabidopsis thaliana*–the role in zinc tolerance. New Phytol 190(1):125–137. https://doi.org/10.1111/j.1469-8137.2010.03606.x

Shin LJ, Lo JC, Chen GH, Callis J, Fu H, Yeh KC (2013) IRT1 degradation factor1, a ring E3 ubiquitin ligase, regulates the degradation of iron-regulated transporter1 in *Arabidopsis*. Plant Cell 25(8):3039–3051. https://doi.org/10.1105/tpc.113.115212

Stephens BW, Cook DR, Grusak MA (2011) Characterization of zinc transport by divalent metal transporters of the ZIP family from the model legume *Medicago truncatula*. Biometals 24(1):51–58. https://doi.org/10.1007/s10534-010-9373-6

Talke IN, Hanikenne M, Kramer U (2006) Zinc-dependent global transcriptional control, transcriptional deregulation, and higher gene copy number for genes in metal homeostasis of the hyperaccumulator *Arabidopsis halleri*. Plant Physiol 142(1):148–167. https://doi.org/10.1104/pp.105.076232

Tiong J, McDonald G, Genc Y, Shirley N, Langridge P, Huang CY (2015) Increased expression of six ZIP family genes by zinc (Zn) deficiency is associated with enhanced uptake and root-to-shoot translocation of Zn in barley (*Hordeum vulgare*). New Phytol 207(4):1097–1109. https://doi.org/10.1111/nph.13413

Varotto C, Maiwald D, Pesaresi P, Jahns P, Salamini F, Leister D (2002) The metal ion transporter IRT1 is necessary for iron homeostasis and efficient photosynthesis in *Arabidopsis thaliana*. Plant J 31(5):589–599

Vert G, Briat JF, Curie C (2001) *Arabidopsis* IRT2 gene encodes a root-periphery iron transporter. Plant J 26(2):181–189

Vert G, Grotz N, Dedaldechamp F, Gaymard F, Guerinot ML, Briat JF, Curie C (2002) IRT1, an *Arabidopsis* transporter essential for iron uptake from the soil and for plant growth. Plant Cell 14(6):1223–1233

Vert G, Barberon M, Zelazny E, Seguela M, Briat JF, Curie C (2009) *Arabidopsis* IRT2 cooperates with the high-affinity iron uptake system to maintain iron homeostasis in root epidermal cells. Planta 229(6):1171–1179. https://doi.org/10.1007/s00425-009-0904-8

Wintz H, Fox T, Wu YY, Feng V, Chen W, Chang HS, Zhu T, Vulpe C (2003) Expression profiles of Arabidopsis thaliana in mineral deficiencies reveal novel transporters involved in metal homeostasis. J Biol Chem 278(48):47644–47653. https://doi.org/10.1074/jbc.M309338200

Yang X, Huang J, Jiang Y, Zhang HS (2009) Cloning and functional identification of two members of the ZIP (Zrt, Irt-like protein) gene family in rice (*Oryza sativa* L.). Mol Biol Rep 36(2):281–287. https://doi.org/10.1007/s11033-007-9177-0

13

Comparative RNA-Seq analysis on the regulation of cucumber sex differentiation under different ratios of blue and red light

Jiali Song, Yiting Zhang, Shiwei Song, Wei Su, Riyuan Chen, Guangwen Sun, Yanwei Hao* and Houcheng Liu*

Abstract

Cucumber (*Cucumis sativus* L.) is a typical monoecism vegetable with individual male and female flowers, which has been used as a plant model for sex determination. It is well known that light is one of the most important environmental stimuli, which control the timing of the transition from vegetative growth to reproductive development. However, whether light controls sex determination remains elusive. To unravel this problem, we performed high-throughput RNA-Seq analyses, which compared the transcriptomes of shoot apices between R2B1(Red light:Blue light = 2:1)-treated and R4B1(Red light:Blue light = 4:1)-treated cucumber seedlings. Results showed that the higher proportion of blue light in the R2B1 treatment significantly induced the formation of female flowers and accelerated female flowering time in this whole study. The genes related to flowering time, such as *flowering locus T* (*FT*) and *SUPPRESSOR OF OVEREXPRESSION OF CO1* (*SOC1*), were up-regulated after R2B1 treatment. Furthermore, the transcriptome analysis showed that up-regulation and down-regulation of specific DEGs (the differentially expressed genes) were primarily the result of plant hormone signal transduction after treatments. The specific DEGs related with auxin formed the highest percentage of DEGs in the plant hormone signal transduction. In addition, the expression levels of transcription factors also changed after R2B1 treatment. Thus, sex differentiation affected by light quality might be induced by plant hormone signal transduction and transcription factors. These results provide a theoretical basis for further investigation of the regulatory mechanism of female flower formation under different light qualities in cucumber seedlings.

Keywords: RNA-seq, Auxin, Cucumber, Flower, Sex differentiation

Background

Sex differentiation of flower buds is an important developmental process, which directly affects the product yield in plants. Cucumber (*Cucumis sativus* L.) is a typical monoecious plant with distinct male and female flowers. It has been served as a model system for studying physiological and molecular aspects of sex determination and differentiation in plants (Bai and Xu 2013). In the young floral buds of cucumber, both the stamen primordia and carpel primordia are initiated, with sex determination occurring just after the bisexual stage. Subsequently, male or female flowers are formed and become enlarged due to the selective arrestment of carpel or stamen development, respectively (Bai et al. 2004).

Sex differentiation of cucumber can be affected by phytohormones, such as ethylene and gibberellin. Exogenous ethylene treatment induced female flower formation in the cucumber. The ethylene content in gynoecious cucumbers is found to be higher than that of monoecious plants (Trebitsh et al. 1987; Rudich et al. 1972). Ethylene synthesis genes play an important role in the sex differentiation of the cucumber flowers, such as 1-aminocyclopropane-1-carboxylic acid oxidases (*CsACO2*).

*Correspondence: yanweihao@scau.edu.cn; yanweihaogbf@gmail.com; liuhch@scau.edu.cn
College of Horticulture, South China Agricultural University, Guangzhou 510642, China

Organ-specific overexpression of *CsACO2* driven by the organ-specific promoter P (AP3) significantly affected stamen but not carpel development, acting similarly to natural floral development of female cucumber flowers (Duan et al. 2008). A conserved-residue conversion in *CsACS2* induced the formation of bisexual flowers in the cucumber (Li et al. 2009). Furthermore, ethylene signaling pathway also affects the sex differentiation. *CsETR1*, an ethylene receptor involved in ethylene signaling transduction, had been demonstrated to play a key role in stamen arrest in female cucumber flowers (Wang et al. 2010). Ethylene-responsive gene associated with the formation of female flowers (*ERAF17*), a MADS-box gene, could be induced by ethylene and might be involved in formation of female flowers in cucumbers (Ando et al. 2001). Exogenous gibberellic acid (GA_3) application promoted the formation of male flowers in gynoecious plants (Pike and Peterson 1969). GA production in andromonoecious cucumbers was higher than that in gynoecious and monoecious plants (Junior et al. 1972). The GA signaling pathway is involved in stamen development in the cucumber (Zhang et al. 2014b; Fei et al. 2004). Cucumber DELLA Homolog (*CsGAIP*) is predominantly expressed in the male specific organs during cucumber flower development and belongs to the DELLA (the negative regulators of the GA action) family. *CsGAIP* inhibited stamen development through transcriptional repression of B-class floral homeotic genes *APETALA3* (*AP3*) and *PISTILLATA* (*PI*) in Arabidopsis (Zhang et al. 2014a). *CsGAMYB1*, a positive regulator involved in the GA signaling pathway, also mediates the sex expression of cucumbers. Knocking out the *CsGAMYB1* gene in cucumbers resulted in decreased ratios of nodes with male to female flowers (Zhang et al. 2014b). The relationship between ethylene and gibberellin in mediating sex differentiation was interpreted recently, with this study finding that gibberellin mediates sex differentiation via ethylene-dependent and ethylene-independent pathways in cucumbers (Zhang et al. 2017). Apart from ethylene and gibberellin, other hormones are also involved in the flower sex differentiation of plant (Song et al. 2013). This includes the abscisic acid (ABA) induction of male flower formation (Zhu et al. 2010); Jasmonic acid (JA) signaling pathway, which might participate in the abortion of male flowers (Acosta et al. 2009); and auxin (IAA) content, which was found to increase during female flower development and decrease during male flower development (Sakata et al. 2010). Ethylene biosynthesis has probably been induced by IAA to promote female flower formation as well (Trebitsh et al. 1987). The IAA induced *CsACS1* gene expression in cucumbers, which suggested that IAA promoted female flower formation through inducing ethylene synthesis (Trebitsh et al. 1997; Mibus and Tatlioglu 2004).

Environmental cues, such as temperature, photoperiod, nutrition, also affected sex determination in many species (Golenberg and West 2013; Korpelainen 1998). Light is one of the major external factors that influence plant growth and development (Chen et al. 2004). Plants respond to light through mediating light receptors. Until now, known photoreceptors were divided into three classes: the UV-B (280–320 nm light) photoreceptors; the red/far-red reversible photoreceptors, known as the phytochromes PhyA–PhyE (Mathews and Sharrock 1997) and blue UVA photoreceptors. These blue UVA photoreceptors have three classes, which are the cryptochromes (CRY1, CRY2 and CRY3) (Mathews and Sharrock 1997; Brudler et al. 2003), phototropins (PHOT1 and PHOT2) (Briggs et al. 2001) and aureochromes (AUREO1 and AUREO2) (Ishikawa et al. 2009; Takahashi et al. 2007). There is less evidence examining the effects of light on plant sex differentiation. In the gametophytes of the *Her1* mutant, blue light induced male development and red light suppressed male development (Kamachi et al. 2007). Our previous studies showed that light quality affected cucumber flower formation and sex differentiation. In our previous study, we used 4R1B (Red:Blue = 4:1), R2B1 (Red:Blue = 2:1), R6G2B1 (Red:Green:Blue = 6:2:1) ratios supplemental light to treat the cucumber seedlings. Comparing to natural light, R2B1 treatment was the best light quality ratio for improving the female flower formation, location and flowering time, but none of them has effects on male flower formation (our unpublished data). However, the mechanism of how light quality regulates flower sex differentiation still remains unclear. In our study, we aimed to understand the mechanism of light regulating cucumber sexual development. We performed RNA-Seq analyses to compare the transcriptomes of shoot apices between cucumber seedlings after R2B1 treatment and R4B1 treatment. Our results built a foundation for dissecting the molecular mechanism of flower sex differentiation in cucumbers.

Results

Effects of different LED light quality treatments on flower formation and flowering time

Comparing to natural light, R2B1 treatment was the best light quality ratio for improving the female flower formation, location and flowering time (Additional file 1: Table S1). After 20 days lighting under R2B1 and R4B1 light treatment, the formation of female flowers occurred mainly at the lower node positions under the R2B1 irradiation, as compared with R4B1 irradiation (Table 1). The number of female flowers in the 15 nodes of the cucumber plant increased markedly and flowered earlier (Table 1). All of these results implied that blue light is responsible for the high number of female flowers, early

Table 1 Effects of different light quality on the formation of female flowers and flowering time

Treatment	First female flower located node	Total female flower numbers in 15 nodes	Days from transplanting to first female flowering
R2B1 (Red light:blue light = 2:1)	7.20 ± 0.23b	2.70 ± 0.13a	21.85 ± 0.17b
R4B1 (Red light:blue light = 4:1)	8.03 ± 0.19a	2.32 ± 0.10b	22.53 ± 0.12a

a,b: Statistically significant variations of mean values at different sampling points (ANOVA, $p < 0.05$) were indicated with different letters. Means denoted by the same letter did not differ significantly at $p < 0.05$

Table 2 Statistical results of expressed genes in different libraries

library	Numbers of known genes detected	Numbers of detected new genes
R2B1-5	18,619 (80.09%)	603
R4B1-5	18,831 (81.00%)	614
R2B1-10	18,943 (81.48%)	609
R4B1-10	19,099 (82.15%)	618
R2B1-15	19,075 (82.05%)	611
R4B1-15	19,306 (83.04%)	618

flowering time and reduction in the first female flower node.

Analysis of differentially expressed genes (DEGs) under treatments with different ratios of blue and red light

Based on deep sequencing, 23,248 unigenes were collected in 12 libraries. Genes expressed at differential stages under two treatments were listed in Table 2. After R2B1 treatment, the detected number of genes in the 5 days, 10 days and 15 days library were 18,619, 18,943 and 19,075. The mapping ratios of these libraries according to cucumber genome were 80.09, 81.48 and 82.5%. The detected new genes in these libraries were 603, 609 and 611. The relevant data for R4B1 treatment were shown in Table 2. Using |log2 (fold change)| > 1 and false discovery rate (FDR) < 0.05 as the cut-offs for significance, we identified DEGs at differential stages under two treatments (Fig. 1a). The results showed that there were 470 DEGs including 11 up-regulated genes and 459 down-regulated genes in pairwise R4B1-5-vs-R2B1-5 (5 days library comparison after R4B1 treatment and R2B1 treatment). Furthermore, we also found that 2697 DEGs were differentially expressed, with 704 genes being up-regulated and 1993 genes down-regulated in pairwise R4B1-10-vs-R2B1-10 (10 days library comparison after R4B1 treatment and R2B1 treatment). Moreover, 3 DEGs containing 1 up-regulated gene and 2 down-regulated genes were found in pairwise R4B1-15-vs-R2B1-15 (15 days library comparison after R4B1 treatment and R2B1 treatment). Most DEGs were found in the 10 days samples, which might implicate that the light quality affected flower sex differentiation mostly on the 10th day.

In addition, there were more DEGs in the 5 days, 10 days and 15 days library after R2B1 irradiation compared to R4B1 irradiation. We used a Venn diagram to analyze all these up-regulated or down-regulated DEGs collected from all these treatments at different time points. The results showed that 423 DEGs were found in all three stages after both R4B1 and R2B1 treatment as well as more up-regulated DEGs (1795 DEGs) at all three stages under both treatments (Fig. 1b).

Gene ontology (GO) analysis of DEGs after different ratios of blue and red light treatments

Gene ontology (GO) mainly includes three categories: molecular function, cellular component and biological process. A total of 110 DEGs in pairwise R4B1-5-vs-R2B1-5 and 745 DEGs in pairwise R4B1-10-vs-R2B1-10 were divided into cell components, which were related to the cell membrane, cytoplasm and organelle (Figs. 2 and 3). In pairwise R4B1-5-vs-R2B1-5, a total of 207 DEGs were classified into molecular functions: catalytic activity, binding, transporter activity and oxidoreductase activity (Fig. 2). A total of 195 DEGs were termed into the biology pathway, which mainly included ion transport, primary metabolic process, localization establishment of localization, biological regulation and response to stimulus (Fig. 2). In pairwise R4B1-10-vs-R2B1-10, 1060 DEGs were termed into molecular functions, which were mainly related to purine nucleoside binding, carbohydrate derivative binding, heterocyclic compound binding, organic cyclic compound binding, catalytic activity, hydrolase activity and transferase activity (Fig. 3). In pairwise R4B1-10-vs-R2B1-10, a total of 1060 genes were classified into the biology pathway. These genes mainly participated in a cellular process, cellular metabolic process, single-organism process, organic substance metabolic process, metabolic process and protein metabolic process (Fig. 3). In pairwise R4B1-15-vs-R2B1-15, only 3 DEGs were found and cannot be classified into any category.

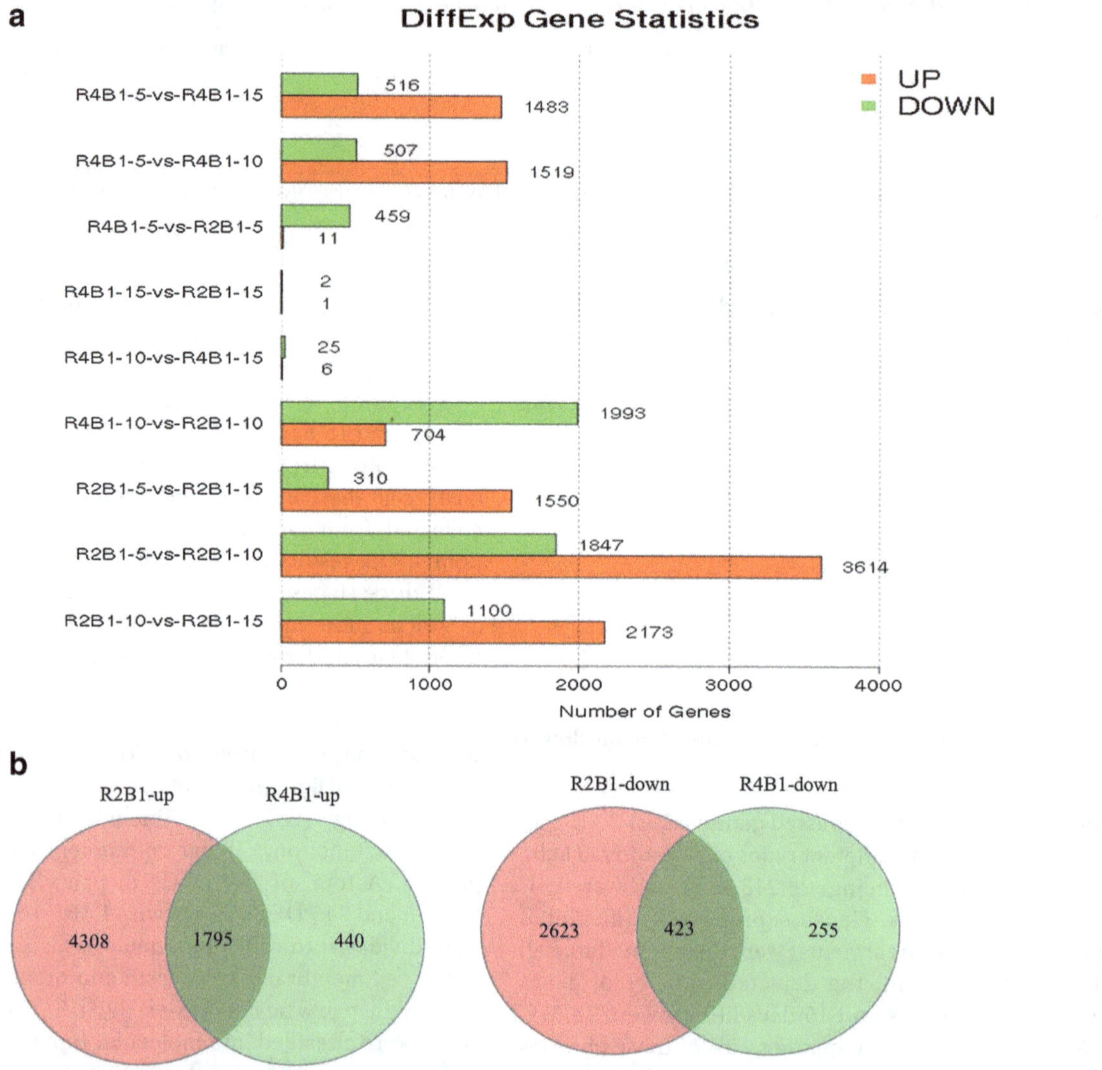

Fig. 1 **a** Statistical results of differentially expressed genes (DEGs) among different stages and treatments and **b** Venn diagrams of DEGs that were significantly (left) up-regulated and (right) down-regulated after R2B1 and R4B1 treatment

Kyoto encyclopedia of genes and genomes (KEGG) pathway analysis of DEGs after different ratios of blue and red light treatments

To further explore the biological pathways associated with sex differentiation development, DEGs in each comparison were mapped to the reference canonical pathways in the KEGG database. In pairwise R4B1-5-vs-R2B1-5, 108 DEGs were assigned into 52 KEGG pathways. They were significantly enriched in the following pathways: metabolic pathways, biosynthesis of secondary metabolites, starch and sucrose metabolism and plant hormone signal transduction (Fig. 4a). In pairwise R4B1-10-vs-R2B1-10, a total of 798 DEGs were assigned into 106 KEGG pathways. They were significantly enriched in the following pathways: metabolic pathways, biosynthesis of secondary metabolites and plant hormone signal transduction (Fig. 4b).

All specifically up-regulated or down-regulated DEGs after R2B1 and R4B1 treatment were also mapped to the reference canonical pathway in the KEGG database. The results showed that 543 specifically up-regulated DEGs from R2B1 were mostly enriched in the following KEGG pathways: plant hormone signal transduction, secondary protein processing in endoplasmic reticulum, RNA transport and spliceosome pathway (Table 3). A total of 382 specifically down-regulated DEGs from R2B1 were grouped into the following KEGG pathways: biosynthesis of amino acids, protein

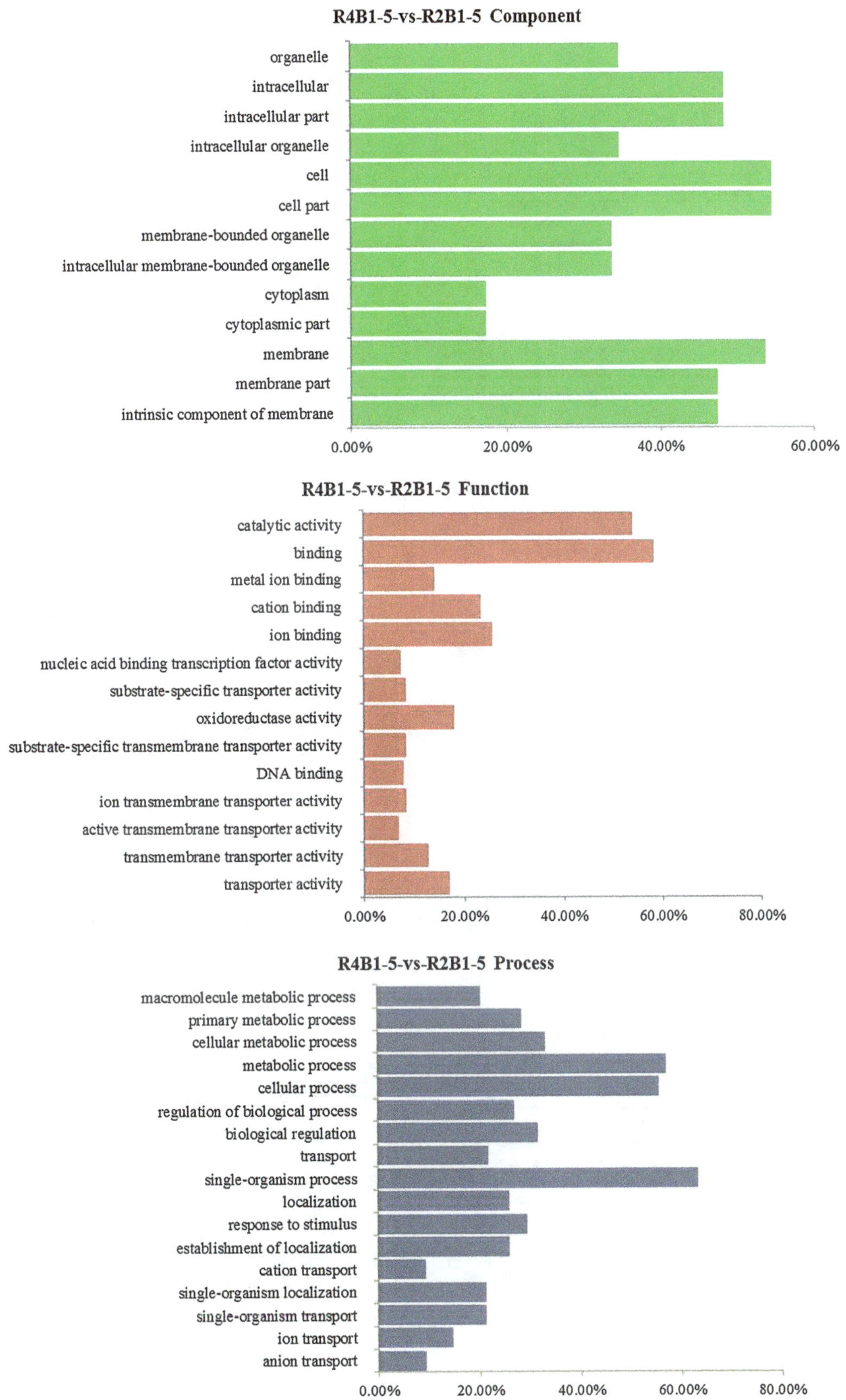

Fig. 2 Gene ontology (GO) categories of differently expressed genes (DEGs) in the pairwise R4B1-5-vs-R2B1-5

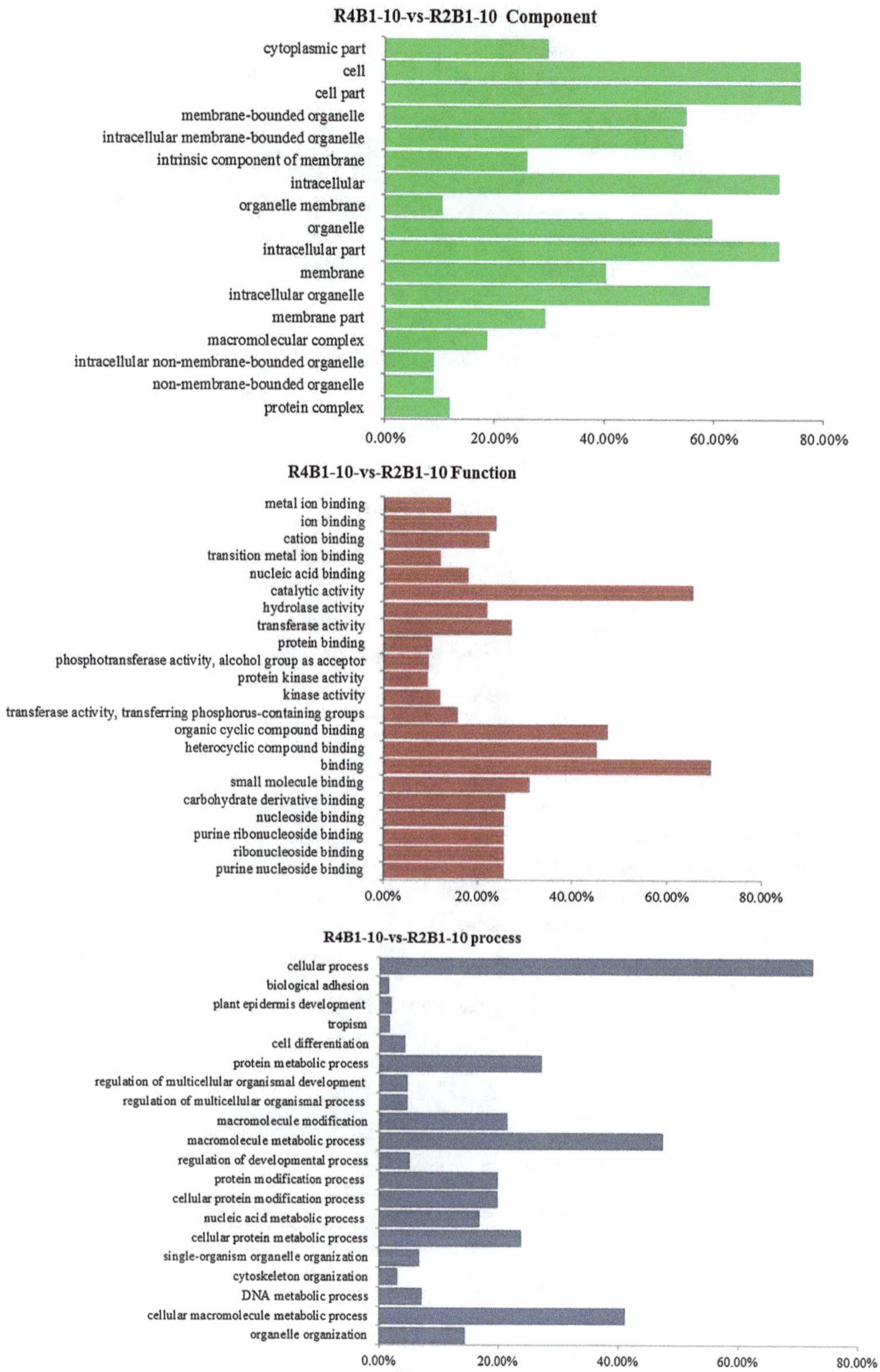

Fig. 3 Gene ontology (GO) categories of differently expressed genes (DEGs) in the pairwise R4B1-10-vs-R2B1-10

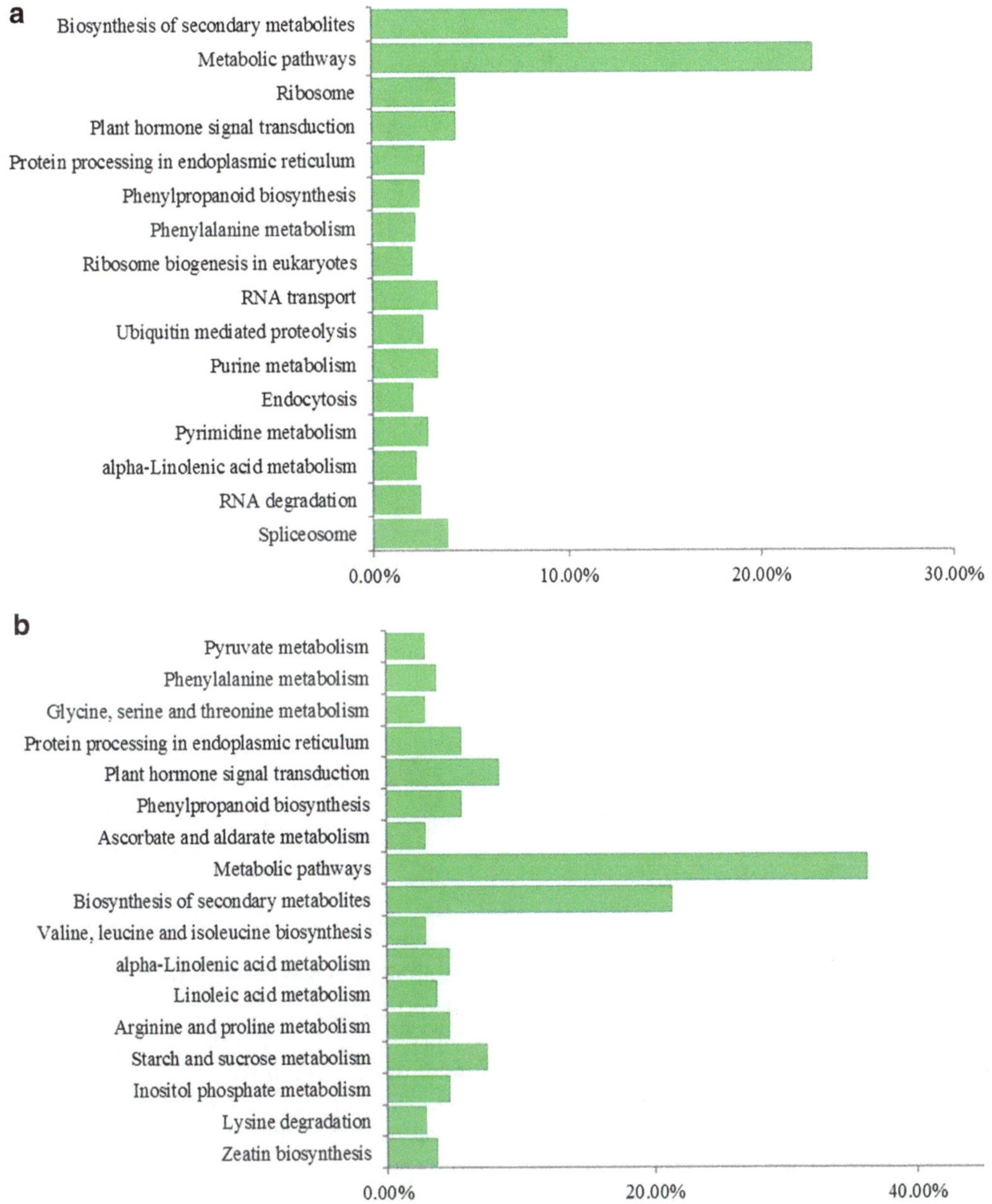

Fig. 4 Kyoto encyclopedia of genes and genomes (KEGG) pathway enrichment of differently expressed genes (DEGs) in **a** pairwise R4B1-5-vs-R2B1-5 and **b** pairwise R4B1-10-vs-R2B1-10

processing in endoplasmic reticulum and RNA transport pathway (Table 3). The relevant data for R4B1 treatment were listed in Table 3.

Auxin plays an important role in the regulation of cucumber sex differentiation under different ratios of blue and red light treatments

Since cucumber sex differentiation is closely related to all types of endogenous hormones, we further explored the role of endogenous hormones regulating the sex differentiation process under different light treatments. Thus, DEGs enriched in plant hormone signal transduction were further analyzed. In pairwise R4B1-5-vs-R2B1-5, 33.33% of the DEGs were related to auxin, 33.33% to cytokines, 22.22% to SA and 11.11% to ABA. In pairwise R4B1-10-vs-R2B1-10, 55.88% of the DEGs were related to auxin, 14.71% to ethylene, 11.76% to CTK, 8.82% to ABA, 5.88% to JA and 2.94% to SA (Fig. 5a, b). In addition, there were 49 up-regulated DEGs related to the plant hormone signaling pathway in R2B1, including

Table 3 Kyoto encyclopedia of genes and genomes (KEGG) pathway enrichment of significantly up/down-regulated DEGs in R2B1/R4B1

Pathway	DEGs with pathway annotation (543)	P-value	
Plant hormone signal transduction	49	0.000917	Up-regulated DEGs in R2B1
Protein processing in endoplasmic reticulum	40	0.000665	
RNA transport	33	0.012593	
Spliceosome	31	0.078185	
Endocytosis	25	0.010854	
Pyrimidine metabolism	24	0.029783	
RNA degradation	22	0.001054	
Ubiquitin mediated proteolysis	21	0.0133	
Starch and sucrose metabolism	6	0.024124	
Alanine, aspartate and glutamate metabolism	4	0.002149	Up-regulated DEGs in R4B1
Diterpenoid biosynthesis	3	0.005284	
Insulin resistance	3	0.010482	
Biosynthesis of amino acids	34	0.000934	Down-regulated DEGs in R2B1
Protein processing in endoplasmic reticulum	29	0.002805	
RNA transport	29	0.000742	
Spliceosome	28	0.004517	
Purine metabolism	25	0.00642	
Aminoacyl-tRNA biosynthesis	23	0	
Pyrimidine metabolism	22	0.001743	
RNA polymerase	10	0.034698	
Plant hormone signal transduction	16	0.955624	
Ribosome	16	0	Down-regulated DEGs in R4B1
Porphyrin and chlorophyll metabolism	5	0.000222	
Photosynthesis	4	0.002138	

53.06% related to IAA, 8.16% to ETH, 12.24% to CTK and 14.29% to ABA (Fig. 5c). IAA plays an important role during the flower sex differentiation of cucumber seedlings under light irradiation. As most of the DEGs were related to plant hormone, 16 DEGs were found to be involved in auxin biosynthesis and signaling pathway, with most of them showing a higher expression level at R2B1-10 stage (Fig. 6). Five DEGs involved in CTK signaling pathway, eight DEGs in ETH signal transduction pathway and three DEGs in ABA signaling pathway were also identified from the transcriptomes of cucumber shoot apices (Fig. 6). There was no DEGs involved in the GA signal transduction, while only two genes, *CsCPS1* (*Csa6G410650*) and *CsGA2ox2* (*Csa7G413380*), involved in GA biosynthesis were identified. *CsGA2ox2* (*Csa7G413380*) accumulated more transcripts at the R2B1-10 stage, while *CsCPS1* (*Csa6G410650*) expressed more at the R4B1-15 stage (Fig. 6). In conclusion, light regulated cucumber flower sex differentiation through meditating multiple hormone signals: auxin plays an important role; CTK, ABA and ETH function as supplementary regulators; and gibberellin is not involved.

Validation of RNA-Seq results by real-time quantitative PCR (qRT-PCR)

Twenty-six genes with different expression profiles were randomly selected to verify the RNA-Seq results by qRT-PCR. *α-Tubulin* (*TUA*) was used as the internal reference control to standardize the results (Wan et al. 2010). As shown in Additional file 2: Figure S1, twenty-three genes had the same expression pattern. The Pearson correlation coefficient between the RNA-Seq and qRT-PCR data were high, confirming the accuracy of the RNA-Seq data (Additional file 3: Figure S2 and Additional file 4: Figure S3).

Effect of different ratios of blue and red light treatments on light signal transduction of cucumber seedling leaves

Photoperiodic induction affects flower meristem differentiation and number of flowers. As leaves are the site for photoperiodic induction, genes related to light signal transduction and flowering were checked by qRT-PCR on 5-day, 10-day and 15-day cucumber seedling leaves after R2B1 and R4B1 light irradiation. There were two blue light receptors, named CRY1 (Csa3M889670)

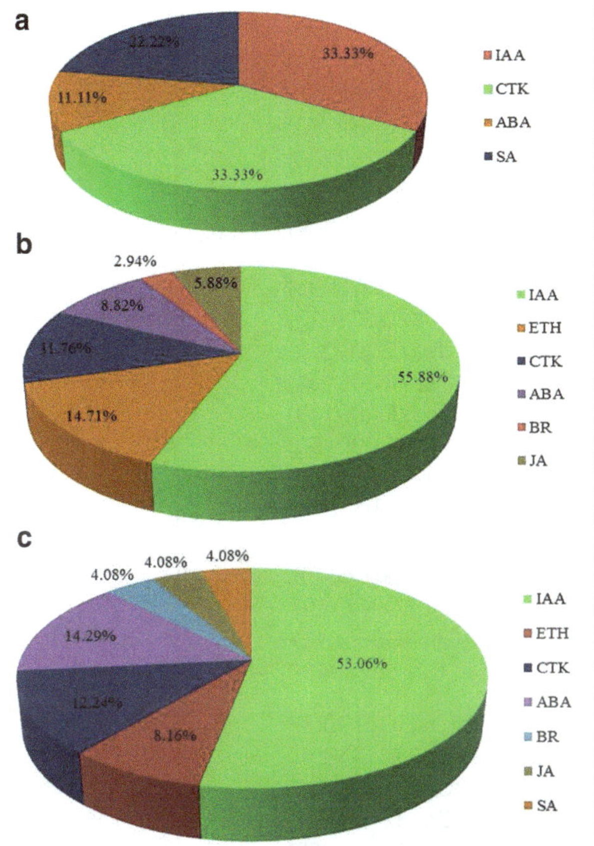

Fig. 5 The proportion of differently expressed genes (DEGs) related to plant hormone signaling pathways in **a** pairwise R4B1-5-vs-R2B1-5, **b** pairwise R4B1-10-vs-R2B1-10 and **c** R2B1 library. *IAA* auxin, *CTK* cytokine, *ABA* abscisic acid, *SA* salicylic acid, *ETH* ethylene, *BR* brassinolide, *JA* jasmonic acid

and CRY2 (Csa3M88160). Only CRY1(Csa3M889670) was found in cucumber leaves. In the 10 days treatment samples, the expression of *CRY1* was higher in R2B1 comparing to R4B1 (Fig. 7). Four red light receptors (PHYA, PHYB, PHYC and PHYE) were found in cucumber leaves. Compared to R2B1 treatment, there was a greater expression of *PHYA* in 10 days samples under R4B1 treatment, while the others showed a higher expression in the 15 days samples after R4B1 treatment (Fig. 7). *CONSTANS* (*CO*) is responsible for turning the light transduction into a flowering signal. It showed a higher expression in 15 days samples after R4B1 treatment than after R2B1 treatment (Fig. 7). *FLOWERING LOCUST* (*FT*) is a key gene in regulating flowering time. *FT* was expressed more in 10 and 15 days samples after R2B1 treatment than after R4B1 treatment (Fig. 7). All of these results suggested that the high proportion of blue light promoted the early flowering of cucumbers by regulating these flowering genes.

Analysis of differentially expressed genes related to flower sex differentiation of cucumber seedlings under different ratios of blue and red light treatments

Plant sex differentiation is closely related to endogenous hormone regulation, while flowering time is related to the photoperiod pathway, vernalization and gibberellin pathway. Aiming to identify genes related to flower sex differentiation and promoting flowering under light irradiation, the expression pattern analysis of DEGs was performed in pairwise R4B1-5-vs-R2B1-5, R4B1-10-vs-R2B1-10 and R4B1-15-vs-R2B1-15. Four light receptors (*Csa6G403570* (*PHYA*), *Csa3G166340* (*PHYB*), *Csa7G031720* (*PHYC*) and *Csa1G461000* (*PHYE*)) responding to red light were identified from the shoot apices of cucumber seedlings. These genes showed a higher expression level after R4B1 treatment compared to R2B1 treatment (Fig. 8a). The (basic helix-loop-helix) *bHLH* transcription factor is involved in light signal transduction, hormone signal transduction and flowering pathway. Seven *bHLH* DEGs were identified in the RNA-seq transcriptome, including 6 up-regulated and 1 down-regulated genes in R2B1-10 samples. *PIF*, phytochrome interaction factor, initiates light signal transduction through interacting with the activated phytochrome. *PIF* belongs to *bHLH* family. Its expression levels increased at the R4B1-10 stage, while these levels decreased at the R2B1-10 stage (Fig. 8a). The *WKRY* (transcription factors with WRKYGQK peptide) family transcription factors participate in multiple metabolic pathways. Six differently expressed *WKRY* transcription factors were identified in cucumber shoot apices, including 3 (*Csa4G003720, Csa6G526230* and *Csa7G236280*) up-regulated genes and 2 (*Csa1G547510* and *Csa3G061560*) down-regulated genes at R2B1-10 stage (Fig. 8c). The *MYB* transcription factor is related to plant hormone and flower organ development. Nine *MYB* DEGs were identified in shoot apices transcriptome of cucumber seedlings, including 2 new genes (*XLOC_008111* and *XLOC_012211*), 7 (*Csa1G586860, Csa2G035350, Csa2G100560, Csa2G355030, Csa6G303240, Csa6G519620* and *XLOC_012211*) up-regulated genes and 2 (*Csa4G338950* and *XLOC_012211*) down-regulated genes at the R2B1-10 stage (Fig. 8c) (AGAMOUS [AG]-like MADS box protein) *AGL* and (*SUPPRESSOR OF OVEREXPRESSION OF CO1*) *SOC1* belong to MADS-box gene, which both could regulate flowering time. Seven *AGLs* and two *SOC1 s* were identified from the shoot apices of cucumber seedlings. *AGL* (*Csa5G156170*) and *SOC1* (*Csa3G124870, Csa6G076720*) accumulated more transcripts at the R2B1-10 stage (Fig. 8b, c).

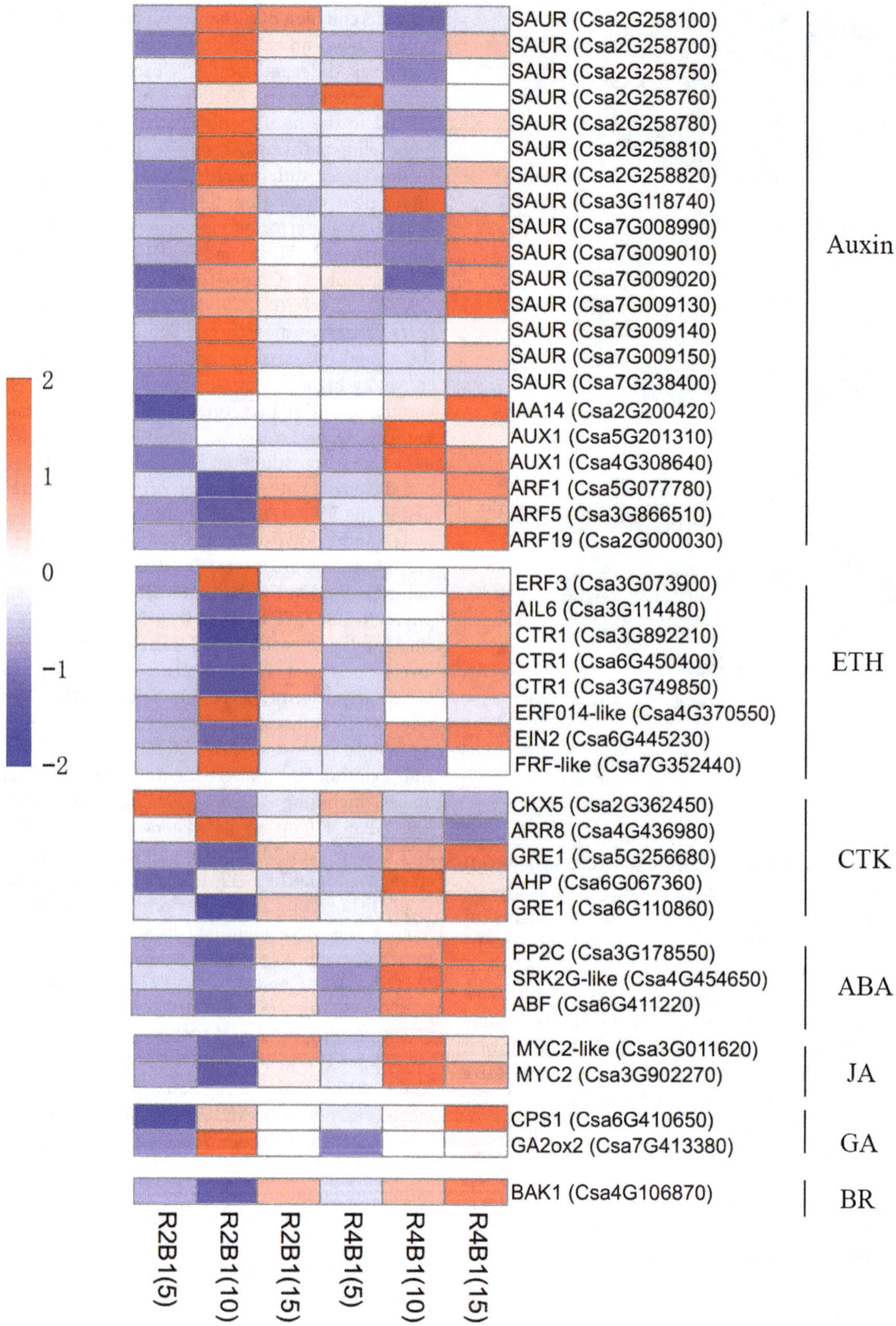

Fig. 6 The expression of a number of genes altered in cucumber shoot apices under different light quality treatments: genes involved in plant hormone signaling pathway

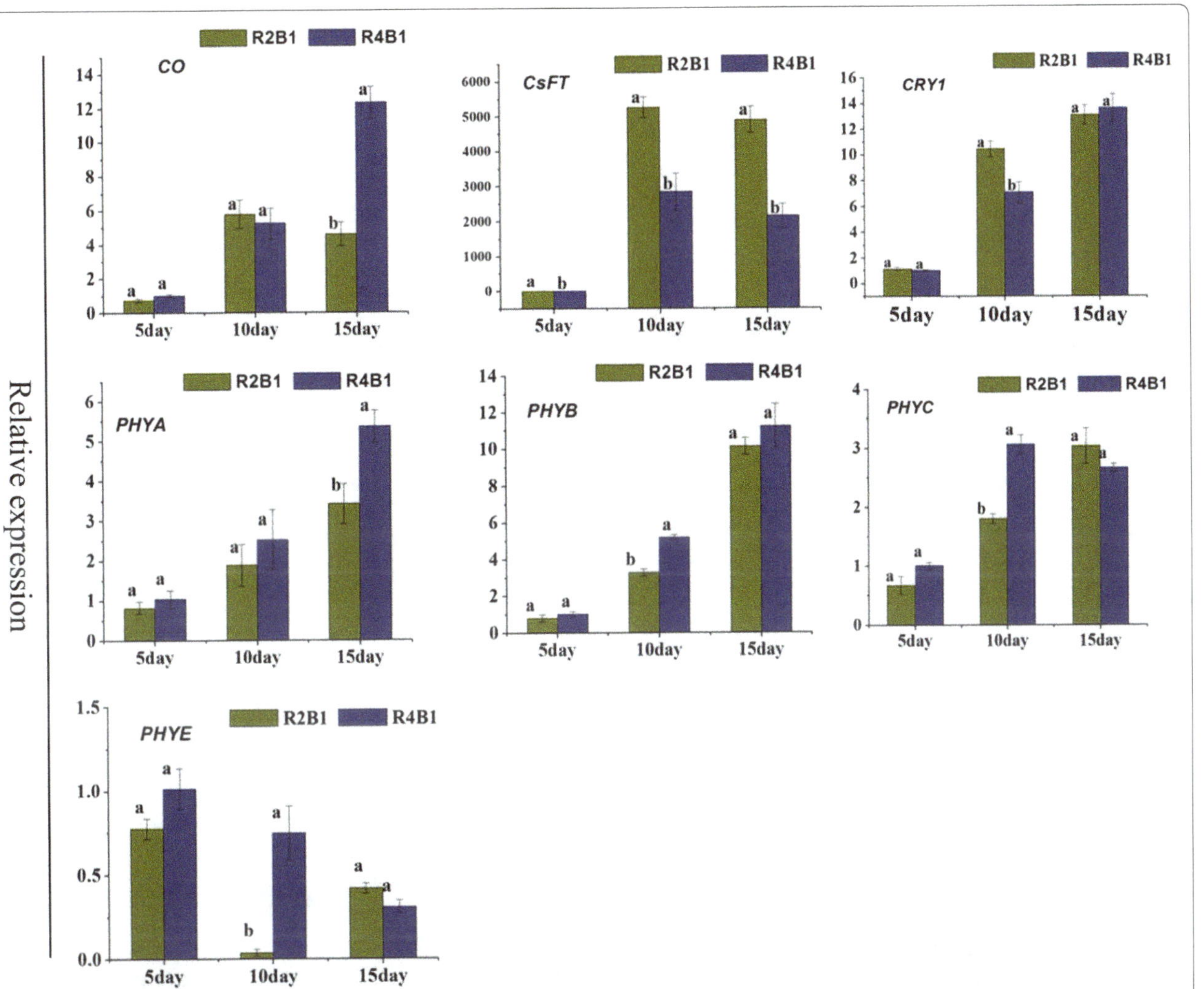

Fig. 7 The expression of a number of flowering and light signal transduction related genes were altered in cucumber seedlings leaves under different light quality treatments. Error bars represented standard deviations of the means of three independent replicates. Statistically significant variations of expression and mean values at different sampling points (ANOVA, $p < 0.05$) were indicated with different letters

Discussion

Light quality affects cucumber flowering time

Blue light is a strong signal in floral bud formation. In Chrysanthemum, flower budding was formed even after a longer photoperiod than a critical day length under blue light illumination (Jerzy et al. 2011). In petunia plants, floral bud formation and flowering occurred earlier under blue light treatment, while no floral buds were observed under low red irradiance. High red irradiance and temporal switching to blue light during long-term low red irradiance induced floral development (Fukuda et al. 2016). In our experiment, we found that the first female flower opened earlier and occurred mainly at the lower node positions under a higher proportion of blue light irradiance (Table 1). This suggested that blue light promoted flower bud formation and accelerated cucumber flowering.

Cryptochromes (CRY) are flavo-proteins that direct a diverse array of developmental processes in response to blue light in plants (Liu et al. 2016). *CRY1* and *CRY2* function as major blue light receptors regulating blue light-induced de-etiolation and photo-periodic flowering (Guo and Li 1998). In Arabidopsis, *CRY1* and *CRY2* serve both distinct and partially overlapping functions in regulating photomorphogenic responses and photoperiodic flowering. The gain-of-function mutant alleles of *CRY1* exhibited an early flowering phenotype after several days (Exner et al. 2010). In our study, we detected a higher expression of *CRY1* in cucumber seedling leaves after R2B1 treatment than R4B1 treatment (Fig. 7), which was consistent with the earlier flowering phenotype after R2B1 irradiation. The phytochrome (phy) family of sensory photoreceptors (phyA to phyE in *Arabidopsis thaliana*) also responds to inducing floral budding. Low red/

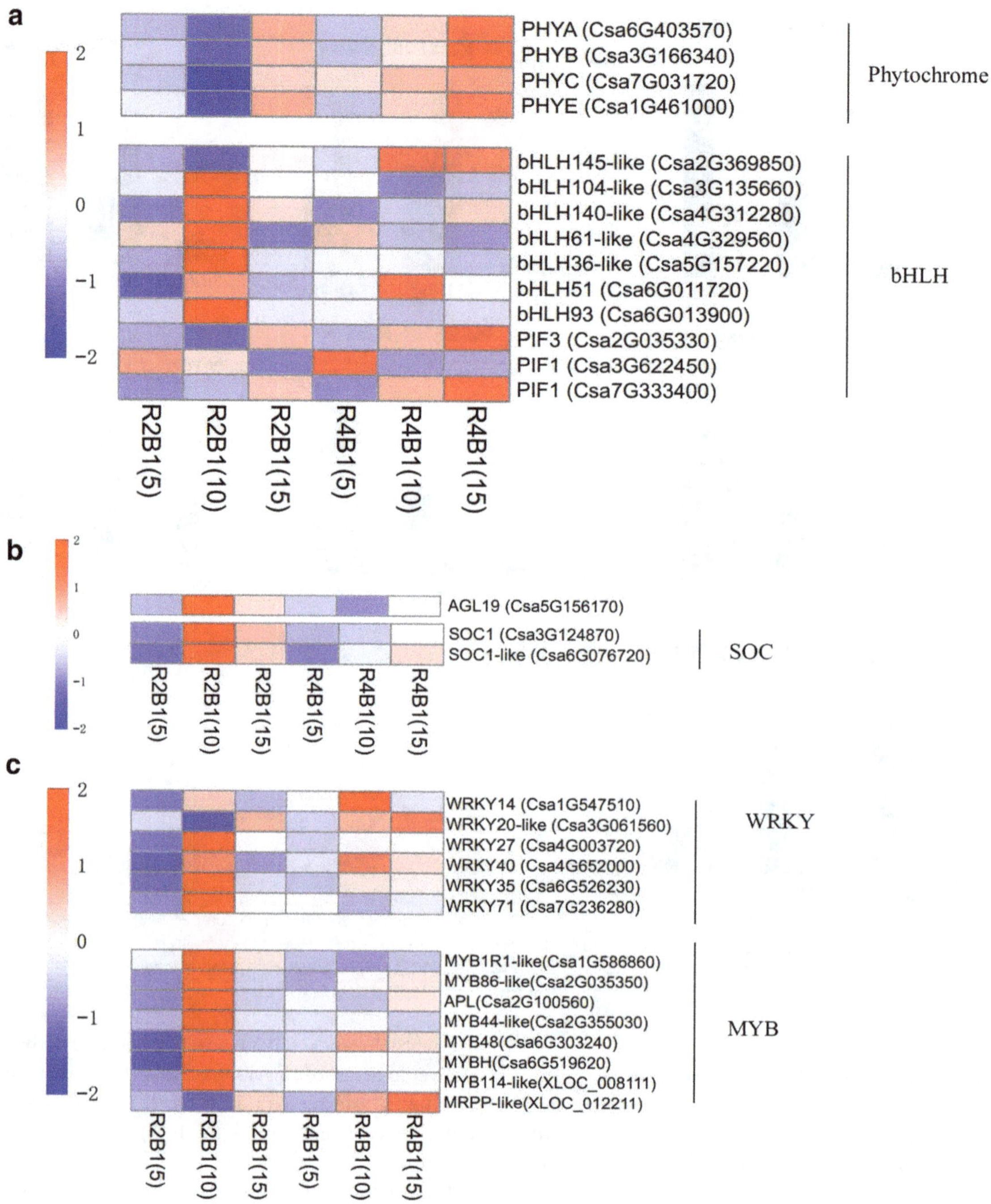

Fig. 8 The expression of a number of genes altered in cucumber shoot apices under different light quality treatments: **a** genes involved in light signaling transduction; **b** genes related to flowering time regulation and **c** transcription factors involved in plant hormone and flower organ development

far-red ratio promotes flowering in Arabidopsis through *PHYA* inhibiting *PHYB* (Chory 2003; Halliday et al. 2003). In our study, we detected 4 phytochrome orthologous genes in cucumber seedling leaves and shoot apices. Expression analysis showed that *PHYA* accumulated more in leaves at the R4B1-15 stage than R2B1-15 stage (Fig. 7). In comparison, the other genes displayed a higher expression in leaves at R4B1-10 stage than R2B1-10 stage (Fig. 7). Similar results were found in shoot apices, with *PHYA*, *PHYB*, *PHYC*, *PHYE* expressed more in the shoot apices of R4B1-10 sample than R2B1-10 sample (Fig. 7). The expression of *CRY2* and *PHYD* were not detected in both cucumber leaves and shoot apices.

CONSTANS (*CO*) is a key transcription factor regulating flowering time. *CO* positively regulates two floral integrators. In Arabidopsis, *CO* activated *SOC1* through *FT* to promote flowering (Rosas et al. 2014; Lee and Lee 2010). In this study, compared to R2B1-15, *CO* showed a higher expression in R4B1-15 (Fig. 7). Although *CO* transcription regulation was affected by blue light treatment, *CO* was not only positively regulated by *CRY2* and *PHYA*, but also negatively regulated by *PHYB* (Valverde et al. 2004). The lower expression under a high proportion of blue irradiation might result from the co-regulation of the cryptochrome receptor and phytochrome receptor. In Arabidopsis, blue light could induce *SOC1* and *FT* expression to promote flowering (Hori et al. 2011). In petunia plants, blue light induces *FBP28* expression, the orthologous gene of *SOC1* (Fukuda et al. 2011). *SOC1* is one of the direct targets of *AGL24* and is up-regulated by *AGL24*. In Arabidopsis, *SOC1* and *AGL24* are able to up-regulate each other's expressions, both of them are MADS-domain-containing transcription factors that determine flowering time (Michaels et al. 2003). In this study, *FT* showed a higher expression in R2B1-10 and R2B1-15 than in R4B1-10 and R4B1-15 (Fig. 8b), which suggested that higher proportion of blue light irradiation accelerated cucumber flowering through inducing *FT* expression. Two *SOC1* orthologous genes and seven *AGLs* (only *AGL19* showed different expression) orthologous genes were found in cucumber shoot apices. *SOC1* and *AGL19* displayed the same expression pattern, with increased expression in R2B1-10 and R2B1-15 treatments (Fig. 8b). All these results suggested that a higher proportion of blue light induced cucumber flower bud formation and promoted flowering time through up-regulating expression of these flowering regulation factors.

Light quality affects flower sex differentiation of cucumber seedlings through regulating plant hormone signaling transduction

In this study, we found that under a higher proportion of blue light (R2B1) treatment, the first female flower node declined significantly and the number of female flowers within fifteen plant nodes increased markedly (Table 1). Transcriptome analysis of cucumber seedlings under different light treatment by RNA-seq technology showed that there were more DEGs under R2B1 treatment. These DEGs were mostly involved in the hormone signaling pathway (Table 3). This suggested that a higher proportion of blue light regulated female flower formation through mediating plant hormone signaling pathways.

Plant sex differentiation is closely related to plant hormones, such as ethylene, which is considered as a potent sex hormone in cucumbers that can induce formation of female flowers (Trebitsh et al. 1987; Rudich et al. 1972; Li et al. 2017). Another example is the gibberellins, which can promote the formation of male flowers via ethylene-dependent and ethylene-independent pathways in cucumber (Hao et al. 2003). ABA can promote male flower development (Zhu et al. 2010), while auxin is able to promote the female tendency of vegetables and induce female flower development through triggering ethylene synthesis (Atsmon and Tabbak 1979; Galun et al. 1962). Exogenous CTK application can turn the grape male flower into a female flower (Chang et al. 1999; Negi and Olmo 1966). JA can regulate gynoecium development (Figueroa and Browse 2015), while BR induces androecium development in maize and inhibits gynoecium development (Hartwig et al. 2011; Makarevitch et al. 2012). In this study, we found that among these DEGs related to plant hormone transduction pathways, the specific DEGs related with IAA formed the higher proportion of DEGs in plant hormone signal transduction, which was followed by the specific DEGs related with CTK, ABA, ETH, BR, JA and SA (Fig. 5).

Ethylene is considered as a potent sex hormone in cucumber that can induce formation of female flowers (Malepszy and Niemirowicz-Szczytt 1991). Ethylene content in shoot apices of gynoecious cucumbers is higher than that of monoecious plants (Rudich et al. 1972; Fujita and Fujieda 1981; Trebitsh et al. 1987). Treatment with exogenous ethylene or ethylene-releasing reagents can increase the numbers of female and bisexual flowers in monoecious and andromonoecious lines, respectively (Mcmurray and Miller 1968; Iwahori et al. 1969). Until now, the molecular mechanism of ethylene-regulated sex determination of cucumber has been well understood. Ethylene biosynthetic genes and ethylene signal transduction genes are involved in the sex expression of cucumber (Wang et al. 2010). In the ethylene signal transduction pathway, the receptors, such as ETRs, function as negative regulators, while ERFs (downstream components of receptors) act as positive transcription factors to regulate sex determination (Tao et al. 2018; Prescott et al. 2016). In our study, the expression levels of some *ERFs* were increased under R2B1 treatment compared to R4B1 treatment (Fig. 6a). They were probably involved in cucumber sex expression through inhibiting the differentiation into males and promoting the differentiation into females. However, this understanding was based on bioinformatics analysis. The precise roles of *ERFs* in cucumber flower development remained unclear and should be verified in future studies using advanced physiological and molecular techniques. *CTR1* is a negative regulator of ethylene signaling, while *CsCTR1* expression gradually declined during male flower development and increased during female flower development (Bie et al. 2014). In our study, *CsCTR1* expression increased on the 5th day after a higher proportion of blue light treatment.

Ethylene plays a key role in plant sex determination, with several research results proving that there is communication between auxin and ethylene during plant development (Trebitsh et al. 1987; Gao et al. 2015; Wang et al. 2018). In our study, we found a considerably number of transcription factors involving in auxin signaling pathway altered their expression levels under R2B1 treatment (Fig. 6). IAA may play a key role in effect of light quality on sex differentiation in cucumber. IAA has been considered as a transcription factor activated by phytochrome, which participates in signaling transduction. During the early development of plants, there are several reactions between phytochrome and auxin, such as auxin homeostasis promoted by light (Zhang et al. 2014c). Auxin is an important regulator in the development of male flowers (Sakata et al. 2010). The auxin synthesis pathway can be regulated by aldehyde oxidases, which have been identified in cucumber (Brown and Purves 1978). Aldehyde oxidases are involved in the conversions of indole-3-acetaldehyde to IAA (Mano and Nemoto 2012). Higher activity of aldehyde oxidase has also been measured in the auxin-overproducing superroot1 (sur1) mutant of Arabidopsis (Seo et al. 1998). *Aldehyde oxidase* was reported to regulate auxin and ABA biosynthesis pathway (Koshiba et al. 1996; Seo et al. 1998, 2000). In our study, the expression of *Aldehyde oxidase* (*Csa4G269130*) was down-regulated in R2B1 samples (Additional file 5: Table S2), which suggested that the auxin synthesis or ABA synthesis in cucumber seedlings flowers was low under a higher proportion of blue light. This result was contradicted against the finding that more female flower were formed in the R2B1 samples which might result from higher auxin content. Meanwhile, it was also contrary to the most up-regulated auxin responsive genes *SAURs* (Fig. 6). While the expression of auxin transporting gene LAX1 and PIN1 were also altered indicating that the auxin transport was changed inside the cucumber flower. Even though the low biosynthesis might be found in the flowers, the higher transportation might also cause a high auxin content in the flower tissues. However, auxin content detection and gene function analysis experiments should be performed to confirm this hypothesis.

Gibberellins, a class of tetracyclic diterpenoid phytohormones, can promote the differentiation of cucumbers into male flowers. GA production in andromonoecious cucumbers is higher than that in gynoecious and monoecious plants (Junior et al. 1972). Exogenous GA_3 application can increase the ratio of males to females in monoecious cucumbers and induce the formation of male flowers in gynoecious plants (Wittwer and Bukovac 1962; Pike and Peterson 1969). In our study, we identified two DEGs involved in the GA synthesis pathway. Ent-copalyl diphosphate synthase (CPS) is limiting for ent-kaurene production in the first portion of the GA synthesis pathway (Prisic and Peters 2007). *CsCPS1* was down-regulated in R2B1-5 and up-regulated in R4B1-15 (Fig. 6), which suggested that a high proportion of blue light reduces levels of GA in cucumber seedling flowers through breaking down the *CPS1* gene expression. GA2-oxidase (GA2ox) is also a key gene for GA synthesis pathway. It catalyzes catabolism and inactivation of bioactive GAs or their precursors (Schomburg and Amasino 2003; Lester et al. 1999; Thomas et al. 1999). In our study, *CsGA2ox2* was up-regulated in R2B1-10 samples (Fig. 6), which suggested that a high proportion of blue light reduces active GA levels through inducing the expression of *CsGA2ox2*. In addition, the GA signaling pathway is involved in stamen and anther development in hermaphroditic plants, such as Arabidopsis and rice (*Oryza sativa*) (Aya et al. 2009; Plackett et al. 2011; Sun 2011). However, in this study, we did not identify any DEGs involved in GA signaling pathway.

PIF, phytochrome interaction factor, plays a central role in light signal transduction mediated by phytochrome (De and Prat 2014). PIF3 can activate the transcription factor *EIN3* directly, which was involved in the ethylene signaling pathway (Zhong et al. 2012). *COI1*, *JAZ*, *MYC2* and *JAR1* involved in the JA signaling pathway affects the light signal response (Kazan and Manners 2011). *PHYB* negatively regulated ABA accumulation through mediating light signals, while *PHYA* positively regulated ABA signaling pathway under far-red light treatment (Gu et al. 2012; Mach 2014). Thus, it was seen that light could regulate plant sex differentiation through mediating hormones by light receptors. The phytohormones, auxin, BR, CK, ETH, GA and JA, influenced sex differentiation in flowering plants (Durand and Durand 2010; Louis and Durand 1978; Chailakhyan 1979; Irish and Nelson 1989). Our results showed that the expression of some key genes involved in ABA, auxin, CK, ETH, GA and JA biosynthesis pathways significantly changed during development and sex expression, which suggesting that these hormones might participate in these processes. Thus, it is interesting to investigate how these hormones interact with one another to regulate the abortion of male flowers in gynoecious plants.

Light quality affects flowering time of cucumber seedlings through regulating transcription factors

A total of 433 transcription factors have been identified by RNA-seq technology. Among them, ethylene transcription factor accounted for 24.02%, *MYB* transcription factor accounted for 20.79%, *bHLH* transcription factor accounted for 16.17% and *WRKY* transcription factor accounted for 10.62%.

Most studies have shown that *MYB*, *bHLH* and *WRKY* transcription factors play an important role in biotic stress and abiotic stress (Oh et al. 2012; Li et al. 2010; Jiang and Yu 2009), but some studies proved that they were also related to flower development (Wang et al. 2011; Luo et al. 2013; Zhang et al. 2014b). In Arabidopsis, JA mediates the transcription factors *MYB21* and *MYB24* to regulate stamen development (Song et al. 2011), while *MYC5* regulated stamen development by activating JA signaling pathway through inducing *MYB21* expression. *PIFs* belongs to *bHLH* family of transcription factors, which interact with activated Pr to initiate phytochrome signaling transduction. In Arabidopsis, *PIF4* accelerated flowering by activating the *FT* gene (Kumar et al. 2016). *WRKY* family members are key factors for the ABA response pathway (Rushton et al. 2012). It had been shown that *WRKY* transcription factors regulated plant flowering. *WRKY25* accelerated flowering through negatively regulating *AP1* directly or indirectly (Wang et al. 2011). In pea, *GsWRKY20* promoted flowering by regulating the flowering related genes *FT*, *SOC1* and *CO* (Luo et al. 2013).

Materials and methods

Plant materials and treatments

LED light produced by Kedao technology corporation (Huizhou, China) (Red light: 650–660 nm and Blue light: 450–460 nm) was chosen for this study. Cucumber (*Cucumis sativus* L. cv. Huaqing No. 5) seeds were cultivated on a sponge block with a Yamasaki culture solution. When cotyledon appeared, we placed these seedlings under light conditions as follows: 300 μmol/m^2/s, 12/12 (light dark) at 25 °C and a relative humidity of 70–80%. There were 2 treatments with ratios of red to blue being 4:1 (R4B1) and 2:1 (R21B), respectively. Each treatment lasted 20 days. As the flower primordium appeared 10 days after light treatment, the shoot apices and leaves after 5, 10 and 15 days of light treatments were collected immediately flash frozen in liquid nitrogen and stored at − 80 °C until use. Two biological replicates of shoot apices were used for RNA-seq analysis and real-time RT-PCR. After 20 days of treatment, 30–40 seedlings from each treatment were transplanted into the greenhouse of South China Agriculture University with a relative humidity ranging from 70 to 85% and 28 °C/20 °C (14 h/10 h) day/night temperature for further flowering time analysis.

cDNA library construction and illumina sequencing

The cucumber shoot apices stored at − 80 °C were provided to Guangzhou Gene Denovo Biological Technology Co., Ltd. (Guangzhou, China). RNA isolation and RNA-Seq library preparation and sequencing were carried out by Guangzhou Gene Denovo Biological Technology Co., Ltd. (Guangzhou, China). In brief, the workflow of library construction for transcriptome analysis was as follows: After RNA was collected, poly(A)-containing mRNA was purified using oligo(dT) magnetic beads. Following this, the mRNAs were fragmented and cDNA was synthesized using a random hexamer, DNA polymerase I and RNase H. The double-stranded cDNAs were purified and ligated to adaptors for Illumina paired-end sequencing. PCR amplication was done on the purified cDNAs with ligated adaptors. PCR products were separated by 2% agarose gel. The final PCR product chosen for sequencing is 400–500 bp band, which were cut and recovered from the gel. The quality and quantity of the library were verified using an Agilent 2100 Bioanalyzer and ABI real time RT-PCR, respectively (Additional file 6: Table S3 and Additional file 2: Figure S1). The cDNA libraries were sequenced using Illumina HiSeqTM 2500 by Gene Denovo Co. (Guangzhou, China). The libraries of two biological replicates of shoot apices were prepared independently.

Sequence read mapping and assembly

The original image data produced by sequencing were transferred into sequences (raw reads). The calculating methods for raw reads was conducted according to Cock (2010). To obtain clean reads for further analysis, the raw reads were filtered by removing adaptor sequences, low-quality reads and reads with a percentage of unknown bases (N) of more than 10%. Following this, we used Bowtie to remove the ribosome reads (Langmead et al. 2009). The clean reads were mapped to the cucumber genome assembly (Chinese long) v2 (http://www.icugi.org/cgi-bin/ICuGI/index.cgi) according to Kim et al. (2013) and Trapnell et al. (2012).

Quantification and differential expression analysis of transcripts

The expression level of each gene was determined by the numbers of reads, which were uniquely mapped to the specific gene and the total number of uniquely mapped reads in the sample. Reads per kilobase of exon model per million mapped reads (RPKM) were calculated according to Mortazavi to estimate gene expression levels. The FDR was used to determine the threshold of the P-value in multiple tests. In our study, a FDR < 0.05 and an absolute value of |log2 (fold change)| > 1 were used as the threshold to determine the significant DEGs.

GO and KEGG enrichment analysis

To identify putative biological functions and pathways for the DEGs, the Gene ontology (GO) and Kyoto encyclopedia of gene and genomes (KEGG) database were searched for annotation. The methods for GO and KEGG enrichment analysis were according to Young and Mao respectively (Young et al. 2010; Mao et al. 2005).

Real-time RT-PCR

Total RNA was isolated from 100 mg cucumber leaves and shoot apices using Huayueyang reagent kit (with DNase I step: Pipet the DNase I onto the membrane containing RNA. Incubate at 37 °C for 30 min) according to the manufacturer's instructions. The RNA quality and purity were verified by Nanodrop 2000 and electrophoresis on 1.0% agarose gels. The first-strand cDNA was synthesized from aliquots of 1 μg of total RNA using PrimeScriptTM RT Reagent Kit with gDNA Eraser (TaKaRa, RR047A) in a reaction volume of 20 μL. The synthesized cDNA was diluted 20 times with sterile water and used as the template for real-time PCR. The reactions were carried out in a Roche LightCycler 480 system with SYBR Premix Ex Taq II (Tli RNaseH Plus) (Takara Bio, Dalian, China). The reactions reagent mix was 5 μL SYBR Premix Ex Taq II, 1.5 μL cDNA template, 0.4 μL each primer (10 μmol/μL), and 2.7 μL nuclease-free water. The amplification program was 95 °C for 30 s and 40 cycles of 95 °C for 5 s and 60 °C for 30 s. Melting curve analyses were performed at the end of 40 cycles (95 °C for 5 s followed by a constant increase from 60 to 95 °C). In addition, TUA was used to normalize expression levels (Wan et al. 2010), and the relative expression of genes was calculated using the $2^{-\Delta\Delta Ct}$ method (Livak and Schmittgen 2001). The results were analyzed by Excel 2010. qPCR reactions in leaves were prepared in biological triplicate. qPCR reactions in shoot apices for RNA-seq validation were prepared in three biological replicates.

Genes used for Real-Time RT-PCR were designed on the NCBI primer blast website (https://www.ncbi.nlm.nih.gov/tools/primer-blast/). The primer quality and efficiency were checked by qPCR on series of diluting cDNA. The expression of 26 genes chosen from RNA-seq were analyzed by real-time RT-PCR in shoot apex samples (R2B1 and R4B1 at 5 days, 10 days and 15 days). Meanwhile, 9 genes were used for checking light signal transduction in leaves (Additional file 7: Table S4).

Authors' contributions

JS and YZ carried out the experiment, participated in the analysis. YH drafted the manuscript. WS carried out the qRT-PCR analysis. SS and GS performed the statistical analysis. HL and YH conceived of the study, and participated in its design. RC and HL acquired of funding and helped to draft the manuscript. All authors read and approved the final manuscript.

Acknowledgements

Not applicable.

Competing interests

The authors declare that they have no competing interests.

Fundings

This work was supported by Teamwork Projects Funded by Guangdong Natural Science Foundation (No. S2013030012842), and the Guangzhou Science & Technology Project (201605030005).

References

Acosta IF, Laparra H, Romero SP, Schmelz E, Hamberg M, Mottinger JP, Moreno MA, Dellaporta SL (2009) Tasselseed1 is a lipoxygenase affecting jasmonic acid signaling in sex determination of maize. Science 323:262–265

Ando S, Sato Y, Kamachi S, Sakai S (2001) Isolation of a MADS-box gene (*ERAF17*) and correlation of its expression with the induction of formation of female flowers by ethylene in cucumber plants (*Cucumis sativus* L.). Planta 213:943–952

Atsmon D, Tabbak C (1979) Comparative effects of gibberellin, silver nitrate and aminoethoxyvinyl glycine on sexual tendency and ethylene evolution in the cucumber plant (*Cucumis sativus* L.). Plant Cell Physiol 20:1547–1555

Aya K, Ueguchitanak M, Kondo M, Hamada K, Yano K, Nishimura M, Matsuoka M (2009) Gibberellin modulates anther development in rice via the transcriptional regulation of GAMYB. Plant Cell 21:1453–1472

Bai N, Xu H (2013) Chapter one-unisexual cucumber flowers, sex and sex differentiation. Int Rev Cel Mol Blo 304:1–55

Bai L, Peng B, Cui X, Gu T, Xu Y, Li Q, Xu H, Bai N (2004) Developmental analyses reveal early arrests of the spore-bearing parts of reproductive organs in unisexual flowers of cucumber (*Cucumis sativus* L.). Planta 220:230–240

Bie B, Sun J, Pa J, He H, Cai R (2014) Ectopic expression of *CsCTR1*, a cucumber CTR-like gene, attenuates constitutive ethylene signaling in an Arabidopsis *ctr1-1* mutant and expression pattern analysis of *CsCTR1* in cucumber (*Cucumis sativus*). Int J Mol Sci 15:16331–16350

Briggs WR, Beck CF, Cashmore AR, Christie JM, Hughes J, Jarillo JA, Kagawa T, Kanegae H, Liscum E, Nagatani A (2001) The phototropin family of photoreceptors. Plant Cell 13:993–997

Brown HM, Purves WK (1978) Cucumber seedling indoleacetaldehyde oxidase. Plant Physiol 61:107–110

Brudler R, Hitomi K, Daiyasu H, Toh H, Kucho KI, Ishiura M, Kanehisa M, Roberts VA, Todo T, Tainer JA, Getzoff ED (2003) Identification of a new cryptochrome class: structure, function, and evolution. Mol Cell 11:59–67

Chailakhyan MK (1979) Genetic and hormonal regulation of growth, flowering, and sex expression in plants. Am J Bot 66:717–736

Chang T, Chen S, Hsu Y, Yu C, Du S, Huang L (1999) Changes in cytokinin activities before, during and after floral initiation in *Polianthes tuberosa*. Plant Physiol Bioch 37:679–684

Chen M, Chory J, Fankhauser C (2004) Light signal transduction in higher plants. Annu Rev Genet 38:87–117

Chory J (2003) Regulation of flowering time by light quality. Nature 423:881–885

Cock J, Fields J, Goto N, Heuer L, Rice M (2010) The sanger fastq file format for sequences with quality scores, and the solexa/illumina fastq variants. Nucleic Acids Res 38:1767–1771

De M, Prat S (2014) Pifs get BR right: phytochrome interacting factors as integrators of light and hormonal signals. New Phytol 202:1126–1141

Duan H, Wang H, Xu H, Bai N (2008) Stamen development in *Arabidopsis* is arrested by organ-specific overexpression of a cucumber ethylene synthesis gene *CsACO2*. Planta 228:537–543

Durand R, Durand B (2010) Sexual differentiation in higher plants. Physiol Plantarum 60:267–274

Exner V, Alexandre C, Rosenfeldt G, Alfarano P, Nater M, Caflisch A, Gruissem W, Batschauer A, Hennig L (2010) A gain-of-function mutation of *Arabidopsis* cryptochrome1 promotes flowering. Plant Physiol 154:1633–1645

Fei H, Zhang R, Pharis P, Sawhney K (2004) Pleiotropic effects of the male sterile33 (ms33) mutation in *Arabidopsis* are associated with modifications in endogenous gibberellins, indole-3-acetic acid and abscisic acid. Planta 219:649–660

Figueroa P, Browse J (2015) Male sterility in *Arabidopsis* induced by overexpression of a MYC5-SRDX chimeric repressor. Plant J 81:849–860

Fujita Y, Fujieda K (1981) Relation between sex expression types and cotyledon etiolation of cucumber in vitro I. on the role of ethylene evolved from seedlings. Plant Cell Physiol 22:667–674

Fukuda N, Ishii Y, Ezura H, Olsen E (2011) Effects of light quality under red and blue light emitting diodes on growth and expression of FBP28 in petunia. Acta Hortic 907:361–366

Fukuda N, Ajima C, Yukawa T, Olsen JE (2016) Antagonistic action of blue and red light on shoot elongation in petunia depends on gibberellin, but the effects on flowering are not generally linked to gibberellin. Environ Exp Bot 121:102–111

Galun E, Jung Y, Lang A (1962) Culture and sex modification of male cucumber buds in vitro. Nature 194:596–598

Gao P, Sheng Y, Luan F (2015) RNA-Seq transcriptome profiling reveals differentially expressed genes involved in sex expression in melon. Crop Sci 55:1686–1695

Golenberg EM, West NW (2013) Hormonal interactions and gene regulation can link monoecy and environmental plasticity to the evolution of dioecy in plants. Am J Bot 100:1022–1037

Gu J, Fang Z, Jie Z, Zhou J, Qian F, Yan L (2012) Light signals mediated by phytochrome B affect abscisic acid pathway in rice. Chin Sci Bull 57:2371

Guo H, Li C (1998) Regulation of flowering time by *Arabidopsis* photoreceptors. Science 279:1360–1363

Halliday KJ, Salter MG, Thingnaes E, Whitelam GC (2003) Phytochrome control of flowering is temperature sensitive and correlates with expression of the floral integrator FT. Plant J 33:875–885

Hao J, Wang H, Peng B, Bai L, Xu Y, Li Q, Xu H, Bai N (2003) DNA damage in the early primordial anther is closely correlated with stamen arrest in the female flower of cucumber (*Cucumis sativus* L.). Planta 217:888–895

Hartwig T, Chuck GS, Fujioka S, Klempien A, Weizbauer R, Potluri DP, Choe S, Johal GS, Schulz B (2011) Brassinosteroid control of sex determination in *maize*. Proc Natl Acad Sci USA 108:19814–19819

Hori Y, Nishidate K, Nishiyama M, Kanahama K, Kanayama Y (2011) Flowering and expression of flowering-related genes under long-day conditions with light-emitting diodes. Planta 234:321–330

Irish EE, Nelson T (1989) Sex determination in monoecious and dioecious plants. Plant Cell 1:737–744

Ishikawa M, Takahashi F, Nozaki H, Nagasato C, Motomura T, Kataoka H (2009) Distribution and phylogeny of the blue light receptors aureochromes in eukaryotes. Planta 230:543–552

Iwahori S, Lyons JM, Sims WL (1969) Induced femaleness in cucumber by 2-chloroethanephosphonic acid. Nature 222:271–272

Jerzy M, Zakrzewski P, Schroeterzakrzewska A (2011) Effect of colour of light on the opening of inflorescence buds and post-harvest longevity of pot chrysanthemums (chrysanthemum × grandiflorum (ramat.) kitam). Acta Agrobot 64:13–18

Jiang W, Yu D (2009) Arabidopsis WRKY2 transcription factor mediates seed germination and postgermination arrest of development by abscisic acid. BMC Plant Biol 9:20

Junior H, Baker R, Sell HM (1972) Different sex phenotypes of *Cucumis sativus* L. and *C. melo* L. and their endogenous gibberellin activity. Euphytica 21:285–291

Kamachi H, Iwasawa O, Hickok LG, Nakayama M, Noguchi M, Inoue H (2007) The effects of light on sex determination in gametophytes of the fern *Ceratopteris richardii*. J Plant Res 120:629–634

Kazan K, Manners JM (2011) The interplay between light and jasmonate signalling during defence and development. J Exp Bot 62:4087–4100

Kim D, Pertea G, Trapnell C, Pimentel H, Kelley R, Salzberg SL (2013) Tophat2: accurate alignment of transcriptomes in the presence of insertions, deletions and gene fusions. Genome Biol 14:R364

Korpelainen H (1998) Labile sex expression in plants. Biol Rev 73:157–180

Koshiba T, Saito E, Ono N et al (1996) Purification and properties of flavin- and molybdenum-containing aldehyde oxidase from coleoptiles of Maize. Plant Physiol 110:781–789

Kumar SV, Lucyshyn D, Jaeger KE, Alós E, Alvey E, Harberd NP, Wigge PA (2016) Phytochrome interacting factor 4 controls the thermosensory activation of flowering. Nature 484:242–245

Langmead B, Trapnell C, Pop M, Salzberg SL (2009) Ultrafast and memory-efficient alignment of short DNA sequences to the human genome. Genome Biol 10:1–10

Lee J, Lee I (2010) Regulation and function of SOC1, a flowering pathway integrator. J Exp Bot 61:2247–2254

Lester DR, Ross JJ, Smith JJ, Elliott RC, Reid JB (1999) Gibberellin 2-oxidation and the *SLN* gene of *Pisum sativum*. Plant J 19:65–73

Li Z, Huang S, Liu S, Pan J, Zhang Z, Tao Q, Shi Q, Jia Z, Zhang W, Chen H, Si L, Zhu L, Cai R (2009) Molecular isolation of the *M* gene suggests that a conserved-residue conversion induces the formation of bisexual flowers in cucumber plants. Genetics 182:1381–1385

Li F, Guo S, Zhao Y, Chen D, Chong K, Xu Y (2010) Overexpression of a homopeptide repeat-containing bHLH protein gene (*OrbHLH001*) from dongxiang wild rice confers freezing and salt tolerance in transgenic *Arabidopsis*. Plant Cell Rep 29:977–986

Li SF, Zhang GJ, Zhang XJ, Yuan JH, Deng CL, Gao WJ (2017) Comparative transcriptome analysis reveals differentially expressed genes associated with sex expression in garden asparagus (*asparagus officinalis*). BMC Plant Biol 17:143

Liu B, Yang Z, Gomez A, Liu B, Lin C, Oka Y (2016) Signaling mechanisms of plant cryptochromes in *Arabidopsis thaliana*. J Plant Res 129:137–148

Livak KJ, Schmittgen TD (2001) Analysis of relative gene expression data using real-time quantitative PCR and the $2^{-\Delta\Delta CT}$ method. Methods 25:402–408

Louis JP, Durand B (1978) Studies with the dioecious angiosperm *Mercurialis annua* L. (2n = 16): correlation between genic and cytoplasmic male sterility, sex segregation and feminizing hormones (cytokinins). Mol Gen Genet 165:309–322

Luo X, Sun X, Liu B, Zhu D, Bai X, Cai H, Ji W, Cao L, Wu J (2013) Ectopic expression of a WRKY homolog from glycine soja alters flowering time in *Arabidopsis*. PLoS ONE 8:e73295

Mach J (2014) Walk into the light response: direct targets of phytochrome A include genes that respond to light, stress, and hormones. Plant Cell 26:1832

Makarevitch I, Thompson A, Muehlbauer GJ, Springer NM (2012) *Brd1* gene in *maize* encodes a brassinosteroid C-6 oxidase. PLoS ONE 7:e30798

Malepszy S, Niemirowicz-Szczytt K (1991) Sex determination in cucumber (*Cucumis sativus*) as a model system for molecular biology. Plant Sci 80:39–47

Mano Y, Nemoto K (2012) The pathway of auxin biosynthesis in plants. J Exp Bot 63:2853–2872

Mao X, Cai T, Olyarchuk JG, Wei L (2005) Automated genome annotation and pathway identification using the KEGG Orthology (KO) as a controlled vocabulary. Bioinformatics 21:3787–3793

Mathews S, Sharrock RA (1997) Phytochrome gene diversity. Plant, Cell Environ 20:666–671

Mcmurray AL, Miller CH (1968) Cucumber sex expression modified by 2-chloroethanephosphonic acid. Science 162:1397–1398

Mibus H, Tatlioglu T (2004) Molecular characterization and isolation of the *f/f* gene for femaleness in cucumber (*Cucumis sativus* L.). Theor Appl Genet 109:1669–1676

Michaels SD, Ditta G, Gustafson-Brown C, Pelaz S, Yanofsky M, Amasino RM (2003) AGL24 acts as a promoter of flowering in *Arabidopsis* and is positively regulated by vernalization. Plant J 33:867–874

Negi SS, Olmo HP (1966) Sex conversion in a male *Vitis vinifera* L. by a Kinin. Science 152:1624–1625

Oh E, Zhu JY, Wang ZY (2012) Interaction between BZR1 and PIF4 integrates brassinosteroid and environmental responses. Nat Cell Biol 14:802–809

Pike LM, Peterson CE (1969) Gibberellin A4/A7, for induction of staminate flowers on the gynoecious cucumber (*Cucumis sativus* L.). Euphytica 18:106–109

Plackett AR, Thomas SG, Wilson ZA, Hedden P (2011) Gibberellin control of stamen development: a fertile field. Trends Plant Sci 16:568–578

Prescott AM, Mccollough FW, Eldreth BL, Binder BM, Abel SM (2016) Analysis of network topologies underlying ethylene growth response kinetics. Front Plant Sci 7:1308

Prisic S, Peters RJ (2007) Synergistic substrate inhibition of ent-copalyl diphosphate synthase: a potential feed-forward inhibition mechanism limiting gibberellin metabolism. Plant Physiol 144:445–454

Rosas U, Mei Y, Xie Q, Banta JA, Zhou RW, Seufferheld G, Gerard S, Chou L, Bhambhra N, Parks JD, Flowers JM, Mcclung CR, Hanzawa Y, Purugganan MD (2014) Variation in *Arabidopsis* flowering time associated with cis -regulatory variation in CONSTANS. Nat Commun 5:3651

Rudich J, Halevy AH, Kedar N (1972) The level of phytohormones in monoecious and gynoecious cucumbers as affected by photoperiod and ethephon. Plant Physiol 50:585–590

Rushton DL, Tripathi P, Rabara RC, Lin J, Ringler P, Boken AK, Langum J, Smidt L, Boomsma DD, Emme NJ, Chen X, Finer JJ, Shen J, Rushton PJ (2012) WRKY transcription factors: key components in abscisic acid signalling. Plant Biotechnol J 10:2–11

Sakata T, Oshino T, Miura S, Tomabechi M, Tsunaga Y, Higashitani N, Miyazawa Y, Takahashi H, Watanabe M, Higashitani A (2010) Auxins reverse plant male sterility caused by high temperatures. Proc Natl Acad Sci USA 107:8569–8574

Schomburg FM, Amasino RM (2003) Overexpression of a novel class of gibberellin 2-oxidases decreases gibberellin levels and creates dwarf plants. Plant Cell 15:151–163

Seo M, Akaba S, Oritani T et al (1998) Higher activity of an aldehyde oxidase in the auxin-overproducing superroot1 mutant of *Arabidopsis thaliana*. Plant Physiol 116:687–693

Seo M, Peeters AJM, Koiwai H et al (2000) The arabidopsis aldehyde oxidase 3 (AAO3) gene product catalyzes the final step in abscisic acid biosynthesis in leaves. Proc Natl Acad Sci USA 97:12908–12913

Song S, Qi T, Huang H, Ren Q, Wu D, Chang C, Peng W, Liu Y, Peng J, Xie D (2011) The jasmonate-ZIM domain proteins interact with the R2R3-MYB transcription factors MYB21 and MYB24 to affect jasmonate-regulated stamen development in *Arabidopsis*. Plant Cell 23:1000–1013

Song S, Qi T, Huang H, Xie D (2013) Regulation of stamen development by coordinated actions of jasmonate, auxin, and gibberellin in *Arabidopsis*. Mol Plant 6:1065–1073

Sun TP (2011) The molecular mechanism and evolution of the GA-GID1-DELLA signaling module in plants. Curr Biol Cb 21:R338–R345

Takahashi F, Yamagata D, Ishikawa M, Fukamatsu Y, Ogura Y, Kasahara M, Kiyosue T, Kikuyama M, Wada M, Kataoka H (2007) AUREOCHROME, a photoreceptor required for photomorphogenesis in stramenopiles. P Natl Acad Sci USA 104:19625–19630

Tao Q, Niu H, Wang Z, Li Z (2018) Ethylene responsive factor ERF110 mediates ethylene-regulated transcription of a sex determination-related orthologous gene in two cucumis species. J Exp Bot 69:12

Thomas SG, Phillips AL, Hedden P (1999) Molecular cloning and functional expression of gibberellin 2- oxidases, multifunctional enzymes involved in gibberellin deactivation. P Natl Acad Sci USA 96:4698–4703

Trapnell C, Roberts A, Goff L, Pertea G, Kim D, Kelley DR, Pimentel H, Salzberg SL, Rinn JL, Pachter L (2012) Differential gene and transcript expression analysis of RNA-seq experiments with TopHat and Cufflinks. Nat Protoc 7:562–578

Trebitsh T, Rudich J, Riov J (1987) Auxin, biosynthesis of ethylene and sex expression in cucumber (*Cucumis sativus*). Plant Growth Regul 5:105–113

Trebitsh T, Staub JE, O'Neill SD (1997) Identification of a 1-aminocyclopropane-1-carboxylic acid synthase gene linked to the female (f) locus that enhances female sex expression in cucumber. Plant Physiol 113:987–995

Valverde F, Mouradov A, Soppe W, Ravenscroft D, Samach A, Coupland G (2004) Photoreceptor regulation of CONSTANS protein in photoperiodic flowering. Science 629:1003–1006

Wan H, Zhao Z, Qian C, Sui Y, Malik AA, Chen J (2010) Selection of appropriate reference genes for gene expression studies by quantitative real-time polymerase chain reaction in cucumber. Anal Biochem 399:257–261

Wang DH, Li F, Duan QH, Han T, Xu ZH, Bai SN (2010) Ethylene perception is involved in female cucumber flower development. Plant J 61:862–872

Wang FX, Li SJ, Yu DQ (2011) Overexpression of WRKY25 causes early flowering in *Arabidopsis* under long-day conditions. Plant Divers 33:653–659

Wang C, Xin M, Zhou X (2018) Transcriptome profiling reveals candidate genes associated with sex differentiation induced by night temperature in cucumber. Sci Hortic 232:162–169

Wittwer SH, Bukovac MJ (1962) Staminate flower formation on gynoecious cucumbers as influenced by the various gibberellins. Sci Nat 49:305–306

Young MD, Wakefield MJ, Smyth GK, Oshlack A (2010) Method gene ontology analysis for RNA-seq: accounting for selection bias. Genome Biol 11:R14

Zhang Y, Liu B, Yang S, An J, Chen C, Zhang X, Ren H (2014a) A cucumber *DELLA* homolog *CsGAIP* may inhibit staminate development through transcriptional repression of B class floral homeotic genes. PLoS ONE 9:e91804

Zhang Y, Zhang X, Liu B, Wang W, Liu X, Chen C, Liu X, Yang S, Ren H (2014b) A *GAMYB* homologue *CsGAMYB1* regulates sex expression of cucumber via an ethylene-independent pathway. J Exp Bot 65:3201–3213

Zhang Z, Ji R, Li H, Zhao T, Liu J, Li C, Liu B (2014c) CONSTANS-LIKE 7 (COL7) is involved in phytochrome B (phyB)-mediated light-quality regulation of auxin homeostasis. Mol Plant 7:1429–1440

Zhang Y, Zhao G, Li Y, Mo N, Zhang J, Liang Y (2017) Transcriptomic analysis implies that GA regulates sex expression via ethylene-dependent and ethylene-independent pathways in cucumber (*Cucumis sativus* L.). Front. Plant Sci 8:10

Zhong S, Shi H, Xue C, Wang L, Xi Y, Li J, Quail PH, Deng XW, Guo H (2012) A molecular framework of light-controlled phytohormone action in *Arabidopsis*. Curr Biol Cb 2:1530–1535

Zhu Y, Dun X, Zhou ZF, Xia SQ, Yi B, Wen J, Shen J, Ma C, Tu J, Fu T (2010) A separation defect of tapetum cells and microspore mother cells results in male sterility in *Brassica napus*: the role of abscisic acid in early anther development. Plant Mol Biol 72:111–123

14

Nuclear and chloroplast DNA phylogeography suggests an Early Miocene southward expansion of *Lithocarpus* (Fagaceae) on the Asian continent and islands

Chih-Kai Yang[1,2†], Yu-Chung Chiang[3†], Bing-Hong Huang[1], Li-Ping Ju[4] and Pei-Chun Liao[1*]

Abstract

Background: Most genera of Fagaceae are thought to have originated in the temperate regions except for the genus *Lithocarpus*, the stone oaks. *Lithocarpus* is distributed in subtropical and tropical Asia, and its ancestral population is hypothesized to be distributed in tropical regions in Borneo and Indochina. Borneo and the nearby islands (the Greater Sunda Islands) were connected to the Malay Peninsula and Indochina prior to the Pliocene epoch and formed the former Sundaland continent. The Southeast Asian *Lithocarpus*, is thought to have dispersed between continental Asia and the present Sundaland. The drastic climate changes during the Pliocene and Pleistocene epochs which caused periodic sea-level changes is often used to explain the cause of its diversity. The aim of this study was to establish phylogenetic relationships by analyzing nuclear (nrDNA) and chloroplast (cpDNA) DNA in order to describe and analyze the origin, causes of diversification and historical biogeography of *Lithocarpus*.

Results: Phylogeny reconstructed through the multiple-species coalescent method with nrDNA and cpDNA revealed that the continental-Asian taxa were clustered at the basal lineages. The derived lineages of tropical *Lithocarpus*, with the inference of a subtropical ancestral state, imply a southward migration in the Early Miocene period with subsequent in situ diversification in the Greater Sunda Islands. The gradual decrease in temperature since the Middle Miocene period is proposed as a cause of the increase in the net diversification rate.

Conclusions: The historical ancestral origin of *Lithocarpus* has been suggested to be mainland Asia. Southward migration in the Early Miocene period with subsequent in situ diversification could explain the current diversity of stone oaks in Southeast Asia. This study also considered the multiple origins of stone oaks currently indigenous to the subtropical islands offshore and near mainland China. Our results provide phylogenetic evidence for a subtropical origin of Asian stone oaks and reveal the process of diversification and how it fits into the timeline of major geologic and climatic events rather than local, episodic, rate-shifting events.

Keywords: Stone oak, Continental Asia, Dispersal–extinction–cladogenesis (DEC), Diversification rate, Endemism, Greater Sunda Islands, Historical biogeography, Indochina, *Lithocarpus*, Phylogeny

*Correspondence: pcliao@ntnu.edu.tw
†Chih-Kai Yang and Yu-Chung Chiang contributed equally to this study and share first authorship
[1] Department of Life Science, National Taiwan Normal University, 88, Ting-Chow Rd, Sec 4, Taipei 116, Taiwan
Full list of author information is available at the end of the article

Background

Dispersal and vicariance are the main biogeographic factors that contribute to species distribution (Donoghue et al. 2001; Manos and Stanford 2001). Quaternary glacial cycles that caused topographic changes and climatic fluctuations should have affected the distribution of species by both dispersal and vicariance (Hewitt 2004). For example, the well-known retreat of European white oaks into the southern refugia during glacial periods and their interglacial northward expansion explains the reshuffling of species distribution (Petit et al. 2002a, b). In Southeast Asia, the emergence of ancient Sundaland, the former continent connecting the Thai-Malay peninsula to Borneo and the Philippine archipelagos, was suggested as a land bridge for species dispersal between Indochina and the Malay Archipelago, but the submergence of the Sunda shelf was the prime barrier for vicariant events between the Malay Archipelago and continental Asia (Lohman et al. 2011). Glacially-dependent demographic changes leading to genetic differentiation of populations were also reported for several eastern and Southeast Asian land plants, such as *Ceriops tagal* (Rhizophoraceae) (Liao et al. 2011), *Castanopsis carlesii* (Fagaceae) (Cheng et al. 2005), *Dysosma versipellis* (Berberidaceae) (Fu et al. 2009), *Machilus* spp. (Lauraceae) (Wu et al. 2006), and *Pinus massoniana* (Pinaceae) (Ge et al. 2012).

In addition to the topographic changes in Southeast Asia (e.g., Sundaland), past climate changes also affected the flora of subtropical China and adjacent areas (Qiu et al. 2011). In their report, Qiu et al. (2011) summarized the evidence that Southwest China, which is adjacent to the Indochinese Peninsula, has the greatest floral biodiversity. The complex geohistory and physiographical heterogeneity of Southwest China provided multiple refuges for plant populations (e.g., Nanling Mts., Yungui Plateau, Hengduan Mts., etc.) during Quaternary glacial periods in terms of inter- and postglacial northward, eastward, and westward recolonization (Qiu et al. 2011). Past climate changes also severely affected the biogeographic patterns of Southeast Chinese, Japanese, Korean, and Taiwanese plants (reviewed by Chiang and Schaal 2006; Qiu et al. 2011) such as *Cunninghamia konishii* (Cupressaceae) (Chung et al. 2004), *Ainsliaea* spp. (Asteraceae) (Mitsui and Setoguchi 2012), *Picea jezoensis* (Pinaceae) (Aizawa et al. 2007), *Pinus luchuensis* complex (Pinaceae) (Chiang et al. 2006), and *Ligularia hodgsonii* (Asteraceae) (Wang et al. 2013). The distribution of these plants was probably due to dispersal/vicariance events driven by the emergence/submergence of the last glacial East China Sea land bridges (Qiu et al. 2011); similar events occurred among fauna (Millien-Parra and Jaeger 1999).

Stone oaks (*Lithocarpus* Bl., Fagaceae) are the most widely distributed trees in East and Southeast Asia ranging from northeast India to South China, southern Japan and Taiwan, Indochina, Borneo, and Papua New Guinea (Nixon 1989). Across this range, severe topography changes took place during the Quaternary glacial/interglacial cycles. In older studies, *Lithocarpus* was supposed to have had a tropical origin in Southeast Asia and an Asian migration which occurred at least twice: a westward dispersal to Europe identified based on the fossil records (no extant species) and an eastward colonization of western North America (one extant species) (Cannon and Manos 2003). However, the American species has been identified as a novel genus *Notholithocarpus* because of its distant phylogenetic relationship to the Asian species (Manos et al. 2008); therefore, the rest of the Asian *Lithocarpus* species formed a monophyly (Manos et al. 2001, 2008; Oh and Manos 2008). *Lithocarpus* sensu lato is composed of approximately 300 species and is the second largest genus of Fagaceae (Nixon 1989).

Manos and Stanford (2001) claimed that the most ancient haplotypes were distributed in northern Borneo and Indochina, which then separated into the Borneo population and the Indochina population in the middle Miocene, during which Sundaland was narrowed and the Gulf of Thailand appeared. After vicariance, the prolonged in situ diversification increased the frequency of endemics (Cannon and Manos 2003). Cannon and Manos (2003) also proposed a recent diversification of the genus *Lithocarpus* according to unresolved gene trees of cpDNA (chloroplast DNA) and nrITS (nuclear ribosomal internally transcribed spacers). However, their inference was mainly based on a study using very biased sampling in Borneo with only a few samples from continental Asia. According to several case studies regarding the isolation and migration of plant species between continental Asia and the Greater Sunda Islands, the diversification of Southeast Asian *Lithocarpus* might be related to recurrent separation and dispersal between geographical-transitional species congeners. Such recurrent secondary contacts were also highlighted for insular species diversity (e.g., the origin of Taiwanese endemic species, Chiang et al. 2012; Chung et al. 2007). The relatively high diversity and endemism of *Lithocarpus* in continental islands near mainland China provide opportunities to test the hypotheses of the origin of insular forest species, either multiple origins by continent-island colonization or single origins with in situ radiation.

In this study, we used East and Southeast Asian stone oaks (*Lithocarpus*) as subjects to determine whether phylodiversity was related to biogeographic events regarding climatic-related geographic timelines. Phylogenetic analyses of *Lithocarpus* (Fagaceae) based on maternally inherited plastid markers and biparentally inherited nuclear markers were assessed to explore the biogeographic patterns via phylogeographical analyses. Based on surveys of topological and temporal analyses, we proposed answers to the following questions: (1) Are there any diversification rate shifts in *Lithocarpus* in different geographical regions? (2) Do the rate-shift events correspond to any geographical event? (3) How is the insular diversity generated?

Methods

Plant material sampling and molecular techniques

Leaf samples were collected from stone oak populations covering the areas of Nanling (E113°01′15″, N24°58′51″), Mangdang Shan (E118°07′33″, N26°41′40″), Emei (E103°16′27″, N29°34′52″), Chinfo Shan (E107°09′43″, N29°01′04″), Kunming Golden Temple (E102°46′10″, N25°05′15″), and botanical gardens (Additional file 1: Table S1). Total genomic DNA was extracted from leaf samples using a modified cetyl trimethylammonium bromide (CTAB) method (Doyle and Doyle 1987). The DNA samples were amplified by polymerase chain reaction (PCR) using primers specific for the regions of the *atp*B-*rbc*L spacer of the chloroplast (cp) genome and the nuclear ribosomal internal transcribed spacer (nrITS) of the nuclear genome. PCR products of the *atp*B-*rbc*L spacer were directly sequenced. The nrITS amplicons were cloned using the yT&A cloning kit (Yeastern Biotech, Taipei, Taiwan) and sequenced with M13F and M13R primers. Sequencing reactions were performed using the ABI BigDye 3.1 Terminator Cycle Sequencing Kit (Applied Biosystems, Foster City, CA, USA) and the fragments were sequenced from both directions with the ABI PRISM®3730XL DNA Sequencer (Perkin-Elmer, Foster City, CA, USA). All sequence polymorphisms were visually rechecked from the chromatograms.

Sequence alignment

Chromatograms of DNA sequences were inspected by SeqMan implemented in DNASTAR ver. 7.0 (Lasergene, Germany). Genetic confirmation of each sequence was done by querying the nucleotide database in NCBI using the BLASTN program and NetPlantGene server in the Center for Biological Sequence Analysis (http://www.cbs.dtu.dk/services/NetPGene/). All sequences were deposited in the NCBI nucleotide sequence database under the following accession numbers: KF992718–KF992796 (*atp*B-*rbc*L) and KJ685163–KJ685212 (nrITS). To enrich the sampling number, 108 sequences of *atp*B-*rbc*L and 77 sequences of nrITS from NCBI were incorporated for analyses. Sequence alignments were performed by Clustal X (Thompson et al. 1997) and manually edited using the BioEdit ver.7.0.9.0 (Hall 1999). Long fragments of sequences were trimmed to equal alignment lengths for subsequent phylogenetic analyses.

Phylogenetic analyses

As the taxa we collected for chloroplast and nuclear phylogenetic analyses were not exactly identical, multispecies coalescent phylogenetic trees (Drummond and Rambaut 2007; Heled and Drummond 2010) were reconstructed based on combinations of genomic markers and the two target markers separately. Consequently, three trees were developed as data structures for further analysis: the cpDNA + nrITS tree (64 species), the cpDNA tree (81 species), and the nrITS tree (70 species). Since extinction events in *Lithocarpus* are rare (Cannon and Manos 2003; Xing et al. 2014), it is likely that the rare event is overestimated from incomplete taxon sampling. Therefore the extinctions were deliberately excluded from species tree reconstruction to explain the species diversification, and we considered only cladogenesis and biogeographical factors. Yule's pure-birth model that ignores the death (extinction) rate parameter was adopted for phylogeny reconstruction. The general-time-reversible substitution model (GTR) together with a gamma and invariant site heterogeneity model by BEAST ver. 1.7.3 (Drummond and Rambaut 2007) were used. Since the ucld.stdev parameter estimate deviated from zero, to account for the rate of heterogeneity among lineages in both markers (95% HPD of ucld.stdev = 0.849 ~ 1.289 and 0.759 ~ 1.407 in ITS and cpDNA, respectively) we have used the uncorrelated lognormal (UCLN) relaxed clock model instead of the strict clock model to estimate divergence times of about 40 Mya from the outgroup genus *Chrysolepis* (Kvaček and Walther 1989). Five independent pre-runs of ten-million generations of the length of MCMC were performed to obtain better parameter priors for the following three independent 50-million generations of the MCMC process. Genealogies for each lineage were sampled every 10,000 generations and the initial 10–25% of the sample run was discarded as burn-in according to the simulation trajectory. The output data containing all the statistics were summarized by TRACER ver. 1.5 (Rambaut and Drummond 2003) and both log and tree files of the last five runs were combined using LogCombiner ver. 1.6.1 (Drummond and Rambaut 2007). We resampled the originally simulated trees with a sampling frequency of 8000 and 6000 trees for ITS and cpDNA, respectively, to reduce the number of trees by LogCombiner ver. 1.6.1 and the posterior samples of trees were summarized by TreeAnnotator ver. 1.6.1 (Drummond and Rambaut 2007). FigTree ver. 1.3.1 (Rambaut 2008) was used to display the summarized trees.

Biogeographic inferences

We applied two strategies to infer the biogeographic events of Asian stone oaks: the statistical dispersal-vicariance (S-DIVA) model and the dispersal–extinction–cladogenesis (DEC) model, performed by RASP (Yu et al. 2015) and LAGRANGE (Matzke 2014) software, respectively. As described above, the extinction rate has probably been misidentified and overestimated by incomplete taxon coverage. However, Yule's pure-birth model is a better fit to our data than the birth-and-death process as suggested by ΔAIC_{RC} test statistics (Table 2). Therefore 'extinction' was not considered in this study when reconstructing the phylogenetic tree for subsequent DEC analysis.

Ancestral area reconstruction was performed using both S-DIVA and DEC models. The study species were mapped into four ancestral areas: (A) continental Asia (including mainland China, northern India, and the islands off continental Asia), (B) Indochina, (C) the Malay Peninsula, and (D) the Greater Sunda Islands (including Borneo, the Philippines, Sumatra, Java, and other Indonesian archipelagos) (Fig. 1). The Greater Sunda Islands and Indochina were Cannon and Manos's (2003) hypothetical ancestral areas; extensions of the Malay Peninsula connected to the Greater Sunda Islands and Indochina form ancient Sundaland. Continental Asia is the major section where most of the sympatric speciation occurs in *Lithocarpus* with other species of Fagaceae, and it is an alternative ancestral area for *Lithocarpus*.

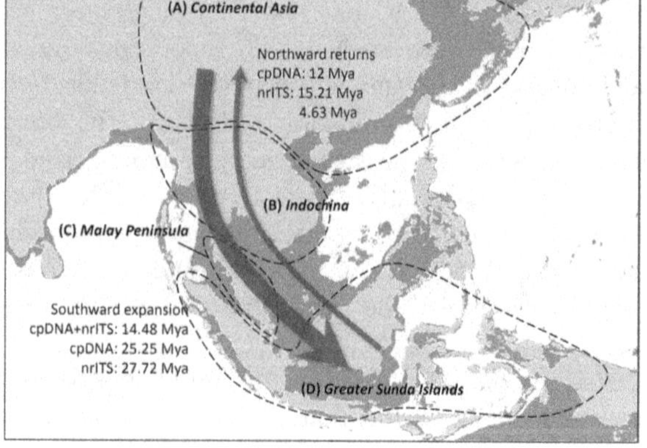

Fig. 1 Geographic distribution and the probable directions of species dispersal. Geographic distribution of the Asian *Lithocarpus* is separated into **A** continental Asia, **B** Indochina, **C** the Malay Peninsula, and **D** the Greater Sunda Islands. Arrows point to the putative dispersal routes and the migration times estimated by cpDNA and nrITS are denoted. Light gray regions are the present landmass; dark gray regions are the shallow seabed, representing the hypothetical coastlines of continental Asia and Sundaland during the Miocene and Quaternary glacials

Homogeneity tests for whole-tree diversification rates

The asymmetric rate shift of nodes of the species tree was estimated using the taxon-size insensitive (TSI) model and equal-rate Markov (ERM) random branching model of the SymmeTree program (Moore and Chan 2005). The Colless's tree imbalance index (I_C), the nodal probability product (M_Π and the modified version M_Π*), and sum (M_Σ and M_Σ*) were used to display the diversification rate variation of the whole tree (Chan and Moore 2002; Moore et al. 2004). Homogeneous evolutionary rates of descendant clades from the common ancestor (the node) were tested with delta-shift statistics (Δ1 and Δ2) at all nodes (Moore et al. 2004). One million random resolutions with 1000 TSI/ERM-resolved trees under one-million ERM simulations were performed to identify significant diversification rate shifts of each node. We selected the best diversification rate model (rate-constant, RC, or rate-variable, RV) through ΔAIC_{RC} test statistics using the *fitdAICrc* function of R package LASER ver. 2.3 (Rabosky 2006). The *fitdAICrc* tested the Yule's pure-birth (pureBirth) and birth–death (bd) models of RC, and the exponential and logistic variants of the density-dependent speciation rate models (DDX and DDL) and multi-rate variants of the pure-birth model (yule2rate and yule3rate) of RV.

Temporal-based diversification-rate-shift analyses

A temporal analysis that accumulates speciation events (lineages) through time was performed to infer the emergence time of species using lineage-through-time (LTT) analysis. The LTT analysis was performed in R using APE package (Paradis et al. 2004). We also resampled 1000 trees from the simulated-trees file by LogCombiner to display the LTT distribution of simulated results using the function *mltt.plot* in APE. Chronograms of the species tree reconstructed by BEAST were used as input trees. Effective sizes of lineages over time were also estimated using the reversible-jump Markov chain Monte Carlo computation (rjMCMC) (Opgen-Rhein et al. 2005) by the function *mcmc.popsize* of APE. One-million-step Markov-chain simulations were performed for the rjMCMC estimation, and the first 1000 steps were dropped as burn-in, with a setting thinning factor = 1000, for both whole tree topologies and subclades.

Diversification rate estimation by Bayesian analysis of macroevolutionary mixtures (BAMM)

Since the number of species used for reconstructing the species tree (63 species) is roughly 20% of the total number of *Lithocarpus* species recognized in the world (~ 300 species), we recalculated the net diversification rate with 20% sampling fractions by the BAMM approach

(Rabosky et al. 2014). For estimating the net diversification rate, 5,000,000 Markov chain Monte Carlo simulations with 1/1000 thinning and 10% burn-in were set. The phylorate plot that shows the speciation rate along branches of the phylogeny was drawn with the Jenks natural breaks method. A histogram of rate heterogeneity was also drawn to illustrate proportions of lineages of rate clusters in the phylogenetic tree. All of the diversification rate analyses were conducted with the assistance of the BAMM program and R package BAMMtools (Rabosky et al. 2014).

Results

Phylogenetic inferences

A total of 187 sequences of cp *atp*B-*rbc*L fragment from 81 species and 132 sequences of nrITS from 70 species were determined for phylogenetic, biogeographic, and diversification rate shift analyses. The tree topologies inferred from individual markers were significantly different (Additional file 1: Figure S1), and a strong heterogeneity and incomplete lineage sorting was found under different evolutionary histories (Hoelzer 1997; Page and Charleston 1998). Therefore, divergence time estimates for biogeographic events were calculated with a strategy that analyzed both markers via the multiple-species coalescent analysis implemented in *BEAST.

The splitting time of *Lithocarpus* estimated by multi-species coalescent methods with a combined cpDNA and nrITS tree (hereafter: combined cpDNA and nrITS data) was about 33.91 Mya [95% highest posterior density interval (95% HPD): 25.98–41.35 Mya], which was probably around the time of the Eocene–Oligocene boundary (known as the Grande Coupure; it is marked by the Eocene–Oligocene extinction event), by rooting with the outgroup *Chrysolepis chrysophylla* at *c.* 40 Mya based on the relaxed-clock model (Fig. 2a). The

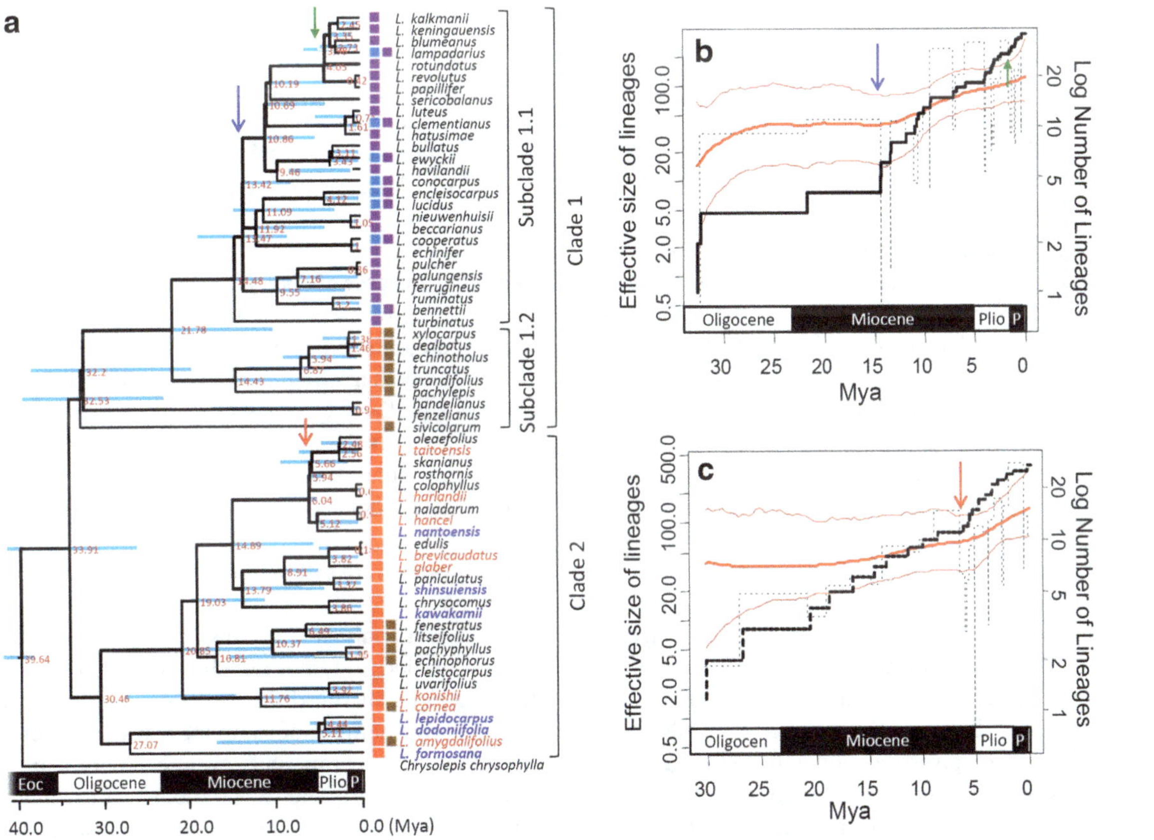

Fig. 2 Species tree and temporal analysis of diversification rates inferred by both cpDNA *atp*B-*rbc*L and nrITS. **a** Species tree reconstructed under the Yule's pure-birth speciation model. Bold lines indicate lineages grouping with posterior probability > 80%; node labels are the splitting time (mya); the node bar is the 95% highest posterior density interval (HPD) of the splitting time. Geographic distribution areas are displayed as colored boxes. **b** and **c** are skylines of diversification rates of clade 1 and clade 2, respectively, inferred by rjMCMC and LTT. Dashed lines represent the classic skyline plots, while the bold and thin red lines indicate the skylines estimated with the rjMCMC and the corresponding 95% confidence intervals, respectively. The bold black line is the LTT plot. The blue, green and red arrows indicate the phylogenetic locations and times of diversification rate shifts

coalescent estimation is consistent with the estimates formed from individual markers cpDNA (33.44 Mya, 95% HPD: 28.23–41.27 Mya) and nrITS (37.19 Mya (95% HPD: 25.89–41.28 Mya). Since that time, *Lithocarpus* has been divided into two clades; the species that make up the first clade are widespread in continental Asia, the continental islands (Japan, Taiwan, and Hainan), the Southeast Asian islands (including the Greater Sunda Islands and the Philippines), and these are shown as clade 1 (cpDNA + nrITS), clade cp-1 (cpDNA), and clade ITS-I (nrITS) respectively (Figs. 2, 3, 4). The second clade is composed of species that are restricted in their geographic distribution to China and the nearby islands (Japan, Taiwan, and Hainan), and these are displayed as clade 2, clade cp-2, and clade ITS-II (Figs. 2a, 3, and 4) respectively. The splitting times of other species that were isolated from their geographic distribution in the Malay Peninsula, the Greater Sunda islands, and the Philippines grouped in subclade 1.1 might have coalesced back to 14.48 Mya (95% HPD: 7.66–19.22 Mya) (Fig. 2a), roughly at the end of the Middle Miocene disruption (14.8 ~ 14.5 Mya). Within clade cp-1 of the cpDNA tree, subclades cp-1a and cp-1b are evident at 25.25 Mya (95% HPD: 14.77–35.54 Mya), during the Late Oligocene period (Fig. 3). However, no sub-clades can be distinguished from clade ITS-I of the nrITS tree (Fig. 4).

Whole-tree diversification rate homogeneity tests

Testing the rate of homogeneity of the whole-tree diversification is the first step in evaluating the cladogenesis imbalance of the independent clades (Ricklefs 2007). Two analytical strategies, the topological analysis of the tree imbalance and the ΔAIC_{RC} statistic, were used for the whole-tree rate homogeneity tests. Table 1 summarizes the distribution of ERM simulations for five tree-imbalance indices. Tail probabilities of all indices are < 0.001 in the tree inferred from combined cpDNA and nrITS data sets, indicating the rejection of rate homogeneity of the whole trees (Table 1). However, no significant imbalance in the number of descendants was derived from any node when tested by delta-shift statistics (data not shown), suggesting a gradual change in diversification rates but not a punctuated diversification (Moore et al. 2004).

The inference of diversification rate heterogeneity of *Lithocarpus* was also supported by positive ΔAIC_{RC} values of the trees assessed from combined marker sets (Table 2). The yule3rate of the RV model had a higher likelihood than other RC and RV models and showed the highest net diversification rate at stage 2 (r2) in all trees (r2 = 9.936 in Fig. 2a and Table 2). A very short time span (stage 1–stage 2 [st1–st2]) of diversification rate increase was estimated in spite of quite different estimated times for each tree (st1 = 5.939 Mya, st2 = 5.936 Mya in Fig. 2a and Table 2).

Biogeographic inferences

The ancestral area of the Asian stone oaks inferred from the DEC model is suggested to be continental Asia (or mainland China) by cpDNA + nrITS markers (Fig. 5 and Additional file 1: Figure S2A, -lnL = 76.94) in the early Oligocene period. The geographical distribution of extant *Lithocarpus* species is mostly a result of dispersal events and joined events of dispersal + vicariance, inferred by S-DIVA using combined cpDNA + nrITS markers (Fig. 5a, Additional file 1: Figure S2). Dispersal events are mostly inferred at clade 1, indicating that the species distributed in southern areas were mostly contributed by the Miocene southward expansions.

The southward colonization history of the common ancestor of these southern species from continental Asia could be traced back to 21.78 ~ 14.48 Mya (Figs. 2a and 5). This timing also matched the Miocene Climatic Optimum (MCO). Furthermore, the dispersal-related events followed by vicariance explained the far-distant close relatives. The S-DIVA inference from combined data showed a southward expansion followed by a vicariance that separated the Greater Sunda Islands from the areas of Indochina + continental Asia at 21.78 Mya (95% HPD: 9.92–25.28 Mya, posterior prob. = 0.4755) (Figs. 2a and 5).

Temporal analyses of diversification rate shifts

In the temporal analyses, the LTT of the tree inferred from the combined markers cpDNA and nrITS (Fig. 6) showed two sharp increases in diversification rates at *c.* 35 Mya (the Early Oligocene) and 15 Mya (Middle Miocene) (see gray area of LTT plot simulated from 1000 post-convergence trees), which roughly correspond to the end of the Grande Coupure and the MCO, respectively.

Diversification rate estimation considering incomplete sampling

The BAMM analysis that considered 20% sampling fractions revealed a temporal change rather than spatial heterogeneity in the diversification rate of *Lithocarpus*. The effective number of sampling sizes (ESS) for the log-likelihood and the rate-shift-event estimation are 2249.43 and 4217.825, verifying the credibility of the inference of simulation. From 4501 simulation samples, the shift posterior distribution of 0, 1, 2, and 3 times rate-shift events are 0.97, 0.032, 0.0011, and 0.00022, respectively, indicating that a no-diversification rate shift between clades is the best evolutionary scenario for the diversification rate shift. We further estimated the speciation rates of tropical lineages (the southern lineages: GSI + Malay

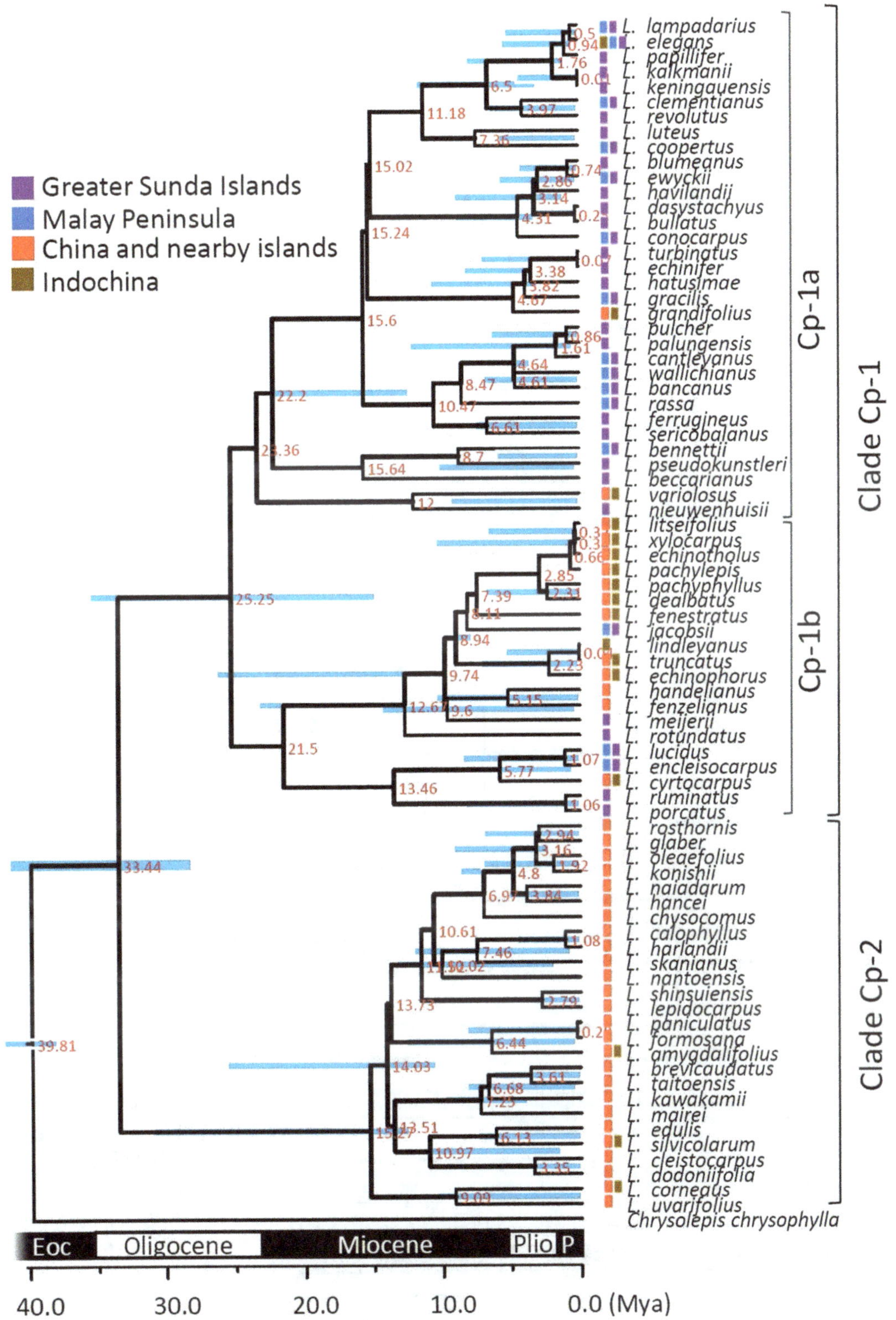

Fig. 3 Species tree and temporal analysis of diversification rates inferred by cpDNA *atp*B-*rbc*L spacer reconstructed under Yule's pure-birth speciation model. Bold lines indicate the lineages grouping with posterior probability > 80%; node labels are the splitting time (unit: mya); node bar is the 95% highest posterior density interval (HPD) of the splitting time. Geographic distribution areas were displayed as the colored box

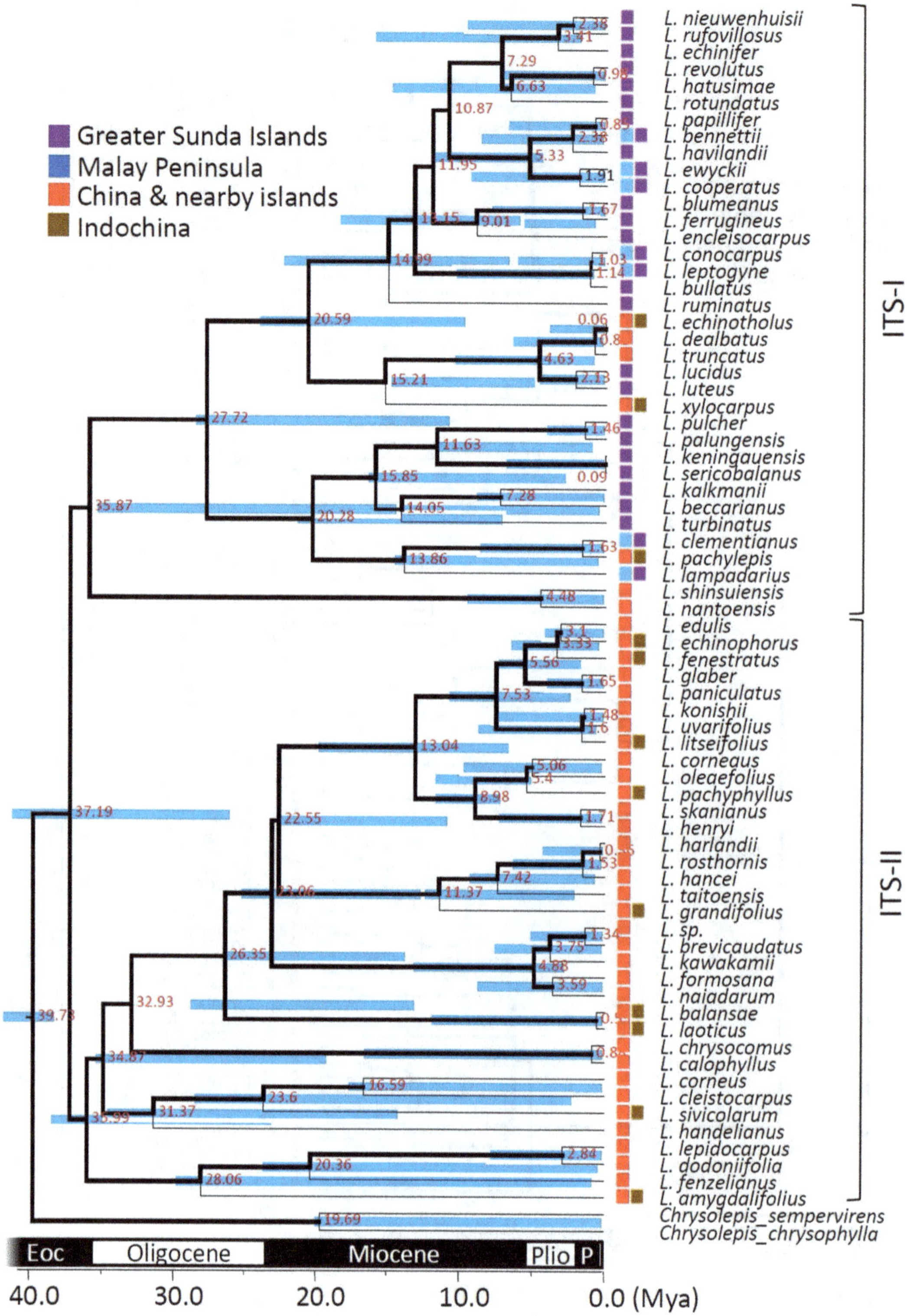

Fig. 4 Species tree and temporal analysis of diversification rates inferred by nuclear ITS spacer reconstructed under the Yule's pure-birth speciation model. Bold lines indicate the lineages grouping with posterior probability > 80%; node labels are the splitting time (mya); the node bar is the 95% highest posterior density interval (HPD) of the splitting time. Geographic distribution areas are displayed as the colored boxes

Table 1 Tail probabilities of asymmetric values for the among-lineage diversification rate variation in the phylogenetic topologies inferred from the species tree (BEAST)

	I_C	M_Π^*	M_Π	M_Σ^*	M_Σ
Both cpDNA and nrITS (Fig. 2a)					
Observed	251	−0.5855	1.50E−10	0.6747	50.2261
Min ERM	507	−0.9566	1.93E−17	0.5302	44.0535
Max ERM	64	−0.0856	0.0472	0.9256	61.3567
0.025 frequentile RR	368	−0.8315	1.08E−13	0.5918	47.8035
0.975 frequentile RR	233	−0.5207	4.01E−08	0.7284	54.2252
TailPr	0.0006	0.0003	3.00E−05	0.0001	6.00E−05
cpDNA (Fig. 3)					
Observed	324	−0.5129	4.33E−10	0.7196	66.2184
Min ERM	651	−0.9533	5.95E−19	0.5418	55.7200
Max ERM	104	−0.1478	0.0010	0.8833	73.5864
0.025 frequentile RR	440	−0.7518	1.98E−15	0.6262	60.3685
0.975 frequentile RR	227	−0.3930	2.91E−08	0.7900	68.394
TailPr	0.0051	0.0005	8.00E−05	0.0002	0.0002
nrITS (Fig. 4)					
Observed	294	−0.5483	1.43E−10	0.7035	58.2993
Min ERM	589	−1.0168	2.51E−19	0.5244	50.1407
Max ERM	79	−0.1281	0.0020	0.8990	67.3080
0.025 frequentile RR	386	−0.7540	3.50E−14	0.6198	55.2110
0.975 frequentile RR	215	−0.4135	9.37E−08	0.7744	62.4962
TailPr	0.0044	0.0023	0.0037	0.0059	0.0048
cpDNA (Additional file 1: Figure S1A)					
Observed	599	−0.5573	3.72E−19	0.6878	103.0830
Min ERM	1112	−0.8805	1.77E−28	0.5576	94.1847
Max ERM	220	−0.2034	2.54E−07	0.8536	117.8260
0.025 frequentile RR	854	−0.7510	2.40E−23	0.6348	100.578
0.975 frequentile RR	463	−0.4276	8.03E−14	0.7606	110.5650
TailPr	0.0007	3.00.E−05	0.0001	0.0007	0.0022
nrITS (Additional file 1: Figure S1B)					
Observed	724	−1.0221	2.02E−21	0.5400	60.0154
Min ERM	754	−1.0083	5.14E−20	0.5637	60.8534
Max ERM	120	−0.1426	0.0004	0.8893	80.0882
0.025 frequentile RR	652	−0.9238	9.54E−20	0.5688	62.8712
0.975 frequentile RR	351	−0.5370	1.53E−12	0.6977	70.1858
TailPr	0.0021	0.0005	0.0019	0.0017	0.0144

I_C is the Colless's tree imbalance index; M_Π and M_Σ are the nodal probability product and nodal probability sum of the tree, respectively; M_Π^* and M_Σ^* are modified versions of M_Π and M_Σ obtained through differential weighting of the individual equal-rate Markov nodal probabilities according to their species diversity. These indices display the diversification rate variation of the whole tree

Peninsula), subtropical lineages (the northern lineages: China + Indochina), and the whole tree (genus *Lithocarpus*). The rates were 0.551 (5%–95% quantile: 0.357–0.799), 0.540 (5%–95% quantile: 0.342–0.806), and 0.543 (5%–95% quantile: 0.348–0.803), respectively, revealing rate homogeneity among clades and a slight, congruent positive skewness of diversification rate (i.e., speciation rate > extinction rate, Fig. 6). The phylo-plot and density-plot of the net diversification rate revealed a congruent trend towards a slight increase of the net diversification rate among different clades, suggesting a homogeneous diversification rate in space and temporal change of diversification rates in whole *Lithocarpus* lineages (Fig. 5). These estimates suggested that although the species composition of *Lithocarpus* differed among localities, the speciation and net diversification rates were unrelated to geographical distribution.

Table 2 Testing for diversification rate variation models by ΔAIC$_{RC}$ test statistic

Model	Log-likelihood	AIC	Parameter
Both cpDNA and nrITS (Fig. 2a)			
PureBirth	−2.3419	*6.684*	r1 = 0.101
bd	−1.682121	7.364	r1 = 0.076, a = 0.405
DDX	−1.972169	7.944	r1 = 0.065, x = −0.136
DDL	−2.342	8.684	r1 = 0.101, k = 1,168,951
yule2rate	−1.083	8.166	r1 = 0.0687, r2 = 0.112, st1 = 14.894
yule3rate	1.871	*6.258*	r1 = 0.087, r2 = 9.936, r3 = 0.115, st1 = 5.939, st2 = 5.936
ΔAIC$_{RC}$		0.426	
cpDNA (Fig. 3)			
PureBirth	25.97718	*−49.95435*	r1 = 0.120
bd	26.15159	−48.30319	r1 = 0.105, a = 0.220
DDX	26.19223	−48.38446	r1 = 0.091, x = −0.082
DDL	25.97713	−47.95426	r1 = 0.120, k = 1,428,290
yule2rate	27.806	−49.612	r1 = 0.063, r2 = 0.130, st1 = 15.636
yule3rate	31.18926	*−52.37852*	r1 = 0.063, r2 = 0.665, r3 = 0.124, st1 = 15.636, st2 = 15.017
ΔAIC$_{RC}$		2.424	
nrITS (Fig. 4)			
PureBirth	−5.870249	13.7405	r1 = 0.086
bd	−4.303511	*12.60702*	r1 = 0.058, a = 0.532
DDX	−5.695618	15.39124	r1 = 0.061, x = −0.105
DDL	−5.870443	15.74089	r1 = 0.086, k = 1,284,297
yule2rate	−2.426486	10.85297	r1 = 0.073, r2 = 0.148, st1 = 2.380
yule3rate	1.189585	*7.62083*	r1 = 0.075, r2 = 0.471, r3 = 0.112, st1 = 1.713, st2 = 1.457
ΔAIC$_{RC}$		4.986	
cpDNA (Additional file 1: Figure S1A)			
PureBirth	103.309	−204.617	r1 = 0.121
bd	107.116	*−210.231*	r1 = 0.076, a = 0.574
DDX	106.169	−208.338	r1 = 0.046, x = −0.251
DDL	103.308	−202.616	r1 = 0.121, k = 2,195,621
yule2rate	108.500	−211.000	r1 = 0.084, r2 = 0.153, st1 = 7.286
yule3rate	112.757	*−215.514*	r1 = 0.095, r2 = 0.184, r3 = 0.024, st1 = 4.059, st2 = 0.328
ΔAIC$_{RC}$		5.283	
nrITS (Additional file 1: Figure S1B)			
PureBirth	27.403	−52.806	r1 = 0.110
bd	31.284	*−58.568*	r1 = 0.060, a = 0.667
DDX	29.731	−55.463	r1 = 0.038, x = −0.306
DDL	21.496	−38.991	r1 = 0.140, k = 174
yule2rate	31.106	−56.213	r1 = 0.103, r2 = 0.368, st1 = 0.227
yule3rate	34.049	*−58.096*	r1 = 0.093, r2 = 9.384, r3 = 0.172, st1 = 2.055, st2 = 2.053
ΔAIC$_{RC}$		−0.471	

r1, r2, r3, net diversification rates at stages 1, 2, and 3; st1 and st2, the first and the second rate-shift times; a, the extinction fraction extinction rate/speciation rate; x, the x parameter in the density-dependent exponetial model; k, the K parameter in the logistic density dependent model

PureBirth pure birth (Yule) model, *bd* rate-constant birth–death model, *DDX and DDL* exponential and logistic variants of the density-dependent speciation rate models, respectively, *yule2rate and yule3rate* multi-rate variants of the pureBirth model, *ΔAIC$_{RC}$* the difference in AIC score between the best rate-constant (AIC$_{RC}$) and rate-variable (AIC$_{RV}$) models. The overall best-fit models were indicated with italic AIC value

The dynamics of effective lineage sizes estimated by rjMCMC sampling represent changes in genetic diversity over time (Opgen-Rhein et al. 2005), which revealed a constant rate of cladogenesis on the logarithmic axis by combined cpDNA and nrITS markers (Fig. 6b). The skylines of effective sizes of lineages inferred by the rjMCMC approach revealed more smooth curves of diversification rates, whereas the classic skylines revealed the intuitional

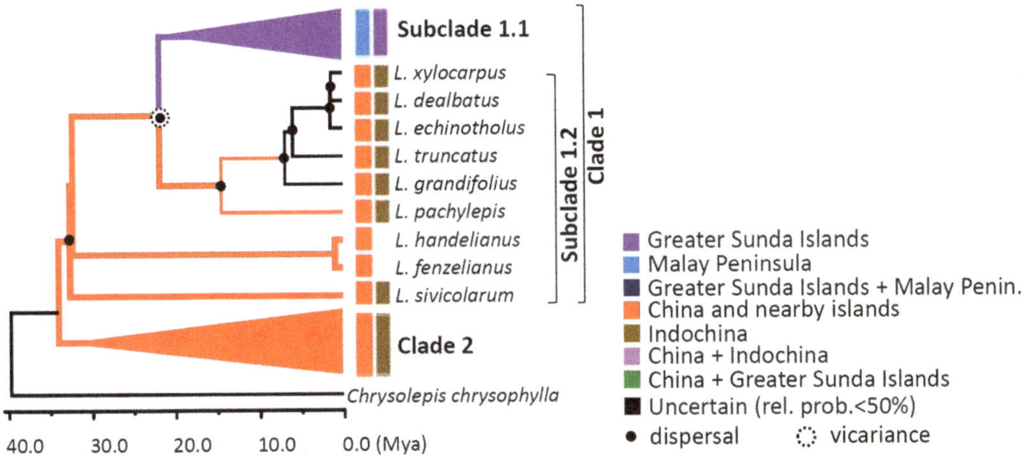

Fig. 5 The species tree reconstructed by combined cpDNA *atpB-rbcL* spacer and nrITS data. Geographic distribution areas are displayed as colored boxes. Colored internodes indicate the ancestral geographic distribution areas reconstructed by DEC model. Lineages with bold and thin lines were derived from nodes (ancestral areas) with likelihood > 0.7 and > 0.5, respectively. Those with likelihoods < 0.5 are treated as unknown (black lineages). Dispersal and vicariance events inferred by S-DIVA are denoted

estimates of lineage sizes at each time interval. Severe fluctuations of effective sizes of lineages at 15 ~ 13 Mya in the Middle Miocene followed by the Late Miocene explosion of lineages at *c.* 9 ~ 7 Mya and another fluctuation of lineage sizes at the Miocene-Pliocene boundary (6 ~ 5 Mya) revealed in the classic skyline plot suggest the conspicuous late-shifts of diversification rates in *Lithocarpus* (Fig. 6b).

Multiple originations of the indigenous stone oaks on mainland China's offshore islands

According to the species tree, the island species were paraphylies or polyphylies, suggesting that these species might have colonized the island multiple times and could have evolved independently. The twelve *Lithocarpus* species on the tropical island of Hainan are sparsely distributed in the phylogeny (Fig. 2a). Two of the three endemic species in Hainan (*L. handelianus* and *L. fenzelianus*) are sister-grouped with short TMRCA at *c.* 0.95 Mya (95% HPD: 0–17.35 Mya), but they are distinct from another endemic species *L. naiadarum* by a long coalescent history [about 33.91 Mya (95% HPD: 25.98–41.35 Mya)], suggesting at least two origins for the Hainan endemic species. The inference of multiple colonization events is also shown on the subtropical island of Taiwan. The Taiwanese species are phylogenetically located in one of the major clades but still sparsely distributed around the tree (Fig. 2a). The coalescent history of Taiwanese indigenous species could be traced back to 30.46 Mya (95% HPD: 9.74–27.74 Mya) and the TMRCA of any two species is longer than 5 Mya (Fig. 2a), which is older than the formation time of Taiwan Island [< 5 Mya, (Chi et al. 1981; Teng 1996; Wang et al. 2002)], indicating multiple colonization times instead of in situ radiation. The temperate island of Japan has only two species, *L. glaber* and *L. edulis*. The former species is also widely distributed in China and Taiwan, while the latter is endemic to Japan. These two Japanese species could have coalesced back to 3.82 Mya (95% HPD: 0.27–6.31 Mya), but the endemic one, *L. edulis*, is sister-grouped with another species, *L. brevicaudatus* (distributed in China and Taiwan, but not in Japan), with very short TMRCA at 0.18 Mya (95% HPD: 0–5.01 Mya) (Fig. 2a). The geological isolation could have differentiated ancestral populations of the common ancestor of *L. brevicaudatus* and *L. edulis* and resulted in the recent speciation of the island species *L. edulis*.

Discussion

Intricate patterns of phylogenetic relationships as inferred from chloroplast and nuclear markers

The different evolutionary trajectories for nuclear and plastid markers may lead to different phylogenetic inferences. The cpDNA tree (Fig. 3, Additional file 1: Figure S1A) and the nrITS tree (Fig. 4, Additional file 1: Figure S1B) share a similar pattern for the early-derived major groupings but are incongruent in the recently derived lineages, suggesting a major influence of biogeographic effects on *Lithocarpus* evolution and less possibly for independent evolutionary trajectories based on chloroplast and nuclear genomes. Conflicts between these two genomes were often attributed to different evolutionary rates and reflect different biogeographic events such as pollen flow and seed dispersal (Ennos 1994; Mccauley 1995; Petit et al. 2005).

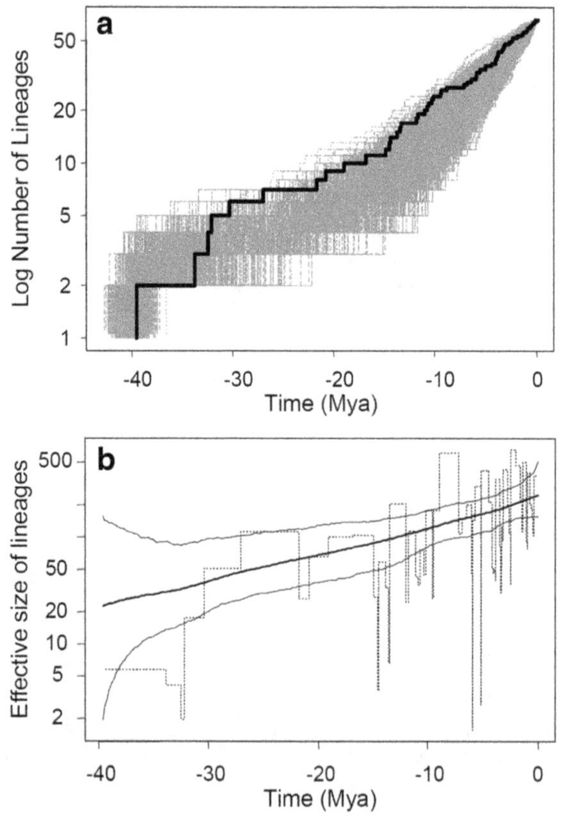

Fig. 6 Skylines of the lineage-through-time (LTT) plots (**a**) and the effective size of lineages (**b**) of Asian *Lithocarpus* species from the cpDNA + nrITS tree. Grey regions of **a** indicate simulated LTT plots based on 1000 post-convergence Bayesian trees computed by BEAST. In **b**, the dashed lines represent the classic skyline plots, while the bold and thin solid lines indicate the skylines estimated with the rjMCMC and the corresponding 95% confidence intervals, respectively

The broad and overlapping confidence intervals in the recently derived clades (Figs. 3 and 4, Additional file 1: Figure S1) indicate uncertainty in phylogenetic inference among closely related taxa. Phylogenetic discrimination between closely related species or between populations may be difficult because of the slow rate of lineage sorting (Hollingsworth et al. 2011), which may result in incorrect phylogenetic inferences in the terminal branches of tree topology. Therefore, we will focus the discussion in terms of global diversification and biogeographic patterns instead of fine-scaled inferences.

Using DNA sequence-based phylogeny assessed from only two DNA fragments to infer the biogeographic pattern of *Lithocarpus* is a trade-off between sample size and phylogenetic resolution. We used a large number of samples resourced mostly from published data, but the availability of relatively few markers might result in a bias of phylogeny. The major concern for the use of limited markers is that most biogeographic analyses are tree-dependent, and the tree quality determines the accuracy of the biogeographic inferences. A biased phylogenetic hypothesis can lead to an erroneous biogeographic inference. Therefore, in this study, integrated biogeographic inferences were obtained by comparing the analyses from combined marker sets and two single loci of different genomes and the contrasting results from different trees are further discussed.

Biogeographic inferences

The ancestral area of *Lithocarpus* inferred from the combined-marker tree is located in mainland China (Fig. 5, Additional file 1: Figures S2 and S4), implying a continental-Asian origin of the genus. Except for the genera *Lithocarpus* and *Trigonobalanus* (cf. Manos and Stanford 2001), most extant species of Fagaceae are distributed in temperate areas, suggesting a temperate origin of Fagaceae, which was also evidenced by the fossil distribution (Crepet and Nixon 1989). Given the apparent ecological niche conservatism throughout the evolutionary history of *Fagaceae*, it seems more likely that the subtropical and tropical genus *Lithocarpus* originated in a temperate region and expanded southward with adaptive divergence (Manos and Stanford 2001). The southward expansion was extended to the Greater Sunda Islands (Fig. 5, Additional file 1: Figures S2–S4). The estimated time of the southward migration places it in the Early Miocene, during which the sea level had not begun to rise and Sundaland was not yet submerged (Hall 2002), showing no or less relationship between the southward range expansion and recent dramatic climate changes, such as the Quaternary glacial oscillations. In fact, the early long-distance dispersal events did not occur only in Southeast Asia. Oligocene leaf fossils of the extant *Lithocarpus* analogs, *L. saxonicus*, discovered in Europe (East Germany) suggest a westward expansion of ancestors of extant *Lithocarpus* (Kvaček and Walther 1989). However, another example of out-of-Asia dispersal of *L. densiflora* into North America in the mid-Eocene (Manos and Stanford 2001) is paradoxical because this species is now recognized as a novel genus *Notholithocarpus* which is sister-grouped with *Quercus* instead of *Lithocarpus* (Manos et al. 2008). The Late-Miocene and Early Pliocene sea-level fluctuations that accompanied the disjunction and connection of Sunda Islands (Susilohadi et al. 2009) might have caused continual divergences and secondary contact with *Lithocarpus* species and resulted in rapid diversification. The following Pliocene fragmentation of island blocks (Susilohadi et al. 2009) further accelerated species divergence (e.g., Cyrtudru, Atkins et al.

2001; murine rodents, Gorog et al. 2004; crabs, Klaus et al. 2013).

The MCO provided suitable environmental conditions for broad spreading, and the following Middle Miocene climatic transition that caused the ice sheet expansion and formation of land bridges (Holbourn et al. 2005) might have facilitated southward colonization of these continental species into the Sundaland region. In addition to this southward colonization, we found at least two independent and synchronized northward returns from the Greater Sunda Islands into continental Asia at roughly 12 Mya, the Middle Miocene, by cpDNA (Fig. 3) and at least two serial northward returns at 15.21 Mya (95% HPD: 1.8–14.94 Mya) and 4.63 Mya (95% HPD: 0.72–10.58 Mya) by nrITS (Fig. 4) analysis.

The colonization date inference (21.78 ~ 14.48 Mya, Figs. 2a and 5a) of Southeast Asian stone oaks supports Cannon and Manos's (2003) diversification hypothesis of Middle-Miocene vicariance between Indochina and Borneo. Manos and Stanford (2001) suggested that the most ancient haplotypes were distributed in northern Borneo and Indochina. The S-DIVA analysis with consideration of the scenario of vicariance rather than the DEC model implied serial dispersal events from the ancestral area of mainland China south towards Indochina and Borneo, indicating that the Borneo species were descended from the continental-Asian immigrants (Fig. 5). Our biogeographic inferences support Cannon and Manos's (2003) hypothesis of Middle-Miocene vicariance between Indochina and Borneo, but the ancestral area of *Lithocarpus* is probably in mainland China, further north than Cannon and Manos's (2003) inference of Borneo and Indochina. Furthermore, the Middle-Miocene long-distance dispersals also support the later in situ diversification inferred by Cannon and Manos (2003).

Diversification rate shift of Asian stone oaks

The rejection of constant diversification rate models (Tables 1 and 2) and the gradual increase of certain diversification rate shifts (Fig. 6a and d) supported the rejection of the stochastic model of ecological drift at long-term time scales (Benedetti-Cecchi et al. 2008)—i.e., that the Asian stone oaks are a group of non-neutral species responsive to heterogeneous environments. The maternally inherited cpDNA that reflects seed-mediated biogeographic events revealed a more obvious diversification rate shift during the Miocene (*c.* 15 Mya, Fig. 6b and e) than the biparentally inherited nrITS marker (Fig. 6c and f). The Miocene southward colonization by long-distance seed dispersal plays an important role in the extant species diversity of Asian stone oaks (Fig. 2a and b; the blue arrows). The lineages of clade 1 experienced southward colonization and an increased diversification rate at the Middle-Miocene (ca. 15 Mya, Fig. 2b), while lineages of the continental Asian clade 2 revealed a later increase of diversification rate since the early Pliocene (*c.* 5 Mya, Fig. 2c). Ancient Sundaland did not separate during the Miocene but narrowed from the Middle Miocene (Hall 2002). Restricted routes for long-distance dispersal in addition to climatic deterioration might have obstructed the range expansion and accelerated population differentiation (or species divergence) in Borneo. In contrast, the topography of mainland China did not vary as did that of Sundaland (Hall 2002; Scotese 2014), so the diversification rate of clade 2 in Fig. 2 did not reveal a rate-transition in the Middle Miocene (Fig. 2c).

However, the single-marker analyses revealed a constant growth of diversification rate in subclades cp-1, cp-2 (Fig. 3), and ITS-I (Fig. 4), but showed a rate-decrease pattern during the Miocene in subclade ITS-II (*c.* 23 ~ 5 Mya, Fig. 4). The constant rate increase in subclades of cpDNA (Fig. 3), and the shift in the whole tree topology (Fig. 6b), were caused by different diversification rates of subclades cp-1 (higher rate increase) and cp-2 (lower rate increase) (Fig. 3). Before the Middle Miocene (ca. 15 Mya), the numbers of continental Asian species were higher than those in southern Sundalands, but the southward colonization of stone oaks and the more severe geotopological changes to Sundaland caused a faster rate of species accumulation in Sundaland (the Greater Sunda Islands) than in continental Asia. Clade ITS-II might reflect a counteraction of frequent interspecific pollen flow and restricted seed dispersal in continental Asia since the Middle Miocene. The early Middle Miocene had an optimal climate (i.e., the MCO) with subsequent Middle Miocene disruption, during which the mid-latitude arid belt replaced the humid climate in East Asia (Liu et al. 2009), which caused an expansion of grasslands and restriction of forested areas (Kurschner et al. 2008). This explains the seed-mediated differentiation. In addition, the abundant insects with rapid evolutionary rates during the Miocene (Nyman et al. 2012) could have created opportunities for successful interspecific pollen flow of entomophilic stone oaks.

In addition to the principal diversification rate shift in the Middle Miocene (*c.* 15 Mya, Fig. 2b), we assessed two minor rate shifts at the Miocene-Pliocene boundary (*c.* 5 Mya) and Pliocene–Pleistocene boundary (*c.* 2.5 Mya). The former minor diversification rate shift was caused by continental Asian lineages of subclade 2 (Fig. 2c), while the latter was caused by lineages of subclade 1 (Figs. 2b and 4b). The climate since the end of the MCO gradually became colder until drastic climatic changes occurred at the end of the Miocene (Hodell and

Kennett 1986) that restricted forest expansion on the Asian landmass. Restricted habitats (refugia) obstructed the gene flow with other populations and increased population differentiation, which accelerated speciation rates of the continental stone oaks in Asia, although the risk of extinction was also increased. Climatic oscillations provided suitable corridors for species survival and dispersal (Nakamura et al. 2012) that may have facilitated species diversification by recurrent extinctions and abrupt speciation from open niches (Dynesius and Jansson 2000).

In contrast, the species distributed in relatively warm Sundaland were less affected until the Pliocene–Pleistocene boundary, during which Sundaland started to break up. The second minor diversification rate shift since the Pliocene–Pleistocene boundary (*c.* 2.5 Mya) reflects the drastic Quaternary climatic change and periodic glacial periods. Quaternary fluctuations in biota transitions were commonly adopted to explain the current biodiversity (Alexeeva and Erbajeva 2005; Graham et al. 2003). Climatic changes at the Quaternary that predominantly attended dispersal, isolation, and local extinction to foster speciation could explain the increased diversification rates (Gavrilets and Losos 2009; Graham et al. 2003).

Multiple originations of insular stone oaks

After the southward colonization, the expansion and migration of *Lithocarpus* led to colonization of islands around continental Asia and Sundaland. Because of the distinct phylogenetic positions of island species in Japan, Taiwan, and Hainan, the hypothesis of multiple colonization events for each island is supported rather than in situ diversification. The deep branches of the cpDNA tree suggest long histories of seed colonization (Fig. 3), implying that the high species diversity of stone oaks in Taiwan and Hainan could be due to multiple origins instead of rapid speciation from single lineages (i.e., radiation). Furthermore, the formation time of Taiwan and Hainan was younger (since the Pliocene) than the formation of the landmass of Sundaland, i.e., Borneo (since at least the end of the Jurassic, Cracraft 1988), and the area of those two subtropical islands was also much more smaller than the landmass of Sundaland. These factors led to fewer species (or taxa) in Taiwan (16 spp.) and Hainan (24 spp.) than in Borneo (73 spp.) (Huang et al. 1999; Liao 1996; Soepadmo 1972). However, the endemism in Taiwan (8/16) was larger than in Borneo (23/73), other tropical islands (Hainan: 6/24, Philippines: 13/32, Sumatra: 4/33, Java: 4/17, Sulawesi: 0/4), and the Malay Peninsula (8/48) (Soepadmo 1972). Diversity of stone oaks in the Greater Sunda Islands was thought to be a consequence of prolonged in situ diversification (Cannon and Manos 2003) and is supported by our biogeographic analysis (Fig. 5, Additional file 1: Figures S2–S4), while the rapidity of speciation seems lower than the multiple-source colonization over a short span in Taiwan. The short distance between Taiwan and China (180 km on average) and the shallow continental shelf of the Taiwan Strait (100 m depth on average) joined to continental Asia repeatedly during glacial epochs, and led to multiple colonization between China and Taiwan, which resulted in periodic dispersal and isolation between species congeners (compared to Hainan Island: 18 km from China, 44 m depth in the Qiongzhou Strait, which reduced the isolation effect). The synergy of genetic isolation and adaptation to the rugged topography could have accelerated the speciation process (Bittkau and Comes 2005), which could be the cause of high endemism of Taiwanese stone oaks. In Taiwan, *Lithocarpus* species are highly diverged, with 14 species recognized and 7 species as endemic (*L. formosana, L. dodonaeifolius, L. shinsiensis, L. nantoensis, L. konishii, L. kawakamii,* and *L. lepidocarpus*) (Yang et al. 2014) usually in central and southern parts of the island, mixing with other species of Fagaceae, Lauraceae and coniferous forests at altitudes of 150–2700 m. Ecologically, seven species usually grow on wind-facing slopes in tropical or subtropical forests (Yang et al. 2014).

Temporal change in diversification rate of Asian stone oaks

Both topological (Table 1) and temporal analyses (Fig. 4) of diversification rates suggest that *Lithocarpus* species were slowly accumulated and formed persistent lineages through time, supporting Thorne's (1999) theory that East Asia serves as a living museum (refugium) to preserve archaic flora (cf. Favre et al. 2016). Such temporal increases in the synchronous diversification rate in subtropical and tropical lineages indicates that (1) the southward founder event did not accelerate the diversification rate and (2) the diversification rate shift of Asian stone oaks was spatially independent. Since the species composition and climate differ between the subtropics and tropics, the congruent patterns of diversification rates of stone oaks in these two areas suggest that the evolution of *Lithocarpus* may fit the stochastic model of ecological drift (Benedetti-Cecchi et al. 2008) rather than the ecological opportunity model (Yoder et al. 2010). The ecological opportunity theory posits an increased rate of morphological or species diversification due to a reduction in interspecific competitive pressures (i.e., ecological release) when novel habitats were colonized. In contrast, under the stochastic model, species assemblages in an environment are the consequence of random variation in the fundamental biological processes of birth, death, and migration (Bell 2000), in which environmental heterogeneity is not responsible for species diversity (Maurer and McGill 2004). One of the characteristics of the ecological opportunity theory is adaptive radiation (Losos and Mahler 2010; Yoder et al. 2010). Species that meet the ecological opportunity model have shorter life histories and are more sensitive to environmental changes—e.g., birds (Norman and Christidis 2016; Schweizer et al. 2014), fishes (Arbour and Lopez-Fernandez 2016; Silva et al. 2016), ferns (Sundue et al. 2015), and Gentianaceae (von Hagen and Kadereit 2003). Stone oaks with longer life histories could not instantaneously respond to environmental variations spatially, but in the long process of time, diversity could accumulate from different archaic sources (refugia). Such a "living museum" model of the diversification patterns of genus *Lithocarpus* could be explained by the hypothesis of multiple origins with in situ diversification of insular species.

Conclusions

The diversification of Asian stone oaks is related to dispersal events in association with past climatic changes. The diversification rate heterogeneity across phylogenetic trees instead of punctuated rate shifts across clades signifies that the diversification events were related to global geohistory rather than episodic local events. Based on the coalescent species tree inferred by nuclear and chloroplast genome markers, we propose to revise the former hypothesis of a tropical origin to a subtropical origin and modify the vicariance hypothesis to one of dispersal with variance for explaining the in situ diversification of tropical stone oaks. The boundaries of geological times (i.e., the Oligocene–Miocene, Miocene-Pliocene, and Pliocene–Pleistocene) and the occurrence of drastic climate changes during the Middle-Miocene matched perfectly with the diversification rate shifts in *Lithocarpus*. Such drastic climatic changes at chronological times which led to restricted gene flow within continents might have accelerated species divergence and promoted diversification. The genus *Lithocarpus* serves as a good model for tropical-subtropical forest trees in reflecting environmental changes. Furthermore, our results suggest multiple origins of the newly originated indigenous species on subtropical continental islands, leading to high levels of insular biodiversity. An excess of high frequency endemics on Taiwan Island suggests the synergistic consequences of adaptive divergence and experienced secondary contacts between (isolated) specific congeners. However, we still need genomic data from a larger population to test the hypothesis of adaptive-divergence-driven in situ diversification.

Additional file

Additional file 1: Table S1. Collecting sites and distribution. **Table S2.** Best substitution models for the *atpB-rbcL* and ITS used in the Bayesian analyses. **Table S3.** T Statistical summary of asymmetric values for the among-lineage diversification rate variation in the phylogenetic topologies. **Table S4.** Testing for diversification rate variation models by ΔAIC_{RC} test statistic. **Figure S1.** The phylogenetic tree reconstructed by cpDNA *atpB-rbcL* spacer and nrITS under the Yule's pure-birth speciation model. Samples that were identified as the same species but represented at different clades in the gene trees (supplementary) were separated to different OTUs. Bold lines indicate the lineages grouping with posterior probability > 80%; node labels are the splitting time (unit: mya); node bar is the 95% highest posterior density interval (HPD) of the splitting time. The nodes probably have diversification rate shift were labeled by nodes *i* and *ii* of cpDNA tree and nodes *a* and *b* of ntITS tree. The testing results of diversification-rate shift of descendents of nodes *i*, *ii*, *a*, and *b* inferred by delta-statistics were indicated in the inner table. **Figure S2.** Biogeographic inferences by cpDNA+nrITS (Fig. 2A). (A) Biogeographic inference under the DEC model. Lineages with bold and thin lines indicate the lineages derived from nodes (ancestral areas) with likelihood > 0.7 and > 0.5, respectively. Likelihood less than 0.5 are treated as unknown (black lineages). (B) Biogeographic inference under the S-DIVA model. The dispersal (d) or vicariance (v) events with a posterior probability > 0.5 are marked in the node. The yellow dots indicate the events of range transition inferred by both dispersal and vicariance events. The open dots indicate that thebiogeographic distribuion of deriving lineages were explained as consequences of dispersal events. **Figure S3.** Biogeographic inferences by cpDNA (Fig. 3). (A) Biogeographic inference under the DEC model. (B) Biogeographic inference under the S-DIVA model. **Figure S4.** Biogeographic inferences by nrITS (Fig. 4). (A) Biogeographic inference under the DEC model. (B) Biogeographic inference under the S-DIVA model.

Abbreviations

DEC: dispersal–extinction–cladogenesis; MCO: Miocene climatic optimum; CTAB: cetyl trimethylammonium bromide; PCR: polymerase chain reaction; nrITS: nuclear ribosomal internal transcribed spacer; LTT: lineage-through-time; rjMCMC: reversible jump Markov chain Monte Carlo; GTR: general-time-reversible substitution model; UCLN: uncorrelated lognormal.

Authors' contributions

CYC and PCL conceived the idea and designed the project. CKY, BHH, LPJ, and PCL collect plant materials. CYC, BHH and PCL conducted data cleaning and data analysis. CKY, CYC, BHH, and PCL wrote the manuscript. All authors read and approved the final manuscript.

Author details

[1] Department of Life Science, National Taiwan Normal University, 88, Ting-Chow Rd, Sec 4, Taipei 116, Taiwan. [2] The Experimental Forest, College of Bio-Resources and Agriculture, National Taiwan University, 12, Sec. 1, Chien-Shan Rd., Nantou 55750, Taiwan. [3] Department of Biological Sciences, National Sun Yat-sen University, 70, Lien-Hai Rd., Kaohsiung 80424, Taiwan. [4] Botanical Garden Division, Taiwan Forestry Research Institute, 53, Nan-Hai Rd., Taipei 10066, Taiwan.

Funding

Supported by the Ministry of Science and Technology, Taiwan (MOST 105-2628-B-003 -002 -MY3) (to P-CL).

References

Aizawa M, Yoshimaruth H, Saito H, Katsuki T, Kawahara T, Kitamura K, Shi F, Kaji M (2007) Phylogeography of a northeast Asian spruce, *Picea jezoensis* inferred from genetic variation observed in organelle DNA markers. Mol Ecol 16:3393–3405

Alexeeva NV, Erbajeva MA (2005) Changes in the fossil mammal faunas of Western Transbaikalia during the Pliocene-Pleistocene boundary and the Early-Middle Pleistocene transition. Quatern Int 131:109–115

Arbour JH, Lopez-Fernandez H (2016) Continental cichlid radiations: functional diversity reveals the role of changing ecological opportunity in the Neotropics. Proc R Soc B Biol, Sci, p 283

Atkins H, Preston J, Cronk QCB (2001) A molecular test of Huxley's line: Cyrtandra (Gesneriaceae) in Borneo and the Philippines. Biol J Linn Soc 72:143–159

Bell G (2000) The distribution of abundance in neutral communities. Am Nat 155:606–617

Benedetti-Cecchi L, Bertocci I, Vaselli S, Maggi E, Bulleri F (2008) Neutrality and the response of rare species to environmental variance. PLoS ONE 3:e2777

Bittkau C, Comes HP (2005) Evolutionary processes in a continental island system: molecular phylogeography of the Aegean *Nigella arvensis* alliance (Ranunculaceae) inferred from chloroplast DNA. Mol Ecol 14:4065–4083

Cannon CH, Manos PS (2003) Phylogeography of the Southeast Asian stone oaks (*Lithocarpus*). J Biogeogr 30:211–226

Chan KMA, Moore BR (2002) Whole-tree methods for detecting differential diversification rates. Syst Biol 51:855–865

Cheng YP, Hwang SY, Lin TP (2005) Potential refugia in Taiwan revealed by the phylogeographical study of *Castanopsis carlesii* Hayata (Fagaceae). Mol Ecol 14:2075–2085

Chi W-R, Namson J, Suppe J (1981) Stratigraphic record of plate interactions in the Coastal Range of eastern Taiwan. Mem Geol Soc China 4:155–194

Chiang TY, Schaal BA (2006) Phylogeography of plants in Taiwan and the Ryukyu archipelago. Taxon 55:31–41

Chiang YC, Hung KH, Schaal BA, Gest XJ, Hsu TW, Chiang TY (2006) Contrasting phylogeographical patterns between mainland and island taxa of the *Pinus luchuensis* complex. Mol Ecol 15:765–779

Chiang YC, Huang BH, Liao PC (2012) Diversification, biogeographic pattern, and demographic history of Taiwanese *Scutellaria* species inferred from nuclear and chloroplast DNA. PLoS ONE 7:e50844

Chung JD, Lin TP, Tan YC, Lin MY, Hwang SY (2004) Genetic diversity and biogeography of *Cunninghamia konishii* (Cupressaceae), an island species in Taiwan: a comparison with *Cunninghamia lanceolata*, a mainland species in China. Mol Phylogen Evol 33:791–801

Chung JD, Lin TP, Chen YL, Cheng YP, Hwang SY (2007) Phylogeographic study reveals the origin and evolutionary history of a *Rhododendron* species complex in Taiwan. Mol Phylogen Evol 42:14–24

Cracraft J (1988) From Malaysia to New Guinea: evolutionary biogeography within a complex continent-island arc contact zone. In: Ouellet H (ed) 19 Congressus Internationalis Ornithologici, 2581-2593. National Museum of Natural Sciences by University of Ottawa Press, Ottawa, pp 2581–2593

Crepet WL, Nixon KC (1989) Earliest megafossil evidence of Fagaceae: phylogenetic and biogeographic implications. Am J Bot 76:842–855

Donoghue MJ, Bell CD, Li JH (2001) Phylogenetic patterns in Northern Hemisphere plant geography. Int J Plant Sci 162:S41–S52

Doyle J, Doyle J (1987) Genomic plant DNA preparation from fresh tissue-CTAB method. Phytochem Bull 19:11–15

Drummond AJ, Rambaut A (2007) BEAST: bayesian evolutionary analysis by sampling trees. BMC Evol Biol 7:214

Dynesius M, Jansson R (2000) Evolutionary consequences of changes in species' geographical distributions driven by Milankovitch climate oscillations. Proc Natl Acad Sci USA 97:9115–9120

Ennos RA (1994) Estimating the relative rates of pollen and seed migration among plant populations. Heredity 72:250–259

Favre A, Michalak I, Chen CH, Wang JC, Pringle JS, Matuszak S, Sun H, Yuan YM, Struwe L, Muellner-Riehl AN (2016) Out-of-Tibet: the spatio-temporal evolution of *Gentiana* (Gentianaceae). J Biogeogr 43:1967–1978

Fu CX, Qiu YX, Guan BC, Comes HP (2009) Did glacials and/or interglacials promote allopatric incipient speciation in East Asian temperate plants? Phylogeographic and coalescent analyses on refugial isolation and divergence in *Dysosma versipellis*. Mol Phylogen Evol 51:281–293

Gavrilets S, Losos JB (2009) Adaptive radiation: contrasting theory with data. Science 323:732–737

Ge XJ, Hsu TW, Hung KH, Lin CJ, Huang CC, Huang CC, Chiang YC, Chiang TY (2012) Inferring multiple refugia and phylogeographical patterns in *Pinus massoniana* based on nucleotide sequence variation and DNA fingerprinting. PLoS ONE 7:e43717

Gorog AJ, Sinaga MH, Engstrom MD (2004) Vicariance or dispersal? Historical biogeography of three Sunda shelf murine rodents (*Maxomys surifer, Leopoldamys sabanus* and *Maxomys whiteheadi*). Biol J Linn Soc 81:91–109

Graham MH, Dayton PK, Erlandson JM (2003) Ice ages and ecological transitions on temperate coasts. Trends Ecol Evol 18:33–40

Hall TA (1999) BioEdit: a user-friendly biological sequence alignment editor and analysis program for Windows 95/98/NT. Nucleic Acids Symp Ser 41:95–98

Hall R (2002) Cenozoic geological and plate tectonic evolution of SE Asia and the SW Pacific: computer-based reconstructions, model and animations. J Asian Earth Sci 20:353–431

Heled J, Drummond AJ (2010) Bayesian inference of species trees from multilocus data. Mol Biol Evol 27:570–580

Hewitt GM (2004) Genetic consequences of climatic oscillations in the Quaternary. Philos T Roy Soc B 359:183–195

Hodell DA, Kennett JP (1986) Late Miocene-Early Pliocene stratigraphy and paleoceanography of the South Atlantic and southwest Pacific Oceans: a synthesis. Paleoceanography 1:285–311

Hoelzer GA (1997) Inferring phylogenies from mtDNA variation: mitochondrial-gene trees versus nuclear-gene trees revisited. Evolution 51:622–626

Holbourn A, Kuhnt W, Schulz M, Erlenkeuser H (2005) Impacts of orbital forcing and atmospheric carbon dioxide on Miocene ice-sheet expansion. Nature 438:483–487

Hollingsworth PM, Graham SW, Little DP (2011) Choosing and using a plant DNA barcode. PLoS ONE 6:e19254

Huang C, Zhang Y, Bruce B (1999) Fagaceae Flora of China, vol 4, pp 314–400. http://www.eFloras.org. Accessed 28 Oct 2018

Klaus S, Selvandran S, Goh JW, Wowor D, Brandis D, Koller P, Schubart CD, Streit B, Meier R, Ng PKL, Yeo DCJ (2013) Out of Borneo: neogene diversification of Sundaic freshwater crabs (Crustacea: Brachyura: Gecarcinucidae: *Parathelphusa*). J Biogeogr 40:63–74

Kurschner WM, Kvacek Z, Dilcher DL (2008) The impact of Miocene atmospheric carbon dioxide fluctuations on climate and the evolution of terrestrial ecosystems. Proc Natl Acad Sci USA 105:449–453

Kvaček Z, Walther H (1989) Paleobotanical studies in Fagaceae of the European Tertiary. Plant Syst Evol 162:213–229

Liao JC (1996) Fagaceae. In: Boufford DE, Hsieh CF, Huang TC, Ohashi H, Yang YP, Lu SY (eds) Flora of Taiwan, vol 2, 2nd edn. Editorial Committee of the Flora of Taiwan, Department of Botany, National Taiwan University, Taipei, pp 51–123

Liao PC, Hwang SY, Huang S, Chiang YC, Wang JC (2011) Contrasting demographic patterns of *Ceriops tagal* (Rhizophoraceae) populations in the South China Sea. Aust J Bot 59:523–532

Liu L, Eronen JT, Fortelius M (2009) Significant mid-latitude aridity in the middle Miocene of East Asia. Palaeogeogr., Palaeoclimatol. Palaeoecol 279:201–206

Lohman DJ, de Bruyn M, Page T, von Rintelen K, Hal R, Ng PKL, Shih HT, Carvalho GR, von Rintelen T (2011) Biogeography of the Indo-Australian Archipelago. Annu Rev Ecol Evol S 42:205–226

Losos JB, Mahler DL (2010) Adaptive radiation: The interaction of ecological opportunity, adaptation, and speciation. In: Bell MA, Futuyma DJ, Eanes WF, Levinton JS (eds) Evolution Since Darwin: The First 150 Years. Sinauer Associates Inc, Sunderland, pp 381–420

Manos PS, Stanford AM (2001) The historical biogeography of Fagaceae: tracking the tertiary history of temperate and subtropical forests of the Northern Hemisphere. Int J Plant Sci 162:S77–S93

Manos PS, Zhou ZK, Cannon CH (2001) Systematics of Fagaceae: phylogenetic tests of reproductive trait evolution. Int J Plant Sci 162:1361–1379

Manos PS, Cannon CH, Oh S-H (2008) Phylogenetic relationships and taxonomic status of the paleoendemic Fagaceae of western North America: recognition af a new genus, *Notholithocarpus*. Madrono 55:181–190

Matzke NJ (2014) Model selection in historical biogeography reveals that founder-event speciation is a crucial process in island clades. Syst Biol 63:951–970

Maurer BA, McGill BJ (2004) Neutral and non-neutral macroecology. Basic Appl Ecol 5:413–422
Mccauley DE (1995) The use of chloroplast DNA polymorphism in studies of gene flow in plants. Trends Ecol Evol 10:198–202
Millien-Parra V, Jaeger JJ (1999) Island biogeography of the Japanese terrestrial mammal assemblages: an example of a relict fauna. J Biogeogr 26:959–972
Mitsui Y, Setoguchi H (2012) Recent origin and adaptive diversification of *Ainsliaea* (Asteraceae) in the Ryukyu Islands: molecular phylogenetic inference using nuclear microsatellite markers. Plant Syst Evol 298:985–996
Moore BR, Chan KMA (2005) SymmeTREE: whole-tree analysis of differential diversification rates. Bioinformatics 21:1709–1710
Moore BR, Chan KMA, Donoghue MJ (2004) Detecting diversification rate variation in supertrees. In: Bininda-Emonds ORP (ed) Phylogenetic supertrees: combining information to reveal the tree of life. The Netherlands Kluwer Academic, Dordrecht, pp 487–533
Nakamura K, Denda T, Kokubugata G, Forster PI, Wilson G, Peng CI, Yokota M (2012) Molecular phylogeography reveals an antitropical distribution and local diversification of *Solenogyne* (Asteraceae) in the Ryukyu Archipelago of Japan and Australia. Biol J Linn Soc 105:197–217
Nixon KC (1989) Origins of Fagaceae. In: Crane PR, Blackmore S (eds) Evolution systematics and fossil history of the Hamamelidae: higher Hamamelidae, vol 2. Clarendon, Oxford, pp 436–506
Norman JA, Christidis L (2016) Ecological opportunity and the evolution of habitat preferences in an arid-zone bird: implications for speciation in a climate-modified landscape. Sci Rep 6:19613
Nyman T, Linder HP, Pena C, Malm T, Wahlberg N (2012) Climate-driven diversity dynamics in plants and plant-feeding insects. Ecol Lett 15:889–898
Oh S-H, Manos PS (2008) Molecular phylogenetics and cupule evolution in Fagaceae as inferred from nuclear *CRABS CLAW* sequences. Taxon 57:434–451
Opgen-Rhein R, Fahrmeir L, Strimmer K (2005) Inference of demographic history from genealogical trees using reversible jump Markov chain Monte Carlo. BMC Evol Biol 5:6
Page RDM, Charleston MA (1998) Trees within trees: phylogeny and historical associations. Trends Ecol Evol 13:356–359
Paradis E, Claude J, Strimmer K (2004) APE: analyses of phylogenetics and evolution in R language. Bioinformatics 20:289–290
Petit RJ, Brewer S, Bordacs S, Burg K, Cheddadi R, Coart E, Cottrell J, Csaikl UM, van Dam B, Deans JD, Espinel S, Fineschi S, Finkeldey R, Glaz I, Goicoechea PG, Jensen JS, Konig AO, Lowe AJ, Madsen SF, Matyas G, Munro RC, Popescu F, Slade D, Tabbener H, de Vries SGM, Ziegenhagen B, de Beaulieu JL, Kremer A (2002a) Identification of refugia and post-glacial colonisation routes of European white oaks based on chloroplast DNA and fossil pollen evidence. For Ecol Manage 156:49–74
Petit RJ, Csaikl UM, Bordacs S, Burg K, Coart E, Cottrell J, van Dam B, Deans JD, Dumolin-Lapegue S, Fineschi S, Finkeldey R, Gillies A, Glaz I, Goicoechea PG, Jensen JS, Konig AO, Lowe AJ, Madsen SF, Matyas G, Munro RC, Olalde M, Pemonge MH, Popescu F, Slade D, Tabbener H, Taurchini D, de Vries SGM, Ziegenhagen B, Kremer A (2002b) Chloroplast DNA variation in European white oaks—phylogeography and patterns of diversity based on data from over 2600 populations. For Ecol Manage 156:5–26
Petit RJ, Duminil J, Fineschi S, Hampe A, Salvini D, Vendramin GG (2005) Comparative organization of chloroplast, mitochondrial and nuclear diversity in plant populations. Mol Ecol 14:689–701
Qiu YX, Fu CX, Comes HP (2011) Plant molecular phylogeography in China and adjacent regions: tracing the genetic imprints of Quaternary climate and environmental change in the world's most diverse temperate flora. Mol Phylogen Evol 59:225–244
Rabosky DL (2006) Likelihood methods for detecting temporal shifts in diversification rates. Evolution 60:1152–1164
Rabosky DL, Grundler M, Anderson C, Title P, Shi JJ, Brown JW, Huang H, Larson JG (2014) BAMMtools: an R package for the analysis of evolutionary dynamics on phylogenetic trees. Methods Ecol Evol 5:701–707
Rambaut A (2008). FigTree ver. 1.3.1. Institute of Evolutionary Biology, University of Edinburgh, Edinburgh. http://tree.bio.ed.ac.uk/software/FigTree/. Accessed 28 Oct 2018
Rambaut A, Drummond AJ (2003) Tracer ver. 1.4.1. Institute of Evolutionary Biology, University of Edinburgh, Edinburgh. http://evolve.zoo.ox.ac.uk/software/. Accessed 28 Oct 2018
Ricklefs RE (2007) Estimating diversification rates from phylogenetic information. Trends Ecol Evol 22:601–610
Schweizer M, Hertwig ST, Seehausen O (2014) Diversity versus disparity and the role of ecological opportunity in a continental bird radiation. J Biogeogr 41:1301–1312
Scotese CR (2014) Atlas of neogene paleogeographic maps (Mollweide projection), Maps 1–7, vol 1, The Cenoaoic, PALEOMAP Atlas for ArcGIS, PALEOMAP Project, Evanston, Illinois
Silva GSC, Roxo FF, Lujan NK, Tagliacollo VA, Zawadzki CH, Oliveira C (2016) Transcontinental dispersal, ecological opportunity and origins of an adaptive radiation in the Neotropical catfish genus *Hypostomus* (Siluriformes: Loricariidae). Mol Ecol 25:1511–1529
Soepadmo E (1972) Fagaceae. In: van Steenis CGGJ (ed) Flora Malesiana, vol 7. Noordhoff International Publishing, Leyden, pp 265–403
Sundue MA, Testo WL, Ranker TA (2015) Morphological innovation, ecological opportunity, and the radiation of a major vascular epiphyte lineage. Evolution 69:2482–2495
Susilohadi S, Gaedicke C, Djajadihardja Y (2009) Structures and sedimentary deposition in the Sunda Strait, Indonesia. Tectonophysics 467:55–71
Teng LS (1996) Extensional collapse of the northern Taiwan mountain belt. Geology 24:949–952
Thompson JD, Gibson TJ, Plewniak F, Jeanmougin F, Higgins DG (1997) The CLUSTAL_X windows interface: flexible strategies for multiple sequence alignment aided by quality analysis tools. Nucleic Acids Res 25:4876–4882
Thorne RF (1999) Eastern Asia as a living museum for archaic angiosperms and other seed plants. Taiwania 44:413–422
von Hagen KB, Kadereit JW (2003) The diversification of *Halenia* (Gentianaceae): ecological opportunity versus key innovation. Evolution 57:2507–2518
Wang CS, Huang CP, Ke LY, Chien WJ, Hsu SK, Shyu CT, Cheng WB, Lee CS, Teng LS (2002) Formation of the Taiwan island as a solitary wave along the Eurasian continental plate margin: magnetic and seismological evidence. Terr Atmos Ocean Sci 13:339–354
Wang JF, Gong X, Chiang YC, Kuroda C (2013) Phylogenetic patterns and disjunct distribution in *Ligularia hodgsonii* Hook. (Asteraceae). J Biogeogr 40:1741–1754
Wu SH, Hwang CY, Lin TP, Chung JD, Cheng YP, Hwang SY (2006) Contrasting phylogeographical patterns of two closely related species, *Machilus thunbergii* and *Machilus kusanoi* (Lauraceae), in Taiwan. J Biogeogr 33:936–947
Xing Y, Onstein RE, Carter RJ, Stadler T, Peter Linder H (2014) Fossils and a large molecular phylogeny show that the evolution of species richness, generic diversity, and turnover rates are disconnected. Evolution 68:2821–2832
Yang CK, Chiou LR, Chang LF, Wang CC (2014) Fagaceae in Taiwan Forest Bureau. Council of Agriculture, Executive Yuan, Taipei
Yoder JB, Clancey E, Des Roches S, Eastman JM, Gentry L, Godsoe W, Hagey TJ, Jochimsen D, Oswald BP, Robertson J, Sarver BAJ, Schenk JJ, Spear SF, Harmon LJ (2010) Ecological opportunity and the origin of adaptive radiations. J Evol Biol 23:1581–1596
Yu Y, Harris AJ, Blair C, He X (2015) RASP (Reconstruct Ancestral State in Phylogenies): a tool for historical biogeography. Mol Phylogen Evol 87:46–49

15

Identification of phosphorylation proteins in response to water deficit during wheat flag leaf and grain development

Fei Luo[1†], Xiong Deng[1†], Yue Liu[1] and Yueming Yan[1,2*]

Abstract

Background: Wheat (*Triticum aestivum* L.) serves as important grain crop widely cultivated in the world, which is often suffered by drought stress in natural conditions. As one of the most important post translation modifications, protein phosphorylation widely participates in plant abiotic stress regulation. In this study, we performed the first comparative analysis of phosphorylated protein characterization in flag leaves and developing grains of elite Chinese bread wheat cultivar Zhongmai 175 under water deficit by combining with proteomic approach and Pro-Q Diamond gel staining.

Results: Field experiment showed that water deficit caused significant reduction of plant height, tiller number, ear length and grain yield. 2-DE and Pro-Q Diamond gel staining analysis showed that 58 proteins were phosphorylated among 112 differentially accumulated proteins in response to water deficit, including 20 in the flag leaves and 38 in the developing grains. The phosphorylated proteins from flag leaves mainly involved in photosynthesis, carbohydrate and energy metabolism, while those from developing grains were closely related with detoxification and defense, protein, carbohydrate and energy metabolism. Particularly, water deficit resulted in significant downregulation of phosphorylated modification level in the flag leaves, which could affect photosynthesis and grain yield. However, some important phosphorylated proteins involved in stress defense, energy metabolism and starch biosynthesis were upregulated under water deficit, which could benefit drought tolerance, accelerate grain filling and shorten grain developing time.

Conclusions: The modification level of those identified proteins from flag leaves and grains had great changes when wheat was suffered from water deficit, indicating that phosphoproteins played a key role in response to drought stress. Our results provide new insights into the molecular mechanisms how phosphoproteins respond to drought stress and thus reduce production.

Keywords: Bread wheat, Phosphorylated proteins, Water deficit, Flag leaves, Developing grains

Background

As one of the most important grain crops, wheat (*Triticum aestivum* L.) is widely cultivated in the world due to its value as a stable source of saccharides and proteins, and its ability to adapt to different surviving environment. However, most agricultural areas where wheat mainly spread over strongly depend on natural rainfall, including almost all arid and semiarid areas (Cai et al. 2012). In recent years, climate changing, especially global warming, has exacerbated the effects of drought stress on crop production. In general, temperature rising 1 °C can produce a decrease in yield of up to 10% (Lobell et al. 2011).

Drought is one of the most common abiotic stresses, which significantly affects crop production (Fedoroff et al. 2010). Drought stress induces a series of physiological and biochemical changes in plants, strongly interferes cellular homeostasis and causes morphological changes,

*Correspondence: yanym@cnu.edu.cn
†Fei Luo and Xiong Deng contributed equally to this work
[1] College of Life Science, Capital Normal University, Beijing 100048, China
Full list of author information is available at the end of the article

e.g. repression of cell growth and photosynthesis, and activation of respiration. Moreover, when two experimental groups are respectively exposed to water sufficiency and water deficiency conditions during grain filling, the results of stored carbon remobilization in wheat are apparently different: one group is senescence and the other accelerates grain filling (Yang et al. 2000, 2001). However, telluric plants have developed specific mechanisms to response and tolerance to short- and long-term adverse environments, particularly to drought (Harb et al. 2010). For example, plants can accumulate reactive oxygen species (ROS) and proteins to enhance drought tolerance (Zhu 2002). In addition, when encountered drought stress, plant root caps synthesize abscisic acid (ABA) to trigger a signaling cascade in guard cells, and then leads to stomatal closure and reduction of water loss (MacRobbie 1998).

Plant leaves are the largest organ of photosynthesis with highest photosynthetic efficiency, providing an important source of carbohydrate for developing grains. Therefore they serve as the ultimate yield-limiting factor (Slafer et al. 1990). As reported, the contribution of wheat flag leaves to grain yield is up to 41–43% (Araus and Tapia 1987). However, photosynthesis is particularly sensitive to water deficit and the photosynthetic rate is negatively related to water content (Lawlor and Cornic 2002). Meanwhile, drought causes foliar stoma limitation and reduces air exchanges (Cornic 2000). In addition, metabolic repair can be suppressed by drought, causing the limitation of photosynthesis and significant reduction of carbon assimilation and utilization capacity (Reddy et al. 2004). Grain endosperm in wheat consists of about 70% starch and 14% proteins, which are the major determinant of yield and quality (Johansson et al. 2001; Donner and Mesdag 2000). These reserve substances are rapidly synthesized and accumulated after flowering, and this process involves lots of genes and enzymes (Cao et al. 2015; Yu et al. 2016). At least four types of enzymes, phosphorylated by other kinases in the amyloplast of higher plant endosperm, are involved in starch biosynthesis such as branching enzymes, ADP glucose pyrophosphorylase (AGPase), starch synthases (SS) and debranching enzymes (DBE).

Protein phosphorylation, as one of the most important protein posttranslational modifications (PTMs), is often reversible and transient (Hunter and Karin 1992). Serine, threonine, and tyrosine (Ser/Thr/Tyr) are the key phosphorylated modification sites in proteins (Stock et al. 1989; Cohen 2002; Aivaliotis et al. 2009), Plenty of enzymes perform their functions in signaling pathways, protein abundance or activity regulation via protein phosphorylation (Engelsberger and Schulze 2012). The phosphorylation of some key proteins involved in ABA signal transduction pathway contributes to cellular growth and abiotic stress response (Thingholm et al. 2009; Zhang et al. 2014a). Protein phosphorylation also regulates many essential biochemical processes such as DNA transcription, protein translation, and energy metabolism (Kersten et al. 2009).

In recent years, extensive phosphoproteomic analyses in different plant species were performed to explore the molecular basis of plant growth and abiotic stress responses such as *Arobidopsis thaliana* (Umezawa et al. 2013), *Brachypodium distachyon* (Lv et al. 2014a, b; Yuan et al. 2016), rice (Chang et al. 2012), wheat (Zhang et al. 2014a, b; Dong et al. 2015; Lv et al. 2016), maize (Hu et al. 2015a, b) and barley (Horie et al. 2011). Particularly, Pro-Q Diamond phosphoprotein staining can specially bind to the part of phosphate of phosphoproteins as well as phosphoamino acids. Therefore it serves as an efficient mean to rapidly detect phosphorylated proteins (Silva-Sanchez et al. 2015). Through Pro-Q Diamond staining, protein phosphorylated modification was widely present in germinating seeds (Dong et al. 2015), developing grains (Guo et al. 2012; Zhang et al. 2014b), starch granules (Chen et al. 2014, 2017; Cao et al. 2015) and seedling leaves response to salt stress (Lv et al. 2016). Various abiotic stresses such as drought and salt can induce significant changes in protein phosphorylated levels (Zhang et al. 2014a; Bian et al. 2017). Thus, protein phosphorylation plays key roles in wheat growth and development, and starch biosynthesis as well as in response to various adverse environments. However, to our knowledge, the phosphorylated protein characterization in wheat developing flag leaves and grains and their potential roles in regulating grain development and yield formation are still not clear.

In this study, we aim to characterize the phosphorylated proteins in wheat developing flag leaves and grains at 20 days postanthesis (DPA) under field water-deficit treatment and to reveal the potential roles of phosphorylated proteins involved in drought stress response and grain development. Our results provide new evidence for further understanding the molecular mechanisms of flag leaves and grains synergistically respond to drought stress.

Materials and methods

Wheat materials, field drought treatments and sampling

Chinese elite bread wheat cultivar Zhongmai 175 (*Triticum aestivum* L.) was used as material and planted in the experimental station of China Agricultural University (CAU), Wuqiao, Hebei Province (116°37′23″E and 37°16′02″N) during 2016–2017 wheat growing season. Basic fertility of the experiment plot was measured before sowing. The organic matter, total nitrogen, hydrolysable nitrogen, available phosphorus and available potassium in the topsoil (0–20 cm) of the experimental plots were 12.1 g kg^{-1}, 1.0 g kg^{-1}, 106.7 mg kg^{-1}, 33.8 mg kg^{-1} and 183.4 mg kg^{-1}, respectively. Total precipitation was 128 mm at 2016–2017 wheat growing season, lower than the annual mean amount (130–180 mm).

Field experiments contained two irrigation treatment groups in three biological replicates: drought treatment group without irrigation after sowing and control group with two times of irrigation after sowing watered 75 mm at the jointing and anthesis stage, respectively. Each experimental plot was 8 × 4 m with rows spaced at 0.16 m and 1 m interval between control and treatment groups was maintained to minimize the effects of adjacent plots. The watering amount was measured by a flow meter. Soil samples at 20 DPA were collected from 0 m increments to a depth of 2 m with a soil corer and soil water content was determined by using the oven-drying method (Gardner 1986).

Before sowing, relative water content of target field was irrigated to 80.5% of the field water capacity at 0–200 cm soil layer, and then seeds were sowed when soil water content got to 80% (Chu et al. 2016). The Zadoks scale was used to categorize crop developmental stages (Zadoks et al. 1974). The plants were marked after flowering, and then flag leaves and developing grains at 20 DPA in three biological replicates were harvested, and immediately immerged into liquid nitrogen prior to use. After maturity, main agronomic traits and grain yield were tested.

Protein extraction

Proteins from flag leaves and grains were extracted according to the previous report (Zhang et al. 2014b) with minor modifications. Leaves and grain samples (each 0.5 g) from three biological replicates were ground into fine powder by liquid nitrogen, then mixed with 1 mL of extraction buffer consisting of 50 mM Tris–HCl (pH 8.0), 0.1 M KCl, 5 mM EDTA and 30% sucrose. PhosSTOP phosphatase inhibitor cocktail (1 tablet/10 mL, Roche, Basel, Switzerland) was added. After placing for 15 min, the samples were shaken vigorously for 30 min at room temperature, and then centrifuged to divide mixture. Protein supernatants were precipitated with a one-quarter volume of cold 10% trichloroacetic acid at − 20 °C for 4 h. The precipitated proteins were rinsed with cold acetone (− 80 °C) and then centrifuged three times at 13,000*g* for 10 min. After frozen dryly, 300 μL solubilization buffer was added and placed at room-temperature for 4 h. The protein concentrations were quantitated by 2-D Quant Kit (Amersham Bioscience, USA) and the protein solution was stored at − 80 °C for later use.

2-DE

The differentially accumulated proteins (DAPs) from three biological replicates were separated by two-dimensional electrophoresis (2-DE) based on Lv et al. (2014b). The isoelectric focusing (IEF) was performed using 18 cm linear gradient IPG strip (GE Healthcare, Little Chalfont, UK). After IEF, the equilibration solution (1% DTT) was applied to equilibrate the strips for 15 min and then the second equilibration was performed with 2.5% w/v iodoacetamide. Subsequently, the strips were loaded on the top of 12% SDS–polyacrylamide gels for SDS–PAGE, and 2-DE gels were stained by Coomassie blue. The ImageMaster 2D Platinum 7.0 (GE Healthcare, USA) was used to analyze the images and only those with significant and biological reproducible changes (abundance variation at least twofold, Student's *t* Test, $p < 0.05$) were used as DAP spots.

Detection of phosphorylated proteins by Pro-Q Diamond gel staining

2-DE gels were stained with Pro-Q Diamond (Invitrogen, USA) to detect the different level of phosphoproteins by the previous method (Zhang et al. 2014a). The gels were fixed twice for 30 min/each time and washed three times with ddH_2O for 10 min/each time. Subsequently, the gels were incubated in Pro-Q Diamond staining in darkness for 2 h and destained with 20% acetonitrile in 50 mM sodium acetate (pH 4.0) four times (30 min each). The gels were scanned on a TyphoonTM 9400 scanner (GE Healthcare, USA) with a 532 nm excitation laser and a 610 nm long pass filter. The gels were stained with CBB to visualize total phosphoproteins after fluorescent image acquisition. The phosphorylated proteins were identified through comparison between 2-DE gels and Pro-Q Diamond gel staining results.

MALDI-TOF/TOF-MS

The DAP and phosphorylated protein spots identified were manually excised from the 2-DE gels and digested with trypsin as the reported method (Lv et al. 2016). Matrix-assisted laser desorption/ionization time-of-flight/time-of-flight mass spectrometer (MALDI-TOF/TOF-MS) was used to identify the DAPs by ABI 4800 Proteomics Analyzer. The MS/MS spectra were searched in Viridiplantae (green plant) sequences in the nonredundant National Center for Biotechnology Information (NCBI) database and Triticum NCBI database. MASCOT software (ver. 2.1; Matrix Science, London, UK) was used and all searches were evaluated based on the significant scores obtained from MASCOT. A significance threshold of $p < 0.05$ was used, and the protein score CI% and total ion score CI% were both set to > 95%.

Validation of phosphorylated proteins by Western blotting

Western blotting was performed to further verify the phosphorylated proteins identified by Pro-Q Diamond staining by using Anti-Phosphoserine/threonine/tyrosine monoclonal antibody from Abcam (Cat. No. SPM101, MA, USA) according to the previous report (Chen et al. 2014).

Bioinformatics analysis

Protein function was classified based on the annotation from UniProt (Wang et al. 2015). The subcellular localization was predicted according to FUEL-mLoc Server (http://bioinfo.eie.polyu.edu.hk/FUEL-mLoc/), WoLF PSORT (http://www.genscript.com/wolf-psort.html), CELLO version 2.5 (http://cello.life.nctu.edu.tw/), Plant-mPLoc (http://www.csbio.sjtu.edu.cn/bioinf/plant-multi/) and UniProtKB. Biological statistic analyses were performed by SPSS statistics software (ver. 19.0; SPSS Inc., Chicago, IL, USA) to intuitively view the results. The relative expression quantity of proteins between well-watered and water-deficit group was represented as the average of three replicates and compared by one-way ANOVA. Cluster 3.0 and visualized with Java TreeView were used for protein clustering analysis. Hierarchical clustering of the expression profiles was carried out based on the log transformed fold change values of protein spots and average linkage clustering method with Euclidean distance similarity metric was used (Wang et al. 2017). Meanwhile, NetPhos 3.1 Server (Blom et al. 1999) (http://www.cbs.dtu.dk/services/NetPhos/) was used to predict phosphorylation sites among the identified phosphorylated proteins.

Results

Soil water content, agronomic trait and yield changes under water deficit

The changes of relative soil water content between 0 and 2 m depths in the control group and drought treatment groups at 20 DPA were shown in Fig. 1. The results showed that the relative soil water content of 0 to 1.4 m depth had obvious discrepancy between the control and drought treatment. Thus, severe drought in the treatment group occurred in the 0–60 cm soil layer, and mild drought in the 60–140 cm soil layer based on the grade of agricultural drought in GB/T 32136-2015 (Lv et al. 2015).

Water deficit resulted in significant changes of main agronomic traits and grain yield. In the field, plant leaves became yellower and plant height was more lethargic (Fig. 2a, b) and wheat ears and grains were smaller and yellower in the treatment group than the control group (Fig. 2c, d). Thus, plant height, ear length, tiller numbers, grain number per spike and 1000-grain weight were significantly decreased. Ultimately, drought treatment caused significant reduction of grain yield, up to 19.23% (Additional file 1: Table S1).

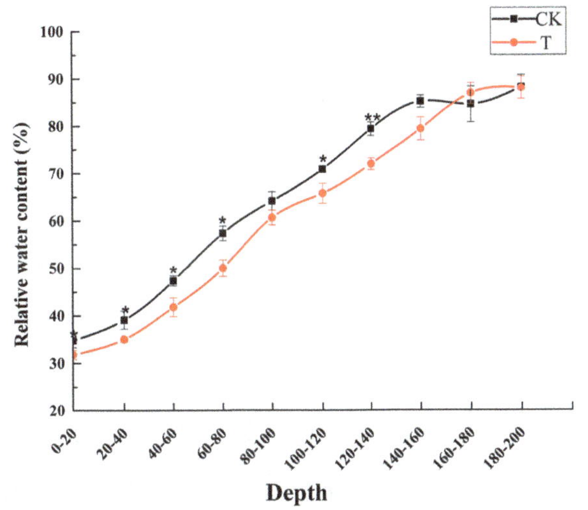

Fig. 1 Soil relative water content at 20 DPA. Soil relative water content was measured at 2 m depth of soil in the control (CK) and drought treatment (T) groups. Statistically significant differences compared to the control were calculated based on an independent Student's t-tests. Asterisks indicate significant different (*$p < 0.05$; **$p < 0.01$). Error bars indicate standard errors of three biological replicates

Identification of differentially accumulated proteins (DAPs) responsive to water deficit in flag leaves and developing grains

2-DE and tandem mass spectrometry identified 41 and 71 DAP spots in response to water deficit, which respectively contained 36 and 64 unique proteins at 20 DPA in flag leaves and grains (Fig. 3). The detailed information and peptide sequences of the identified proteins are listed in Additional file 1: Tables S2A, B and S3A, B. The function annotation from UniProt showed that 100 unique proteins were classified into 7 functional categories: carbon metabolism, stress defense, energy metabolism, photosynthesis, protein metabolism, storage substance biosynthesis. In the flag leaves, the DAPs were mainly involved in energy metabolism (34.17%), photosynthesis (31.70%) and protein metabolism (14.63%) while those in the developing grains were mainly related to stress defense (23.94%), protein metabolism (16.90%) and storage substance biosynthesis (16.90%) (Fig. 4a). Subcellular localization showed that most of the DAPs in flag leaves and grains were located in chloroplast (73.17%) and cytoplasm (45.07%), respectively (Fig. 4b, c). These results indicate that the foremost function of

Fig. 2 Plant phenotypes under stress treatment and CK. **a, b** Plant height compared at 20 DPA between CK and drought treatment groups. **c** Ear length at 20 DPA under drought treatment. **d** Grain appearance at 20 DPA under drought treatment

flag leaves is photosynthesis while grain proteins play important roles in abiotic stress defense and reserve substance synthesis.

Characterization of phosphorylated proteins in the developing flag leaves and grains under water deficit

Through comparison of 2-DE (Fig. 3), Pro-Q Diamond staining (Fig. 5), and tandem mass spectrometry results, 20 and 38 phosphorylated DAP spots in response to water deficit were respectively identified in flag leaves and developing grains, which were individually represented 18 and 38 unique phosphoproteins in flag leaves and developing grains (Table 1). According to the functional categories (Fig. 4a), the phosphorylated proteins in flag leaves were mainly involved in energy metabolism (35%) and photosynthesis (35%), while those in the developing grains mainly participated in stress defense (26.31%), protein metabolism (23.68%), and energy metabolism (18.42%). Most of the phosphorylated proteins in flag leaves were present in chloroplast (80%) (Fig. 4b), and 52.63% grain phosphorylated proteins were located in cytoplasm (Fig. 4c). Thus, drought stress mainly affected photosynthesis and energy metabolism in the flag leaves,

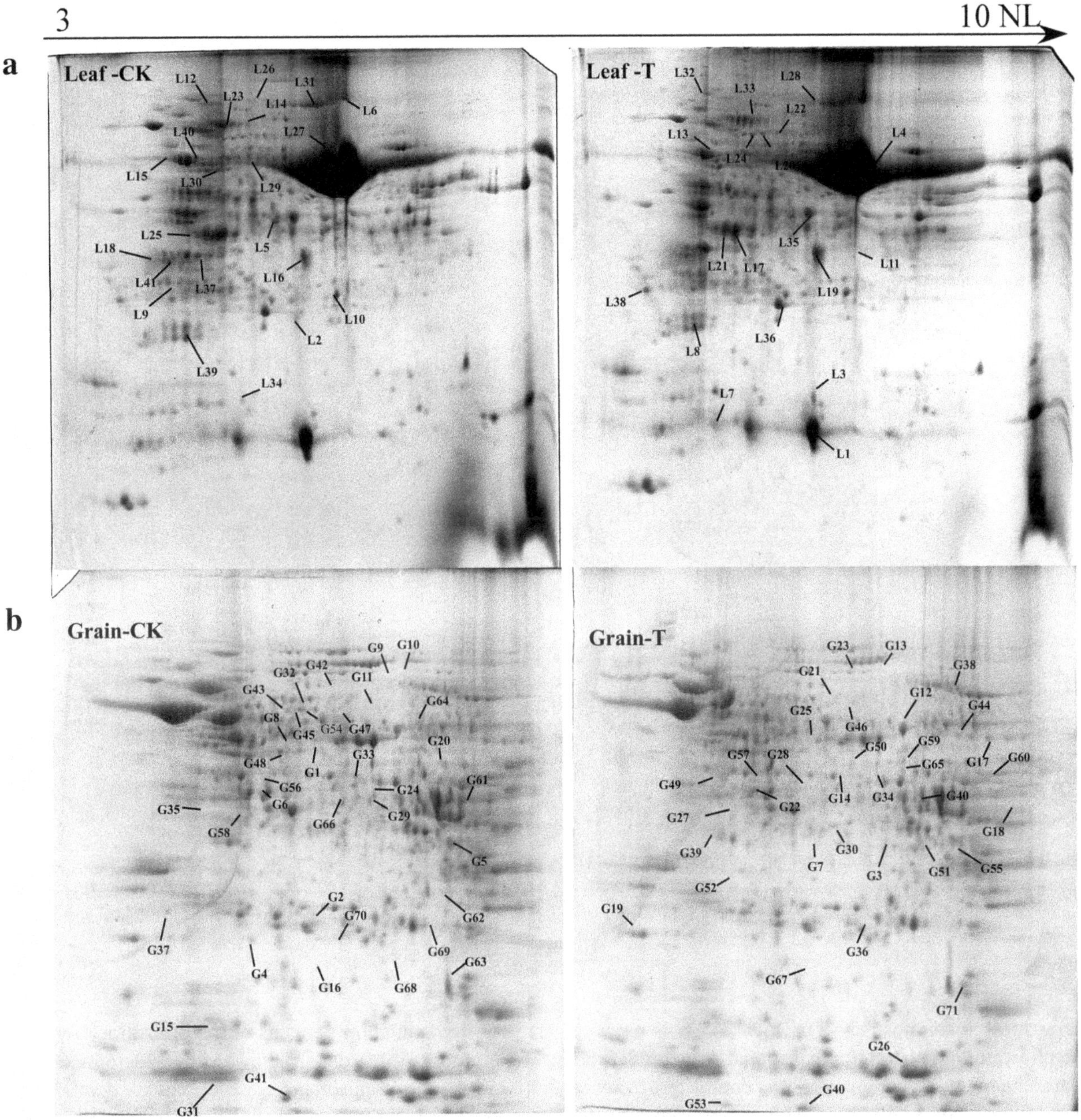

Fig. 3 2-DE gel images of flag leaves and developing grains at 20 DPA in Zhongmai 175 under drought stress. **a** Flag leaf 2-DE gels of CK and drought treatment. **b** Grain 2-DE gels of CK and drought treatment

and stress/defense, energy metabolism, protein trafficking and degradation in the developing grains.

Associated with 2-DE gels and Pro-Q Diamond gel staining, some isoforms of DAPs resulted from phosphoprotein were found. In the flag leaves, ribulose-1,5-bisphosphate carboxylase/oxygenase large subunit had three different isoforms (L4, L5 and L6). Meanwhile, the phosphorylation of both L4 and L5 was detected by Pro-Q Diamond. In the developing grains, ribulose-1,5-bisphosphate carboxylase/oxygenase large subunit (G1 and G59) and fructose-bisphosphate aldolase cytoplasmic isozyme (G2 and G3) also had different isoforms, of which G2 and G3 was individually phosphorylated (Table 1).

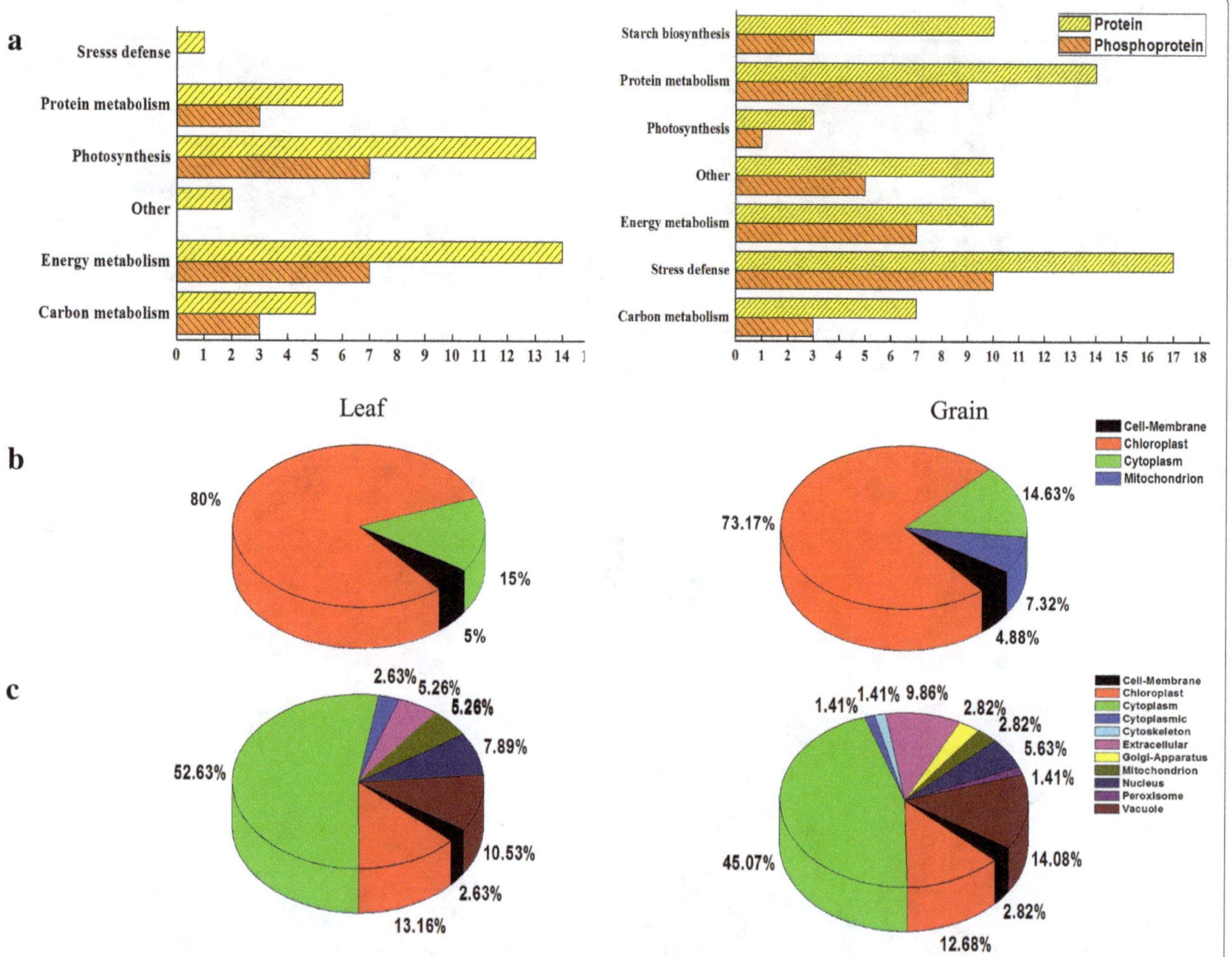

Fig. 4 Functional classification, subcellular localization and Venn diagram analysis of DAPs from flag leaves and developing grains in Zhongmai 175. **a** Functional classification of proteins in flag leaf and developing grain. **b** Subcellular localization of all proteins (right) and phosphoproteins (left) that we had identified in leaves. **c** Subcellular localization of all proteins (right) and phosphoproteins (left) that we had identified in developing grains

Hierarchical clustering analysis of DAPs and phosphorylated proteins in response to water deficit

Hierarchical clustering analysis was conducted to reveal the accumulation patterns of DAPs and phosphorylated proteins under water deficit (Fig. 6). The results showed that 24 (58.54%) and 45 (63.38%) DAPs from flag leaves and developing grains were significantly upregulated, respectively (Fig. 6a). In the flag leaves, the upregulated DAPs were mainly related to energy metabolism while the downregulated DAPs were majorly involved in photosynthesis. In the developing grains, stress defense and energy metabolism related DAPs were drastically increased whereas those related to protein metabolism were downregulated.

The accumulation patterns of the phosphorylated proteins under water deficit showed that 13 (65%) and 25 (65.78%) phosphoproteins were upregulated in the flag leaves and grains, respectively (Fig. 6b). We found that most phosphoproteins involved in carbon metabolism and energy metabolism were upregulated in flag leaves, but some important phosphoproteins participating in photosynthesis were downregulated. In the developing grains, the modification levels of those related to energy metabolism, stress defense and storage substance biosynthesis increased. Thus, most phosphoproteins involved in photosynthesis were both downregulated whereas those related to energy metabolism, drought stress response and storage substance biosynthesis were generally

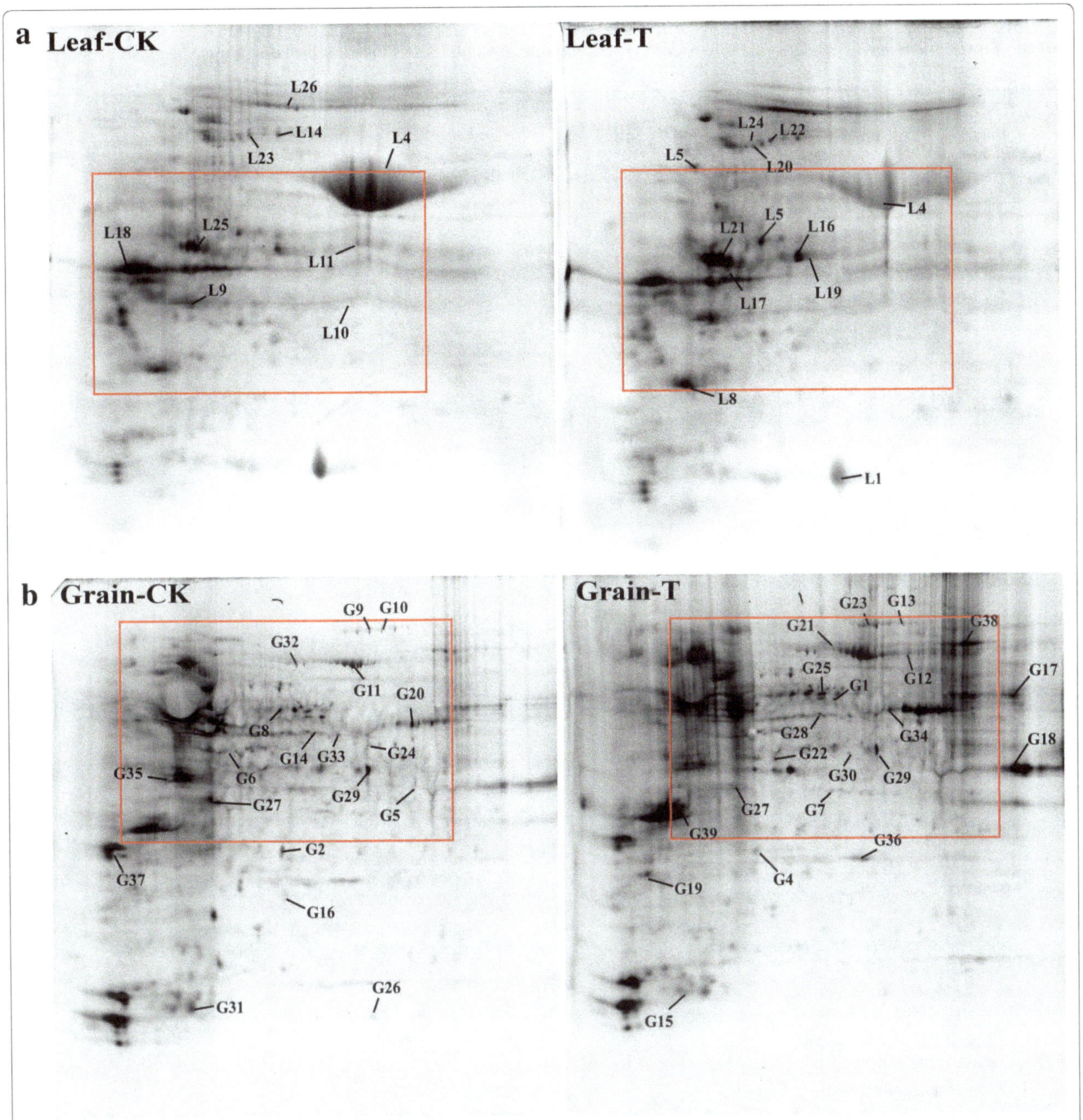

Fig. 5 Detection of phosphorylated proteins in 2-DE gels by Pro-Q diamond staining. **a** The image of Pro-Q diamond staining in flag leaves. **b** The image of Pro-Q diamond staining in grains

upregulated in both phosphorylated modification level and protein level. Particularly, ribulose bisphosphate carboxylase/oxygenase activase A (L10 and L11) was only identified in flag leaves under well-watered group and translationally-controlled tumor protein (G19) was only identified in the developing grains under drought treatment group. This suggests that water deficit caused the dephosphorylation of ribulose bisphosphate carboxylase/oxygenase activase A and specific phosphorylation of translationally-controlled tumor protein.

Verification of phosphorylated proteins by phosphorylated site prediction and Western blotting

To provide further supporting for the presence of the phosphorylated proteins identified by Pro-Q Diamond staining, the phosphorylated sites of all 58

Table 1 Representative phosphorylated proteins identified by MALDI-TOF/TOF-MS and Pro-Q staining

Spot no.	Protein name	Protein score	Peptide count	*p*-value	Subcellular localization	Protein level	Phosphorylation level
Photosynthesis							
L1	Ribulose-1,5-bisphosphate carboxylase/oxygenase small subunit	300	18	0.021	Chloroplast	Down	Down
L4	Ribulose-1,5-bisphosphate carboxylase/oxygenase large subunit	1120	30	0.03	Chloroplast	Down	Down
L5	Ribulose-1,5-bisphosphate carboxylase/oxygenase large subunit	695	29	0.024	Chloroplast	Up	Up
L8	Chlorophyll a–b binding protein 8, chloroplastic	111	5	0.041	Chloroplast	Up	Up
L9	Chlorophyll a–b binding protein, chloroplastic	399	10	0.028	Chloroplast	Up	Up
L10	Ribulose bisphosphate carboxylase/oxygenase activase A, partial	363	17	0.029	Chloroplast	Down	Down–
L11	Ribulose bisphosphate carboxylase/oxygenase activase A, partial	402	17	0.042	Chloroplast	Down	Down–
L14	RuBisCO large subunit-binding protein subunit beta	1030	20	0.013	Chloroplast	Down	Down
G1	Ribulose-1,5-bisphosphate carboxylase/oxygenase large subunit, partial	166	10	0.038	Chloroplast	Up	Down
Energy metabolism							
L13	Ribulose bisphosphate carboxylase activase B	539	18	0.031	Chloroplast	Up	Up
L16	Fructose-bisphosphate aldolase, cytoplasmic isozyme 1	606	24	0.019	Cytoplasm	Up	Up
L17	Chloroplast fructose-1,6-biphosphate aldolase	490	17	0.035	Chloroplast	Up	Up
L19	ATP synthase subunit	255	20	0.043	Chloroplast	Down	Down
L21	Phosphoglycerate kinase, chloroplastic	916	17	0.028	Chloroplast	Up	Up
L25	V-type proton ATPase subunit B 1	538	36	0.019	Chloroplast	Up	Up
L26	ATP-dependent Clp protease ATP-binding subunit clpA-like protein CD4B	383	38	0.02	Chloroplast	Down	Up
G2	Fructose-bisphosphate aldolase cytoplasmic isozyme	118	12	0.019	Cytoplasm	Up	Up
G4	Triosephosphate-isomerase	619	3	0.018	Cytoplasm	Up	Up
G5	Glyceraldehyde-3-phosphate dehydrogenase	75	14	0.032	Cytoplasm	Down	Up
G6	Enolase	366	18	0.018	Cytoplasm	Down	Down
G7	Pyruvate, phosphate dikinase 1	491	39	0.018	Chloroplast	Up	Up
G8	Dihydrolipoyl dehydrogenase	253	23	0.015	Mitochondrion	Up	Up
G9	Aconitate hydratase	163	25	0.043	Cytoplasm	Up	Up
Storage substance biosynthesis							
G10	Sucrose synthase type 2	809	31	0.002	Cytoplasm	Up	Up
G11	Sucrose synthase 2	432	35	0.028	Cytoplasm	Up	Up
G12	Phosphoglucomutase	476	18	0.019	Cytoplasm	Up	Up
Stress defense							
G13	Heat shock protein 101	916	40	0.039	Nucleus	Up	Up

Table 1 (continued)

Spot no.	Protein name	Protein score	Peptide count	*p*-value	Subcellular localization	Protein level	Phosphorylation level
G14	17.9 kDa class I heat shock protein-like	224	21	0.017	Cytoplasm	Up	Up
G15	Oxalate oxidase 2	97	5	0.021	Cell-Membrane	Up	Up
G16	Dehydroascorbate reductase	812	15	0.035	Cytoplasm	Up	Down
G17	Disease resistance protein RPP8	48	14	0.022	Cytoplasmic	Up	Up
G18	Peroxidase 1	340	13	0.032	Vacuole	Down	Up
G19	Translationally-controlled tumor protein	552	16	0.022	Cytoplasm	Down	Up+
G31	Hypothetical protein TRIUR3_21260	632	8	0.006	Cytoplasm	Down	Down
G35	Cold regulated protein	210	6	0.021	Cytoplasm	Down	Down
G39	Translationally controlled tumor protein	264	7	0.021	Cytoplasm	Up	Up
Carbon metabolism							
L20	Arabinoxylan arabinofuranohydrolase isoenzyme AXAH-II	251	12	0.029	Cell-Membrane	Up	Up
L24	Predicted protein	316	14	0.039	Chloroplast	Down	Up
G24	Basic endochitinase C	130	3	0.029	Extracellular	Up	Down
G28	Formate dehydrogenase	196	3	0.026	Mitochondrion	Up	Up
G30	Glucose and ribitol Dehydrogenase-like protein	264	11	0.032	Cytoplasm	Up	Down
Protein biosynthesis and degradation							
L18	Peptidyl-prolylcis-trans isomerase CYP38	371	21	0.015	Chloroplast	Down	Down
L22	Methionine synthase 1 enzyme	391	27	0.032	Cytoplasm	Down	Up
L23	Methionine synthase 1 enzyme	391	27	0.006	Cytoplasm	Down	Up
G20	Globulin-1 S allele	416	16	0.024	Vacuole	Up	Up
G22	Serpin-Z2B	712	18	0.034	Chloroplast	Down	Down
G23	ATP-dependent zinc metalloprotease FTSH 2	53	12	0.027	Chloroplast	Up	Up
G25	Alanine aminotransferase 2	304	22	0.017	Cytoplasm	Down	Up
G27	Serpin 1	500	17	0.041	Extracellular	Up	Down
G32	Putative methionine synthase	157	1	0.014	Cytoplasm	Up	Up
G33	Aspartate aminotransferase	194	9	0.021	Cytoplasm	Down	Up
G34	Globulin 3	489	20	0.017	Vacuole	Down	Up
G37	Elongation factor 1-beta	159	4	0.042	Vacuole	Down	Down

Up: upregulated in phosphorylation level; Down: downregulated in phosphorylation level. Up+: the phosphoproteins were only identified in drought treatment. Down−: the phosphoproteins were only identified in CK

phosphorylated proteins were predicted by NetPhos 3.1 Server. High scores (≥0.90) were set to increase the accuracy of prediction and the results are shown in Additional file 1: Table S4A, B. In total, 58 phosphorylated proteins contained 543 phosphorylated peptides, of which 191 were from flag leaves and 352 from developing grains (Additional file 2: Fig. S1a). Combined with functional categories, we found that phosphorylated proteins related to energy metabolism contained the most phosphorylated sites, including 73 (38.22%) from flag leaves and 76 (21.59%) from grains (Fig. 7a). The second largest category was those related to protein metabolism, which accounted for 26.70% (51) in flag leaves and 21.02% (74) in developing grains. Generally, both flag leaves and grains had similar percentages of serine (74%), threonine (14%) and tyrosine (12%) phosphorylation (Additional file 2: Fig. S1b, d). Thus, consistent with the previous reports (Zhang et al. 2014a, b), serine phosphorylation had the highest percentage among three phosphorylated amino acid residues.

Western blotting was performed to further verify the identified phosphorylated proteins using an

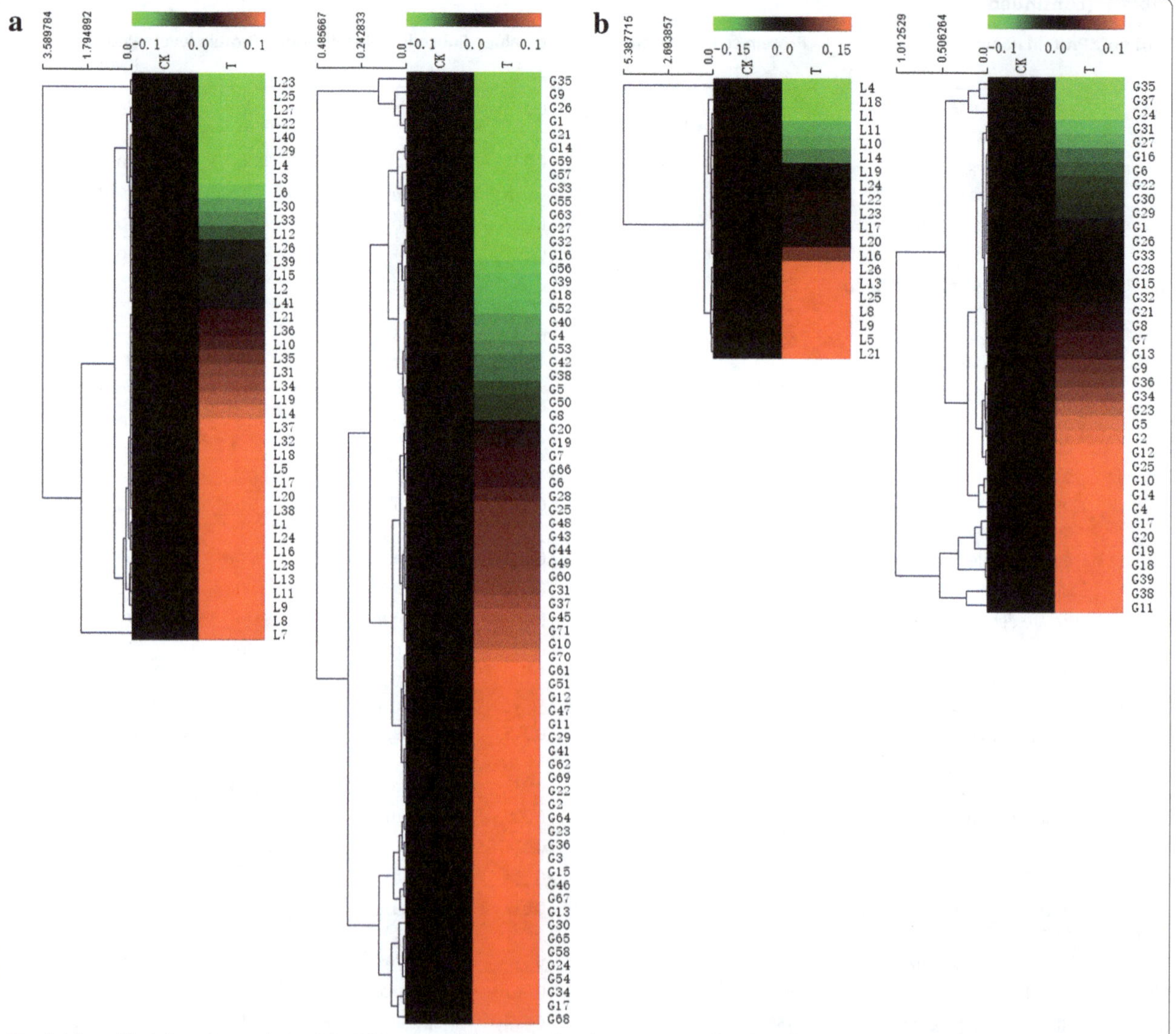

Fig. 6 Hierarchical clustering analysis of the DAP spots and phosphorylated proteins identified from flag leaves and developing grains. **a** Hierarchical clustering of all DAP spots from flag leaves and developing grains in protein level; **b** hierarchical clustering of phosphorylated proteins from flag leaves and developing grains in phosphorylation level. The upregulation and downregulation are indicated are indicated in red and green, respectively

Anti-Phosphoserine/threonine/tyrosine monoclonal antibody (Fig. 7b, c). Through comparison of Western blotting results with 2-DE and Pro-Q Diamond gels, the results showed that 16 phosphorylated proteins, including 7 from flag leaves (Fig. 7b) and 9 from developing grains (Fig. 7c) were confirmed to be phosphorylated, such as ribulose-1,5-bisphosphate carboxylase/oxygenase large subunit, phosphoglycerate kinase, chloroplastic and formate dehydrogenase etc. These phosphorylated proteins shown in Fig. 7b, c were well corresponding to Pro-Q Diamond staining results listed in Table 1.

Discussion

Phosphorylated proteins participating in photosynthesis regulation

Photosynthesis is an essential metabolic process that directly impacts carbohydrate synthesis, grain development and yield formation. Phosphorylation of chloroplast membrane proteins is ultimately responsible in plants, responding to changes in incident light and redox poise (Allen 1983, 1992a). Light-harvesting complex II (LHCII), the chloroplast light-harvesting chlorophyll a/b-binding complex, is a major substrate protein of phosphorylation, which binds perhaps half of the chlorophyll in nature. It is

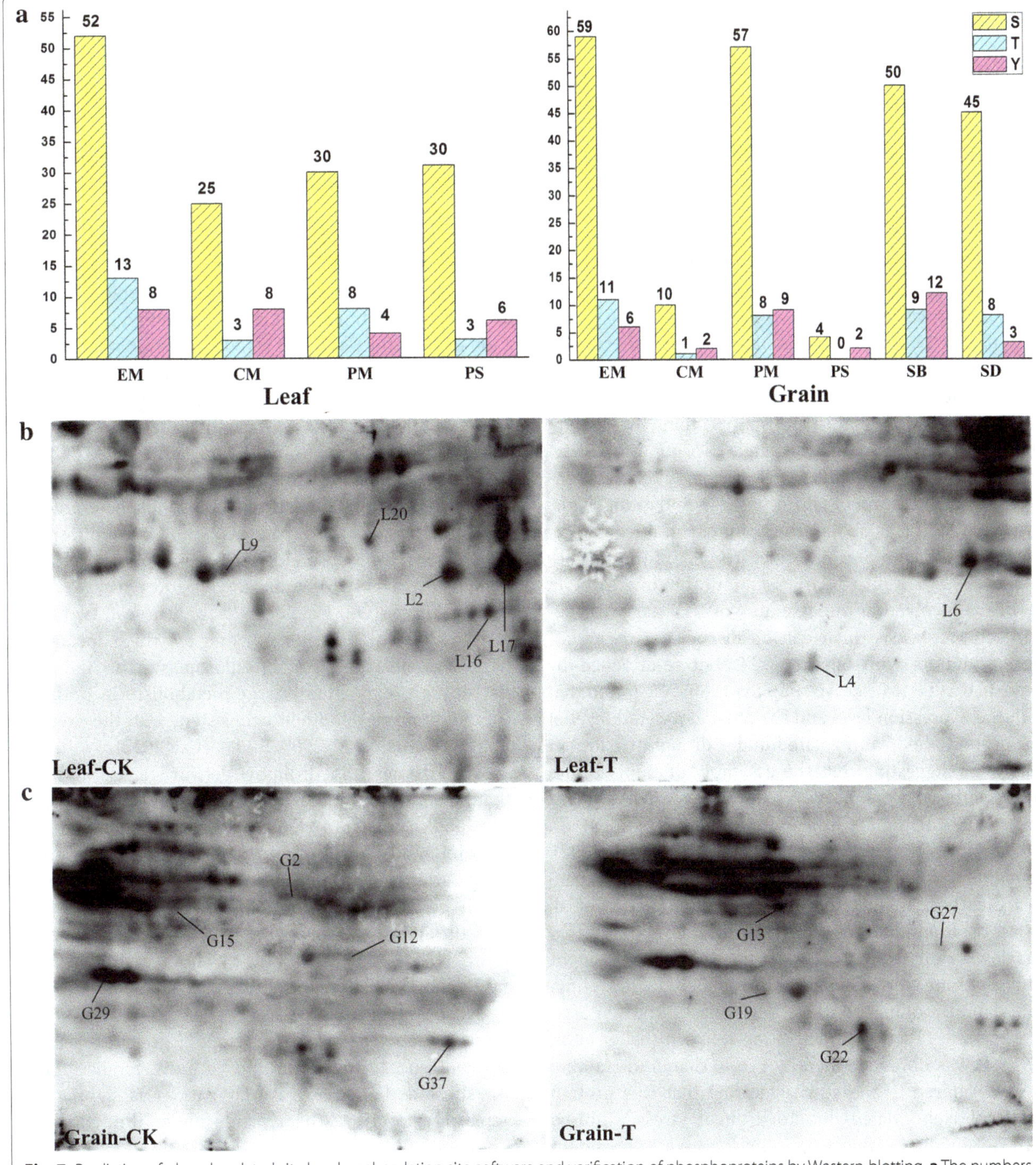

Fig. 7 Prediction of phosphorylated site by phosphorylation site software and verification of phosphoproteins by Western-blotting. **a** The numbers of serine/threonine/tyrosine predictive phosphorylation sites in main functional groups in flag leaf and grain. S: phosphosites in serine residues, T: phosphosites threonine residues, Y: phosphosites tyrosine residues. EM: energy metabolism; CM: carbon metabolism; PM: protein metabolism; PS: photosynthesis; SB: Starch biosynthesis; SD: stress defense. **b** Western-blotting analysis in leaves of CK and drought treatment group, respectively. **c** Western-blotting analysis in the grains of CK and drought treatment group. All the gels used to perform Western blotting were chosen from red box in Fig. 5

known that the light-harvesting complex changes its allegiance upon phosphorylation. Phosphorylated proteins have a great effect on enzyme structure, specificity or activities such as glycogen phosphorylase and isocitrate dehydrogenase and therefore play a key role in photosynthesis (Allen 1992b). In this study, we identified several phosphorylated proteins associated with photosynthesis in flag leaves, including a series of ribulose-1,5-bisphosphate carboxylase/oxygenase (Rubisco) proteins, chlorophyll a–b binding proteins (CBPs, L8–L9) and ribulose bisphosphate carboxylase/oxygenase activase (L10–L13). One phosphorylated Rubisco protein in response to water deficit was also identified in the developing grains (G1) (Table 1). This indicates that photosynthesis also happened in grains when subjected to drought stress, which could be beneficial for grain development. Meanwhile, we found that most phosphoproteins involved in photosynthesis were downregulated in protein modification level as well as protein level, which could be responsible for the reduction of photosynthesis and grain yield.

Rubisco initiates both the photosynthetic carbon reduction and photorespiratory carbon oxidation cycles (Bowes 1991). Rubisco large subunit binding protein (RSBL) is involved in the assembly of Rubisco in higher plant chloroplasts, which can be dissociated to monomeric subunits with binding ATP, but re-attached by removal of ATP (Musgrove and Ellis 1986). We found that both the expression level and the phosphorylated level of this protein are downregulated under drought treatment (Table 1), suggesting that the assembly of Rubisco and photosynthesis were heavily restrained under drought stress. Rate of photosynthesis is heavily dependent on the activity of Rubisco proteins (Chaitanya et al. 2002). Drought normally diminishes the biochemical capacity for carbon assimilation and utilization (Reddy et al. 2004). We found that the phosphorylated level of Rubisco was downregulated in flag leaves, suggesting that drought inhibits photosynthesis mainly through reducing phosphorylation modification level of Rubisco. Rubisco activase has the function of activating Rubisco (Feller et al. 1998). It was only identified by Pro-Q diamond staining in well-watered flag leaves, suggesting that this protein was dephosphorylated duo to drought stress. Rubisco activase mediates Rubisco, which is ATP-dependent (Feller et al. 1998). We found that drought heavily inhibited phosphorylation of Rubisco activase and affected the activity of Rubisco activase. Thus, the function of photosynthesis was severely disturbed, ultimately leading to grain yield decrease.

CBP, a light-harvesting complex serves as a light receptor and has diverse functions (Bassi et al. 1992). It captures and delivers excitation energy to photosystems. In this study, both phosphorylation and protein levels of CBP and CBP 8 were upregulated under drought stress in the leaf, consistent with the recent report (Bian et al. 2017). This suggests that a few phosphorylated proteins such as CBP were upregulated to alleviate the influences on photosynthesis when suffered from drought stress.

Phosphorylated proteins involved in energy metabolism

Energy metabolism is basic radical cellular activity to maintain normal growth and development, which involves in three respiratory pathways: glycolysis, themitochondrial electron transport chain, and TCA cycle. These pathways are essential for energy supply to numerous cellular functions and greatly depend on phosphorylation of proteins (Fernie et al. 2004). In this study, energy metabolism related phosphorylated proteins accounted for a great proportion in function categories in both flag leaves and grains. Among them, 5 DAPs involved in glycolysis with an upregulation at 20 DPA of grain development in both protein and phosphorylaed modification levels, including fructose-bisphosphate aldolase (FBA) cytoplasmic isozyme (G2-G3), triosephosphate isomerase (G4), glyceraldehyde-3-phosphate dehydrogenase (GAPDH, G5), enolase (G6) and pyruvate phosphate dikinase (G7) (Table 1). In the flag leaf, two FBAs (L16–L17) were upregulated, one of which located in cytoplasm and the other in chloroplast. This indicated that drought expedited energy metabolism in both flag leaf and grain development, consistent with the previous reports (Caruso et al. 2009; Budak et al. 2013).

Two enzymes (dihydrolipoyl dehydrogenase and aconitate hydratase, G8 and G9) involved in TCA cycle were identified in this study. Both of them were up-regulated at protein and phosphoprotein level at 20 DPA of developing grains. The pyruvate dehydrogenase complex (PDC), consisting of dihydrolipoyl dehydrogenase, catalyzes the irreversible conversion of pyruvate, coenzyme A and NAD+ into CO_2, NADH and acetyl-CoA (Patel and Roche 1990). Obviously, TCA cycle has similar trend to glycolysis, and plants increased TCA cycle and glycolysis metabolism in the developing grains to provide sufficient energy for starch biosynthesis. Similar report also showed that the ATP/ADP ratio was significantly increased in spring wheat plants under drought stress (Chen et al. 2004).

Starch biosynthesis regulation via protein phosphorylation

Starch biosynthesis needs abundant triosephosphates provided by photosynthesis in the early grain development stages (Tschiersch et al. 2011). A large number of phosphorylated enzymes related to starch biosynthesis were needed to transform intermediates of photosynthesis into starch (Tschiersch et al. 2011). In this study, we identified two key enzymes involved in starch biosynthesis: sucrose synthase (G10, G11) and phosphoglucomutase (PGM, G12). Phosphorylation of these proteins could be induced by drought stress (Zhang et al. 2014a) and high-nitrogen (Zhen et al. 2017). At the beginning of starch biosynthesis, ADP glucose pyrophosphorylase (ADPase) catalyzes the first committed step of the starch biosynthetic pathway, converting glucose 1-phosphate and ATP to glucosyl moiety of ADP (ADPG) and pyrophosphate (Tschiersch et al. 2011). Then, ADPG is metabolized by sucrose synthase (SS) as it arrives in the cytosol of endosperm cells (Tomlinson and Denyer 2003). SS catalyzes starch biosynthesis by transferring the ADPG glucose to the non-reducing end of an existing α-1,4-glucan chain. It has reported that loss of SSII in *T. aestivum* and *H. vulgare* endosperms causes a drastic reduction in amylopectin synthesis and abolishes binding of SSI, SBEIIa, and SBEIIb to the starch granules (Yamamori et al. 2000). We found that the phosphorylation level as well as protein level of SS was upregulation under drought stress, which could enhance enzyme activity and protein–protein interactions (Tetlow et al. 2004), and expedite starch biosynthesis and grain development under water deficit environment.

PGM has function of interconverting glucose-1-phosphate and glucose-6-phosphate, with glucose 1,6-bisphosphate as a cofactor (Ray et al. 1983). It has function that the transfer of a phosphoryl group from the itself to glucose 1-phosphate, forming glucose 1,6-bisphosphate (Najjar and Pullman 1954). The glucose-1-phosphate can be supplied in chloroplast phosphoglucomutase through the reductive pentose phosphate pathway (Smith and Martin 1993). However, it could be directly supplied by cytosol (Tyson and Rees 1988) or synthesized with the catalysis of a plastidial phosphoglucomutase in nonphotosynthetic tissues (Hill and Smith 1991). The biochemistry reaction catalyzed by PGM is a principal distinction in carbon and starch synthesis pathways. According to our results, PGM was unregulated in both phosphorylation and protein levels at 20 DPA of developing grains under drought stress conditions, which might boost to transform glucose-6-phosphate to glucose-1-phosphate. Thus, grain filling was advanced by drought stress due to acceleration of starch biosynthesis, consistent with energy metabolism in the developing grains.

Protein phosphorylation regulating plant drought stress response

During grain development, many hydrophobic proteins were gradually synthesized and accumulated to protect cells from various adverse environments (Zhang et al. 2014a). Drought has a severe limitation on wheat growth and yield formation. In this study, several proteins under drought conditions were found to be phosphorylated and up-regulated as well as in protein level, including 2 heat shock proteins (G13 and G14), one manganese ion binding protein (oxalate oxidase 2, G15), dehydroascorbate reductase (DHAR, G16), disease resistance protein RPP8 (G17), peroxidase 1 (G18) and translationally-controlled tumor protein (TCTP, G19). Particularly, the phosphorylation of TCTP was specifically induced by drought stress, suggesting its important roles in resistance to drought stress. TCTP is a calcium-binding protein (Sanchez et al. 1997) as well as tubulin-binding protein that associates with microtubules (Vandre et al. 1991). The expression level of plant TCTP could increase in response to abiotic stresses such as salt (Santa Brígida et al. 2014) and drought (Kim et al. 2012). TCTP could be phosphorylated at serine residues 46 and 64 by a polo-like protein kinase that regulates spindle function (Jung et al. 2004).

Plants are easily damaged by accumulated ROS under drought stress, including singlet oxygen, superoxide radical (O^{2-}), hydrogen peroxide (H_2O_2), and hydroxyl radical (OH) (Smith and Martin 1993). Many biological processes such as stress responses, hormone signaling, cell growth, and development, are related to ROS (Pei et al. 2000). Therefore, it is highly important for plants to scavenge high levels of ROS when suffered from drought. It is known that the ascorbate–glutathione (AsA–GSH) cycle is particularly important in plant antioxidant defense mechanism. Four anti-oxidative enzymes participate in ascorbate–glutathione cycle, including ascorbate peroxidase (APX), monodehydroascorbate reductase, dehydroascorbate reductase (DHAR), and glutathione reductase. In this study, a DHAR was identified in the drought treatment group at 20 DPA of developing grains, which is an indispensable enzyme functioning in the regeneration of ascorbate and glutathione that scavenges hydrogen peroxide and reduces it to water, with concomitant oxidation of NADPH (Eltayeb et al. 2006). Meanwhile, its phosphorylation and protein levels were up-regulated under drought stress, in accordance with our conjecture that drought stress may transform into oxidative stress, and proteins phosphorylated modification plays a key role to respond drought stress. The other anti-oxidative enzymes peroxidase 1 was also upregulated in grains, which is a key capable enzyme of removing ROS (De Pinto et al. 2006). Their phosphorylation could enhance ROS scavenging and drought tolerance.

Under abiotic stresses, maintaining proteins in their functional conformations and preventing protein misfolding are highly important for cell survival (Kim et al. 2013). Heat-shock proteins (Hsps) function in protein folding, assembly, translocation and degradation as well as assistance in protein refolding under stress conditions (Boston et al. 1996). Thus, Hsp family plays an imperative role in protecting plants against stress by re-establishing normal protein structure and cellular homeostasis (Bukau and Horwich 1998). Up-regulation of two Hsps was identified with enhancement in modification and expression level in this study. Hsp 101 can augment *Arabidopsis* resistance to thermo stress (Queitsch 2000). To our knowledge, the mechanism of HSP 101 functioned in adverse response is still not clear. Its phosphorylation was also found in wheat developing grains (Guo et al. 2012). The 17.9 kDa class I heat shock protein-like was found to be upregulated under drought stress in both phosphorylation and protein accumulation levels. We speculate that the phosphorylation of Hsp family may be an important mechanism in response to drought stress.

Conclusion

This study identified 58 phosphorylated proteins among 112 differentially accumulated proteins in response to water deficit at 20 days of postanthesis, including 20 in the flag leaves and 38 in the developing grains. The phosphorylated proteins from flag leaves mainly involved in photosynthesis, carbohydrate metabolism and energy metabolism while those from developing grains mainly participated in detoxification and defense, protein metabolism, carbohydrate metabolism and energy metabolism. Most phosphoproteins related to photosynthesis in flag leaves were significantly downregulated such as ribulose-1,5-bisphosphate carboxylase/oxygenase large subunit, ribulose bisphosphate carboxylase/oxygenase activase, leading to significant decrease in starch biosynthesis and grain yield. On the contrary, most phosphoproteins associated with drought stress response as well as energy metabolism were upregulated in the developing grains, demonstrating the important roles of protein phosphorylation in resisting to drought, and expediting starch biosynthesis and grain filling. Our results provide new evidence for protein phosphorylated modification how to regulate plant drought tolerance and grain development.

Abbreviations

2-DE: two-dimensional electrophoresis; AGPase: ADP glucose pyrophosphorylase; APX: ascorbate peroxidase; CBPs: chlorophyll a–b binding proteins; DAPs: differentially accumulated proteins; DHAR: dehydroascorbate reductase; DHAR: dehydroascorbate reductase; DPA: days postanthesis; FBA: fructose-bisphosphate aldolase; GAPDH: glyceraldehyde-3-phosphate dehydrogenase; Hsps: heat-shock proteins; MALDI-TOF/TOF-MS: matrix-assisted laser desorption/ionisation time-of-flight/time-of-flight mass spectrometry; NCBI: National Center for Biotechnology Information; PDC: pyruvate dehydrogenase complex; PGM: phosphoglucomutase; PTMs: protein posttranslational modifications; ROS: reactive oxygen species; Rubisco: ribulose-1,5-bisphosphate carboxylase/oxygenase; SDS–PAGE: sodium dodecyl sulfate–polyacrylamide gel electrophoresis; SS: starch synthases; TCTP: translationally-controlled tumor protein.

Authors' contributions

YY designed the experiments. FL witted this paper, conducted the experiment and analyzed experimental data. XD collected materials and measured agronomic trait. YL reviewed the manuscript. All authors read and approved the final manuscript.

Author details

[1] College of Life Science, Capital Normal University, Beijing 100048, China.
[2] Hubei Collaborative Innovation Center for Grain Industry (HCICGI), Yangtze University, Jingzhou 434025, China.

Competing interests

The authors declare that they have no competing interests.

Funding

This research was financially supported by the Grant from the National Natural Science Foundation of China (31771773).

References

Aivaliotis M, Macek B, Gnad F, Reichelt P, Mann M, Oesterhelt D (2009) Ser/Thr/Tyr protein phosphorylation in the archaeon *Halobacterium salinarum*—a representative of the third domain of life. PLoS ONE 4:e4777. https://doi.org/10.1371/journal.pone.0004777

Allen JF (1983) Protein phosphorylation—carburettor of photosynthesis? Trends Biochem Sci 8:369–373. https://doi.org/10.1016/0968-0004(83)90364-X

Allen JF (1992a) Protein phosphorylation in regulation of photosynthesis. Biochim Biophys Acta Bioenerg 1098:275–335. https://doi.org/10.1016/S0005-2728(09)91014-3

Allen JF (1992b) How does protein phosphorylation regulate photosynthesis? Trends Biochem Sci 17:12–17. https://doi.org/10.1016/0968-0004(92)90418-9

Araus JL, Tapia L (1987) Photosynthetic gas exchange characteristics of wheat flag leaf blades and sheaths during grain filling: the case of a spring crop grown under mediterranean climate conditions. Plant Physiol 85:667–673. https://doi.org/10.1104/pp.85.3.667

Bassi R, Soen Su Yin, Frank G, Zuber H, Rochaix JD (1992) Characterization of chlorophyll a/b proteins of photosystem I from *Chlamydomonas reinhardtii*. J Biol Chem 267:25714–25721

Bian Y, Deng X, Yan X, Zhou J, Yuan L, Yan Y (2017) Integrated proteomic analysis of *Brachypodium distachyon* roots and leaves reveals a synergistic network in the response to drought stress and recovery. Sci Rep 7:46183. https://doi.org/10.1038/srep46183

Blom N, Gammeltoft S, Brunak S (1999) Sequence and structure-based prediction of eukaryotic protein phosphorylation sites. J Mol Biol 294:1351–1362. https://doi.org/10.1006/jmbi.1999.3310

Boston RS, Viitanen PV, Vierling E (1996) Molecular chaperones and protein folding in plants. Plant Mol Biol 32:191–222. https://doi.org/10.1007/BF00039383

Bowes G (1991) Growth at elevated CO_2: photosynthetic responses mediated through Rubisco. Plant Cell Environ 14:795–806. https://doi.org/10.1111/j.1365-3040.1991.tb01443.x

Budak H, Akpinar BA, Unver T, Turktas M (2013) Proteome changes in wild and modern wheat leaves upon drought stress by two-dimensional electrophoresis and nanoLC-ESI–MS/MS. Plant Mol Biol 83:89–103. https://doi.org/10.1007/s11103-013-0024-5

Bukau B, Horwich AL (1998) The Hsp70 and Hsp60 chaperone machines. Cell 92:351–366. https://doi.org/10.1016/S0092-8674(00)80928-9

Cai W, Cowan T, Thatcher M (2012) Rainfall reductions over Southern Hemisphere semi-arid regions: the role of subtropical dry zone expansion. Sci Rep 2:702. https://doi.org/10.1038/srep00702

Cao H, Yan X, Chen G, Zhou J, Li X, Ma W, Yan Y (2015) Comparative proteome analysis of A- and B-type starch granule-associated proteins in bread wheat (*Triticum aestivum* L.) and *Aegilops crassa*. J. Proteomics 112:95–112. https://doi.org/10.1016/j.jprot.2014.08.002

Caruso G, Cavaliere C, Foglia P, Gubbiotti R, Samperi R, Laganà A (2009) Analysis of drought responsive proteins in wheat (*Triticum durum*) by 2D-PAGE and MALDI-TOF mass spectrometry. Plant Sci 177:570–576. https://doi.org/10.1016/J.PLANTSCI.2009.08.007

Chaitanya KV, Masilamani S, Jutur PP, Reddy AR (2002) Variation in photosynthetic rates and biomass productivity among four mulberry cultivars. Photosynthetica 40:305–308. https://doi.org/10.1023/A:1021318412861
Chang I-F, Hsu J-L, Hsu P-H, Sheng W-A, Lai S-J, Lee C, Chen C-W, Hsu J-C, Wang S-Y, Wang L-Y, Chen C-C (2012) Comparative phosphoproteomic analysis of microsomal fractions of *Arabidopsis thaliana* and *Oryza sativa* subjected to high salinity. Plant Sci 185–186:131–142. https://doi.org/10.1016/j.plantsci.2011.09.009
Chen K-M, Gong H-J, Chen G-C, Wang S-M, Zhang C-L (2004) Gradual drought under field conditions influences the glutathione metabolism, redox balance and energy supply in spring wheat. J Plant Growth Regul 23:20–28. https://doi.org/10.1007/s00344-003-0053-4
Chen G, Zhu J, Zhou J, Subburaj S, Zhang M, Han C, Hao P, Li X, Yan Y (2014) Dynamic development of starch granules and the regulation of starch biosynthesis in *Brachypodium distachyon*: comparison with common wheat and *Aegilops peregrina*. BMC Plant Biol 14:198. https://doi.org/10.1186/s12870-014-0198-2
Chen G-X, Zhen S-M, Liu Y-L, Yan X, Zhang M, Yan Y-M (2017) *In vivo* phosphoproteome characterization reveals key starch granule-binding phosphoproteins involved in wheat water-deficit response. BMC Plant Biol 17:168. https://doi.org/10.1186/s12870-017-1118-z
Chu P, Zhang Y, Yu Z, Guo Z, Shi Y (2016) Winter wheat grain yield, water use, biomass accumulation and remobilisation under tillage in the North China Plain. F Crop Res 193:43–53. https://doi.org/10.1016/J.FCR.2016.03.005
Cohen P (2002) The origins of protein phosphorylation. Nat Cell Biol 4:E127–E130. https://doi.org/10.1038/ncb0502-e127
Cornic G (2000) Drought stress inhibits photosynthesis by decreasing stomatal aperture—not by affecting ATP synthesis. Trends Plant Sci 5:187–188. https://doi.org/10.1016/S1360-1385(00)01625-3
De Pinto MC, Paradiso A, Leonetti P, De Gara L (2006) Hydrogen peroxide, nitric oxide and cytosolic ascorbate peroxidase at the crossroad between defence and cell death. Plant J 48:784–795. https://doi.org/10.1111/j.1365-313X.2006.02919.x
Dong K, Zhen S, Cheng Z, Cao H, Ge P, Yan Y (2015) Proteomic analysis reveals key proteins and phosphoproteins upon seed germination of wheat (*Triticum aestivum* L.). Front Plant Sci 6:1–14. https://doi.org/10.3389/fpls.2015.01017
Donner DA, Mesdag J (2000) Bread-making quality of wheat: a century of breeding in Europe. Kluwer Academic Publishers, Wageningen
Eltayeb AE, Kawano N, Badawi GH, Kaminaka H, Sanekata T, Morishima I, Shibahara T, Inanaga S, Tanaka K (2006) Enhanced tolerance to ozone and drought stresses in transgenic tobacco overexpressing dehydroascorbate reductase in cytosol. Physiol Plant 127:57–65. https://doi.org/10.1111/j.1399-3054.2006.00624.x
Engelsberger WR, Schulze WX (2012) Nitrate and ammonium lead to distinct global dynamic phosphorylation patterns when resupplied to nitrogen-starved Arabidopsis seedlings. Plant J 69:978–995. https://doi.org/10.1111/j.1365-313X.2011.04848.x
Fedoroff NV, Battisti DS, Beachy RN, Cooper PJM, Fischhoff DA, Hodges CN, Knauf VC, Lobell D, Mazur BJ, Molden D, Reynolds MP, Ronald PC, Rosegrant MW, Sanchez PA, Vonshak A, Zhu J-K (2010) Radically rethinking agriculture for the 21st century. Science 327:833–834. https://doi.org/10.1126/science.1186834
Feller U, Crafts-Brandner SJ, Salvucci ME (1998) Moderately high temperatures inhibit ribulose-1,5-bisphosphate carboxylase/oxygenase (Rubisco) activase-mediated activation of Rubisco. Plant Physiol 116:539–546. https://doi.org/10.1104/pp.116.2.539
Fernie AR, Carrari F, Sweetlove LJ (2004) Respiratory metabolism: glycolysis, the TCA cycle and mitochondrial electron transport. Curr Opin Plant Biol 7:254–261. https://doi.org/10.1016/j.pbi.2004.03.007
Gardner WH (1986) Water content. Methods of soil analysis. In: Klute A (ed) Part 1. Agronomy monographs 9, 2nd edn. Verlag, American Society of Agronomy and Soil Science Society of America Madison, pp 493–544
Guo G, Lv D, Yan X, Subburaj S, Ge P, Li X, Hu Y, Yan Y (2012) Proteome characterization of developing grains in bread wheat cultivars (*Triticum aestivum* L.). BMC Plant Biol 12:147. https://doi.org/10.1186/1471-2229-12-147
Harb A, Krishnan A, Ambavaram MMR, Pereira A (2010) Molecular and physiological analysis of drought stress in Arabidopsis reveals early responses leading to acclimation in plant growth. Plant Physiol 154:1254–1271. https://doi.org/10.1104/pp.110.161752
Hill L, Smith A (1991) Evidence that glucose 6-phosphate is imported as the substrate for starch synthesis by the plastids of developing pea embryos. Planta 185:91–96. https://doi.org/10.1007/BF00194519
Horie T, Kaneko T, Sugimoto G, Sasano S, Panda SK, Shibasaka M, Katsuhara M (2011) Mechanisms of water transport mediated by PIP aquaporins and their regulation via phosphorylation events under salinity stress in barley roots. Plant Cell Physiol 52:663–675. https://doi.org/10.1093/pcp/pcr027
Hu X, Li N, Wu L, Li C, Li C, Zhang L, Liu T, Wang W (2015a) Quantitative iTRAQ-based proteomic analysis of phosphoproteins and ABA-regulated phosphoproteins in maize leaves under osmotic stress. Sci. Rep 5:15626. https://doi.org/10.1038/srep15626
Hu X, Wu L, Zhao F, Zhang D, Li N, Zhu G, Li C, Wang W (2015b) Phosphoproteomic analysis of the response of maize leaves to drought, heat and their combination stress. Front Plant Sci 6:298. https://doi.org/10.3389/fpls.2015.00298
Hunter T, Karin M (1992) The regulation of transcription by phosphorylation. Cell 70:375–387. https://doi.org/10.1016/0092-8674(92)90162-6
Johansson E, Prieto-Linde ML, Jönsson JÖ (2001) Effects of wheat cultivar and nitrogen application on storage protein composition and breadmaking quality. Cereal Chem 78:19–25. https://doi.org/10.1094/CCHEM.2001.78.1.19
Jung J, Kim M, Kim MJ, Kim J, Moon J, Lim JS, Kim M, Lee K (2004) Translationally controlled tumor protein interacts with the third cytoplasmic domain of Na, K-ATPase α subunit and inhibits the pump activity in HeLa cells. J Biol Chem 279:49868–49875. https://doi.org/10.1074/jbc.M400895200
Kersten B, Agrawal GK, Durek P, Neigenfind J, Schulze W, Walther D, Rakwal R (2009) Plant phosphoproteomics: an update. Proteomics 9:964–988. https://doi.org/10.1002/pmic.200800548
Kim YM, Han YJ, Hwang OJ, Lee SS, Shin AY, Kim SY, Il Kim J (2012) Overexpression of arabidopsis translationally controlled tumor protein gene *AtTCTP* enhances drought tolerance with rapid ABA-induced stomatal closure. Mol Cells 33:617–626. https://doi.org/10.1007/s10059-012-0080-8
Kim YE, Hipp MS, Bracher A, Hayer-Hartl M, Ulrich Hartl F (2013) Molecular chaperone functions in protein folding and proteostasis. Annu Rev Biochem 82:323–355. https://doi.org/10.1146/annurev-biochem-060208-092442
Lawlor DW, Cornic G (2002) Photosynthetic carbon assimilation and associated metabolism in relation to water deficits in higher plants. Plant Cell Environ 25:275–294. https://doi.org/10.1046/j.0016-8025.2001.00814.x
Lobell DB, Schlenker W, Costa-Roberts J (2011) Climate trends and global crop production since 1980. Science 333:616–620. https://doi.org/10.1126/science.1204531
Lv DW, Li X, Zhang M, Gu AQ, Zhen SM, Wang C, Li XH, Yan YM (2014a) Large-scale phosphoproteome analysis in seedling leaves of *Brachypodium distachyon* L. BMC Genomics 15:1–18. https://doi.org/10.1186/1471-2164-15-375
Lv D-W, Subburaj S, Cao M, Yan X, Li X, Appels R, Sun D-F, Ma W, Yan Y-M (2014b) Proteome and phosphoproteome characterization reveals new response and defense mechanisms of *Brachypodium distachyon* leaves under salt stress. Mol Cell Proteomics 13:632–652. https://doi.org/10.1074/mcp.M113.030171
Lv HQ, Zhang YS, Li MS et al (2015) [GB/T 32136-2015] Grade of agricultural drought. China Standards Press, Beijing
Lv D-W, Zhu G-R, Zhu D, Bian Y-W, Liang X-N, Cheng Z-W, Deng X, Yan Y-M (2016) Proteomic and phosphoproteomic analysis reveals the response and defense mechanism in leaves of diploid wheat *T. monococcum* under salt stress and recovery. J Proteomics 143:93–105. https://doi.org/10.1016/j.jprot.2016.04.013
MacRobbie EA (1998) Signal transduction and ion channels in guard cells. Philos Trans R Soc B 353:1475–1488. https://doi.org/10.1098/rstb.1998.0303
Musgrove J, Ellis R (1986) The rubisco large subunit binding protein. Philos Trans R Soc 313:419–428. https://doi.org/10.1007/BF00039488
Najjar VA, Pullman ME (1954) The occurrence of a group transfer involving enzyme (phosphoglucomutase) and substrate. Science 119(3097):631–634. https://doi.org/10.1126/science.119.3097.631
Patel MS, Roche TE (1990) Molecular biology and biochemistry of pyruvate dehydrogenase complexes. FASEB J 4:3224–3233. https://doi.org/10.1096/FASEBJ.4.14.2227213

Pei Z-M, Murata Y, Benning G, Thomine S, Klüsener B, Allen GJ, Grill E, Schroeder JI (2000) Calcium channels activated by hydrogen peroxide mediate abscisic acidsignalling in guard cells. Nature 406:731–734. https://doi.org/10.1038/35021067

Queitsch C (2000) Heat shock protein 101 plays a crucial role in thermotolerance in Arabidopsis. Plant Cell 12:479–492. https://doi.org/10.1105/tpc.12.4.479

Ray WJ, Hermodson MA, Puvathingal JM, Puvathingal JM (1983) The complete amino acid sequence of rabbit muscle phosphoglucomutase. J Biol Chem 258:9166–9174

Reddy AR, Chaitanya KV, Vivekanandan M (2004) Drought-induced responses of photosynthesis and antioxidant metabolism in higher plants. J Plant Physiol 161:1189–1202. https://doi.org/10.1016/j.jplph.2004.01.013

Sanchez J-C, Schaller D, Ravier F, Golaz O, Jaccoud S, Belet M, Wilkins MR, James R, Deshusses J, Hochstrasser D (1997) Translationally controlled tumor protein: a protein identified in several nontumoral cells including erythrocytes. Electrophoresis 18:150–155. https://doi.org/10.1002/elps.1150180127

Santa Brígida AB, dos Reis SP, de Nazaré Monteiro Costa C, Cardoso CMY, Lima AM, de Souza CRB (2014) Molecular cloning and characterization of a cassava translationally controlled tumor protein gene potentially related to salt stress response. Mol Biol Rep 41:1787–1797. https://doi.org/10.1007/s11033-014-3028-6

Silva-Sanchez C, Li H, Chen S (2015) Recent advances and challenges in plant phosphoproteomics. Proteomics 15:1127–1141. https://doi.org/10.1002/pmic.201400410

Slafer GA, Andrade FH, Satorre EH (1990) Genetic-improvement effects on pre-anthesis physiological attributes related to wheat grain-yield. Field Crop Res 23:255–263. https://doi.org/10.1016/0378-4290(90)90058-J

Smith AM, Martin C (1993) Starch biosynthesis and the potential for its manipulation. Topics Biosynth Manip Plant Prod. https://doi.org/10.1007/978-94-011-2142-2_1

Stock JB, Ninfa AJ, Stock AM (1989) Protein phosphorylation and regulation of adaptive responses in bacteria. Microbiol Rev 53:450–490

Tetlow IJ, Wait R, Lu ZX, Akkasaeng R, Bowsher CG, Esposito S, Kosar-Hashemi B, Morell MK, Emes MJ (2004) Protein phosphorylation in amyloplasts regulates starch branching enzyme activity and protein-protein interactions. Plant Cell 16:694–708. https://doi.org/10.1105/tpc.017400

Thingholm TE, Jensen ON, Larsen MR (2009) Analytical strategies for phosphoproteomics. Proteomics 9:1451–1468. https://doi.org/10.1002/pmic.200800454

Tomlinson K, Denyer K (2003) Starch synthesis in cereal grains. Adv Bot Res 40:1–61. https://doi.org/10.1016/S0065-2296(05)40001-4

Tschiersch H, Borisjuk L, Rutten T, Rolletschek H (2011) Gradients of seed photosynthesis and its role for oxygen balancing. Biosystems 103:302–308. https://doi.org/10.1016/J.BIOSYSTEMS.2010.08.007

Tyson RH, Rees T (1988) Starch synthesis by isolated amyloplasts from wheat endosperm. Planta 175:33–38. https://doi.org/10.1007/BF00402879

Umezawa T, Sugiyama N, Takahashi F, Anderson JC, Ishihama Y, Peck SC, Shinozaki K (2013) Genetics and phosphoproteomics reveal a protein phosphorylation network in the abscisic acid signaling pathway in *Arabidopsis thaliana*. Sci Signal 6:8. https://doi.org/10.1126/scisignal.2003509

Vandre DD, Centonze VE, Peloquin J, Tombes RM, Borisy GG (1991) Proteins of the mammalian mitotic spindle: phosphorylation/dephosphorylation of MAP-4 during mitosis. J Cell Sci 98:577–588

Wang H, Wu K, Liu Y, Wu Y, Wang X (2015) Integrative proteomics to understand the transmission mechanism of Barley yellow dwarf virus-GPV by its insect vector *Rhopalosiphum padi*. Sci Rep 5:10971. https://doi.org/10.1038/srep10971

Wang Y, Xu C, Zhang B, Wu M, Chen G (2017) Physiological and proteomic analysis of rice (*Oryza sativa* L.) in flag leaf during flowering stage and milk stage under drought stress. Plant Growth Regul 82:201–218. https://doi.org/10.1007/s10725-017-0252-9

Yamamori M, Fujita S, Hayakawa K, Matsuki J, Yasui T (2000) Genetic elimination of a starch granule protein, SGP-1, of wheat generates an altered starch with apparent high amylose. Theor Appl Genet 101:21–29. https://doi.org/10.3389/fpls.2017.00067

Yang J, Zhang J, Huang Z, Zhu Q, Wang L (2000) Remobilization of carbon reserves is improved by controlled soil-drying during grain filling of wheat. Crop Sci 40:1645. https://doi.org/10.2135/cropsci2000.4061645x

Yang J, Zhang J, Wang Z, Zhu Q, Liu L (2001) Water deficit-induced senescence and its relationship to the remobilization of pre-stored carbon in wheat during grain filling. Agron J 93:196. https://doi.org/10.2134/agronj2001.931196x

Yu Y, Zhu D, Ma C, Cao H, Wang Y, Xu Y, Zhang W, Yan Y (2016) Transcriptome analysis reveals key differentially expressed genes involved in wheat grain development. Crop J 4:92–106. https://doi.org/10.1016/j.cj.2016.01.006

Yuan LL, Zhang M, Yan X, Bian YW, Zhen SM, Yan YM (2016) Dynamic phosphoproteome analysis of seedling leaves in *Brachypodium distachyon* L. reveals central phosphorylated proteins involved in the drought stress response. Sci Rep 6:1–14. https://doi.org/10.1038/srep35280

Zadoks JC, Chang TT, Konzak CF (1974) A decimal code for the growth stages of cereals. Weed Res 14:415–421. https://doi.org/10.1111/j.1365-3180.1974.tb01084.x

Zhang M, Lv D, Ge P, Bian Y, Chen G, Zhu G, Li X, Yan Y (2014a) Phosphoproteome analysis reveals new drought response and defense mechanisms of seedling leaves in bread wheat (*Triticum aestivum* L.). J Proteomics 109:290–308. https://doi.org/10.1016/j.jprot.2014.07.010

Zhang M, Ma CY, Lv DW, Zhen SM, Li XH, Yan YM (2014b) Comparative phosphoproteome analysis of the developing grains in bread wheat (*Triticum aestivum* L.) under well-watered and water-deficit conditions. J Proteome Res 13:4281–4297. https://doi.org/10.1021/pr500400t

Zhen S, Deng X, Zhang M, Zhu G, Lv D, Wang Y, Zhu D, Yan Y (2017) Comparative phosphoproteomic analysis under high-nitrogen fertilizer reveals central phosphoproteins promoting wheat grain starch and protein synthesis. Front Plant Sci 8:67. https://doi.org/10.3389/fpls.2017.00067

Zhu J-K (2002) Salt and drought stress signal transduction in plants. Annu Rev Plant Biol 53:247–273. https://doi.org/10.1146/annurev.arplant.53.091401.143329

16

Identification of differentially expressed genes in sunflower (*Helianthus annuus*) leaves and roots under drought stress by RNA sequencing

Chunbo Liang[1,2], Wenjun Wang[2], Jing Wang[2], Jun Ma[2], Cen Li[2], Fei Zhou[2], Shuquan Zhang[2], Ying Yu[2], Liguo Zhang[2], Weizhong Li[3] and Xutang Huang[1,2*]

Abstract

Background: Sunflower is recognized as one of the most important oil plants with strong tolerance to drought in the world. In order to study the response mechanisms of sunflower plants to drought stress, gene expression profiling using high throughput sequencing was performed for seedling leaves and roots (sunflower inbred line R5) after 24 h of drought stress (15% PEG 6000). The transcriptome assembled using sequences of 12 samples was used as a reference.

Results: 805 and 198 genes were identified that were differentially expressed in leaves and roots, respectively. Another 71 genes were differentially expressed in both organs, in which more genes were up-regulated than down-regulated. In agreement with results obtained for other crops or from previous sunflower studies, we also observed that nine genes may be associated with the response of sunflower to drought.

Conclusions: The results of this study may provide new information regarding the sunflower drought response, as well as add to the number of known genes associated with drought tolerance.

Keywords: Digital gene expression, Drought stress, Leaf, Sunflower (*Helianthus annuus* L.), Root

Background

Drought is one of the most serious ecological problems challenging mankind. Statistically, of all natural insults, drought stress has the highest impact on crop yields worldwide and its impact may equal the total impact of all other abiotic stresses combined (Chen et al. 2009). Due to the magnitude of drought's influence, improving crop drought tolerance is an urgent priority. Modern molecular biological technologies have greatly facilitated elucidation of plant response mechanisms to water stress, which in turn are relevant to understanding the genetic basis of drought tolerance of plants. Ultimately, identification of drought tolerance genes should inform breeding strategies for eventual creation of drought resistant sunflower hybrids.

Sunflower (*Helianthus annuus*) is the fourth largest source of vegetable oil in the world. In 2016, the global planting area of this crop was greater than 24,970,000 hm^2 and total output approached 45,750,000 tons (the Food and Agriculture Organization of the United Nations, FAO). China is one of the most important producers and consumers of sunflower in the world. Since sunflower plants are somewhat tolerant to drought, poor soil conditions and salt, sunflower is suitable for cultivation in most areas of Heilongjiang Province. Consequently, improved sunflower drought tolerance would allow cultivation of this crop across an even larger area than currently utilized, extending to the mid-western arid and semi-arid middle- and low-yielding fields of this province. By first studying sunflower varieties most resistant to drought, with characterization of drought

*Correspondence: hxthlj@126.com
[1] Heilongjiang Academy of Agricultural Sciences Postdoctoral Programme, Xuefu Road 368, Harbin 150086, People's Republic of China
Full list of author information is available at the end of the article

resistance genes and molecular mechanisms underlying drought tolerance, improved drought-tolerant sunflower might then be effectively developed.

In recent years, researchers have paid increasing attention to the molecular biology underlying drought tolerance mechanisms in order to improve sunflower drought tolerance. For example, Liu et al. (2003) analyzed the expression of five drought-induced genes in roots, stems and leaves of sunflower using differential display PCR and real-time quantitative PCR. Using a different approach, Diaz-Martin et al. (2005) cloned a drought-related DREB transcription factor, HaDREB2, in sunflower using gene interaction analyses. Meanwhile, Roche et al. (2007) analyzed the expression of genes associated with metabolism and signal transduction in leaves and immature embryos of sunflower using a cDNA array. Subsequently, 409 differentially expressed genes (DEGs) were identified, of which 82 were organ specific and induced by drought stress. In addition, Kane and Risesberg (2007) identified candidate sunflower drought and salt tolerance genes using selective screening and discovered 17 genes that were induced by each of these stresses. Subsequently, Cheng et al. (2009) transformed the drought and salt resistance gene P5CS into sunflower and obtained six transformed buds which were resistant to Kan, thus showing that successful gene transfer and expression had occurred; however, no fertile transgenic plants were obtained. Soon thereafter, Sauca et al. (2011) introduced the same drought resistance gene into cultivated sunflower inbred lines using an embryo rescue technique. More recently, Yi et al. (2013) demonstrated that a betaine aldehyde dehydrogenase (BADH) gene is induced by drought and salt stress in sunflower.

Moreover, due to the lack of a complete sunflower transcriptome, the study of molecular mechanisms underlying drought resistance has mainly focused on several independent genes instead of focusing on broader stress-induced genetic response networks. Because tolerance to drought is a complex phenotypic trait controlled by polygenes, it would not be useful to research any single gene or single class of genes to understand molecular mechanisms underlying drought tolerance. Fortunately, transcriptome profiling and digital gene expression (DGE) analysis using high throughput sequence technology have facilitated discovery of DEGs under drought stress without relying on genome sequence information. Although annotation of DEGs is challenging without a genome sequence, the information acquired from DGE profiling could nonetheless describe the stress response mechanism to a certain extent.

In this study, DGE profiling of leaves and roots of sunflower (inbred line R5) in response to drought stress was conducted using Illumina high throughput sequencing technology. The DEGs that were subsequently described here should lay the foundation for research of molecular mechanisms of sunflower drought tolerance. Moreover cloning and expression analysis of DEGs identified in this study may provide several candidate genes and information to improve sunflower drought stress tolerance using genetic engineering technology.

Methods

Plant growth and stress treatments

The sunflower line R5 was used in this study. R5 seeds were provided by the Industrial Crops Institute of Heilongjiang Academy of Agricultural Sciences. Sunflower seeds were rinsed with sterilized distilled water three times and allowed to germinate in Petri dishes for 72 h at 28 °C in darkness. The germinated seeds were transferred into pots containing nutrients, soil and vermiculite (1:1). Seedlings were grown in an artificial climate chamber with 16 h light and 8 h darkness (22 °C) with 70% relative humidity. Seedlings at the four-leaf stage showing appropriate growth status were rinsed with sterilized distilled water and placed into culture bottles filled with 15% PEG in water for 24 h to serve as drought stressed plants. Control plants (distilled water only) were grown in parallel and collected at the same time points. The process was repeated in triplicate, where each replicate was biologically and temporally independent.

Sample preparation and RNA isolation

The leaves and roots of drought-stressed and control groups were harvested for gene expression analysis. The tissue samples were immediately frozen in liquid nitrogen and stored at − 80 °C. Total RNA was isolated separately from the leaves and roots of sunflower seedlings using TRIzol Reagent (Tiangen, Beijing, China) following the manufacturer's protocols. Yield and quality of the RNA samples were determined using a NANODROP 2000 spectrophotometer (Thermo Fisher Scientific, USA). Since no genome sequence has yet been generated for sunflower, a portion of the total RNA from each of 12 samples studied here was pooled for construction of a library for use as a reference to validate transcriptome sequencing results presented here.

Library preparation and sequencing

Total RNA preparations from 13 samples (including one pool of all samples and three samples from each of four sample types: leaves from PEG-treated plants, roots from PEG-treated plants, leaves from control plants, roots from control plants) were used for mRNA isolation and concentration using magnetic oligo (dT) beads. The mRNA was broken into short segments by adding fragmentation buffer. With mRNA as the template, first-strand cDNA was synthesized by adding random hexamers and second-strand cDNA was synthesized by adding buffer, dNTPs, DNA polymerase I and RNase H. Double-stranded cDNA was purified using AMPure XP beads. The ends of purified double-stranded cDNA were repaired, A-tails were added and overlapping sequences were joined. Next, AMPure XP beads were used for fragment size selection. Finally, the cDNA libraries were constructed using PCR amplification and products were purified using AMPure XP beads. Qubit 2.0 was used to perform a preliminary quantitation of the cDNA library. Each library was diluted to 1.5 ng/μL and an Agilent 2100 Bioanalyzer was used to analyze the insert size range of the cDNA library. Quantitative PCR (qPCR) was used to analyze the effective concentration of each library. In order to ensure the quality of each library, the effective concentration of library must be > 2 nM. According to the effective concentration and data quantification, the various libraries were pooled and sequenced using Illumina HiSeq/MiSeq.

Sequencing data analysis

In order to ensure the quality of downstream analyses, the raw reads obtained from sequencing were filtered. Clean reads were obtained after removing from the raw sequence data the reads containing only adapters, reads with a content of N > 10% and low quality reads. All downstream analyses were based on clean reads with high quality.

Trinity software (v2012-10-05) was used for splicing of transcripts (Grabherr et al. 2011). The databases used for gene annotation in this study included: nr (NCBI non-redundant protein sequences), nt (NCBI nucleotide sequences), Pfam (Protein family), KOG/COG (euKaryotic Ortholog Groups/Clusters of Orthologous Groups of proteins), Swiss-Prot (a manually annotated and reviewed protein sequence database), KEGG (Kyoto Encyclopedia of Genes and Genomes) an GO (Gene Ontology). MISA (1.0) software was used to detect simple sequence repeats (SSRs) in the unigenes and Primer 3 (2.3.5) was used to design primers to characterize SSR markers. Clean DEG sequence reads were mapped to the transcripts by RSEM (Li and Dewey 2011) and were used to obtain the read count for each sample that mapped to each gene. Finally, the read count number was converted to FPKM (expected number of Fragments Per Kilobase of transcript sequence per Millions base pairs sequenced) (Trapnell et al. 2010). DESeq (Anders et al. 2010) was used to analyze differential gene expression. The screening threshold was padj < 0.05. This method was based on a negative binomial distribution model; the read count of gene i in the sample of j was designated K_{ij}. The P values were adjusted using the Benjamini and Hochberg method. The corrected *P* value of 0.005 and log2 (fold change) of 1 were set as thresholds for significance for scoring of differential expression. The method of GO enrichment analysis in this study was GOseq (Young et al. 2010). KOBAS (2.0) software was used to test the statistical enrichment of differential expression of genes in the KEGG pathway (Mao et al. 2005).

Quantitative real-time PCR (qRT-PCR) analysis

Five genes (comp9755_c0, comp11967_c0, comp29815_c0, comp36138_c0 and comp72325_c0) which have been identified may be related to drought stress response were validated by quantitative real-time PCR (qRT-PCR) using the same RNA samples as in the DGE library construction. Leaves and roots were removed from the freezer and ground in liquid nitrogen. Six pairs of gene-specific primers (five candidate genes and one actin gene) were designed based on target gene sequences using Primer Express software (Additional file 1: Table S1). First-strand cDNA synthesis and qRT-PCR were carried out using a PrimeScript™ RT reagent Kit with gDNA Eraser (TaKaRa), SYBR Premix Ex Taq™ II (Tli RNaseH Plus) and ROX plus (TaKaRa). Each PCR (18 μL) contained 10 μL 2 × Real-Time PCR Mix, 5 μM of each primer and diluted cDNA. The thermal cycling conditions were 95 °C for 30 s followed by 40 cycles of 5 s at 95 °C and 40 s at 60 °C. All reactions were performed in triplicate, including the non-template controls. The relative expression levels were calculated as $2^{-(\triangle Ct\ of\ treatment\ -\ \triangle Ct\ of\ control)}$.

Result

Results of sunflower transcriptome sequencing

To identify genes that are differentially expressed in response to drought stress, 12 samples [leaves of plants treated with PEG (three samples), roots of plants treated with PEG (three samples), leaves of control plants (three samples) and roots of control plants (three samples)] were mixed for the construction of a library for transcriptome sequencing to serve as a reference DGE sequence library. In this study, a total of 62,252 unigenes were obtained by merging overlapping transcript fragments for each locus. By analyzing the length of transcripts and unigenes, we found that the numbers of unigenes of lengths of 200–500 bp, 0.5–1 kb, 1–2 kb or greater than 2 kb were 37,645, 12,190, 8714 and 3703, respectively. The average length of all gene fragments was 390 bp. All unigenes were annotated using NR, NT, KO, Swiss-Proto, PFAM, GO, KOG databases. The results demonstrated that 39,356 unigenes could be annotated using at least one database, accounting for 62.33% of the total unigenes observed. Of these, 4152 unigenes could be annotated using all seven databases, which accounted for 6.66% of the total number of unigenes (Table 1), the detailed

Table 1 Success rate statistics of gene annotation

Term	Number of unigenes	Percentage
Annotated in NR	36,286	58.28
Annotated in NT	12,683	20.37
Annotated in KO	12,043	19.34
Annotated in Swiss Prot	26,583	42.7
Annotated in PFAM	25,009	40.17
Annotated in GO	28,579	45.9
Annotated in KOG	14,038	22.55
Annotated in all databases	4152	6.66
Annotated in at least one database	39,356	63.22
Total	62,252	100

NR, NT, KO, Swiss Prot, PFAM, GO and DOG indicate the database of NCBI non-redundant protein sequences, NCBI, nucleotide sequences, Kyoto Encyclopedia Genes and Genomes Orthology, a manually annotated and reviewed protein sequence database, Protein family, Gene ontology and euKaryotic Orthology groups

results of annotation is listed in Additional file 2: Table S2. After annotation using the GO database, successfully annotated genes could be categorized into 47 sub-groups from three main categories (biological process, cellular component and molecular function) (Additional file 3: Table S3). The results of this study should provide abundant gene resources for future studies focusing on the molecular biology of sunflower.

In this study, 4,283,598 bp nucleotides belonging to 62,252 unigenes were analyzed for SSR calling. Subsequently, 8808 SSR sequences were found in 7441 unigenes of which 1120 unigenes contained more than one SSR sequence and 507 SSR sequences were present in compound formations. The primers of these SSR sequences are listed in Additional file 4: Table S4.

Identification of DEGs in sunflower leaves and roots in response to drought stress

The results of correlation analysis of three replicates for each treatment showed that the correlation index between the replicates ranged from 0.786 to 0.920 (Fig. 1), which indicated that the sequencing data were reliable and could be used for further analysis.

In this study, 876 and 269 genes were differentially expressed between control and drought-stressed plants in their leaves and roots, respectively. Next, gene expression profiles were compared between leaves and roots under drought stress. Under drought stress, 805 and 198 DEGs were identified to be leaf-specific and root-specific, respectively, and 71 of these genes were differentially expressed both in leaves and roots. Among these DEGs, 616 and 172 genes were up-regulated during drought stress in leaves and roots, respectively, while 260 and 97 genes were down-regulated in leaves and roots, respectively (Fig. 2). Of the 616 up-regulated genes in leaves, 26 and 415 genes could be annotated in sunflower and other plant species, respectively, while 175 genes could not be annotated. Of the 260 down-regulated genes in leaves, 14 and 191 genes could be annotated in sunflower and other plant species, respectively. In the 172 up-regulated genes in roots, 5 and 110 genes could be annotated in sunflower and other plant species, respectively, while 57 genes could not be annotated. Of the 97 down-regulated genes in roots, three and 60 genes could be annotated in sunflower and other plant species, respectively, while 34 genes could not be annotated.

In response to drought, both leaves and roots showed a greater number of up-regulated genes than down-regulated genes. Some stress-induced genes were regulated in a tissue-specific manner, which indicates that these genes may play roles as part of distinct mechanisms for coping with drought stress. Another 71 genes exhibited the same pattern of expression in leaves versus roots, indicating that overlaps exist at the transcriptional level between leaves and roots in their response to drought stress.

Functional categorization of stress-regulated genes

Differentially expressed genes from leaves and roots were annotated using the GO database. Clearly annotated genes belonged to three main categories (cellular component, molecular function and biological process) and the distributions of gene numbers for each category were shown in Fig. 3. Notably, the "cellular component" category was not found within the top 15 categories in leaves.

In this study, we focused on a group of genes known to be involved in stress responses (Table 2). Of 17 genes involved in responses to abiotic stimuli, nine genes are known to be involved to responses to water stimuli and 16 genes were differentially expressed in leaves; only one gene, designated Comp11967_c0, was differentially expressed in both leaves and roots. We also found that each gene responded to more than one abiotic stress. This result indicated that relatively conserved mechanisms may be shared by sunflower responses to various abiotic stresses.

To identify the biological pathways represented within the DGE libraries, all annotated genes were mapped using the KEGG database to identify significantly enriched genes involved in metabolic or signal transduction pathways. Subsequently, 409 and 119 DEGs were assigned to 105 and 47 KEGG pathways in leaves and roots for drought, respectively. The top 20 pathways among all of the KEGG annotations of leaves and roots are listed in Table 3. The top KEGG pathways which contained the highest number of DEGs in leaves and roots were metabolic pathways, biosynthesis of secondary metabolites, plant hormone signal transduction, ribosome and RNA transport.

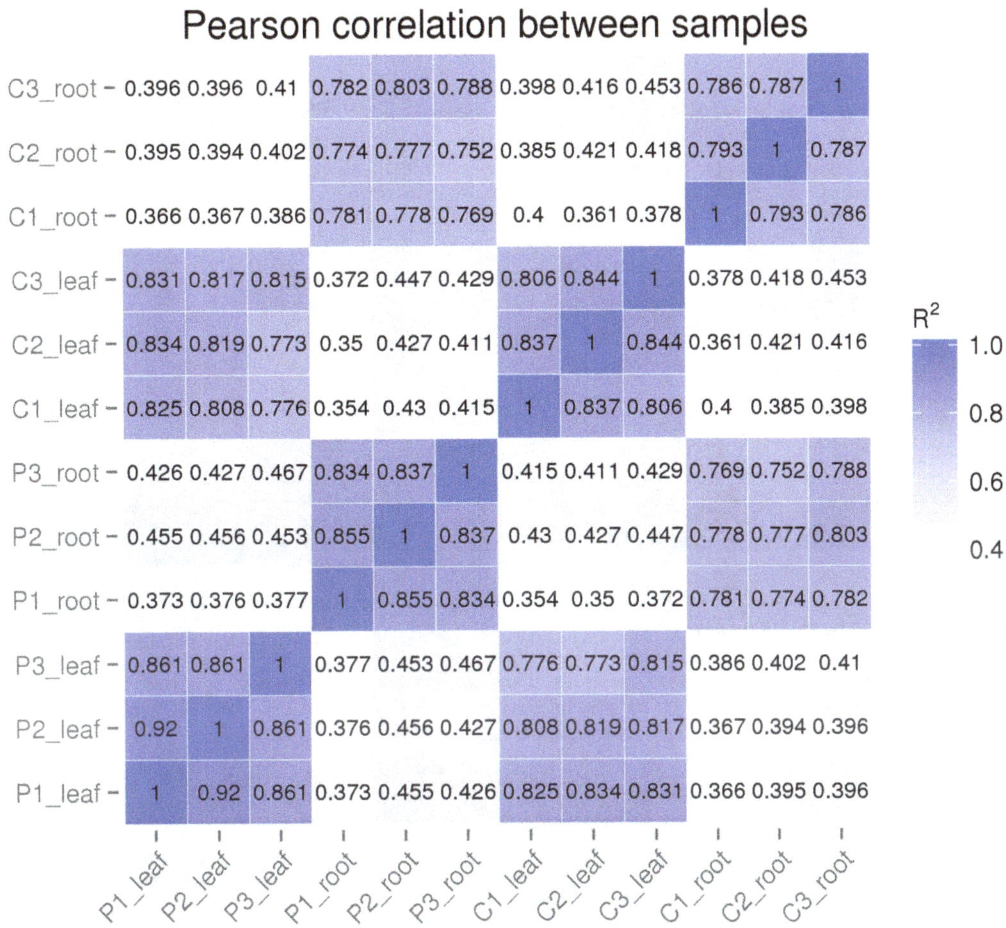

Fig. 1 Pearson correlation between samples. C represents the control group; P represents the PEG-treated group. R^2 represents the correlation coefficient. The darker the blue background, the greater the correlation coefficient

Validation of DGE data by qRT-PCR

To evaluate the validity of DGE sequencing and to further analyze the patterns of differential gene expression, five genes were selected for detection using real-time PCR with gene-specific primers (Additional file 1: Table S1) and the data were compared to DEG sequence data. The results demonstrate that both in leaves and roots, four genes, comp9755_c0, comp11967_c0, comp29815_c0, comp36138_c0, were up-regulated after PEG treatment, but the expression of comp72325_c0 was significantly decreased during drought stress. The results of qRT-PCR are consistent with the trend observed for DEG sequencing (Fig. 4). Comparison of data obtained from DEG sequencing analysis methods to data from qRT-PCR confirmed the reliability of the DEG sequencing method.

Discussion

Numerous studies have shown that adaptation mechanisms of plants in response to drought make up a complex metabolic regulatory network based on the expression of many genes. Although response mechanisms to various stresses are mainly conserved, specific plants often possess some of their own unique response mechanisms. Therefore, expression profiling studies of sunflower at the transcriptional level may benefit from knowledge of molecular drought tolerance mechanisms for other plants.

Before the large-scale application of high-throughput sequencing, cDNA microarrays were the traditional method for conducting gene expression profiling. Using cDNA microarrays, a number of drought response genes have been identified in *Arabidopsis thaliana* (Shinozaki et al. 2007), rice (Wang et al. 2007), maize (Zheng et al. 2006) and other crops. However, accurate expression profiling could only be conducted using a clear genome sequence to guide comparison of samples with identical genetic backgrounds for various conditions. Since no complete genome sequence exists for sunflower, cDNA microarray analysis could only be performed using the genome sequence of another crop. However, differing

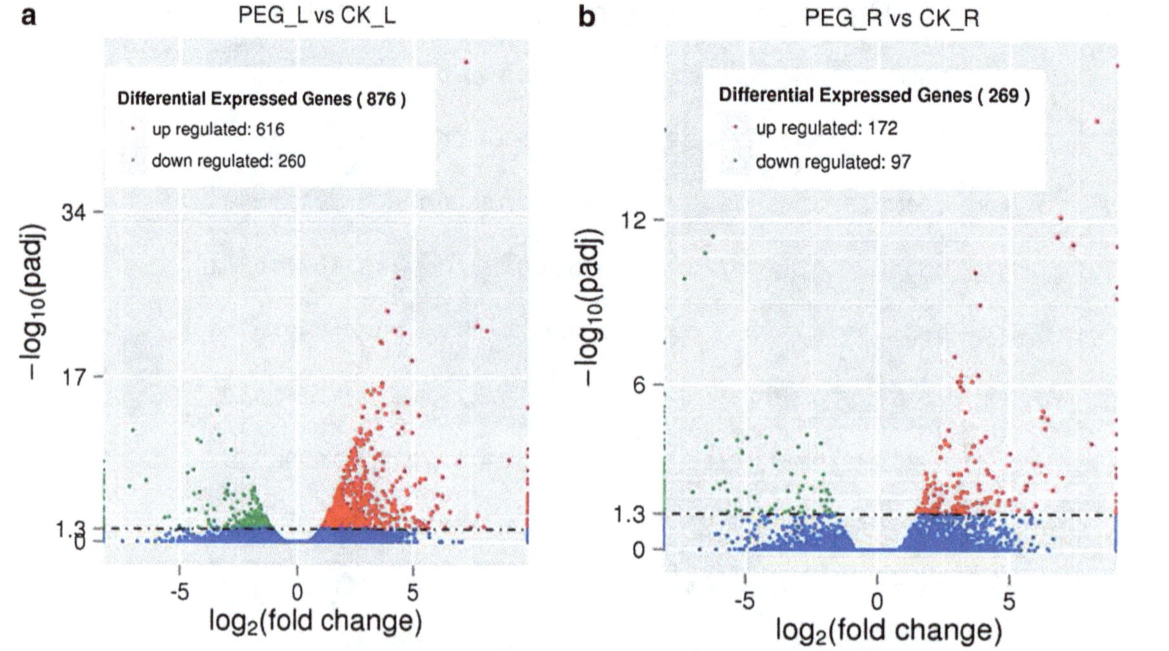

Fig. 2 Volcano plot of significantly differently expressed genes in leaves and roots under drought stress. Red and green dot represent the up-regulated and down-regulated genes, respectively. **a** Differential expressed genes in the leaves under PEG treatment (PEG _L) and control group (CK_L). **b** Differential expressed genes in the roots under PEG treatment (PEG _R) and control group (CK_R)

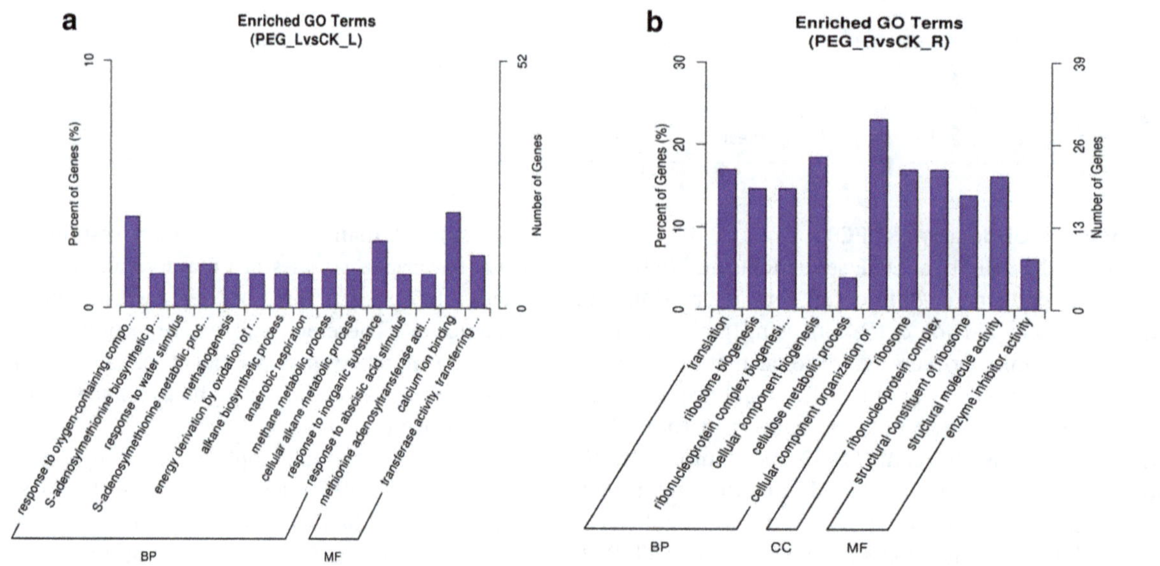

Fig. 3 GO enrich analysis of significantly differently expressed genes in leaves and roots under drought stress. BP, CC and MF indicate biological process, cellular component and molecular function. **a** Enrich GO terms of significantly differently expressed genes in leaves under PEG treatment (PEG_L) and control group (CK_L). **b** Enrich GO terms of significantly differently expressed genes in roots under PEG treatment (PEG_R) and control group (CK_R)

genetic backgrounds between sunflower and reference crops often caused high analytical uncertainty, generating erroneous results. In contrast, the sequences of unigenes obtained using high-throughput sequencing reflected objective changes in actual gene expression, but later annotation of DEGs was limited by the number of

Table 2 Different expressed genes response to stress

Gene id	Annotation	Leaf			Root			Response to stress
		PEG-readcount	CK	pval	PEG	CK	pval	
Comp11967_c0	Dehydrin-like protein	10,175.3	1617.8	4.44E−08	120.8	1.9	7.27E−05	Water stimulus, chemical stimulus
Comp12003_c1	*H. annuus* mRNA for dehydrin related protein	5070.93	887.9	1.47E−07				Water stimulus, chemical stimulus
Comp14403_c0	NAC domain-containing protein	358.3	111.8	1.70E−06				Water deprivation, chemical stimulus, hormone stimulus, endogenous stimulus, organic substance
Comp17955_c0	Indole-3-acetic acid amido synthetase activity	33.43	5.4	1.64E−05				Hormone stimulus, endogenous stimulus, organic substance, chemical stimulus, radiation
Comp21257_c0	*S*-Adenosylmethionine synthetase	712.53	116.3	2.27E−14				Cold, temperature stimulus
Comp22725_c0	Galactinol synthase 2	31.43	6.9	0.0004				Abscisic acid stimulus, cold, oxidative stress, water deprivation, salt stress
Comp25287_c0	Aldehyde dehydrogenase family 7 member A1	2745.1	525.1	5.97E−13				Endogenous stimulus, salt, chemical stimulus, water stimulus, osmotic stress, organic substance, desiccation
Comp26183_c0	CBL-interacting serine/threonine-protein kinase	587.5	226.9	3.16E−05				Osmotic stress, salt, chemical stimulus, water stimulus,
Comp30219_c0	Glycosyl transferase family	1270.1	363.5	8.76E−08				Endogenous stimulus, salt, oxidative stress, chemical stimulus, temperature stimulus, water stimulus, organic substance, cold, osmotic stress
Comp305249_c0	DNA binding	0.3	10.6	0.0258				Radiation
Comp31776_c0	ATP-dependent Clp protease ATP-binding subunit ClpB	755.1	289.3	3.34E−05				Hydrogen peroxide, oxidative stress, chemical stimulus, temperature, light stimulus, reactive oxygen species, radiation
Comp32945_c0	Galactinol synthase 3	1420.8	461.4	0.0004				Chemical stimulus, water deprivation, temperature stimulus, cold
Comp34774_c0	Dehydrin	82412.6	5573.7	9.12E−29				Chemical stimulus, water stimulus
Comp35427_c0	Calcium-binding protein CML	1744.4	383.7	0.0003				Organic substance, endogenous stimulus, hormone stimulus, chemical stimulus, temperature stimulus, radiation
Comp36183_c0	Root phototropism protein 2	1014.1	2130.7	0.0455				Radiation
Comp39610_c0	Protein serine/threonine kinase activity	262.8	59.8	1.45E−08				Osmotic stress, endogenous stimulus, hormone stimulus, chemical stimulus, temperature stimulus, radiation
Comp42687_c0	DNAJ heat shock protein-like protein	216.9	66.0	2.04E−05				Hydrogen peroxide, oxidative stress, chemical stimulus, temperature stimulus, high light intensity, reactive oxygen species, heat, radiation

Gene id represent the sub-component serial number obtained by Trinity stitching

available reference sequences. The results of this study showed that DGE sequencing can achieve near complete coverage of the entire genome sequence with reliable and reproducible sequencing results under various conditions. However, we also found that the lack of genome sequence information may limit gene annotation,

Table 3 Top 20 KEGG pathways of different expression genes in leaves and roots

Organs	Term	Gene number	Background number	P-value	Corrected P-value	Hyperlink
Leaves	Cysteine and methionine metabolism	15	164	7.70E−07	0.000141601	http://www.genome.jp/kegg-bin/show_pathway?ko00270
	Plant hormone signal transduction	19	320	1.77E−05	0.001632768	http://www.genome.jp/kegg-bin/show_pathway?ko04075
	Photosynthesis—antenna proteins	6	37	7.20E−05	0.00441311	http://www.genome.jp/kegg-bin/show_pathway?ko00196
	Other glycan degradation	3	23	0.009632561	0.338207262	http://www.genome.jp/kegg-bin/show_pathway?ko00511
	Porphyrin and chlorophyll metabolism	5	69	0.010937551	0.338207262	http://www.genome.jp/kegg-bin/show_pathway?ko00860
	Biosynthesis of secondary metabolites	43	1546	0.011028498	0.338207262	http://www.genome.jp/kegg-bin/show_pathway?ko01110
	Cyanoamino acid metabolism	4	56	0.02325574	0.523505194	http://www.genome.jp/kegg-bin/show_pathway?ko00460
	Valine, leucine and isoleucine degradation	5	85	0.024892835	0.523505194	http://www.genome.jp/kegg-bin/show_pathway?ko00280
	Biosynthesis of amino acids	16	479	0.025606232	0.523505194	http://www.genome.jp/kegg-bin/show_pathway?ko01230
	Metabolic pathways	70	2916	0.033671031	0.553343443	http://www.genome.jp/kegg-bin/show_pathway?ko01100
	Ascorbate and aldarate metabolism	4	66	0.039317106	0.553343443	http://www.genome.jp/kegg-bin/show_pathway?ko00053
	Pantothenate and CoA biosynthesis	3	41	0.04494819	0.553343443	http://www.genome.jp/kegg-bin/show_pathway?ko00770
	Carotenoid biosynthesis	3	41	0.04494819	0.553343443	http://www.genome.jp/kegg-bin/show_pathway?ko00906
	Lysine degradation	3	41	0.04494819	0.553343443	http://www.genome.jp/kegg-bin/show_pathway?ko00310
	Beta-Alanine metabolism	4	69	0.04510952	0.553343443	http://www.genome.jp/kegg-bin/show_pathway?ko00410
	Novobiocin biosynthesis	1	6	0.110953484	1	http://www.genome.jp/kegg-bin/show_pathway?ko00401
	Chlorocyclohexane and chlorobenzene degradation	1	6	0.110953484	1	http://www.genome.jp/kegg-bin/show_pathway?ko00361
	Phosphonate and phosphinate metabolism	1	7	0.128212836	1	http://www.genome.jp/kegg-bin/show_pathway?ko00440
	Styrene degradation	1	7	0.128212836	1	http://www.genome.jp/kegg-bin/show_pathway?ko00643
	Nitrogen metabolism	3	65	0.13220423	1	http://www.genome.jp/kegg-bin/show_pathway?ko00910
	Terpenoid backbone biosynthesis	3	67	0.141100694	1	http://www.genome.jp/kegg-bin/show_pathway?ko00900
	Benzoate degradation	1	8	0.14513793	1	http://www.genome.jp/kegg-bin/show_pathway?ko00362
Roots	Ribosome	16	979	0.000125245	0.023045141	http://www.genome.jp/kegg-bin/show_pathway?ko03010
	RNA transport	6	257	0.003419288	0.267326383	http://www.genome.jp/kegg-bin/show_pathway?ko03013
	Biotin metabolism	2	22	0.006784043	0.267326383	http://www.genome.jp/kegg-bin/show_pathway?ko00780
	Hippo signaling pathway—fly	3	75	0.008823808	0.267326383	http://www.genome.jp/kegg-bin/show_pathway?ko04391
	Arginine and proline metabolism	4	147	0.00971839	0.267326383	http://www.genome.jp/kegg-bin/show_pathway?ko00330
	Adherens junction	3	79	0.010167104	0.267326383	http://www.genome.jp/kegg-bin/show_pathway?ko04520

Table 3 continued

Organs	Term	Gene number	Background number	P-value	Corrected P-value	Hyperlink
	Regulation of actin cytoskeleton	4	153	0.011134171	0.267326383	http://www.genome.jp/kegg-bin/show_pathway?ko04810
	Biosynthesis of unsaturated fatty acids	3	83	0.011622886	0.267326383	http://www.genome.jp/kegg-bin/show_pathway?ko01040
	Tight junction	3	94	0.016,217,849	0.331,564,911	http://www.genome.jp/kegg-bin/show_pathway?ko04530
	Hippo signaling pathway	3	110	0.024479743	0.438901021	http://www.genome.jp/kegg-bin/show_pathway?ko04390
	Focal adhesion	3	113	0.026238648	0.438901021	http://www.genome.jp/kegg-bin/show_pathway?ko04510
	Fatty acid biosynthesis	2	49	0.031,291,426	0.450,398,707	http://www.genome.jp/kegg-bin/show_pathway?ko00061
	Phagosome	4	211	0.031821648	0.450398707	http://www.genome.jp/kegg-bin/show_pathway?ko04145
	Two-component system	2	52	0.03,490,051	0.458,692,412	http://www.genome.jp/kegg-bin/show_pathway?ko02020
	Benzoate degradation	1	8	0.044298565	0.543395733	http://www.genome.jp/kegg-bin/show_pathway?ko00362
	Nitrogen metabolism	2	65	0.052231368	0.600660726	http://www.genome.jp/kegg-bin/show_pathway?ko00910
	Fatty acid metabolism	3	153	0.055894477	0.60497552	http://www.genome.jp/kegg-bin/show_pathway?ko01212
	Arachidonic acid metabolism	1	25	0.132,075,196	1	http://www.genome.jp/kegg-bin/show_pathway?ko00590
	Alanine, aspartate and glutamate metabolism	2	114	0.135824158	1	http://www.genome.jp/kegg-bin/show_pathway?ko00250
	Pentose and glucuronate interconversions	2	124	0.155261816	1	http://www.genome.jp/kegg-bin/show_pathway?ko00040

subsequently influencing in-depth mining and selection of candidate genes.

Of the total 1074 DEGs identified in this study, nine unigenes were associated with water stress, as demonstrated using annotation through sequence alignment. These DEGs included three dehydrin-like genes, one NAC transcription factor, two galactosidase family genes, one aldehyde dehydrogenase family gene, one CBL-interacting serine/threonine-protein kinase gene and one glycosyltransferase gene. Many researchers have observed a positive correlation between the expression and accumulation of plant dehydrin genes and plant resistance to abiotic stress (Hu et al. 2010). Meanwhile, the NAC transcription factor (Fang et al. 2015), galactosidase family gene (Santos et al. 2015), aldehyde dehydrogenase family gene (Singh et al. 2013), calcineurin-B-like (CBL)-interacting serine/threonine protein kinase gene (Chen et al. 2012) and glycosyltransferase gene (Tognetti et al. 2010) have all been shown to play important roles in response to abiotic and biotic stresses. Although 321 DEGs identified in this work could not be annotated, this study provides a rich set of candidate genes for further study of sunflower responses to drought stress. Although the functions and molecular mechanisms of these stress genes have not yet been fully elucidated, powerful genetic engineering methodologies, such as RNA interference, gene overexpression and transgenic technologies hold promise for defining their roles in sunflower drought tolerance.

Conclusions

A total of 62,252 unigenes were obtained by transcript profiling sequencing in this study. Under drought stress, 876 and 269 DEGs were identified by DGE sequencing in sunflower leaves and roots, respectively. A greater number of up-regulated genes than down-regulated genes were identified in both leaves and roots. By analyzing and annotating DEGs, we found 17 genes that play roles in sunflower response to abiotic stimuli and thus may have relevance to drought tolerance. Of these, nine genes may be associated with responses to water-related stimuli. To our knowledge, this study is the first high throughput sequencing study for gene expression profiling analysis of sunflower under drought stress. The results of this study demonstrate that the drought stress response in sunflower leaves and roots comprises a complex mechanism

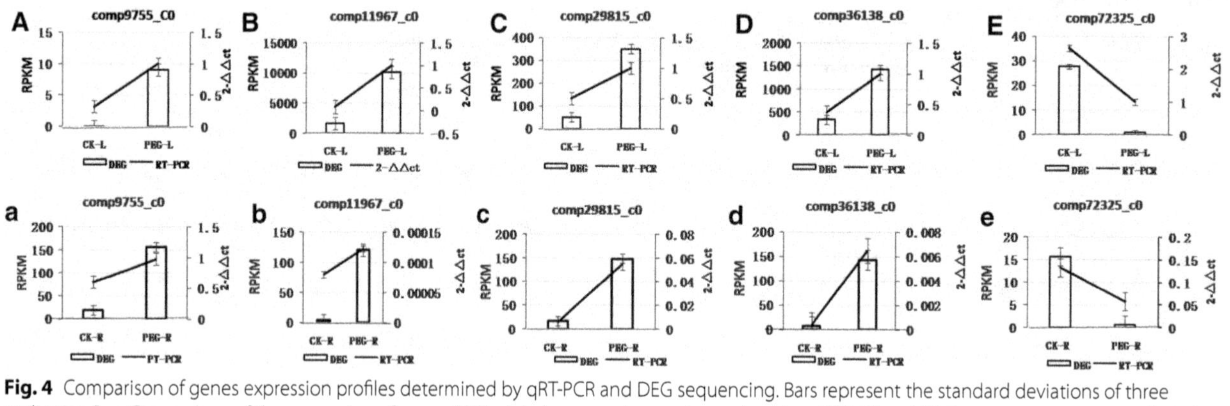

Fig. 4 Comparison of genes expression profiles determined by qRT-PCR and DEG sequencing. Bars represent the standard deviations of three replicates. **A, a** Comparison of the expression of comp9755_c0 in leaves and roots determined by qRT-PCR and DEG sequencing, respectively. **B, b** Comparison of the expression profile of comp11967_c0 in leaves and roots determined by qRT-PCR and DEG sequencing, respectively. **C, c** Comparison of the expression profile of comp29815_c0 in leaves and roots determined by qRT-PCR and DEG sequencing, respectively. **D, d** Comparison of the expression profile of comp36138_c0 in leaves and roots determined by qRT-PCR and DEG sequencing, respectively. **E, e** Comparison of the expression profile of comp72325_c0 in leaves and roots determined by qRT-PCR and DEG sequencing., respectively

involving multiple metabolic pathways. These findings have enriched our understanding of sunflower drought tolerance and provide many candidate genes associated with drought resistance for future development of sunflower varieties better suited to growth in marginal lands.

Abbreviations

BADH: betaine aldehyde dehydrogenase; cDNA: complementary deoxyribonucleic acid; CBL: calcineurin-B-like; CK: control check; DEGs: differentially expressed genes; DGE: digital gene expression; DNA: deoxyribonucleic acid; DREB: dehydration responsive element binding protein; FPKM: Fragments Per Kilobase of exon model per Million mapped reads; KEGG: Kyoto Encyclopedia of Genes and Genomes; mRNA: messenger ribonucleic acid; NAC: no apical meristem; NCBI: National Center for Biotechnology Information; PCR: polymerase chain reaction; PEG: polyethylene glycol; qPCR: real-time quantitative PCR detecting system; RNA: ribonucleic acid; RSEM: RNA-Seq by expectation maximization; SYBR: synergy brands.

Authors' contributions

XH conceived and designed the experiments. CL, WW, JM LZ and CL performed the experiments, SZ, FZ, YY, and WL analyzed the data, CL wrote the paper. All authors read and approved the final manuscript.

Author details

[1] Heilongjiang Academy of Agricultural Sciences Postdoctoral Programme, Xuefu Road 368, Harbin 150086, People's Republic of China. [2] Institute of Industrial Crops, Heilongjiang Academy of Agricultural Sciences, Harbin 150086, People's Republic of China. [3] Crop Research and Breeding Center of Land-Reclamation of Heilongjiang Province, Harbin 150036, People's Republic of China.

Competing interests

The authors declare that they have no competing interests.

Funding

This work was supported by grants from Natural Science Foundation of Heilongjiang Province, China (Grant No. QC2015025), Scientific Research Starting Foundation for Introduced doctorates, Heilongjiang Academy of Agricultural Sciences, China (Grant No. 201507-38) and National Characteristic Oli Plants Industry Technology System (Grant No. CARS-14).

References

Ander S, Huber W (2010) Differential expression analysis for sequence count data. Genome Biol 11(10):R106

Chen K, Jiang Q, Lei YB, Yin CY (2009) Physiological responses of plant to drought stress. J Anhui Agric Sci 37(5):1907–1908 **(Chinese with English abstract)**

Chen L, Ren F, Zhou L, Wang QQ, Zhong H, Li XB (2012) The *Brassica napus* calcineurin B-like 1/CBL-interacting protein kinase 6(CBL1/CIPK6) component is involved in the plant response to abiotic stress and ABA signalling. J Exp Bot 63(17):6211–6222

Cheng JD, An YL, Sun RF (2009) Drought and salt-alkali resistance gene P5CS transformed into sunflower inbred lines. Biotechnol Bull 3:65–69 **(Chinese with English abstract)**

Diaz-Martin J, Almoguera C, Prieto -Dapena P, Espinosa JM, Jordano J (2005) Functional interaction between two transcription factors involved in the developmental regulation of a small heat stress protein gene promoter. Plant Physiol 139:1483–1494

Fang YJ, Liao KF, Du H, Xu Y, Song HZ, Li XH, Xiong LZ (2015) A stress-responsive NAC transcription factor SNAC3 confers heat and drought tolerance through modulation of reactive oxygen species in rice. J Exp Bot 25:2–15

Grabherr MG, Haas BJ, Yassour M, Levin JZ, Thompson DA, Amit I, Adiconis X, Fan L, Raychowdhury R, Zeng QD, Chen ZH, Mauceli E, Hacohen N, Gnirke A, Rhind N, di Palma F, Birren BW, Nusbaum C, Lindblad-Toh K, Friedman N, Regev A (2011) Full-length transcriptome assembly from RNA-Seq data without a reference genome. Nat Biotechnol 29:644–652

Hu LX, Wang ZL, Du HM, Huang BR (2010) Differential accumulation of dehydrins in response to water stress for hybrid and common bermudagrass genotypes differing in drought tolerance. J Plant Physiol 167(2):103–109

Kane NC, Rieseberg LH (2007) Selective sweeps reveal candidate genes for adaptation to drought and salt tolerance in common sunflower, *Helianthus annuus*. Genetics 175:1823–1834

Li B, Dewey CN (2011) RSEM: accurate transcript quantification from RNA-Seq data without a reference genome. BMC Bioinform 12(1):323

Liu XN, Baird WV (2003) Differential Expression of Genes Regulated in Response to Drought or Salinity Stress in Sunflower. Crop Sci 43:678–687 (**Chinese with English abstract**)

Mao XZ, Cai T, Olyarchuk JG, Wei LP (2005) Automated genome annotation and pathway identification using the KEGG Orthology (KO) as a controlled vocabulary. Bioinformatics 21:3787–3793

Roche J, Hewezi T, Bouniols A, Gentzbittel L (2007) Transcriptional profiles of primary metabolism and signal transduction-related genes in response to water stress in field-grown sunflower genotypes using a thematic cDNA microarray. Planta 226:601–617

Santos TBD, de Lima RB, Nagashima GT, de Oliveira Petkowicz CL, Carpentieri-Pipolo V, Pereira LFP, Domingues DS, Vieira LGE (2015) Galactinol synthase

transcriptional profile in two genotypes of *Coffea canephora* with contrasting tolerance to drought. Genet Mol Biol 38(2):182–190
Sauca F, Lazar DA (2011) Scientific results regarding the gene(S) introgression of drought-resistance to *Helianthus annuus* species, using embryo rescue. Rom Biotechnol Lett 16(1):3–8
Shinozaki K, Yamaguchi-Shinozaki K (2007) Gene networks involved in drought stress response and tolerance. J Exp Bot 58:221–227
Singh S, Brocker C, Koppaka V, Ying C, Jackson B, Matsumoto A, Thompson DC, Vasiliou V (2013) Aldehyde dehydrogenases in cellular responses to oxidative/electrophilic stress. Free Radic Biol Med. 56:89–101
Tognetti VB, Van Aken O, Morreel K, Vandenbroucke K, van de Cotte B, Clercq ID, Chiwocha S, Fenske R, Prinsen E, Boerjan W, Genty B, Stubbs KA, Inze D, Breusegem FV (2010) Perturbation of indole-3-butyric acid homeostasis by the UDP-glucosyltransferase UGT74E2 modulates Arabidopsis architecture and water stress tolerance. Plant Cell 22:2660–2679
Trapnell C, Williams BA, Pertea G, Mortazavi A, Kwan G, van Baren MJ, Salzberg SL, Wold BJ, Pachter L (2010) Transcript assembly and quantification by RNA-Seq reveals unannotated transcripts and isoform switching during cell differentiation. Nat Biotechnol 28(5):511–515
Wang H, Zhang H, Gao F, Li J, Li Z (2007) Comparison of gene expression between upland and lowland rice cultivars under water stress using cDNA microarray. Theor Appl Genet 115(8):1109–1126
Yi LF, Zhou KF, Wang P (2013) Development and application of a real-time RT-PCR approach for quantification of *BADH* gene expression in *Helianthus annuus*. Crops 5:70–74 **(Chinese with English abstract)**
Young MD, Wakefield MJ, Smyth GK, Oshlack A (2010) Gene ontology analysis for RNA-seq:accounting for selection bias. Genome Biol 11(2):R14
Zheng J, Zhao JF, Zhang JP, Fu JJ, Gou MY, Dong ZG, Hou W, Huang QS, Wang GY (2006) Comparative expression profiles of maize genes from a water stress-specific cDNA macroarray in response to high-salinity, cold or abscisic acid. Plant Sci 170:1125–1132

17

Transcriptome analysis reveals the genetic basis underlying the biosynthesis of volatile oil, gingerols, and diarylheptanoids in ginger (*Zingiber officinale* Rosc.)

Yusong Jiang, Qinhong Liao, Yong Zou, Yiqing Liu and Jianbin Lan*

Abstract

Background: Ginger (*Zingiber officinale* Rosc.) is a popular flavoring that widely used in Asian, and the volatile oil in ginger rhizomes adds a special fragrance and taste to foods. The bioactive compounds in ginger, such as gingerols, diarylheptanoids, and flavonoids, are of significant value to human health because of their anticancer, anti-oxidant, and anti-inflammatory properties. However, as a non-model plant, knowledge about the genome sequences of ginger is extremely limited, and this limits molecular studies on this plant. In this study, de novo transcriptome sequencing was performed to investigate the expression of genes associated with the biosynthesis of major bioactive compounds in matured ginger rhizome (MG), young ginger rhizome (YG), and fibrous roots of ginger (FR).

Results: A total of 361,876 unigenes were generated by de novo assembly. The expression of genes involved in the pathways responsible for the biosynthesis of major bioactive compounds differed between tissues (MG, YG, and FR). Two pathways that give rise to volatile oil, gingerols, and diarylheptanoids, the "terpenoid backbone biosynthesis" and "stilbenoid, diarylheptanoid and gingerol biosynthesis" pathways, were significantly enriched (adjusted P value < 0.05) for differentially expressed genes (DEGs) (FDR < 0.005) both between the FR and YG libraries, and the FR and MG libraries. Most of the unigenes mapped in these two pathways, including curcumin synthase, phenylpropanoylacetyl-CoA synthase, trans-cinnamate 4-monooxygenase, and 4-hydroxy-3-methylbut-2-en-1-yl diphosphate synthase, were expressed to a significantly higher level ($\log_2$ (fold-change) $\geq$ 1) in FR than in YG or MG.

Conclusion: This study provides the first insight into the biosynthesis of bioactive compounds in ginger at a molecular level and provides valuable genome resources for future molecular studies on ginger. Moreover, our results establish that bioactive compounds in ginger may predominantly synthesized in the root and then transported to rhizomes, where they accumulate.

Keywords: Ginger, Transcriptome sequencing, Rhizome, Fibrous root, Bioactive compounds

Background

Ginger (*Zingiber officinale* Rosc.), which has a cultivar history more than 2000 years in China, is now a popular condiment and medicine all over the world. The distinct flavor of the ginger rhizome makes it popular as a food additive. In traditional Chinese medicine (TCM), ginger is used to relieve nausea and vomiting, eliminate phlegm and relieve cough, improve the internal environment by clearing heat and dampness, and strengthen the immune system (Chang et al. 2011; Li 1985; Qiang et al. 2009; Therkleson 2010; Wang and Wang 2005). In recent years, clinical studies have shown that ginger is an effective medicine for treating degenerative disorders, indigestion, cardiovascular disorders, vomiting, diabetes mellitus, and cancer (Jiang et al. 2006; Mashhadi et al. 2013; Nicoll and Henein 2009; Shukla and Singh 2007; Wei et al. 2005;

*Correspondence: proteomics@163.com
Research Institute for Special Plants, Chongqing University of Arts and Sciences, Chongqing 402160, China

Weng et al. 2010). Moreover, it is generally accepted that ginger has anti-inflammatory properties and functions as an anti-oxidant by scavenging reactive oxygen species (ROS) in cells, which cause human diseases and aging (Dugasani et al. 2010; Mishra et al. 2007; Sekiwa et al. 2000; Torkzadeh-Mahani et al. 2014; Zheng and Wang 2001). Understanding the molecular basis of the biosynthesis of pharmacologically active compounds in ginger is of great practical value for molecular breeding and medicinal use of this plant.

Chemical component analysis has been used to identify hundreds of compounds in ginger. These compounds can generally be classified into three categories: volatile oils, gingerols, and diarylheptanoids (Afzal et al. 2001; Ding and Ding 1988; Jiang et al. 2006). These three categories of compounds contribute to the pungency, biological actions and pharmacological properties. The volatile oil contains sesquiterpenoids and monoterpenoids. Gingerols are a homologous series of phenols that include shogaols, paradols, and gingerone. (Semwal et al. 2015). Diarylheptanoids is a class of derivatives that have a 1,7-diarylheptane skeleton, including curcuminoids, which are compounds found in turmeric (Jiang et al. 2007). In addition to these compounds, ginger contains a large amount of flavonoids, which are natural antioxidants that can lower the risk of cancer, high blood pressure, and heart disease (Ghasemzadeh et al. 2010a, b; Knekt et al. 2002).

Previous studies on ginger have mainly focused on chemical and pharmacological analyses or the products obtained by processing ginger. Little is known about the genome sequences and molecular bases of the biological processes in ginger. In this study, we explored the gene expression profiles of rhizomes and roots, and examined the molecular bases underlying biosynthesis of the major bioactive compounds in ginger. To accomplish this, we *de novo* sequenced the transcriptomes of matured ginger rhizome, young ginger rhizome, and the fibrous roots of the ginger cultivar, Yujiang1. The transcripts obtained in the present study will add to the available genome resources for *Zingiberaceae* plants and provide molecular information for use in molecular breeding and biological study of ginger.

Methods

Ginger material

The ginger variety, Yujiang1, was grown under normal field conditions in Chongqing, China (29°14′ N, 105°52′ E). The sampling were carried out at the vigorous growth and developmental stage, i.e. 90 days after planting. The healthy rhizomes were collected, surface cleaned, and drained. The matured and young ginger from the same plant were than crosscut into slices and frozen in liquid nitrogen, respectively. Matured ginger rhizomes (MG) are ginger rhizomes that harvested at last year and used as planted propagule, and the young ginger rhizomes (YG) are newly formed and immature ginger rhizomes in present year. The fibrous roots in young ginger were immediately frozen in liquid nitrogen after harvest. All samples were stored at −80 °C.

RNA isolation, cDNA library construction and sequencing

Total RNA of matured ginger rhizome (MG, the planted propagule), young ginger rhizome (YG) and fibrous roots (FR) was isolated using RNAprep Pure Plant kits (Bio TeKe, China). The RNA concentration and quality were determined by a Qubit 3.0 (Thermo Scientific, USA) and an Agilent 2100 Bioanalyzer (Agilent Technologies, USA), respectively. For each tissue (MG, YG, and FR), total RNA from five individual plants was pooled for library construction. cDNA libraries were prepared using a NEBNext Ultra Directional RNA Library Prep kit (cat#E7420, NEB, UK) for strand-specific RNA-seq. Briefly, the mRNA was first fragmented and fragments of 300–500 bp were enriched with magnetic beads. The first cDNA strands were then biosynthesized from the fragments. The second cDNA strands were subsequently synthesized by adding dUTP as a marker. Finally, the double strand cDNA was subjected to digestion with uracil-DNA glycosylase (UDG) before PCR reaction and transcriptome sequencing. In this way, only the first strands of cDNA were retained and sequenced. Libraries were sequenced on an Illumina Hiseq 2500 platform using a paired-end run (2 × 150 bp).

Transcriptome assembly and functional annotation

The raw reads generated for each library were quality-filtered by removing adapter sequences and reads with quality scores below Q20. The clean reads of the three libraries were de novo assembled using the Trinity Program (V 2.2.0) with default parameters (Grabherr et al. 2011). The transcriptomes were assembled using the *de Bruijn* strategy, and the longest transcript in each *de Bruijn* graph was defined as a unigene and used as reference sequence for assembly and coding sequence (CDS) prediction. The CDS and protein sequences of unigenes were predicted by TransDecoder (http://transdecoder.github.io/). For functional annotation, BlastX was used with an E-value threshold of 1×10^{-5} to align the unigenes to the following databases: the cluster of orthologous groups of proteins (COG) (https://www.ncbi.nlm.nih.gov/COG/), NCBI non-redundant proteins (NR) (https://www.ncbi.nlm.nih.gov/refseq/about/nonredundantproteins/), Swiss-Prot (http://www.uniprot.org/), gene ontology (GO) (http://www.geneontology.org/) and kyoto encyclopedia of genes and genomes (KEGG) (http://www.genome.jp/kegg/pathway.htkl) databases.

Gene expression measurement, GO and KEGG enrichment analysis

The expression levels of unigenes were calculated and normalized using the fragments per kilobase of exon per million fragments mapped (FPKM) method (Li and Dewey 2011). Significant differentially expressed genes (DEGs) between libraries were restricted according to Benjamini and Yekutieli (2001), with a false discovery rate (FDR) threshold < 0.005 and absolute value of $\log_2$(fold-change) ≥ 1. Significantly enriched GO terms in DEGs were defined as those with corrected P values < 0.05 when compared to the genome background, as described by Benjamini and Yekutieli (2001). KEGG Pathways with significant enrichment for DEGs were defined as those with Q values < 0.05.

Validation of differential expression genes by qRT-PCR

Quantitative RT-PCR (qRT-PCR) was conducted to validate the expression levels of DEGs. cDNA of MG, YG and FR were generated using 1 μg of total RNA treated with an RNase-free DNase I set, followed by reverse transcription with a ReverAid First Strand cDNA Synthesis kit (Thermo Scientific, USA). Quantitative RT-PCR analysis was performed using a LightCycler 96 (Roch, USA) with SYBR Green master mix (BioRed, USA). Genes and primers are listed in Additional file 3: Table S1. Significantly different expression was determined by Tukey's multi-comparison test ($\alpha = 0.05$).

Results and discussion

Transcriptome data and assembly

A total of 18.05 G, 19.22 G, and 16.86 G clean data were generated for MG, YG, and FR, respectively (Table 1). The GC contents ranged from 49.51 to 51.19%, and the Q20 percentage ranged from 97.69 to 97.88% in clean reads. De novo assembly of the pooled cleaned reads was performed using Trinity based on chosen criteria. The assembled sequences were then used as reference for further analysis. The total mapped reads for MG, YG, and FR were 77.44, 75.15, and 73.83%, respectively. A total of 250,969, 240,132, and 225,366 unigenes were identified for the MG, YG, and FR libraries, respectively (Table 1). Finally, the *de novo* assembly generated a total of 361,876 unigenes, with an average length of 515.89 bp, and N50 length of 589 bp. The minimum length and maximum length was 201 bp and the maximum length was 21,690 bp, respectively (Table 2). Length distribution showed that 105,683 (29.2%) unigenes were longer than 500 bp (Additional file 1: Figure S1).

Ginger plays important roles in the food processing and pharmaceutical industries. Many studies have focused on the chemical components and pharmacological effects of ginger (Hoferl et al. 2015; Li et al. 2016; Wang et al. 2015). However, the genome sequence of ginger is still unavailable. Only a limited number of expressed sequence tags (ESTs) for *Zingiber officinale* are available, with only 38,169 ESTs that can be retrieved from NCBI GenBank. The lack of genome information greatly limits molecular biology studies of ginger. The development of next-generation sequencing (NGS) technologies and bioinformatics has greatly facilitated studies on molecular biological changes and regulation at the genome-wide level, especially for non-model plants (Chai et al. 2009; Ward et al. 2012). The N50 value obtained for unigenes in the present study (589 bp) was lower than those reported by Prasath et al. (2014) (943 bp), and Gaur et al. (2016) (1251 bp). N50 length has been used as a criterion of transcriptome assembly quality; however, this value is greatly affected by the genome structure, assembly software, and assembly parameters (Li and Dewey 2011; Postnikova et al. 2013). Therefore, differences between studies in N50 values may be artifacts of the methods used for assembly. Thus, N50 values are not necessarily correlated with transcriptome accuracy. Our study generated 186.69 Mb of transcriptome sequences and the dataset will provide valuable information for future molecular studies of ginger and other Zingiberaceae plants. The raw data from this study are available in the NCBI database (SRA accession number SRP101401).

Functional annotation and classification of unigenes

GO analysis was used to annotate 98,029 (39.06%), 91,092 (37.93%), and 81,695 (36.25%) of the unigenes in the MG, YG, and FR libraries, respectively. The distributions of the GO terms for the unigenes in each library were similar. The majority of the unigenes were contained in metabolic process (GO:0008152), binding (GO:0005488), cellular

Table 1 Throughput and quality of strand specific RNA-seq of the three libraries

Library ID	Clean reads	Clean base (G)	Clean GC%	Clean Q20%	Total mapped reads (%)	Unigenes
MG	120,892,308	18.05	50.41	97.84	93,623,020 (77.44)	250,969
YG	128,809,296	19.22	49.51	97.88	96,789,876 (75.14)	240,132
FR	112,821,264	16.86	51.19	97.69	83,298,040 (73.83)	225,366

Table 2 Overview of the sequencing and assembly

Items	Number
Total unigenes	361,876
Average unigene length (bp)	515.89
Min unigene length (bp)	201
Max unigene length (bp)	21,690
N50 length (bp)	589
Total bases (bp)	186,687,182

process (GO:0009987), catalytic activity (GO:0003824), cell (GO:0005623), cell part (GO:0044464), and single-organism process (GO:0044699) terms (Fig. 1). This suggests that there was no bias in library construction. A large number of the unigenes were classified into the metabolic process term, suggesting high metabolic activities in ginger rhizome and root. This may reflect the biosynthesis and accumulation of bioactive secondary metabolites in ginger.

Of the 361,876 unigenes identified, 184,768 (51.09%) were annotated in at least in one database (Additional file 2: Figure S2A): 99,726 (27.43%) were annotated with COG functions, 79,705 (22.02%) were annotated with KEGG enzymes, 177,366 (49.01%) were mapped to the NR database, and 122,191 (33.77%) were mapped to the Swiss-prot database. The alignment of unigenes against all proteins in the NR database showed that species with the greatest number of matches was *Musa acuminata* (28.38%), followed by *Gossypium* species (Gossypium hirsutum 9.64%, Gossypium raimondii 7.32%, Gossypium arboreum 2.32%; Additional file 2: Figure S2B). The unigenes identified in this study were not associated with any other individual species contained in the databases we examined. This is because of there was no genome data of Zingiberaceae species published to date (by searching the genome assembly information in the NCBI public database). This result indicates that the genome of ginger differs greatly from previously sequenced plant species. The big differences between the genomes of ginger and other species emphasizes the importance of genome and transcriptome sequencing of ginger or other Zingiberaceae plants in order to facilitate molecular studies of the whole genus.

GO and KEGG enrichment of differentially expressed genes (DEGs)

Comparison of unigene expression levels across the three libraries showed that the number of unigenes with medium (3.75 < FPKM < 15) and high (FPKM > 15) expression levels in matured ginger (MG) (29,715, 8.22%) was lower than those in young ginger (YG) (36,789, 10.17%) and fibrous roots (FR) (38,466, 10.63%; Additional file 4: Table S2). The differences in expression levels between these tissues may therefore indicate that metabolic activity is more vigorous in newly generated tissues.

To identify DEGs, pairwise comparisons of unigene expression levels were performed between the FR and

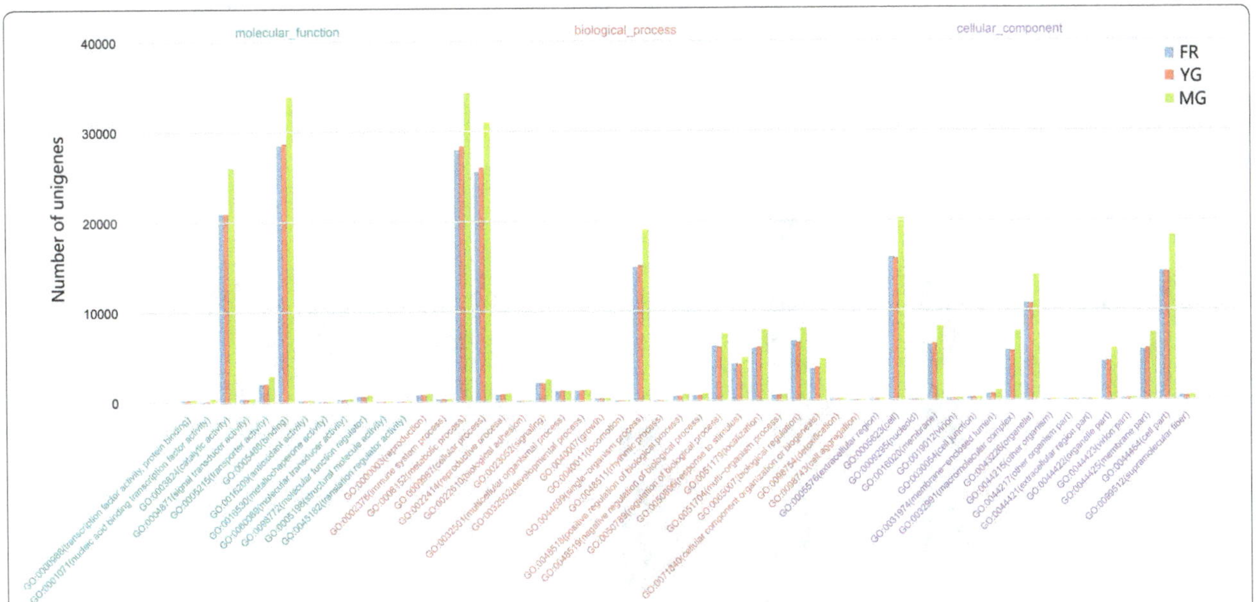

Fig. 1 Histogram of gene ontology (GO) annotations for unigenes in ginger matured rhizome, young rhizome, and fibrous root libraries. Level two GO terms belonging to three categories, molecular function, biological process, and cellular component are along the x-axis and the number of unigenes assigned to each term is represented along the y-axis. Bars with different colors indicate different libraries

YG, FR and MG, and YG and MG libraries, (FDR < 0.005, $\log_2$ (fold change) $\geq$ 1). In comparison between FR and YG libraries, 1080 up-regulated and 2233 down-regulated DEGs were identified. In FR and MG, 1353 up-regulated and 2706 down-regulated DEGs were identified, and in YG and MG, 888 up-regulated and 1173 down-regulated DEGs were identified (Fig. 2a). There were many more DEGs between gingers rhizome tissues and fibrous root tissue (MG versus FR and YG versus FR) than there were between tissues from the different parts of ginger rhizomes (MG versus YG). This suggests a large difference in the expression of genes between ginger tissues. Comparison across the three libraries identified 351 DEGs (Fig. 2b).

GO classification of DEGs between the FR and YG, FR and MG, and YG and MG libraries resulted in assigning them to 52, 54, and 35 level two GO terms, respectively. The most enriched GO terms for DEGs between MG and FR were the same as those for DEGs between YG and FR. However, these terms were notably different from the most enriched GO terms for DEGs between MG and YG (Fig. 3). The "molecular function" (GO: 0009058) term in the molecular function category and many GO terms in the biological process category, which is mainly related to biosynthetic processes and metabolic processes, contained the largest numbers of DEGs between MG and FR and between YG and FR. However, no or few DEGs between MG and YG were assigned to these GO terms (Fig. 3).

KEGG classification identified a total of 49, 32, and 25 significantly enriched pathways (Q value < 0.05) for DEGs between the FR and MG, FR and YG, and MG and YG libraries, respectively (Additional file 5: Table S3). As with the GO term distributions, the majority of significantly enriched KEGG pathways for DEGs between FR and YG were similar to those between FR and MG libraries. Two pathways, the "terpenoid backbone biosynthesis" (ko00900) and "stilbenoid and diarylheptanoid and gingerol biosynthesis" (ko00945) pathways, were significantly enriched for DEGs between FR and MG, and FR and YG. These pathways give rise to the biosynthesis of volatile oil, gingerols, and diarylheptanoids in ginger (http://www.kegg.jp/kegg/pathway.html). Further KEGG enrichment for the 351 DEGs identified across the three libraries resulted in 16 significantly enriched pathways (P value < 0.05; Table 3). Among these pathways, the "phenylalanine metabolism" pathway (ko00360) and the "phenylpropanoid biosynthesis" pathway (ko00940) produces substrates for the "phenylpropanoid biosynthesis" pathway (ko00940), which generates substrates for the biosynthesis of pharmacologically active metabolites in ginger rhizomes (Ramirez-Ahumada et al. 2006).

The study by Zhan et al. (2009) showed that the major components of volatile oil, gingerols, and diarylheptanoids in FG and YG almost the same, while the contents of some of the components such as α-zingiberene, α-curcumene, and 6-gingerol were significant different. Our study provides the molecular evidence for the differences on the contents of bioactive compounds between ginger rhizomes with different mature degree. However, the content and biosynthesis of bioactive compounds in ginger fibrous root had not documented yet. In this study, the DEGs between FR vs. YG, and FR vs. MG libraries in the "terpenoid backbone biosynthesis" and "stilbenoid

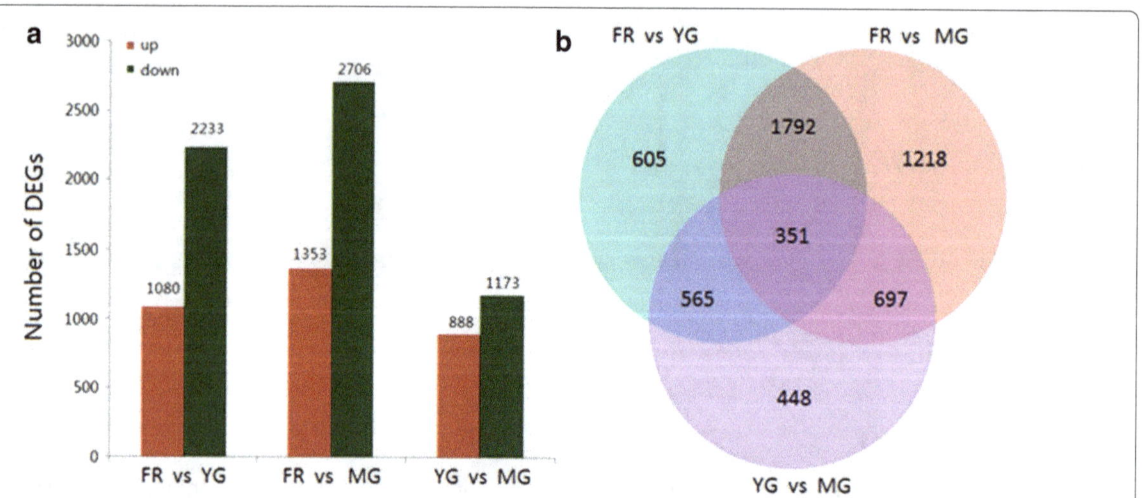

Fig. 2 Pairwise comparisons of unigene expression in ginger matured rhizome, young rhizome, and fibrous root libraries (**a**) and a Venn diagram of the differentially expressed genes (DEGs) among the three libraries (**b**)

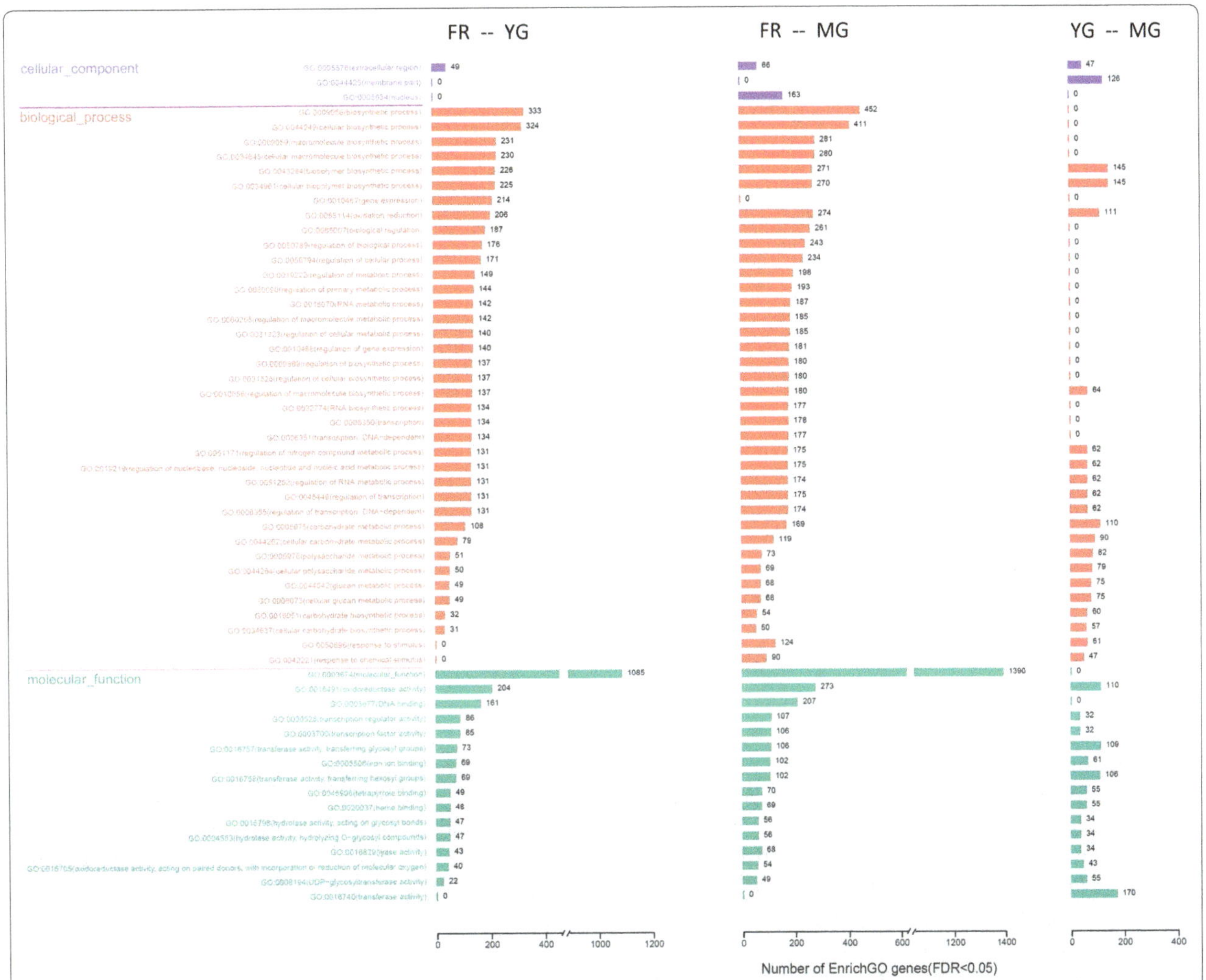

Fig. 3 Gene ontology (GO) enrichment of differentially expressed genes (DEGs) between ginger matured rhizome, young rhizome, and fibrous root libraries. P values are represented along the x-axis and level two GO terms belonging to three categories, molecular function, biological process, and cellular component are represented along the y-axis. Numbers indicate the number of unigenes assigned to each term

and diarylheptanoid and gingerol biosynthesis" pathways brings new insight into the material biosynthesis and metabolism between different ginger parts.

Candidate genes involved in the biosynthesis of bioactive compounds in ginger

Volatile oil, gingerols and diarylheptanoids are three major categories of bioactive chemical compounds in ginger (Afzal et al. 2001; Kuo et al. 2005). In this study, two KEGG pathways that are related to the biosynthesis of these compounds, "terpenoid backbone biosynthesis" (ko00900) and "stilbenoid and diarylheptanoid and gingerol biosynthesis" (ko00945), were significantly enriched. Based on the NR annotation, a total of 33 unigenes were mapped to the "terpenoid backbone biosynthesis", including 11 unigenes for 4-hydroxy-3-methylbut-2-en-1-yl diphosphate synthase (HDS), 5 unigenes for 1-deoxy-D-xylulose 5-phosphate reductoisomerase (DXR), 5 unigenes for 1-deoxy-D-xylulose-5-phosphate synthase (DXS), 4 unigenes for geranyl pyrophosphate synthase (GPPS), 1 unigene for acetyl-CoA acetyltransferase (ACAT), 1 unigenes for mevalonate kinase (MVK), 1 unigenes for phosphomevalonate kinase (MPVK), 1 unigenes for isopentenyl diphosphate isomerase (IPPI), 1 unigenes for 2-C-methyl-D-erythritol 4-phosphate cytidylyltransferase (MEPCT), and 1 unigene for 2-C-methyl-D-erythritol-2, 4-cyclodiphosphate synthase (MECPS). Most of the above unigenes were also identified in the "terpenoid backbone biosynthesis" pathway in Atractylodes lancea (Ahmed et al. 2016), a traditional Chinese medicine, and Cymbopogon flexuosus (lemongrass) (Meena et al. 2016). This indicates that the

Table 3 The significantly enriched KEGG pathways for DEGs across the three cDNA libraries

Pathway ID	Pathway	P value	Adjusted P value	DEG number (70)	All unigenes with pathway annotation (52,831)
ko00960	Tropane, piperidine and pyridine alkaloid biosynthesis	2.22497E−07	1.80223E−05	7 (10%)	316 (0.6%)
ko00360	Phenylalanine metabolism	0.000440777	0.017851483	6 (8.57%)	736 (1.39%)
ko04075	Plant hormone signal transduction	0.000867937	0.019222353	9 (12.86%)	1893 (3.58%)
ko00270	Cysteine and methionine metabolism	0.000949252	0.019222353	6 (8.57%)	853 (1.61%)
ko00940	Phenylpropanoid biosynthesis	0.001677001	0.027167422	6 (8.57%)	954 (1.81%)
ko00062	Fatty acid elongation in mitochondria	0.003175276	0.042866226	3 (4.29%)	220 (0.42%)
ko00130	Ubiquinone and other terpenoid-quinone biosynthesis	0.004387597	0.050770761	3 (4.29%)	247 (0.47%)
ko04626	Plant-pathogen interaction	0.007368338	0.07460442	8 (11.43%)	2145 (4.06%)
ko00592	alpha-Linolenic acid metabolism	0.016010692	0.144096225	3 (4.29%)	398 (0.75%)
ko04740	Olfactory transduction	0.026994683	0.197111092	2 (2.86%)	192 (0.36%)
ko00910	Nitrogen metabolism	0.028187756	0.197111092	3 (4.29%)	495 (0.93%)
ko04910	Insulin signaling pathway	0.030128861	0.197111092	5 (7.14%)	1312 (2.48%)
ko04744	Phototransduction	0.032099363	0.197111092	2 (2.86%)	211 (0.4%)
ko04971	Gastric acid secretion	0.034068584	0.197111092	2 (2.86%)	218 (0.41%)
ko04970	Salivary secretion	0.037551065	0.20277575	2 (2.86%)	230 (0.44%)
ko04745	Phototransduction—fly	0.048435888	0.245206684	2 (2.86%)	265 (0.5%)

major component and the biosynthesis of volatile oil in different aromatic plants are common.

The "terpenoid backbone biosynthesis" pathway includes two biosynthetic pathways, the mevalonate pathway and the non-mevalonate pathway (MEP/DOXP) pathway (KEGG). These two pathways co-exist in higher plants but operate in separate cellular compartments. The mevalonate pathway operates in the cytosol and gives rise to triterpenes, sterols, and most sesquiterpenes. The MEP/DOXP pathway operates in the plastids and mainly synthesizes essential oil monoterpenes and linalyl acetate, some sesquiterpenes, diterpenes, carotenoids, and phytol (http://www.genome.jp/dbget-bin/www_bget?pathway:map00900). Sesquiterpenoids and monoterpenoids are major biochemical constituents of ginger volatile oil (Jeena et al. 2013; Mashhadi et al. 2013). In this study, three unigenes (ACAT, EC:2.3.1.9, MVK, EC:2.7.1.36, and MPVK, EC:2.7.4.2) mapped to the mevalonate pathway, and two of them (ACAT and MVK) expressed to a greater degree in both matured and young ginger rhizomes than in fibrous roots (Fig. 4). In contrast, more unigenes were mapped to the MEP/DOXP pathway than to the mevalonate pathway, and the majority of these unigenes were expressed to a significantly greater degree ($\log_2$ (fold-change) > 1) in fibrous roots than in rhizomes, such as HDS (EC:1.17.7.1 and 1.17.7.3) and GPPS (EC:2.5.1.1, 2.5.1.10, and 2.5.1.29) (Fig. 4). This result suggests that the biological process responsible for the biosynthesis of volatile oil is different between ginger rhizomes and fibrous roots, and the fibrous roots play important roles in the biosynthesis of ginger volatile oil.

In addition, a total of 23 unigenes "stilbenoid, diarylheptanoid and gingerol biosynthesis" pathways (Additional file 6: Table S4), including 12 unigenes for curcumin synthase (CURS), 5 unigenes for trans-cinnamate 4-monooxygenase (CYP73A), 5 unigenes for phenylpropanoylacetyl-CoA synthase (DCS), and 1 unigene for caffeoyl-CoA O-methyltransferase (CCOMT). The "stilbenoid, diarylheptanoid and gingerol biosynthesis" pathway is responsible for the biosynthesis of gingerols and diarylheptanoid. These compounds exhibit a host of biological and pharmacological activities, including anticancer, antioxidant, antimicrobial, anti-inflammatory and anti-allergic activities (Dugasani et al. 2010; Jiang et al. 2007; Mashhadi et al. 2013; Ramirez-Ahumada et al. 2006; Semwal et al. 2015). In this study, none of the 23 unigenes mapped to the "stilbenoid, diarylheptanoid and gingerol biosynthesis" pathway was directly related to the biosynthesis of gingerols (Fig. 5). This is consistent with the previous study by Ramirez-Ahumada et al. (2006), which reported that no gingerol synthase activity could be identified in extracts from ginger rhizomes. However, unigenes involved in the biosynthesis of curcuminoids were expressed at higher levels in fibrous roots than in rhizomes (Fig. 5). Curcuminoids are diarylheptanoid compounds found in turmeric, and represent an important category of biochemical components (Jiang et al. 2007). This indicates that diarylheptanoids are predominantly synthesized in fibrous roots.

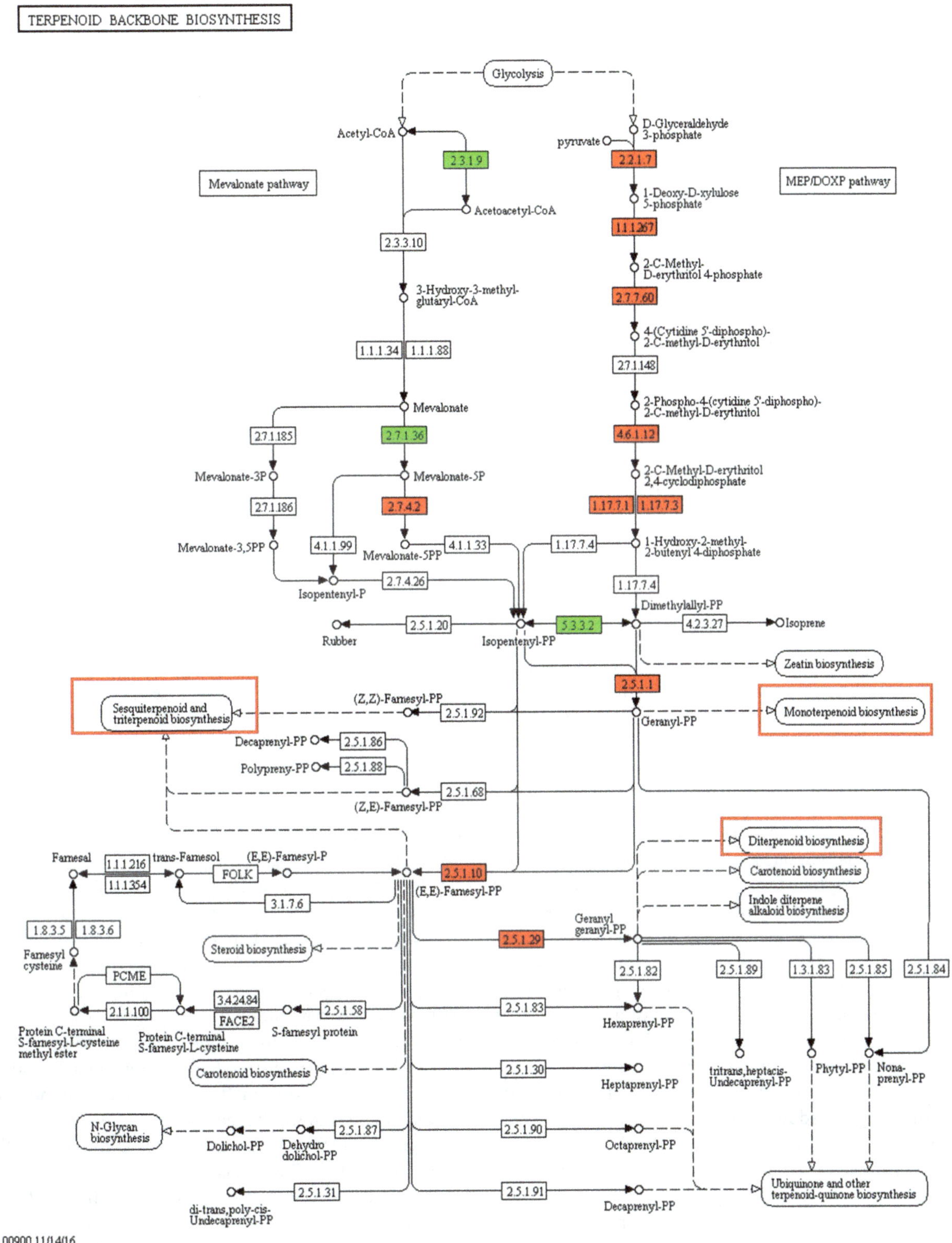

Fig. 4 Differentially expressed genes (DEGs) involved in the "terpenoid backbone biosynthesis" pathway of ginger. The red columns indicate genes expressed at a significantly high level in fibrous roots. The green columns indicate genes expressed at a significantly high level in rhizomes. The red boxes indicate the biosynthesis of the major compounds found in ginger volatile oil. This color-coded map of DEGs corresponds to map00900 in the KEGG database (http://www.genome.jp/dbget-bin/www_bget?pathway:map00900)

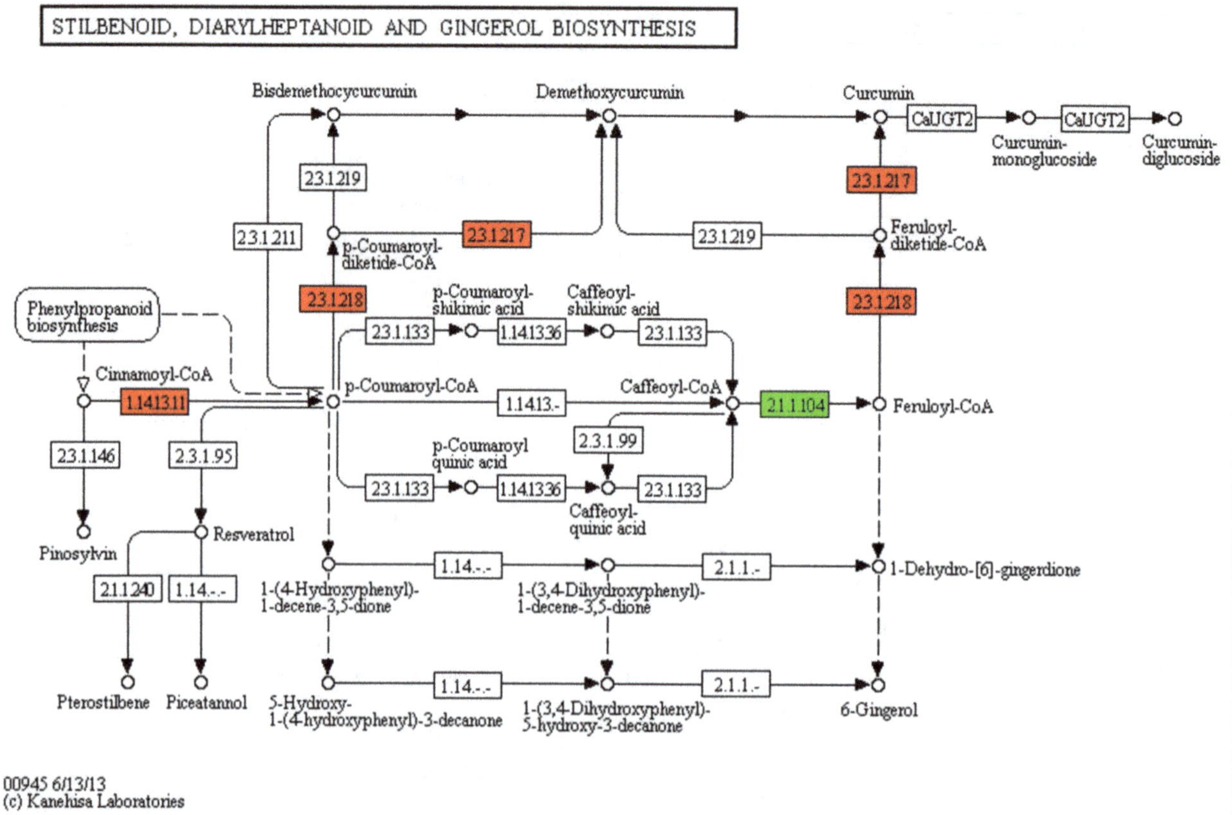

Fig. 5 Differentially expressed genes (DEGs) involved in the "stilbenoid, diarylheptanoid and gingerol biosynthesis" pathway. The red columns indicate genes expressed at a significantly high level in fibrous roots; the green columns indicate genes expressed at a significantly high level in rhizomes. This color-coded map of DEGs corresponds to map00945 in the KEGG database (http://www.genome.jp/dbget-bin/www_bget?pathway:map00945)

In addition to these important bioactive compounds, ginger contains a large amount of flavonoids, which are important pharmacologically active metabolites that display antioxidant, anti-aging, and anti- cancer properties (Ghasemzadeh et al. 2010b; Knekt et al. 2002). In this study, 12 unigenes were mapped to "flavonoid biosynthesis" (ko00941) pathways (Additional file 6: Table S4). In contrast to the expression patterns of unigenes involved in the biosynthesis of volatile oil and diarylheptanoids, most unigenes related to the biosynthesis of flavonoids were expressed most highly in young ginger (Additional file 6: Table S4). The differential expression of genes related to the synthesis of biologically and pharmacologically active compounds in different ginger tissues suggests that, although these compounds accumulate in ginger rhizomes, they may be synthesized in other tissues. The synthesis, transport, and storage of these compounds require further investigation.

Expression validation by qPCR

To evaluate the expression profiles of DEGs obtained by transcriptome sequencing, seven DEGs, including two DEGs (chalcone synthase, CHS, and flavonoid 3′,5′-hydroxylase, F3′H) in the "flavonoid biosynthesis" pathway (ko00941), two (curcumin synthase 2, CURS2, and phenylpropanoylacetyl-CoA synthase, DCS) in the "stilbenoid, diarylheptanoid and gingerol biosynthesis" pathway (ko00945), and three (acetyl-CoA acetyltransferase, ACAT, 1-deoxy-D-xylulose-5-phosphate synthase, DXS, and 4-hydroxy-3-methylbut-2-en-1-yl diphosphate synthase, HDS) in the "terpenoid backbone biosynthesis" pathway (ko00900), were selected for expression validation using qRT-PCR analysis (Additional file 3: Table S1). These three pathways are of the major pathways that are responsible for the biosynthesis of bioactive compounds in ginger such as flavonoid and volatile oil. The expression levels calculated by qRT-PCR were consistent with those from transcriptome sequencing data (Fig. 6). This

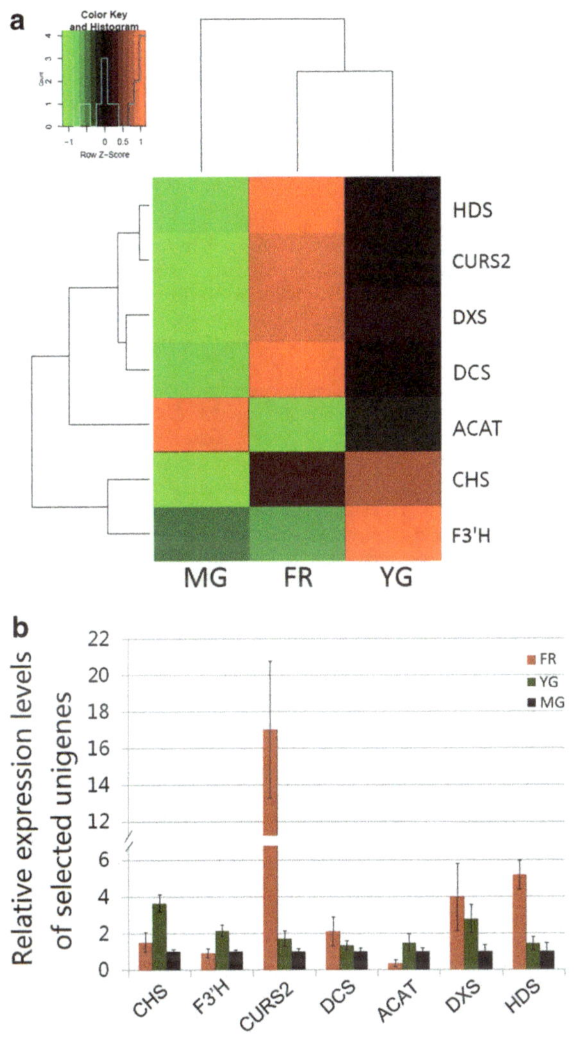

Fig. 6 Heat map of the expression levels of seven ginger unigenes as determined by the fragments per kilobase of exon per million fragments mapped (FPKM) (**a**) and qRT-PCR analysis (**b**)

indicated that the transcriptome sequencing is an effective way to investigate the differences on genes expressions and biological processes between different tissues.

Conclusion

De novo transcriptome sequencing and assembly of three ginger tissues (MG, YG, and FR) generated a total of 361,876 unigenes, with an average length of 515.89 bp. Pairwise comparisons of unigene expression levels indicated that the unigenes' expression pattern were similar between different parts of ginger rhizomes (YG versus MG), but different between ginger rhizomes and fibrous roots (YG versus FR, and MG versus FR). In addition, KEGG pathway enrichment for DEGs between different ginger tissues suggested that pathways responsible for the biosynthesis of volatile oil, diarylheptanoids, and substrates of these compounds, were predominantly represented. Moreover, most of the unigenes mapped in these pathways expressed significantly higher in FR than in YG or MG. Our result established that bioactive compounds in ginger may be predominantly synthesized in fibrous roots and then transported to rhizomes, where they accumulate. This study provides the first insight into the biosynthesis of bioactive compounds in ginger at a molecular level and provides valuable genome resources for future molecular studies on ginger.

Additional files

Additional file 1: Figure S1. The length distribution of the assembled ginger transcriptome.

Additional file 2: Figure S2. A Venn diagram showing annotation of all of the assembled ginger unigenes that were found in different databases (A) and the unigenes that matched the 15 top species in the NR database (B).

Additional file 3: Table S1. Summary of unigene expression levels in ginger matured rhizome, young rhizome, and fibrous root libraries using the fragments per kilobase of exon per million fragments mapped (FPKM) method.

Additional file 4: Table S2. Significantly enriched Kyoto Encyclopedia of Genes and Genomes (KEGG) pathways of differentially expressed genes identified from pairwise comparisons between matured and young rhizome, young rhizome and fibrous root, matured rhizome and fibrous root libraries.

Additional file 5: Table S3. KEGG enrichment of DEGs between libraries of MG and YG.

Additional file 6: Table S4. Unigenes mapped to the pathways that responsible for the biosynthesis of major bioactive compounds in ginger.

Abbreviations

ACAT: accetyl-CoA acetyltransferase; CCOMT: caffeoyl-CoA *O*-methyltransferase; CDS: coding sequence; CHS: chalcone synthase; CURS: curcumin synthase; CYP73A: trans-cinnamate 4-monooxygenase; DCS: phenylpropanoylacetyl-CoA synthase; DXPS: 1-deoxy-D-xylulose-5-phosphate synthase; DXR: 1-deoxy-D-xylulose 5-phosphate reductoisomerase; DXS: 1-deoxy-D-xylulose-5-phosphate synthase; F3′H: flavonoid 3′,5′-hydroxylase; FR: fibrous roots; GPPS: geranyl pyrophosphate synthase; HDS: 4-hydroxy-3-methylbut-2-en-1-yl diphosphate synthase; IPPI: isopentenyl diphosphate isomerase; MECPS: 2-C-methyl-D-erythritol-2, 4-cyclodiphosphate synthase; MEPCT: 2-C-methyl-D-erythritol 4-phosphate cytidylyltransferase; MG: matured ginger rhizome; MPVK: phosphomevalonate kinase; MVK: mevalonate kinase; ROS: reactive oxygen species; TCM: traditional Chinese Medicine; UDG: uracil-DNA glycosylase; YG: young ginger rhizome.

Authors' contributions

Conceived and designed the experiments: YJ and JL. Performed the experiments: QL and YZ. Analyzed the data: YJ, QL, and YL. Wrote the paper: YJ, QL, and JL. All authors read and approved the final manuscript.

Acknowledgements
Not applicable.

Competing interests
The authors declare that they have no competing interests.

Funding
This work was supported by Grants from the National Natural Science Foundation of China (No. 31501273), the Chongqing Research Program of Basic Research and Frontier Technology (No. cstc2015jcyjA80003), the Chongqing social undertakings and the people's livelihood security Program (No. cstc-2016shmszx0483), and the Scientific and Technological Research Program of Chongqing Municipal Education Commission (No. KJ1401104).

References

Afzal M, Al-Hadidi D, Menon M, Pesek J, Dhami MS (2001) Ginger: an ethnomedical, chemical and pharmacological review. Drug Metabol Drug Interact 18:159–190

Ahmed S, Zhan CS, Yang YY, Wang XK, Yang TW, Zhao ZY, Zhang QY, Li XH, Hu XB (2016) The transcript profile of a traditional Chinese medicine, Atractylodes lancea, revealing its sesquiterpenoid biosynthesis of the major active components. PLoS ONE 11:e0151975

Benjamini Y, Yekutieli D (2001) The control of the false discovery rate in multiple testing under dependency. Ann Stat 29(4):1165–1188

Chai YR, Lei B, Huang HL, Li JN, Yin JM, Tang ZL, Wang R, Chen L (2009) TRANSPARENT TESTA 12 genes from Brassica napus and parental species: cloning, evolution, and differential involvement in yellow seed trait. Mol Genet Genom 281:109–123

Chang TT, Chen KC, Chang KW, Chen HY, Tsai FJ, Sun MF, Chen CY (2011) In silico pharmacology suggests ginger extracts may reduce stroke risks. Mol BioSyst 7:2702–2710

Ding AW, Ding QL (1988) Comparison of the contents of the main chemical constituents in various processed preparations of ginger. Zhong Yao Tong Bao 13(17–19):62

Dugasani S, Pichika MR, Nadarajah VD, Balijepalli MK, Tandra S, Korlakunta JN (2010) Comparative antioxidant and anti-inflammatory effects of [6]-gingerol, [8]-gingerol, [10]-gingerol and [6]-shogaol. J Ethnopharmacol 127:515–520

Gaur M, Das A, Sahoo RK, Kar B, Nayak S, Subudhi E (2016) De Novo transcriptome assembly of Zingiber officinale cv Suruchi of Odisha. Genom Data 9:87–88

Ghasemzadeh A, Jaafar HZ, Rahmat A (2010a) Synthesis of phenolics and flavonoids in ginger (Zingiber officinale Roscoe) and their effects on photosynthesis rate. Int J Mol Sci 11:4539–4555

Ghasemzadeh A, Jaafar HZ, Rahmat A (2010b) Identification and concentration of some flavonoid components in Malaysian young ginger (Zingiber officinale Roscoe) varieties by a high performance liquid chromatography method. Molecules 15:6231–6243

Grabherr MG, Haas BJ, Yassour M, Levin JZ, Thompson DA, Amit I, Adiconis X, Fan L, Raychowdhury R, Zeng Q, Chen Z, Mauceli E, Hacohen N, Gnirke A, Rhind N, di Palma F, Birren BW, Nusbaum C, Lindblad-Toh K, Friedman N, Regev A (2011) Full-length transcriptome assembly from RNA-Seq data without a reference genome. Nat Biotechnol 29:644–652

Hoferl M, Stoilova I, Wanner J, Schmidt E, Jirovetz L, Trifonova D, Stanchev V, Krastanov A (2015) Composition and comprehensive antioxidant activity of ginger (Zingiber officinale) essential oil from Ecuador. Nat Prod Commun 10:1085–1090

Jeena K, Liju VB, Kuttan R (2013) Antioxidant, anti-inflammatory and antinociceptive activities of essential oil from ginger. Indian J Physiol Pharmacol 57:51–62

Jiang H, Xie Z, Koo HJ, McLaughlin SP, Timmermann BN, Gang DR (2006) Metabolic profiling and phylogenetic analysis of medicinal Zingiber species: tools for authentication of ginger (Zingiber officinale Rosc.). Phytochemistry 67:1673–1685

Jiang H, Timmermann BN, Gang DR (2007) Characterization and identification of diarylheptanoids in ginger (Zingiber officinale Rosc.) using high-performance liquid chromatography/electrospray ionization mass spectrometry. Rapid Commun Mass Spectrom 21:509–518

Knekt P, Kumpulainen J, Jarvinen R, Rissanen H, Heliovaara M, Reunanen A, Hakulinen T, Aromaa A (2002) Flavonoid intake and risk of chronic diseases. Am J Clin Nutr 76:560–568

Kuo PC, Damu AG, Cherng CY, Jeng JF, Teng CM, Lee EJ, Wu TS (2005) Isolation of a natural antioxidant, dehydrozingerone from Zingiber officinale and synthesis of its analogues for recognition of effective antioxidant and antityrosinase agents. Arch Pharm Res 28:518–528

Li ZW (1985) Application of the breeding of the medicinal ginger family to the systematization and study of Chinese materia medica. Zhong Yao Tong Bao 10:3–6

Li B, Dewey CN (2011) RSEM: accurate transcript quantification from RNA-Seq data with or without a reference genome. BMC Bioinform 12:323

Li Y, Hong Y, Han Y, Wang Y, Xia L (2016) Chemical characterization and antioxidant activities comparison in fresh, dried, stir-frying and carbonized ginger. J Chromatogr B Analyt Technol Biomed Life Sci 1011:223–232

Mashhadi NS, Ghiasvand R, Askari G, Hariri M, Darvishi L, Mofid MR (2013) Antioxidative and anti-inflammatory effects of ginger in health and physical activity: review of current evidence. Int J Prev Med 4:S36–S42

Meena S, Kumar SR, Venkata Rao DK, Dwivedi V, Shilpashree HB, Rastogi S, Shasany AK, Nagegowda DA (2016) De novo sequencing and analysis of lemongrass transcriptome provide first insights into the essential oil biosynthesis of aromatic grasses. Front Plant Sci 7:1129

Mishra C, Sharma S, Kakkar P (2007) A study to evaluate heavy metals and organochlorine pesticide residue in Zingiber officinale Rosc. collected from different ecological zones of India. Bull Environ Contam Toxicol 79:95–98

Nicoll R, Henein MY (2009) Ginger (Zingiber officinale Roscoe): a hot remedy for cardiovascular disease? Int J Cardiol 131:408–409

Postnikova OA, Shao J, Nemchinov LG (2013) Analysis of the alfalfa root transcriptome in response to salinity stress. Plant Cell Physiol 54:1041–1055

Prasath D, Karthika R, Habeeba NT, Suraby EJ, Rosana OB, Shaji A, Eapen SJ, Deshpande U, Anandaraj M (2014) Comparison of the transcriptomes of ginger (Zingiber officinale Rosc.) and mango ginger (Curcuma amada Roxb.) in response to the bacterial wilt infection. PLoS ONE 9:e99731

Qiang LQ, Wang CP, Wang FM, Pan Y, Yi LT, Zhang X, Kong LD (2009) Combined administration of the mixture of honokiol and magnolol and ginger oil evokes antidepressant-like synergism in rats. Arch Pharm Res 32:1281–1292

Ramirez-Ahumada MC, Timmermann BN, Gang DR (2006) Biosynthesis of curcuminoids and gingerols in turmeric (Curcuma longa) and ginger (Zingiber officinale): identification of curcuminoid synthase and hydroxycinnamoyl-CoA thioesterases. Phytochemistry 67:2017–2029

Sekiwa Y, Kubota K, Kobayashi A (2000) Isolation of novel glucosides related to gingerdiol from ginger and their antioxidative activities. J Agric Food Chem 48:373–377

Semwal RB, Semwal DK, Combrinck S, Viljoen AM (2015) Gingerols and shogaols: important nutraceutical principles from ginger. Phytochemistry 117:554–568

Shukla Y, Singh M (2007) Cancer preventive properties of ginger: a brief review. Food Chem Toxicol 45:683–690

Therkleson T (2010) Ginger compress therapy for adults with osteoarthritis. J Adv Nurs 66:2225–2233

Torkzadeh-Mahani S, Nasri S, Esmaeili-Mahani S (2014) Ginger (zingiber officinale roscoe) prevents morphine-induced addictive behaviors in conditioned place preference test in rats. Addict Health 6:65–72

Wang WH, Wang ZM (2005) Studies of commonly used traditional medicine-ginger. Zhongguo Zhong Yao Za Zhi 30:1569–1573

Wang CZ, Qi LW, Yuan CS (2015) Cancer chemoprevention effects of ginger and its active constituents: potential for new drug discovery. Am J Chin Med 43:1351–1363

Ward JA, Ponnala L, Weber CA (2012) Strategies for transcriptome analysis in nonmodel plants. Am J Bot 99:267–276

Wei QY, Ma JP, Cai YJ, Yang L, Liu ZL (2005) Cytotoxic and apoptotic activities of diarylheptanoids and gingerol-related compounds from the rhizome of Chinese ginger. J Ethnopharmacol 102:177–184

Weng CJ, Wu CF, Huang HW, Ho CT, Yen GC (2010) Anti-invasion effects of 6-shogaol and 6-gingerol, two active components in ginger, on human hepatocarcinoma cells. Mol Nutr Food Res 54:1618–1627

Zhan KY, Yin HZ, Zhang XZ, Xu K (2009) Chemical composition analysis on SEF-C02 extracts of seed ginger and fresh ginger. Food Sci 30:33

Zheng W, Wang SY (2001) Antioxidant activity and phenolic compounds in selected herbs. J Agric Food Chem 49:5165–5170

18

Identification of seeds based on molecular markers and secondary metabolites in *Senna obtusifolia* and *Senna occidentalis*

Renjun Mao[1], Pengguo Xia[2], Zhigui He[1], Yan Liu[3], Fenghua Liu[4], Hongguang Zhao[4], Ruilian Han[1] and Zongsuo Liang[1,2*]

Abstract

Background: *Senna obtusifolia* and *Senna occidentalis* (Leguminosae), whose seeds have similar appearance and chemical constituents, are easily confused in using their seeds. To elucidate the similarities and differences between *S. obtusifolia* seeds and *S. occidentalis* seeds, three molecular markers and high performance liquid chromatography (HPLC) were employed to evaluate the seeds characteristics of these two medicinal herbs.

Results: The results showed that selected 3 ISSR and 7 SCoT primers could distinguish *S. obtusifolia* seeds from *S. occidentalis* seeds based on the specific band and UPGMA dendrogram. ITS2 sequence indicated that the intra-specific similarity of 20 *S. obtusifolia* and 16 *S. occidentalis* was 99.79 and 100.0%, respectively, while the inter-specific similarity between *S. obtusifolia* and *S. occidentalis* was 89.58%. Although phylogenetic analysis revealed that these two species had a close relationship, they were assigned to different branches. HPLC fingerprint results showed that seeds of *S. obtusifolia* and *S. occidentalis* shared some secondary metabolites, but aurantio-obtusin was not detected in *S. occidentalis* seeds which could differentiate *S. obtusifolia* seeds from *S. occidentalis* seeds.

Conclusions: The present study not only compared the seeds characters of *S. obtusifolia* and *S. occidentalis* from molecular and secondary metabolites levels, but also provided a convenient method to identify *S. obtusifolia* seeds and *S. occidentalis* seeds effectively.

Keywords: *S. obtusifolia* seeds, *S. occidentalis* seeds, Molecular marker, ITS2 Sequence, HPLC fingerprint

Background

Senna obtusifolia L. (Leguminosae) whose seeds known as Juemingzi in China, is a famous traditional Chinese medicine and has been regarded as a food and drug dual-purpose material by China Food and Drug Administration (CFDA). *S. obtusifolia* seeds owned various pharmaceutical properties such as eyesight improvement (Yang et al. 2012), blood lipid regulation, hypertension regulation (Li and Guo 2002) and hepatoprotective effect (Kim et al. 2009). These properties make *S. obtusifolia* seeds very popular in China and some other Asian countries. Since the wide applications in pharmaceutical and healthcare industries, the demand of *S. obtusifolia* seeds increased rapidly, and the proportion of its adulterant increased at the same time. Among the adulterants, the most common and indistinguishable one is the seeds of *Senna occidentalis* L., which is an ayurvedic medicinal herb also belong to genus *Senna* Leguminosae family. Several investigations had demonstrated that *S. occidentalis* have antibacterial (Jafri et al. 1999; Saganuwan and Gulumbe 2006), antimutagenic (Jafri et al. 1999), antiplasmodial (Tona et al. 2004) and hepatoprotective activities (Yadav et al. 2010).

*Correspondence: liangzs@ms.iswc.ac.cn
[1] State and Local Joint Research Center of TCM Fingerprint and NP Library, College of Life Sciences, Northwest A&F University, Yangling 712100, Shaanxi, China
Full list of author information is available at the end of the article

Unfortunately, as the non-staple traditional medicine, the research level of *S. obtusifolia* and *S. occidentalis* is limited. The published papers about these two species mainly focus on their pharmacological characteristics (Drever et al. 2008; Ju et al. 2010) and chemical structures (Sob et al. 2008; Li et al. 2010). The authenticity of *S. obtusifolia* seeds is still identified by some traditional methods such as appearance characters or microstructures. The traditional identification methods are largely dependent on subjective experience and feelings, which easily bring confusion in identifying of *S. obtusifolia* seeds and *S. occidentalis* seeds. Even more alarming is that improper use of *S. occidentalis* seeds will cause serious symptoms (Panigrahi et al. 2014; Teles et al. 2015). However, there has no systematic research from the molecular and secondary metabolite levels to identify *S. obtusifolia* seeds and *S. occidentalis* seeds. Therefore, it is imperative to elucidate the similarities and differences between *S. obtusifolia* seeds and *S. occidentalis* seeds, establishing a suitable method to distinguish them.

In the present study, 20 *S. obtusifolia* samples and 16 *S. occidentalis* samples were collected mainly from China. ISSR and SCoT markers, ITS2 sequence and HPLC fingerprint were used to compare the differences between *S. obtusifolia* seeds and *S. occidentalis* seeds, and thereby providing an efficient method to identify them.

Methods

Plant materials

20 *S. obtusifolia* samples and 16 *S. occidentalis* samples were collected during 2012–2015. Detailed information of each sample had been list in Table 1. All samples were identified by Professor Yuejin Zhang (College of Life Sciences, Northwest A&F University) and voucher specimens (Specimens number see Table 1) were deposited in the State and Local Joint Research Center of TCM Fingerprint and Natural Product Library, Northwest A&F University.

DNA extraction

30 sterilized seeds of each sample were placed in illumination incubator to germinate under 28 °C, 2500 lx (8 h/day) and 75% relative humidity conditions for 5 days. Young and healthy leaves from ten individuals of each sample were randomly collected and mixed for genomic DNA extraction. Total genomic DNA was extracted by using New NuClean Plant Genomic DNA Kit (CWBIO, China) according to the manufacturer's instructions. The quality and concentration of DNA was examined by 1% agarose gel electrophoresis and spectrophotometer analysis (NanoDrop 1000, Thermo Scientific).

PCR programs and ITS2 sequencing

ISSR and SCoT PCR reactions were conducted in 20 μL volume containing 10 μL 2 × *Taq* Mix (CWBIO, China),

Table 1 Detailed information of 36 samples used in the present study

No.	Name	Haplotypes	GenBank accessions	Voucher no.	No.	Name	Haplotypes	GenBank accessions	Voucher no.
1	HN-DZ-12[a]	A1	MF061316	TCMNP-1241	19	JS-TZ-14	A2	MF061317	TCMNP-1481
2	HN-ZMD-12	A3	MF085547	TCMNP-1242	20	ZJ-HZ-15	A3	MF085555	TCMNP-1571
3	HN-ZMD-13	A3	MF085550	TCMNP-1341	21	SX-YL-O13	B1	MF085530	TCMNP-0301
4	HN-ZMD-14[a]	A3	MF085548	TCMNP-1441	22	SX-YL-O14	B1	MF085531	TCMNP-0401
5	SX-YL-13	A1	MF067418	TCMNP-1301	23	SX-YL-O15	B1	MF085532	TCMNP-0501
6	SX-YL-14	A1	MF067419	TCMNP-1401	24	SX-SL-O14	B1	MF085533	TCMNP-0402
7	SX-YL-15	A1	MF067420	TCMNP-1501	25	SX-SL-O15	B1	MF085534	TCMNP-0502
8	SX-SL-15	A3	MF085549	TCMNP-1502	26	SX-LN-O14	B1	MF085535	TCMNP-0403
9	SX-WN-14	A1	MF067421	TCMNP-1402	27	HN-DZ-O14	B1	MF085536	TCMNP-0441
10	SX-LN-14	A1	MF067422	TCMNP-1503	28	YN-KM-O14	B1	MF085537	TCMNP-0461
11	SX-LN-15	A1	MF067423	TCMNP-1403	29	JS-YZ-O14	B1	MF085538	TCMNP-0481
12	Vietnam-14	A1	MF067424	TCMNP-14F1	30	JS-HA-O14	B1	MF085539	TCMNP-0482
13	Myanmar-14	A2	MF067425	TCMNP-14F2	31	GX-NN-O15	B1	MF085540	TCMNP-0551
14	SC-CD-14	A3	MF061318	TCMNP-1491	32	GX-BS-O15	B1	MF085541	TCMNP-0553
15	SD-HZ-15	A3	MF085551	TCMNP-1531	33	GX-YL-O15	B1	MF085542	TCMNP-0552
16	HB-AG-15	A3	MF085552	TCMNP-1521	34	GD-MM-O14	B1	MF085543	TCMNP-0431
17	AH-BZ-14	A3	MF085553	TCMNP-1461	35	GD-GZ-O14	B1	MF085544	TCMNP-0432
18	GX-YL-15	A3	MF085554	TCMNP-1551	36	JX-PX-O14	B1	MF085545	TCMNP-0493

HN-ZMD-14 indicated that the sample was collected from Zhumadian city, Henan province in 2014

[a] Indicated these samples were collected from wild field and the other samples were collected from the cultivars fields or experimental fields

8 μL ddH_2O, 1 μL primer (1 mM), 1 μL template DNA (20 ng/μL). 100 ISSR and 85 SCoT primers were based on University of British Columbia and Collard and Mackill (Collard and Mackill 2009), respectively. All the primers were synthesized by Sangon Biological Engineering Technology & Services Company (Shanghai, China). ISSR-PCR reaction started at 94 °C for 5 min, followed by 35 cycles of 94 °C for 45 s, 55 °C for 40 s and 72 °C for 1.5 min, the final extension step at 72 °C was held for 7 min. SCoT-PCR reaction started at 94 °C for 5 min, followed by 35 cycles of 30 s at 94 °C, 1 min at 50 °C and then 1.5 min at 72 °C, with the extension at 72 °C for 7 min.

The PCR reaction for ITS2 was conducted in 50 μL volume containing 25 μL 2 × *Taq* Mix, 19 μL ddH_2O, 2 μL primer forward (1 mM), 5′-ATGCGATACTTGGTGTGAAT-3′, 2 μL primer reverse (1 mM), 5′-GACGCTTCTCCAGACTACAAT-3′ and 2 μL template DNA (20 ng/μL). ITS2 PCR reaction started at 94 °C for 5 min, followed by 40 cycles of 94 °C for 30 s, 56 °C for 30 s and 72 °C for 45 s, the final extension step at 72 °C was held for 10 min (Chen et al. 2010). PCR reactions were performed in a Veriti 96 Thermal Cycler (Applied Biosystems, USA). PCR productions were visualized by agarose gel electrophoresis and sequenced by AuGCT Technology & Services Company (Beijing, China).

Determination of secondary metabolites content

The secondary metabolites were determined by HPLC system (Waters, Milford, USA) equipped with a 1525 binary pump and 2487 Dual λ detector. The column was Waters Sunfire C_{18} reserved-phase column (5 μm, 250 mm × 4.6 mm). The secondary metabolites extraction and chromatography condition were according to the method described in Chinese Pharmacopoeia (Chinese Pharmacopoeia Commission 2015). Standard references of aurantio-obtusin (111900-201504), chrysophanol (110796-201520), physcion (110758-200611), emodin (110756-200110) and aloe-emodin (110795-201308) were purchased from the National Institute for the Control of Pharmaceutical and Biological Products (Beijing, China). Every sample was detected with three technical replicates to reduce the error rate.

Data analysis

Every PCR reaction repeated twice and only unambiguous and repeatable bands were taken into account. NTSYS-pc 2.10e was employed to calculate the similarity coefficient (Rohlf 1993). The dendrograms were generated from similarity matrix data by cluster analysis using unweighted pair group method for arithmetic mean (UPGMA). The raw sequences were assembled by Codon Code Aligner V 4.0.3 (Codon Code Co., USA) based on the Hidden Markov model (HMMer) annotation method (Tamura et al. 2013). The sequence divergence and genetic distance were performed with pairwise calculations under Kimura 2-parameter model by using MEGA 6.06 software. Totally 54 samples, including 20 *S. obtusifolia*, 16 *S. occidentalis*, 15 closely related samples and three *Glycine max* samples, i.e., *Senna italica* CIMAP-C026 (Genbank Accession Number: KY492294), *Senna italica* CIMAP-C033 (KY492295), *Senna tora* CIMAP-C044 (KY492308), *Senna alexandrina* CIMAP-C032 (KY492292), *Senna alexandrina* CIMAP-C039 (KY492293), *Senna uniflora* CIMAP-C043 (KY492305), *Senna italica* (JQ301844), *Senna spectabilis* CIMAP-C027 (KY492297), *Senna spectabilis* CIMAP-C034 (KY492298), *Senna spectabilis* CIMAP-C041 (KY492299), *Senna tora* CIMAP-C037 (KY492307), *Senna tora* CIMAP-C044 (KY492308), *Senna auriculata* CIMAP-C028 (KY492300), *Senna auriculata* CIMAP-C035 (KY492301), *Senna auriculata* CIMAP-C042 (KY492302), *Glycine max* SL2 (KP793718), *Glycine max* 001 (AJ009787) and *Glycine max* Williams (KP793718), were used to construct phylogenetic tree by using Neighbor jointing (NJ) and Maximum likelihood (ML) methods, respectively.

Results

Discrimination ability and polymorphism of ISSR and SCoT markers

Among 100 ISSR primers, 3 primers (Table 2) could generate specific band that distinguish *S. obtusifolia* from *S. occidentalis*. The primer UBC80 generated a specific band about 860 bp in *S. occidentalis* samples which did not observe in *S. obtusifolia* samples (Fig. 1). The primers UBC81 and UBC85 yielded the specific bands, which could distinguish *S. obtusifolia* from *S. occidentalis* (Additional file 1: Figure S1). 3 ISSR primers totally yielded 70 bands among 20 *S. obtusifolia* samples, 51 of which were polymorphic bands with 72.9% polymorphism and yielded 54 bands among 16 *S. occidentalis* samples and 35 of which were polymorphic with 64.8% polymorphism.

7 out of 85 SCoT primers could produce specific bands to distinguish *S. obtusifolia* from *S. occidentalis*. The primer SCoT21 produced a unique band about 2800 bp in *S. occidentalis* which could distinguish *S. obtusifolia* from *S. occidentalis* (Fig. 1). The other 6 SCoT primers also yielded different specific bands which could distinguish these two species (Additional file 1: Figure S1). 7 SCoT primers totally yielded 131 bands among 20 *S. obtusifolia* samples, among which 112 (85.5%) were polymorphic, and produced 92 bands among 16 *S. occidentalis* samples, among which 78 (84.8%) were polymorphic.

Table 2 Genetic polymorphism parameters of 36 samples generated by 3 ISSR and 7 SCoT primers

Primer	Primer sequence	Total bands		Polymorphic bands		PPB (%)	
	(5′→3′)	Sob	Soc	Sob	Soc	Sob	Soc
UBC880	(GA)8YG	29	21	20	15	68.9	71.4
UBC881	GGGT(GGGGT)2G	23	17	17	12	73.9	70.6
UBC885	BDB(CA) 7	18	16	14	8	77.8	50.0
Average	–	23.3	18.0	17.0	11.7	73.5	64.0
Total	–	70	54	51	35	72.9	64.8
SCoT 8	ACCATGGCTACCACCGCA	20	15	18	12	90.0	80.0
SCoT 21	CCATGGCTACCACCGCCT	24	17	20	14	83.3	82.4
SCoT 27	ACAATGGCTACCACTGCC	16	11	15	10	93.8	90.9
SCoT 42	ACCATGGCTACCACCGGC	18	11	14	10	77.8	90.9
SCoT 43	ACCATGGCTACCACCGCC	15	12	11	11	73.3	91.7
SCoT 44	ACGACATGGCGACCCACA	17	13	16	11	94.1	84.6
SCoT 66	ACAATGGCTACCACTAGC	21	13	18	10	85.7	76.9
Average	–	18.7	13.1	16.0	11.1	85.4	85.7
Total	–	131	92	112	78	85.5	84.8

Sob: *S. obtusifolia;* Soc: *S. occidentalis;* PPB: percentage of polymorphic bands

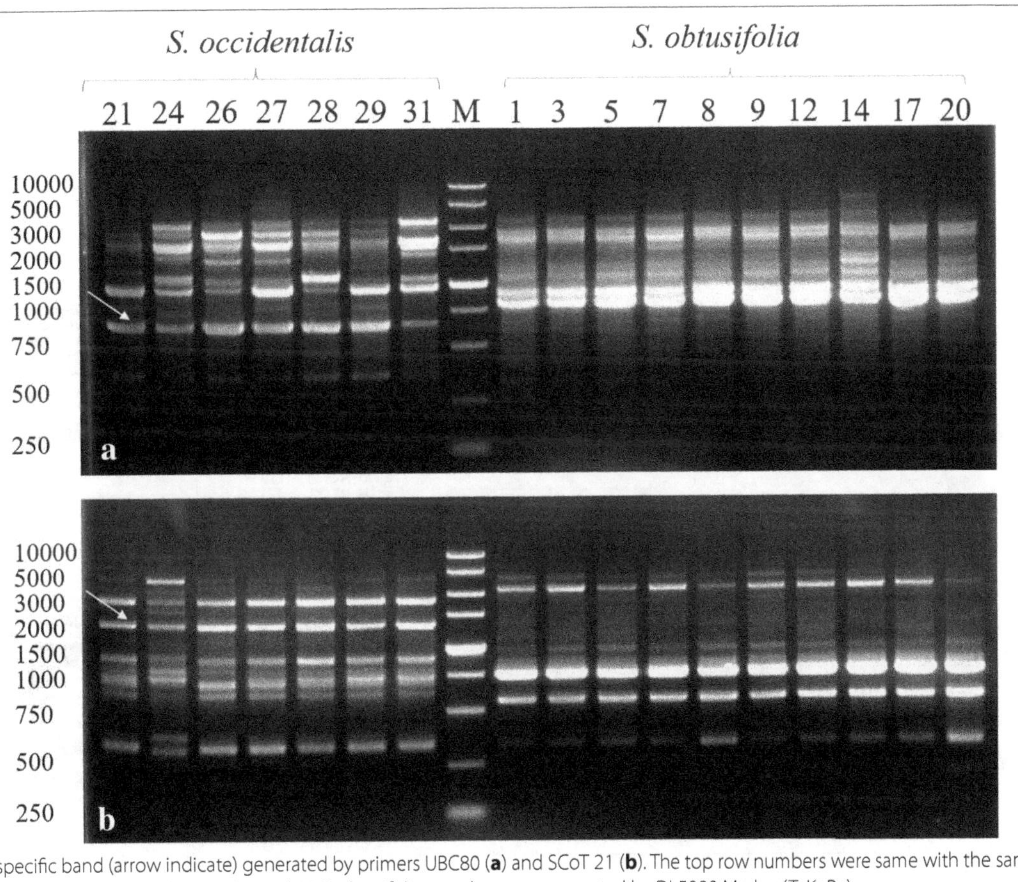

Fig. 1 The specific band (arrow indicate) generated by primers UBC80 (**a**) and SCoT 21 (**b**). The top row numbers were same with the samples number in Table 1. 7 *S. occidentalis* samples and 10 *S. obtusifolia* samples were seperated by DL5000 Marker (TaKaRa)

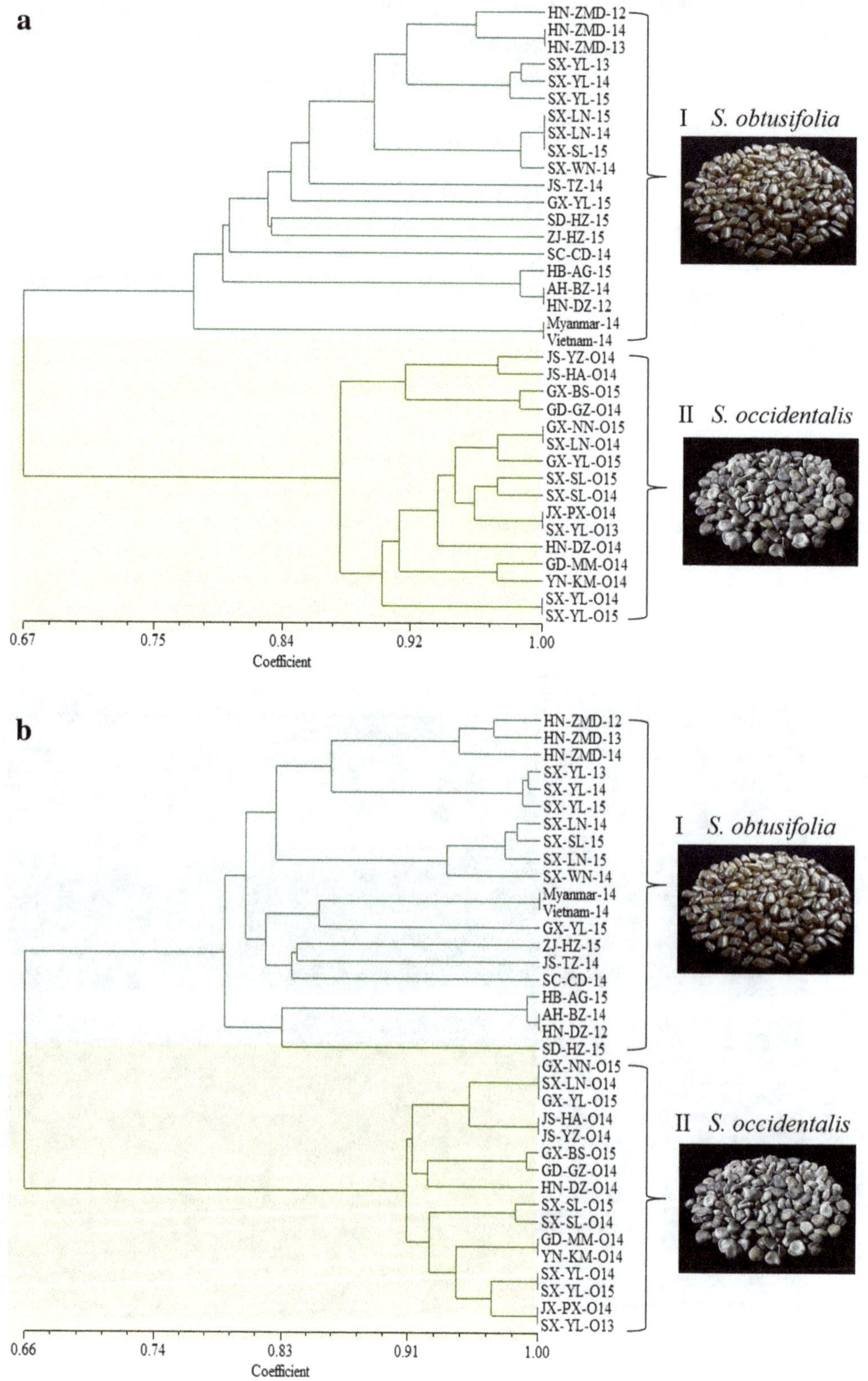

Fig. 2 UPGMA dendrograms generated based on ISSR (**a**) and SCoT (**b**). The cluster I contained 20 *S. obtusifolia* samples and cluster II contained 16 *S. occidentalis* samples, which indicated in blue and pink respectively. The photographs of *S. obtusifolia* and *S. occidentalis* were also exhibited

Clustering analysis

The UPGMA dendrogram generated by ISSR divided 36 samples into two clusters (Fig. 2a). 20 *S. obtusifolia* samples and 16 *S. occidentalis* samples were assigned into cluster I and cluster II, respectively. Most samples were clustered based on their geographic origin. For example, in cluster I, the samples collected from SX-YL and HN-ZMD during 2012–2015 years were clustered together, respectively, and other four samples from Shaanxi province were nested firstly. The samples from different regions did not show a clear cluster pattern. Two samples of Myanmar-14 and Vietnam-14 were clustered and separated from samples of China. Cluster II contained all *S. occidentalis* samples. SX-LN-O14 was clustered with GX-NN-O15, and SX-YL-O13 was clustered with JX-PX-O14. The remaining samples were clustered based on their geographic regions.

The UPGMA dendrogram generated by SCoT also assigned *S. obtusifolia* samples and *S. occidentalis* samples to two different clusters (Fig. 2b). In cluster I, the samples from Shaanxi and Henan showed same cluster pattern with ISSR dendrogram. While the samples of Myanmar-14 and Vietnam-14 were mixed with China samples which were different from ISSR dendrogram. In cluster II, SX-LN-O14 and JX-PX-O14 were grouped with two samples from Guangxi and three samples from Shaanxi, respectively, which was not consist with their origin.

Sequences variations and genetic distance of intra- and inter-specific

ITS2 sequences of 20 *S. obtusifolia* samples and 16 *S. occidentalis* samples were 233 bp. 20 *S. obtusifolia* samples represented 3 haplotypes namely, A1 (included 8 samples, see Table 1), A2 (2 samples) and A3 (10 samples). All 36 ITS2 of *S. obtusifolia* and *S. occidentalis* samples were submitted to GenBank and gained the accession numbers (Table 1). A1 showed the same ITS2 sequences with *Senna obtusifolia* voucher PS1588MT08 (GenBank: GQ434816). A2 and A3 showed the same sequence with voucher PH16960 (KX675057) and PH01988 (KX674796) respectively. Compared to A1, A2 showed a variation at 201 bp with 'A' to 'G' and A3 showed a variation at 216 bp with 'A' to 'C'. ITS2 sequences of 20 *S. obtusifolia* samples exhibited 99.79% similarity (Additional file 2: Figure S2). Compared with the existence *S. occidentalis* ITS2 sequence (JQ301840), no variations were detected among 16 *S. occidentalis* samples (haplotypes B1). Nevertheless, compared to the high similarity value of intra-specific, the inter-specific similarity value of *S. obtusifolia* and *S. occidentalis* was 89.58% (Additional file 2: Figure S2).

The genetic distance of *S. obtusifolia* ranged from 0 to 0.01 with an average of 0.003. This value of *S. occidentalis* was 0. However, the genetic distance between *S. obtusifolia* and *S. occidentalis* ranged from 0.280 to 0.290 with the average of 0.286. The smallest genetic distance was detected between *S. obtusifolia* and *S. tora* (0.06), while the largest genetic distance was observed between *S. obtusifolia* and *Glycine max* (0.80). The results showed that the intra-specific genetic distances (0.003) was lower than the inter-specific genetic distances (0.286) between *S. obtusifolia* and *S. occidentalis*. The largest intra-specific distance was smaller than the distance of inter-specific, which could use as a parameter to distinguish *S. obtusifolia* from *S. occidentalis.*

Phylogenetic analysis

Two phylogenetic trees were constructed using Neighbor Jointing (NJ) and Maximum likelihood (ML) methods, respectively. Two phylogenetic trees showed large similarity with slightly difference (Fig. 3). 15 closely related species samples selected from GenBank were inserted into phylogenetic trees. Two *S. tora* were grouped with *S. obtusifolia* samples in both phylogenetic trees. Another difference was the branch consisted of 3 *Senna spectibilis* samples (KY492297-99) appeared a close evolution relationship with *S. obtusifolia* in NJ-tree (Fig. 3a), while it showed a relative far relationship with *S. obtusifolia* in ML-tree (Fig. 3b). *Senna uniflora* (KY492305) and *Senna italic* (JQ301844) exhibited the different positions in two trees. Despite showing some differences, both phylogenetic trees assigned *S. obtusifolia* samples and *S. occidentalis* samples to different branches (Fig. 3). Being the outgroup, three *Glycine max* samples were clustered together and separated from *Senna* samples. Collectively, the constructed phylogenetic trees could differentiate *S. obtusifolia* from *S. occidentalis* and elucidate the evolution relationship of *Senna* species.

Determination of secondary metabolites

Representative chromatograms of each five samples of *S. obtusifolia* and *S. occidentalis* were shown in Fig. 4. The concentrations of five secondary metabolites of each sample had been listed in Additional file 3: Table S1. The HPLC fingerprints of two species showed some similarities. At the retention times of 29.33, 30.42, 33.11 and 35.23 min, these two species showed the same peaks. However, the peak detected at 12.95 min which was identified as aurantio-obtusin, was observed in *S. obtusifolia* samples only (Fig. 4). The concentrations of five secondary metabolites exhibited a large variation among different samples. The concentration of aurantio-obtusin and chrysophanol among *S. obtusifolia* samples ranged from 0.058 to 0.231% and 0.182 to 0.641%, respectively (Additional file 3: Table S1). Chrysophanol concentration of *S. occidentalis* seeds ranged from 0.125 to 0.315%. Physcion concentration ranged from 0.021 to 0.972 and 0.021 to 0.582% for *S. obtusifolia* and *S. occidentalis*, respectively.

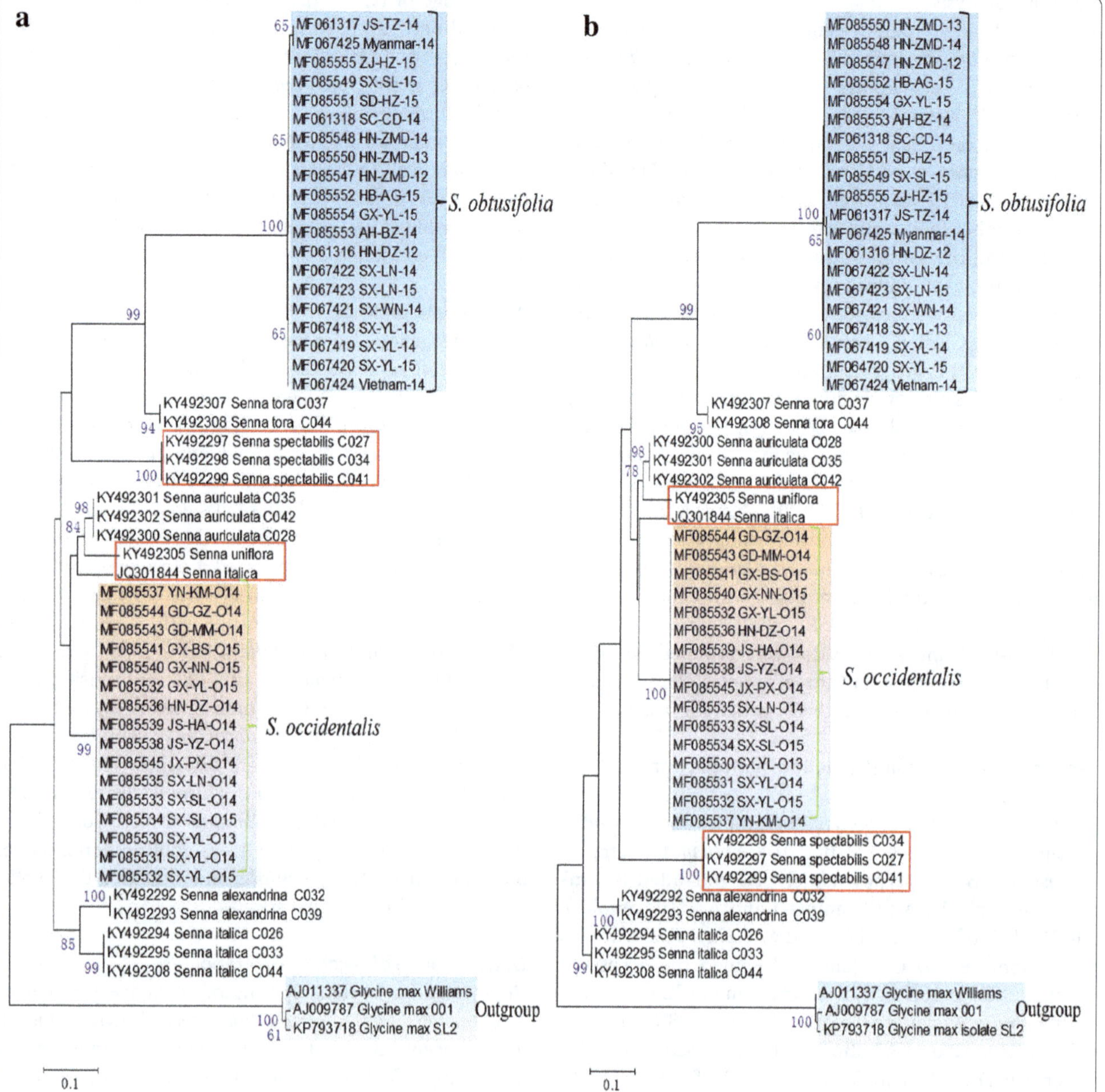

Fig. 3 Neighbour-joining (**a**) and Maximum-likelihood (**b**) phylogenetic trees were constructed based on ITS2 sequence with 1000 replicates. The blue, pink and light blue represented *S. obtusifolia*, *S. occidentalis* and *Glycine max* samples, respectively. The samples differently grouped in two phylogenetic trees were highlighted in the red box

Discussion

Traditional appearance identification offered a direct method to identify medicinal herbs, but this method easily made confusion. It was more difficult to distinguish the real medicinal herb from its adulterants when they were adulterated. In addition, using one identification method only may not be suitable or convenient for all users. Molecular markers, ITS2 sequence and HPLC fingerprint were employed to evaluate the similarities and differences between *S. obtusifolia* seeds and *S. occidentalis* seeds, which could provide a suitable method to identify these two valuable medicinal herbs effectively.

Molecular markers had been widely used to identify the different genotypes of the species or the different ecological types with the same genotype. Several successful studies of plant identification were reported using different molecular marker systems. Based on selected RAPD, ISSR and SSR markers, *Jatropha curcas* samples

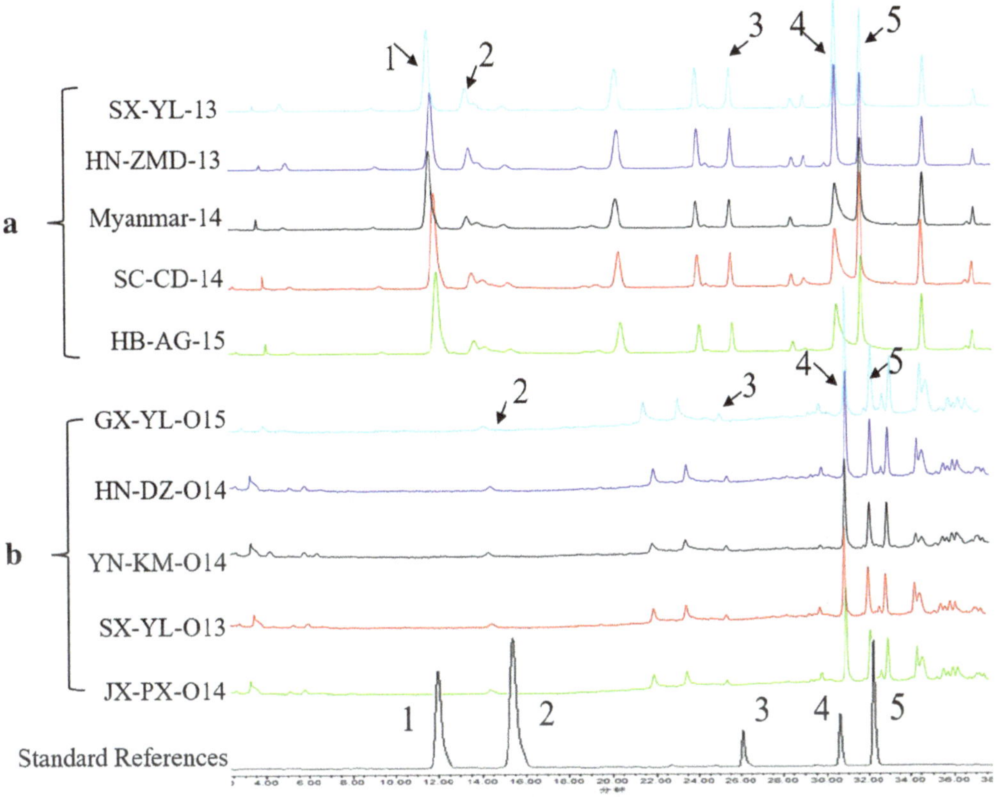

Fig. 4 HPLC fingerprints of five *S. obtusifolia* (**a**) and *S. occidentalis* samples (**b**). Standard references from 1 to 5 represented aurantio-obtusin, emodin, aloe-emodin, chrysophanol and physcion, respectively

from Mexica showed clear distinction with samples from other countries (Pamidimarri et al. 2009). Validation of the primers on *Jatropha curcas* indicated specificity of 3 RAPD primers to toxic genotypes, 5 RAPD and 2 ISSR primers to non-toxic samples and 4 RAPD primers to Mexican samples (Basha and Sujatha 2007). The study on *Olea europaea* indicated that 39 RAPD primers and 12 ISSR primers were able to distinguish 10 cultivars and 6 cultivars, respectively (Brake et al. 2014). In the present study, 3 ISSR primers and 7 SCoT primers could generate specific band to distinguish *S. obtusifolia* from *S. occidentalis*, which present a new method for identifying *S. obtusifolia*. Additionally, the results of molecular markers exhibited that both *S. obtusifolia* and *S. occidentalis* had a high level of genetic diversity and their genetic relationship showed a certain connection with geographical origin. This phenomenon could be explained by the wide cultivation area of *S. obtusifolia* and *S. occidentalis* in China for their medicinal and viewing purposes, which prompted their genetic diversity level at the same time.

The medicinal herbs collected from different regions showed large difference in quality. The adulterants exist is another reason accounting for the unstable quality of medicinal herbs. Identification of medicinal herbs based on morphological characters, microstructure or other traditional methods are lacking of the uniform standard and reliable data, which impede the further use of the medicinal herbs. Chen et al. (2010) proposed that ITS2 is a suitable DNA barcoding for medicinal plants. In recent years, ITS2 is a newly rising method which could effectively solve morphological indistinguishable or complex classification event. Additionally, ITS2 played a pivotal role in promoting the identification authenticity and stability of the medicinal herbs. ITS2 had been successively applied in identifying *Hyoscyamus niger* (Xiong et al. 2016) and *Officinal rhubarb* (Zhou et al. 2017). In this study, 36 samples of *S. obtusifolia* and *S. occidentalis* were collected from different geographic regions and evaluated their differences by ITS2. The sequence similarity of intra-specific was 99.79 and 100.0% for *S. obtusifolia* and *S. occidentalis* accordingly, while the similarity of inter-specific was 89.58%. This results demonstrated that ITS2 was a suitable DNA barcoding for identification of *S. obtusifolia*. NJ-tree separated *S. obtusifolia* and *S. occidentalis* into different branches despite that they had a close relationship. Three *Glycine max* samples formed the outgroup and separated from *Senna* species. To further evaluate the genetic relationship of *Senna* species and offer a more accurate evolution view, ML-tree was also constructed. Showing slight difference with NJ-tree, ML-tree also assigned *S. obtusifolia* and *S. occidentalis* to different branches based on ITS2 sequence. It is no doubt that using two methods to construct phylogenetic trees is helpful in ensuring the accuracy of the results and could provide more information.

Containing different kinds of secondary metabolites was a significant feature to distinguish medicinal herbs from crops or other plants. The concentration of secondary metabolite was also an indispensable parameter to evaluate the quality of the medicinal herbs. The secondary metabolite determination by HPLC can be used as a powerful tool to identify the authenticity of medicinal herbs (Tian et al. 2009; Zhang et al. 2014). In this study, the HPLC fingerprints of *S. obtusifolia* and *S. occidentalis* showed some similarities since they both belong to the genus *Senna* family leguminous. The Chinese pharmacopoeia identified Juemingzi as the dried and ripe seeds of *S. obtusifolia* or *S. tora*, not included *S. occidentalis* seeds. As a famous traditional Chinese medicine, there had already been some HPLC fingerprint reports on *S. obtusifolia* seeds (Wang et al. 2009; Huang et al. 2010). However, to our knowledge, no HPLC fingerprint research was carried out on *S. occidentalis* seeds. In this study, we presented the HPLC fingerprint of *S. occidentalis* seeds for the first time. Notably, aurantio-obtusin was not detected in *S. occidentalis* samples, which was a significant difference from *S. obtusifolia*. Former researches focus on chemical constituents also indicated that *S. occidentalis* seeds did not contain aurantio-obtusin (Yadav et al. 2010). Same to our expectation, the concentration of chrysophanol and several other anthraquinones, i.e., emodin, aloe-emodin and physcion in *S. obtusifolia* seeds were higher than that in *S. occidentalis* seeds. This results implied that *S. occidentalis* seeds maybe not suitable for medicinal use.

There is no doubt that every method has its advantages and deficiencies. Molecular marker is easy to operate and does not need expensive experimental equipment, but the number of available primers is limited. ITS2 has high accuracy and low identification error rate. However, this method requires expensive DNA sequencer and needs to further analyze after obtaining the raw sequences data, which restrict the use of this method more or less. HPLC method is affordable, accurate and could simultaneously determine the concentration of various metabolites in medicinal materials. However, the processing of the sample preparation is cumbersome and the instrument operation has a certain degree of difficulty. In this study, we presented three methods to distinguish *S. obtusifolia* from *S. occidentalis*, aiming to provide a suitable method for users from different industrials to distinguish these two species effectively.

Conclusions

In summary, three approaches employed in this study could effectively identify *S. obtusifolia* seeds and *S. occidentalis* seeds. Selected 3 ISSR and 7 SCoT primers distinguished *S. obtusifolia* from *S. occidentalis* based on UPGMA dendrogram and agarose gel electrophoresis. ITS2 results showed that the intra-specific similarity of *S. obtusifolia* and *S. occidentalis* were 99.79 and 100.0%, respectively. While the inter-specific similarity between *S. obtusifolia* and *S. occidentalis* was 89.58%. Phylogenetic analysis assigned *S. obtusifolia* and *S. occidentalis* to different branches. HPLC fingerprints showed that two species shared some secondary metabolites, but *S. occidentalis* seeds did not contain aurantio-obtusin which could differentiate *S. obtusifolia* from *S. occidentalis*.

Abbreviations

ISSR: inter-simple sequence repeat; SCoT: target start codon polymorphism; ITS2: internal transcribed sequence 2; Sob: *Senna obtusifolia*; Soc: *Senna occidentalis*; PPB: percentage of polymorphic bands; HPLC: high performance liquid chromatography.

Authors' contributions

MRJ and LZS conceived the research; MRJ, XPG, ZHG and LFH collected the samples. MRJ and HZG performed the experimental work; MRJ, HZG, LY and HRL analyzed the data; MRJ, XPG and LZS wrote the manuscript. All authors read and approved the final manuscript.

Author details

State and Local Joint Research Center of TCM Fingerprint and NP Library, College of Life Sciences, Northwest A&F University, Yangling 712100, Shaanxi, China. [2] College of Life Sciences, Zhejiang Sci-Tech University, Hangzhou 310018, Zhejiang, China. [3] Tianjin Tasly Modern TCM Resources Co. Ltd., Tianjin 300400, China. [4] Shaanxi Tasly Plants Pharmaceutical Co. Ltd., Shangluo 726000, Shaanxi, China.

Acknowledgements

We thank Sheng Yang and Fuping Wang (Shaanxi Tasly Plants Pharmaceutical Co. Ltd.) for their help in collecting samples.

Competing interests

The authors declare that they have no competing interests.

Funding

This work was financially supported by China National Twelfth "Five Year" Science and Technology Support Project (2015BAC01303) and the Major Industrial Cluster Project of Shaanxi Province (2012KTCL02-07).

References

Basha SD, Sujatha M (2007) Inter and intra-population variability of *Jatropha curcas* (L.) characterized by RAPD and ISSR markers and development of population-specific SCAR markers. Euphytica 156:375–386

Brake M, Migdadi H, Al-Gharaibeh M, Ayoub S, Haddad N, Oqlah AE (2014) Characterization of Jordanian olive cultivars (*Olea europaea* L.) using RAPD and ISSR molecular markers. Hortic Sci 176:282–289

Chen SL, Yao H, Han JP, Liu C, Song JY, Shi LC, Zhu YJ, Ma XY, Gao T, Pang XH, Luo K, Li Y, Li XW, Jia XC, Lin YL, Leon CJ (2010) Validation of the ITS2 region as a novel DNA barcode for identifying medicinal plant species. PLoS ONE 5:e8613

Chinese Pharmacopoeia Commission (2015) Pharmacopoeia of People's Republic of China, Part 1. Chemical Industry Press, China, p 145 **(In Chinese)**

Collard BCY, Mackill DJ (2009) Start codon targeted (SCoT) polymorphism: a simple, novel DNA marker technique for generating gene-targeted markers in plants. Plant Mol Biol Rep 27:86–93

Drever BD, Anderson WGL, Riedel G, Kim DH, Ryu JH, Choi D, Platt B (2008) The seed extract of *Cassia obtusifolia* offers neuroprotection to mouse hippocampal cultures. J Pharmacol Sci 107:380–392

Huang XP, Zhong GY, Zhang XM (2010) Determination of aurantio-obtusin and chrysophanol in Cassiae Semen by HPLC. Chin J Chin Mater Med 35:2065–2067 **(In Chinese)**

Jafri MA, Subhani MJ, Javed K, Singh S (1999) Hepatoprotective activity of leaves of *Cassia occidentalis* against paracetamol and ethyl alcohol intoxication in rats. J Ethnopharmacol 66:355–361

Ju MS, Kim HG, Choi JG, Ryu JH, Hur J, Kim YJ, Oh MS (2010) Cassiae semen, a seed of *Cassia obtusifolia*, has neuroprotective effects in Parkinson's disease models. Food Chem Toxicol 48:2037–2044

Kim DH, Kim S, Jung WY, Park SJ, Park DH, Kim JM, Cheong JH, Ryu JH (2009) The neuroprotective effects of the seeds of *Cassia obtusifolia* on transient cerebral global ischemia in mice. Food Chem Toxicol 47:1473–1479

Li XE, Guo BL (2002) Effect of protein and anthraquinone glucosides from cassia seed on serum lipid of hyperlipidemia rats. Chin J Chin Mater Med 5:374–376 **(In Chinese)**

Li S, Di Y, Wang Y, Tan C, Fang X, Zhang Y, Zheng Y, Li L, He H, Li S, Hao X (2010) Anthraquinones and lignans from *Cassia occidentalis*. Helv Chim Acta 93:1795–1802

Pamidimarri DV, Singh S, Mastan SG, Patel J, Reddy MP (2009) Molecular characterization and identification of markers for toxic and non-toxic varieties of *Jatropha curcas* L. using RAPD, AFLP and SSR markers. Mol Biol Rep 36:1357–1364

Panigrahi G, Tiwari S, Ansari KM, Chaturvedi RK, Khanna VK, Chaudhari BP, Vashistha VM, Raisuddin S, Das M (2014) Association between children death and consumption of *Cassia occidentalis* seeds: clinical and experimental investigations. Food Chem Toxicol 67:236–248

Rohlf FJ (1993) NTSYS.PC. Numerical taxonomy and multivariate analysis system. Version 2.11Q. Applied biostatistics Inc, New York

Saganuwan AS, Gulumbe ML (2006) Evaluation of in vitro antimicrobial activities and phytochemical constituents of *Cassia occidentalis*. Animal Res Int 3:566–569

Sob SVT, Wabo HK, Tane P, Ngadjui BT, Ma D (2008) A xanthone and a polyketide derivative from the leaves of *Cassia obtusifolia* (Leguminosae). Tetraheron 64:7999–8002

Tamura K, Stecher G, Peterson D, Filipski A, Kumar S (2013) MEGA6: molecular evolutionary genetics analysis version 6.0. Mol Biol Evol 30:2725–2729

Teles AV, Fock RA, Gorniak SL (2015) Effects of long-term administration of *Senna occidentalis* seeds on the hematopoietic tissue of rats. Toxicon 108:73–79

Tian R, Xie P, Liu H (2009) Evaluation of traditional Chinese herbal medicine: Chaihu (*Bupleuri Radix*) by both high-performance liquid chromatographic and high-performance thin-layer chromatographic fingerprint and chemometric analysis. J Chromatogr A 1216:2150–2155

Tona L, Cimanga RK, Mesia K, Musuamba CT, De Bruyne T, Apers S, Hernans N, Miert SV, Pieters L, Totté J, Vlietinck AJ (2004) In vitro antiplasmodial activity of extracts and fractions from seven medicinal plants used in the Democratic Republic of Congo. J Ethnopharmacol 93:27–32

Wang WY, Zhao Q, Zhang TJ, Zhu HJ, Li Y (2009) HPLC fingerprint and chemical pattern recognition of Semen Cassiae. Chin Tradit Herbal Drugs 10:1638–1641 **(In Chinese)**

Xiong C, Hu ZG, Tu Y, Liu HG, Wang P, Zhao MM, Shi YH, Wu L, Sun W, Chen SL (2016) ITS2 barcoding DNA region combined with high resolution melting (HRM) analysis of *Hyoscyami Semen*, the mature seed of *Hyoscyamus niger*. Chin J Nat Med 14:898–903

Yadav JP, Arya V, Yadav S, Panghal M, Kumar S, Dhankhar S (2010) *Cassia occidentalis* L.: a review on its ethnobotany, phytochemical and pharmacological profile. Fitoterapia 81:223–230

Yang J, Ye H, Lai H, Li S, He S, Zhong S, Chen L, Peng A (2012) Separation of anthraquinone compounds from the seed of *Cassia obtusifolia* L. using recycling counter-current chromatography. J Sep Sci 35:256–262

Zhang R, Chen J, Shi Q, Li Z, Peng Z, Zheng L, Wang X (2014) Quality control method for commercially available wild Jujube leaf tea based on HPLC characteristic fingerprint analysis of flavonoid compounds. J Sep Sci 37:45–52

Zhou Y, Du X, Zheng X, Huang M, Li Y, Wang X (2017) ITS2 barcode for identifying the officinal rhubarb source plants from its adulterants. Biochem Syst Ecol 70:177–185

19

Whole plastid transcriptomes reveal abundant RNA editing sites and differential editing status in *Phalaenopsis aphrodite* subsp. *formosana*

Ting-Chieh Chen[1,2], Yu-Chang Liu[1,2], Xuewen Wang[3], Chi-Hsuan Wu[1], Chih-Hao Huang[1,2] and Ching-Chun Chang[1,2*]

Abstract

Background: RNA editing is a process of post-transcriptional level of gene regulation by nucleotide modification. Previously, the chloroplast DNA of Taiwan endemic moth orchid, *P. aphrodite* subsp. *formosana* was determined, and 44 RNA editing sites were identified from 24 plastid protein-coding transcripts of leaf tissue via RT-PCR and then conventional Sanger sequencing. However, the RNA editing status of whole-plastid transcripts in leaf and other distinct tissue types in moth orchids has not been addressed. To sensitively and extensively examine the plastid RNA editing status of moth orchid, RNA-Seq was used to investigate the editing status of whole-plastid transcripts from leaf and floral tissues by mapping the sequence reads to the corresponding cpDNA template. With the threshold of at least 5% C-to-U or U-to-C conversion events observed in sequence reads considered as RNA editing sites.

Results: In total, 137 edits with 126 C-to-U and 11 U-to-C conversions, including 93 newly discovered edits, were identified in plastid transcripts, representing an average of 0.09% of the nucleotides examined in moth orchid. Overall, 110 and 106 edits were present in leaf and floral tissues, respectively, with 79 edits in common. As well, 79 edits were involved in protein-coding transcripts, and the 58 nucleotide conversions caused the non-synonymous substitution. At least 32 edits showed significant (≧20%) differential editing between leaf and floral tissues. Finally, RNA editing in *trnM* is required for the formation of a standard clover-leaf structure.

Conclusions: We identified 137 edits in plastid transcripts of moth orchid, the highest number reported so far in monocots. The consequence of RNA editing in protein-coding transcripts mainly cause the amino acid change and tend to increase the hydrophobicity as well as conservation among plant phylogeny. RNA editing occurred in non-protein-coding transcripts such as tRNA, introns and untranslated regulatory regions could affect the formation and stability of secondary structure, which might play an important role in the regulation of gene expression. Furthermore, some unidentified tissue-specific factors might be required for regulating RNA editing in moth orchid.

Keywords: Plastids, Plastid gene expression, RNA editing, Moth orchids, *Phalaenopsis aphrodite*

Background

RNA editing represents a process of post-transcriptional level of gene regulation by modification of the nucleotide sequence. Two distinct types of RNA editing are identified: insertion/deletion and conversion/substitution. In plant organelles, most of the RNA editing causes C-to-U and rarely U-to-C conversions. The consequence of RNA editing occurring at first or second codon positions can increase proteomic diversity but usually leads to the conservation of amino acid sequences among plant phylogeny. Some RNA editing events could result

*Correspondence: chingcc@mail.ncku.edu.tw
[1] Institute of Biotechnology, National Cheng Kung University, Tainan 701, Taiwan
Full list of author information is available at the end of the article

in the creation of translation initiation or termination codons or repair internal stop codons (Bock 2000; Fiebig et al. 2004; Wolf et al. 2004). Several events occurring in non-coding regions such as tRNA, 5′/3′ untranslated regulatory regions (UTR) or introns might affect RNA stability or splicing by modifying the secondary structure (Drescher et al. 2002; Farre et al. 2012; Zeng et al. 2007). The significance of RNA editing in regulating plant organellar gene expression has been demonstrated, with the alteration of some editing sites having deleteriously effects on plant growth, development, and fertility (Hammani and Giege 2014).

The current model for plant organellar RNA editing requires at least three components: the *cis*-acting elements recognized by trans-acting factors, subsequently recruiting the editing enzyme to catalyze the nucleotide conversion (Takenaka et al. 2012). The *cis*-elements include approximately 20–40 nucleotides (−30/+10) spanning the editing site, and the upstream 20 nucleotides are more important than downstream nucleotides (Bock 2000; Hammani and Giege 2014; Hirose and Sugiura 2001; Yagi et al. 2013); however, the *cis*-elements surrounding organellar editing sites are not generally conserved. The site-specific *cis*-elements are recognized by an editosome that consists of at least four nuclear-encoded protein factors (Sun et al. 2016). The pentatricopeptide repeat (PPR) protein is a large family with more than 400 members in most species of land plants (Barkan and Small 2014). The CRR4 protein, required for site-specific editing at the initiation codon of chloroplast *ndhD* transcripts, was the first identified PPR protein in Arabidopsis (Kotera et al. 2005). In addition, another group of small family proteins, multiple organellar RNA editing factors or RNA editing-interacting proteins (MORF/RIP), is required for plant organelle RNA editing (Bentolila et al. 2012, 2013; Takenaka et al. 2012, 2013, 2014). The MORF/RIP proteins could interact with PPR proteins and might serve as a bridge to connect PPR proteins and editing enzymes, but the actual function is not clear (Bentolila et al. 2012; Takenaka et al. 2013). Furthermore, a zinc finger protein, organelle zinc finger 1 (OZ1), directly interacts with organelle RNA recognition motif-containing protein 1 (ORRM1), which in turn could bind to RIPs, is also required for some C targets (Sun et al. 2013, 2015). The editing enzyme responsible for deamination or amination remains unidentified. Because of a highly conserved motif resembling the zinc-binding active site of nucleotide deaminases and essential for editing, the DYW motif of PPR protein was hypothesized to be responsible for the catalytic reaction but is still debated (Boussardon et al. 2014; Takenaka et al. 2014; Yagi et al. 2013; Wagoner et al. 2015).

The organellar RNA editing sites have been investigated in more than a dozen plant species by traditional Sanger methods, through sequencing cDNA and comparing with corresponding genomic DNA. In plastids of seed plants, the number of editing sites is relatively constant, 21 to 44 which account for less than 0.07% of genome (Sasaki et al. 2003, 2006; Zeng et al. 2007), and mostly occur in protein-coding transcripts. However, the number of mitochondrial editing sites is abundant, from 189 to 635, but varies among species (Grimes et al. 2014; Picardi et al. 2010; Sloan et al. 2010). The editing efficiency of plastid transcripts varies among ecotypes, tissue types, developmental stages and environmental factors (Bentolila et al. 2005; Tseng et al. 2013). Recently, by applying next-generation sequencing (NGS), organellar RNA editing sites have been identified more rapidly and economically and have been extensively examined at the whole transcriptome level (Bentolila et al. 2013; Grimes et al. 2014; Picardi et al. 2010). Besides the fast and large-scale detection of RNA editing sites, NGS technology has provided depth of coverage (DOC) per reference nucleotide and qualities of base call. It could overcome the existing limitations of standard Sanger methodology to sensitively revealing previously undiscovered tissue-specific or low-level partial editing sites. For instance, NGS revealed many new editing sites in the mitochondria of Arabidopsis and tobacco from leaf and/or floral tissues (Bentolila et al. 2013; Grimes et al. 2014; Picardi et al. 2010). In addition, re-addressing RNA editing status in the leaf chloroplasts of Arabidopsis and *pnp* mutant by NGS revealed several novel editing sites (Bentolila et al. 2013; Ruwe et al. 2013).

The Orchidaceae family contains approximately 25,000 species and represents one of the most diverse families in flowering plants (Dressler 2005). The *Phalaenopsis* (moth orchids) genus comprises approximately 66 endemic species (Christenson 2001; Lin et al. 2016). Moth orchids are among the top-traded blooming potted plants in the world, with more than 32,000 hybrids bred and registered in the Royal Horticultural Society (Lin et al. 2015a, b; Liu et al. 2016). The chloroplast DNAs (cpDNAs) of two endemic moth orchids, *P. aphrodite* subsp. *formosana* and *P. equestris*, have been determined (Chang et al. 2006; Jheng et al. 2012); 44 plastid RNA editing sites were identified from 24 protein-coding transcripts of leaf tissues in *P. aphrodite* subsp. *formosana* by conventional Sanger sequencing (Zeng et al. 2007); and 43 editing sites were bioinformatically predicted in *P. equestris* (Jheng et al. 2012). However, the RNA editing status of whole-plastid transcripts in leaf and other distinct tissue types in moth orchids has not been addressed.

In this study, we used NGS for transcriptome analysis to extensively study the RNA editing status in whole-plastid transcripts from leaf and floral tissues of *P. aphrodite* subsp. *formosana*. Our results revealed 93 new plastid edits and significantly differential editing status between the two tissue types.

Methods

Plant material and RNA extraction

The seedlings of 1.5-year-old *P. aphrodite* subsp. *formosana* cultivar TS97 were purchased from TaiSugar Corporation (Tainan, Taiwan). Approximately 50–100 g tissues from leaves or flowers was used to isolate plastids as described (Hrubec et al. 1985; Schwitzguebel and Siegenthaler 1984) with slight modification. Plastid-enriched RNA was extracted with Trizol reagent (Omic Bio, Taiwan). To eliminate DNA contamination, RNA samples were treated with DNase by use of the RapidOut DNA removal kit (Thermo, USA). The RNA quality was checked by PCR and RT-PCR before RNA-sequencing (RNA-Seq).

NGS

Plastid-enriched RNA from leaf and floral tissues underwent NGS with Ion Proton and Illumina Hiseq 2000 platforms, respectively (Yourgene Bioscience, Taiwan). In brief, ribosomal RNA was depleted from total RNA by using RiboMinus™ eukaryote kit (Life technologies, USA), and subsequently RNA was fragmented using RNase III. The fragmented RNA was then hybridized and ligated with adaptor mixture. Subsequently, the reverse transcription was performed. The cDNA was amplified using Platinum PCR SuperMix High Fidelity reaction mix, and then the library was sequenced. In total, two independent libraries from leaf and one from floral tissue were constructed and sequenced (Additional file 1: Table S1). Short-sequenced reads were trimmed with a minimal of 35-bp read and error probability <0.05. Approximately 49,722,976 reads with average 114.9 bp and 29,726,302 reads with average 111.7 bp were obtained from two independent libraries of leaf tissues, respectively, and 21,207,630 reads with average 95.8 bp in length were obtained from floral tissues (Additional file 1: Table S1). The raw sequence reads data were submitted to the Sequence Read Archive (http://trace.ncbi.nlm.nih.gov/Traces/sra/) with BioSample Accession Numbers SRR4996537, SRR4098109 and SRR4098702.

Determination of RNA editing sites and gene expression profile

Transcriptome analysis involved use of CLC Genomic Workbench 7.5.1 (CLC Bio, Aarhus, Denmark) as described with modification (Suzuki et al. 2013). In brief, quality control was adjusted with the mapping parameters of length fraction 0.98 and similarity 0.98 to eliminate undesirable fragment reads. The sequence reads were mapped to the cpDNA (Accession AY916449) template of *P. aphrodite* (Chang et al. 2006). The low frequency variant detection parameters were set to 5% for required significance, and the minimum coverage, minimum count and minimum frequency was set to 10, 2, and 5% respectively. Then, the position of RNA editing sites, total read counts and coverage depth were identified. The level of nucleotide conversion for each site was calculated according to the number of reads with nucleotide conversion divided by number of total reads, and frequency >5% of C-to-U or U-to-C conversions at specific nucleotide positions was defined as an RNA editing site as described in Wang et al. (2015). In addition, we manually inspect for the edits located in the regions of homopolymer runs of more than five to eliminate potentially cause by sequencing errors. Reads per kilobase of exon model per million mapped reads (RPKM) values was measured to estimate the gene expression profile as described (Mortazavi et al. 2008). To compare and validate the obtained RNA editing sites from flower tissue, we alternatively conducted a bioinformatic analysis using RES-scanner (Wang et al. 2016). In brief, the quality of RNA-seq reads from flower tissue was checked, and then trimmed using Trimmomatic (Bolger et al. 2014) version 0.33 with the following parameters: CROP:90 HEADCROP:10 LEADING:30 TRAILING:30 SLIDINGWINDOW:4:20 MINLEN:50. Subsequently, the RNA-seq reads were mapped to the reference cpDNA using RES-scanner (Wang et al. 2016) with the parameters of Phred-scaled base quality score cutoff 30 and mapping quality score cutoff 39. The RNA editing sites were filtered, and then defined the same as described above.

Prediction of RNA structure

The tRNA structure was predicted by tRNA-scanning (Schattner et al. 2005). The secondary structure of non-coding transcripts of 100 nt in length with 50 nt extending from both upstream and downstream of indicated editing sites were predicted by using CLC Genomic Workbench 7.5.1.

Results

Plastid RNA edits in moth orchid

Previously, 44 RNA edits were identified from plastid protein-coding transcripts of leaf tissues in *P. aphrodite* subsp. *formosana* via RT-PCR and then conventional Sanger sequencing (Zeng et al. 2007). To sensitively and extensively examine the plastid RNA editing status of moth orchid, RNA-Seq was used to investigate the editing status of whole-plastid transcripts from leaf and floral tissues by mapping the sequence reads to the corresponding cpDNA in *P. aphrodite* subsp. *formosana*. Two independent libraries from leaf and one from floral tissue were sequenced (Additional file 1: Table S1). We mapped 805,063 (leaf-1), 1,751,673 reads (leaf-2) and 7,177,146 reads (flower) to the chloroplast DNA, which account for 1.5, 5.9 and 33% of total reads, respectively. The average genomic coverage is 591 (leaf-1), 1174 (leaf-2) and 4645 (flower) folds, respectively (Additional file 1: Table S1). The NGS libraries with genome coverage more than 1000 folds were further used for RNA editing analysis. With the threshold of at least 5% C-to-U or U-to-C conversion events observed in sequence reads considered as RNA editing sites, 137 RNA edits with 126 C-to-U and 11 U-to-C conversions, which include 93 newly discovered edits, were identified in plastid transcripts, representing an average of 0.09% of the nucleotides examined in moth orchid (Table 1; Additional file 1: Table S2). Overall, 110 and 106 edits were present in leaf and floral tissue, respectively; 79 edits, including 44 previously identified edits (Zeng et al. 2007), were commonly present in both leaf and floral tissues, whereas 31 and 27 edits were specific to leaf or floral tissues, respectively (Table 1). According to the edited proportion of each nucleotide in sequence reads with a C-to-U or U-to-C conversion after mapping to the corresponding DNA template over the total reads in that nucleotide position, the efficiency of RNA editing for each edit was further classified into five groups. Full editing was defined as >90% efficiency, high partial editing 60–90% efficiency, medium partial editing 40–60% efficiency, low partial editing 20–40% efficiency, and poor partial editing 5–20% efficiency. Overall, 10 (9.1%) and 18 (17%) fully edited sites were present in leaf and floral tissue, respectively; 26 (23.6%) and 30 (28.3%) high partially edited sites; 15 (13.6%) and 11 (10.4%) medium partially edited sites; 14 (12.7%) and 6 (5.7%) low partially edited sites; and 45 (40.9%) and 41 (39.6%) poor partially edited sites (Additional file 1: Figure S1, Table S2).

Table 1 Summary of plastid RNA edits in *Phalaenopsis* orchid

	Leaf	Flower	Common[a]	Total	Reported[b]
Protein coding transcripts	67	69	57	79	43
1st codon position	9	10	7	12	4
2nd codon position	47	51	43	55	39
3rd codon position	11	8	7	12	0
Creation of start codon	2	2	1	3	1
Creation of stop codon	4	5	3	6	0
Nonsynonymous substitution	50	54	46	58	42
Synonymous substitution	11	8	7	12	1
Non-protein coding transcripts	43	37	22	58	1
tRNA	1	1	1	1	0
rRNA	0	0	0	0	0
Intron	4	6	2	8	0
Intergenic spacer (IGS)	38	30	19	49	1
Total	110	106	79	137	44

[a] Common RNA edits in both leaf and floral with editing status >5%

[b] Previously reported in Zeng et al. (2007)

RNA editing in protein-coding transcripts

In total, 42 protein-coding transcripts were responsible for 79 edits, and 67 and 69 edits were present in leaf and floral tissues, respectively, with 57 edits in common (Table 1). In particular, the *psbN* and *petL* transcripts had the highest density of RNA editing sites in both tissues, with more than 1.5% of nucleotides having various levels of RNA editing. In contrast, *ycf1*, *rpoC2* and *psaB* transcripts had the lowest density (Additional file 1: Figure S2). However, 17 edits including two newly identified sites (*rpoB*-55; *rpoC1*-1638) from the *rpo* transcripts showed the highest number among functional gene categories, and three edits (*rpoB*-55, *rpoC1*-203, *rpoC1*-1638) were unique to moth orchids as compared with 17 other species of seed plants (Additional file 1: Tables S2 and S3). Among the 79 edits that involved codons, 12 (15%) were in the first position and 55 (70%) in the second position, which resulted in 12 (15%) synonymous substitutions and 67 (85%) nonsynonymous substitutions (Table 1 and Additional file 1: Figure S3A, B). This result is consistent with previous reports regarding genome-wide RNA editing across widely divergent taxa, showing a bias in favor of second codon-position edits (Zeng et al. 2007). The most frequently edited codon was Ser converted to Leu (28%), followed by Ser to Phe (18%), and Pro to Leu (15%) resulting from nonsynonymous substitution (Additional file 1: Figure S3C). In addition, the consequence of amino acid substitution results in the increase of hydrophobicity (Additional file 1: Figure S3D).

Seven edits in six protein coding transcripts having relatively high editing efficiency (>40%) were previous unidentified in moth orchid (Additional file 1: Table S2). Two edits, *petL*-56 and *rps12*-221 are in high editing level (>83%) both in leaf and floral tissues. Editing of the corresponding orchid *petL*-56 causing the conversion of Ser to Leu was not reported in other eight plant species which they encode Pro or Ser in the genome except rice and cycas already coding for Leu (Fig. 1a). The orchid *rps12*-221 (S → L) edit also occurred in *Amborella* (Hein et al. 2016), and the consequence of nucleotide substitution increase the amino acid conservation among plant species (Fig. 1b). Three edits *psbN*-29, *psbN*-30 and *ccsA*-336 are in medium editing level (45–54%) in leaf, but poor (8–9%)

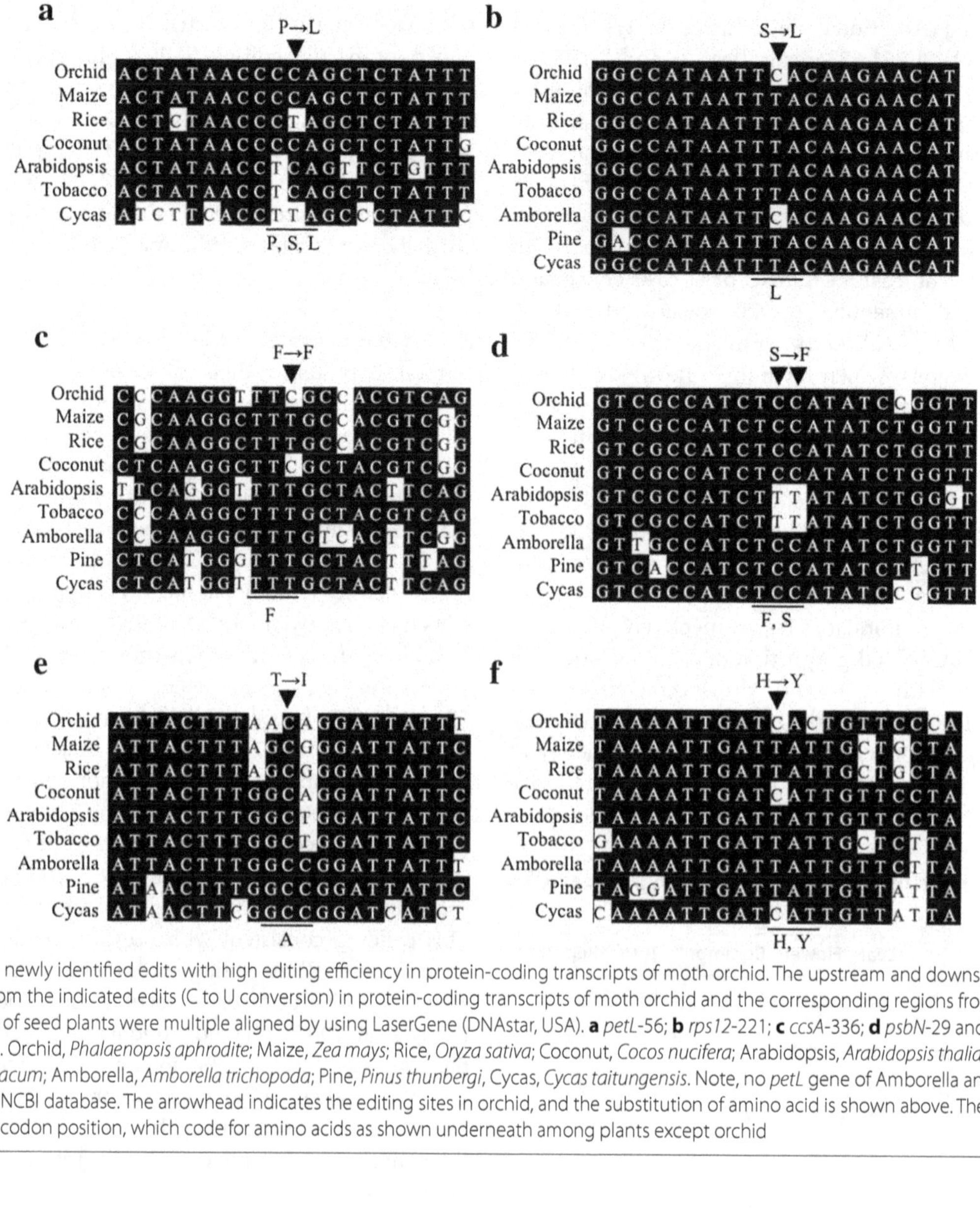

Fig. 1 Seven newly identified edits with high editing efficiency in protein-coding transcripts of moth orchid. The upstream and downstream 10 nucleotide from the indicated edits (C to U conversion) in protein-coding transcripts of moth orchid and the corresponding regions from other eight species of seed plants were multiple aligned by using LaserGene (DNAstar, USA). **a** *petL*-56; **b** *rps12*-221; **c** *ccsA*-336; **d** *psbN*-29 and -30; **e** *petG*-56; **f** *rps3*-583. Orchid, *Phalaenopsis aphrodite*; Maize, *Zea mays*; Rice, *Oryza sativa*; Coconut, *Cocos nucifera*; Arabidopsis, *Arabidopsis thaliana*; Tobacco, *Nicotiana tabacum*; Amborella, *Amborella trichopoda*; Pine, *Pinus thunbergi*, Cycas, *Cycas taitungensis*. Note, no *petL* gene of Amborella and Pine was annotated in NCBI database. The arrowhead indicates the editing sites in orchid, and the substitution of amino acid is shown above. The underline indicates the codon position, which code for amino acids as shown underneath among plants except orchid

in flower. Editing of psbN-29 and psbN-30 (S → F) in moth orchid tend to increase the amino acid conservation with tobacco and Arabidopsis, and ccsA-336 (F → F) is a silent edit (Fig. 1c, d). In contrast, petG-56 and rps3-583 are in high editing level (64–72%) in flower, but relatively low (30–35%) in leaf. Editing of petG-56 (T → I) result in the increase of amino acid hydrophobicity in moth orchid, and the corresponding site encoding Ala was not edited for other eight species of seed plant (Fig. 1e). In contrast, the editing of rps3-583 (H → Y) in moth orchid tends to increase the amino acid conservation among plants except coconut and cycas (Fig. 1f).

One medium (42%) or high partial editing (71%) was involved in the creation of the *rpl2* start codon in leaf and floral tissue, respectively. In contrast, six C-to-U conversions could result in stop codons (Table 1), but their conversion efficiency was poor (Additional file 1: Table S2). For instance, the C-to-U conversion of *ndhE*-106 and *rpl20*-40 resulted in stop codons, but the editing efficiency was poor (5–20%) in both leaf and floral tissues. In addition, *ccsA*-652 and *ycf2*-3868 showed poor partial editing (11–16%) in leaf and much less editing level (0–3%) in floral tissue. In contrast, *ndhB*-106, *rpoB*-55 and *ycf4*-163 showed poor partial editing (5–10%) for the creation of a stop codon in floral tissue, but the level was much less in leaf (1–4%).

RNA editing status in transcripts of two long conserved open reading frames, *ycf1* and *ycf2*, have not been investigated in moth orchid. We found one edit and six edits for *ycf1* and *ycf2* transcripts, respectively (Additional file 1: Table S2), but the editing efficiency was poor (5–23%) in both leaf and floral tissues. *Ndh* genes, encoding the subunits of NADH dehydrogenase complex, have various degrees of nucleotide insertion/deletion and were non-functional pseudogenes in moth orchids (Chang et al. 2006; Jheng et al. 2012). Previously, only one edit was found in the *ndhB* transcripts (1977) (Zeng et al. 2007). In this study, four more edits (*ndhB*-106, *ndhD*-528, *ndhD*-740 and *ndhE*-106) were discovered in *ndh* transcripts. The editing level of *ndhB*-1977 was medium (~57%) in both leaf and floral tissues but the editing efficiency of other four newly identified sites was poor (5–17%).

RNA editing in non-protein coding transcripts

The plastid RNA editing status of tRNA, rRNA and non-coding transcripts was previously unexplored in moth orchid, with the exception of one edit reported in the 5′UTR of *psbH* transcripts (Zeng et al. 2007). In this study, we identified 59 edits with 22 common edits present in non-protein-coding transcripts of moth orchid; 21 and 16 edits were specific to leaf and floral tissues, respectively (Table 1; Additional file 1: Table S2). Among them, one common edit was found in tRNA, seven (four in leaf and five in floral tissue, two in common) in introns, and 51 (38 in leaf and 32 in floral tissue, 19 in common) in intergenic spacer (IGS) or untranslated regions (UTRs), but no edit was found in rRNA. Previously, chloroplast protein-coding transcripts including moth orchid showed a nearest-neighbor bias towards a U_A context immediately before and after RNA editing sites (Zeng et al. 2007). However, the condition surrounding edits in non-protein-coding transcripts have not been analyzed. In this study, we found a similar U_A context bias surrounding plastid edits in both protein-coding and non–protein-coding transcripts in moth orchid (Additional file 1: Figure S4).

Both leaf and floral tissue showed high efficiency editing (>63%) at 52,826 genomic position in *trnM* (cau) transcripts. The consequence of this C-to-U conversion in *trnM* (cau) of moth orchid resulted in increased nucleotide conservation among plant phylogeny (Fig. 2a). In addition, after being edited, despite no difference in free energy change (ΔG), the clover-leaf structure was restored as predicted by tRNAscan-SE 1.21 (Fig. 2b), which implied the functional importance of RNA editing for the *trnM*. Furthermore, we found relatively high-level editing (>48%) in the *rps12* intron (100,611) and *ycf3* intron (44,389; 45,108) (Additional file 1: Table S2). To determine whether the C-to-U conversion in these precursor transcripts affected the RNA secondary structure, the sequence extending 50 nt to both upstream and downstream of the edit in its edited and unedited form was analyzed. The secondary structures and free energy of these two transcripts, *rps12* intron (100,611) and *ycf3* intron (45,108) did not change before and after editing (Additional file 1: Figure S5A, B). In contrast, the secondary structure of *ycf3* intron (44,389) transcript was energetically more stable after editing (Additional file 1: Figure S5C).

Previously, the edit of the *psbT-psbH* (76,156 position) IGS, which also located in the 5′UTR of *psbH* (−30 position), resulted in the formation of an energetically less stable secondary structure (Zeng et al. 2007). We found very high efficiency editing (>82%) for this *psbH* 5′UTR edit in both leaf and floral tissues. In addition, at least a 50% editing level was found in the following seven IGS edits *clpP-rpl20* (70,142), *psaI-ycf4* (60,764), *psbB-psbT* (75,495), *psbM-rpoB* (27,736), *rps16-matK* (3095), *trnD-psbM* (31,129), and *trnN-rps12* (130,345) (Additional file 1: Table S2), which most of them might locate in the 5′ or 3′ UTRs and play an important role in regulation of gene expression. To determine whether the nucleotide substitution in these transcripts affected the RNA secondary structure, the sequence extending 50 nt to each side of the edit was predicted in its edited and unedited form. The secondary structure was energetically less stable in edited than non-edited transcripts of *psaI-ycf4* (60,764) and *rps16-matK* (3095). In contrast, the secondary structure was energetically more stable in edited than non-edited forms of *clpP-rpl20* (70,142), *psbM-rpoB* (27,736) as well as *trnD-psbM* (31,129) transcripts, but with no significant difference in edited and non-edited forms of the *psbB-psbT* (75,495) and *trnN-rps12* (130,345) transcripts were observed (Additional file 1: Figure S5D–J).

Differential status of RNA editing

Among 137 edits, we found significantly differential efficiency (>20%) of RNA editing for at least 32 edits between leaf and floral tissue, with 10 edits upregulated and 22 edits downregulated in comparing leaf and floral tissues (Fig. 3; Additional file 1: Table S2). Among 79 edits in protein-coding transcripts, 4 edits were significantly upregulated from 20 to 46%, and 17 edits were significantly downregulated, from 20 to 72% in comparing leaf and floral tissues (Fig. 3; Additional file 1: Table S2). In particular, two edits (*psbN*-29, 30) in *psbN* transcripts were upregulated to more than 42%, and the *ccsA*-336 edit was upregulated by 36%. In contrast, *rpoB*-614, *rpoB*-629, *rpoB*-686, *atpF*-92 and *rpoC1*-638 were

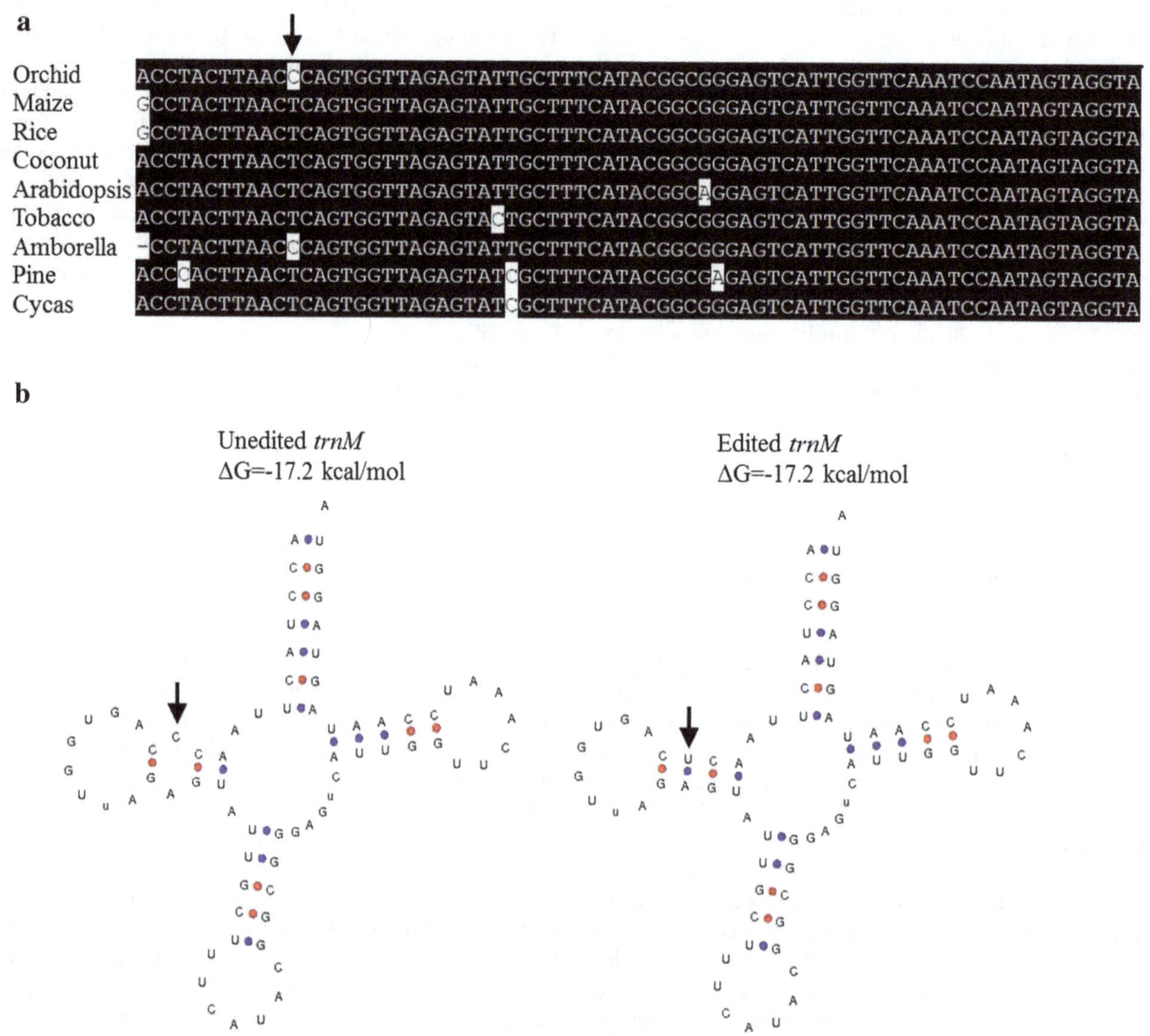

Fig. 2 RNA editing affects the secondary structure of *trnM* transcript in moth orchid. **a** The nucleotide sequences of plastid *trnM* (cau) gene from nine species of seed plants were multiple aligned by using LaserGene (DNAstar, USA). RNA editing at the indicated position is converted from C to U in moth orchid. Orchid, *Phalaenopsis aphrodite*; Maize, *Zea mays*; Rice, *Oryza sativa*; Coconut, *Cocos nucifera*; Arabidopsis, *Arabidopsis thaliana*; Tobacco, *Nicotiana tabacum*; Amborella, *Amborella trichopoda*; Pine, *Pinus thunbergi*, Cycas, *Cycas taitungensis*. **b** The predicted tRNA secondary structures formed by the unedited and edited *trnM* transcripts, by using the tRNAscan-SE 1.21 algorithm. The edited and unedited nucleotides are indicated by arrows

downregulated by 72, 68, 66, 63 and 58%, respectively, in comparing leaf and floral tissues. These results suggested the involvements of unidentified tissue-specific factors in editing of protein-coding transcripts in moth orchid.

The *trnM* (52,826) is the only edit occurred in tRNA transcripts, which it is high partial editing (64%) in floral tissue and was upregulated 29% to nearly fully edited in leaf tissue (Fig. 3; Additional file 1: Table S2). Among 7 edits in introns, one edit at *ycf3* intron (44,389) showed significant differential editing up to 43% between the two tissue types (Fig. 3; Additional file 1: Table S2). Among 51 edits in IGS, 9 differential (>20%) edits were identified: 5 edits were significantly upregulated from 20 to 71%, and 4 edits were significantly downregulated from 20 to 40% in comparing leaf and floral tissues (Fig. 3; Additional file 1: Table S2). For example, the efficiency of C-to-U conversion in the IGSs of *trnT-psbD* (33,566) and *trnN-rps12* (130,345) was upregulated to 33, and 71%, respectively, in comparing leaf and floral tissues, with the editing barely detectable in floral tissue. In contrast, the C-to-U

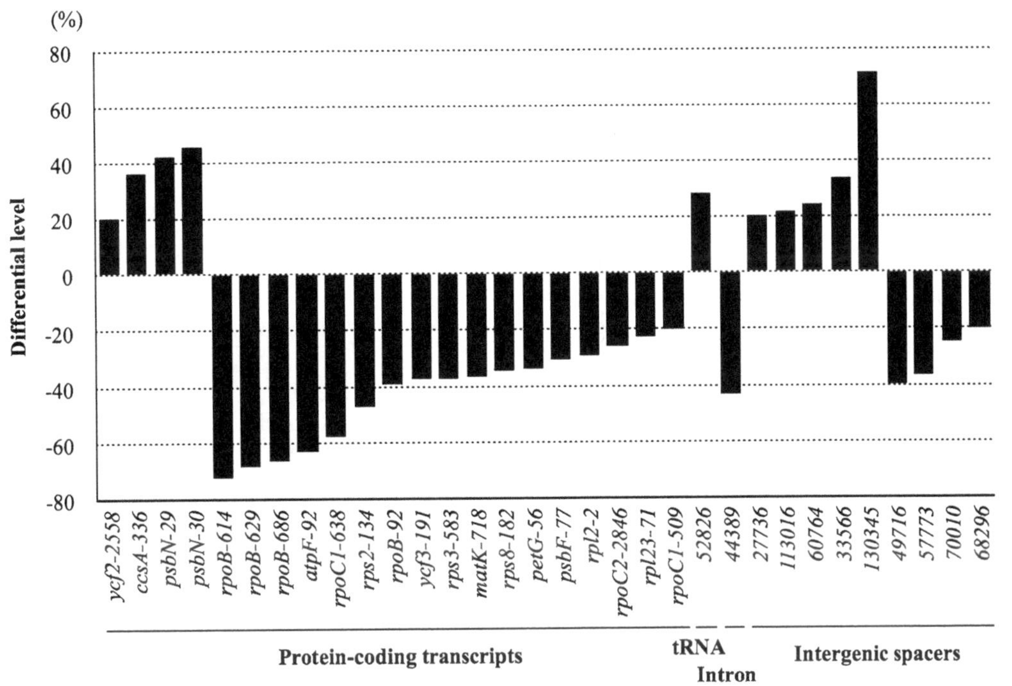

Fig. 3 Differential status of RNA editing between leaf and floral tissues. The differential editing level (>20%) for each edit between leaf and floral tissues is shown. The bars above the x-axis indicate upregulated editing level, and those below indicate downregulated editing level in leaf tissues as compared to floral tissues

conversion efficiency in the IGSs of *ndhJ-trnT* (49,716) and *rbcL-accD* (57,773) was downregulated by 40 and 37%, respectively, in comparing leaf and floral tissues. Meanwhile, the former edit were specific for floral tissue. Those analyses suggested the involvements of unknown tissue-specific factors in editing of none-protein-coding transcripts in moth orchid.

Expression profile of plastid protein-coding genes in moth orchids

RNA-Seq was used to investigate the relative steady-state expression of 68 plastid protein-coding genes from leaf and floral tissues of moth orchid by RPKM analysis. In both leaf and floral tissues, photosynthesis-related genes were most highly expressed. For instance, four genes (*psbA, rbcL, psaC* and *petB*) were the most highly expressed, with RPKM > 20,000 in leaf and four genes (*psbA, psaC, rbcL* and *psbD*) were the most highly expressed, with RPKM > 2250 in flower (Fig. 4a). The expression of *psbA* was highest in both leaf and floral tissues (Fig. 4a). In contrast, the expression of *matK* was the lowest, with RPKM < 2 in leaf and the expression of *petL* and *petN* is extremely low in flower, since the RPKM is barely detectable. Most plastid genes expressed in higher level in leaf than that in flower with the exception of *psbK*. The differential level of gene expression between leaf and floral tissues could range from 0.8-fold (*psbK)* to 80-fold (*psbT*) (Fig. 4b). According to functional categories, the plastid protein-coding genes were classified into seven groups, including CO_2 carboxylation (CO), photosystem I (PSI), photosystem II (PSII), cytochrome $b_{6/f}$ complex (PET), ATP synthase, gene expression machinery, and miscellaneous (Fig. 4). The *rbcL* and genes coding for subunits of electron transport chain complexes had significantly higher (10.6 folds in average) expression level in leaf than floral tissue, although genes involved in other groups such as expression apparatus, ATP synthase and miscellaneous function also had higher (4.6, 3.4 and 2.5 folds, respectively) expression level in leaf than flower (Fig. 4b).

Discussion

Previously, 44 edits were identified from 24 plastid protein-coding transcripts in leaves of *P. aphrodite* subsp. *formosana* by conventional Sanger sequencing (Zeng et al. 2007). In addition, bioinformatic analysis revealed high conservation (43 of 44) of plastid RNA editing sites between *P. equestris* and *P. aphrodite* (Jheng et al. 2012). Recently, RNA-Seq approaches were found to be more rapid, sensitive, extensive and cost-effective for investigating organellar RNA editing events than previously developed methods (Bentolila et al. 2013; Wang et al.

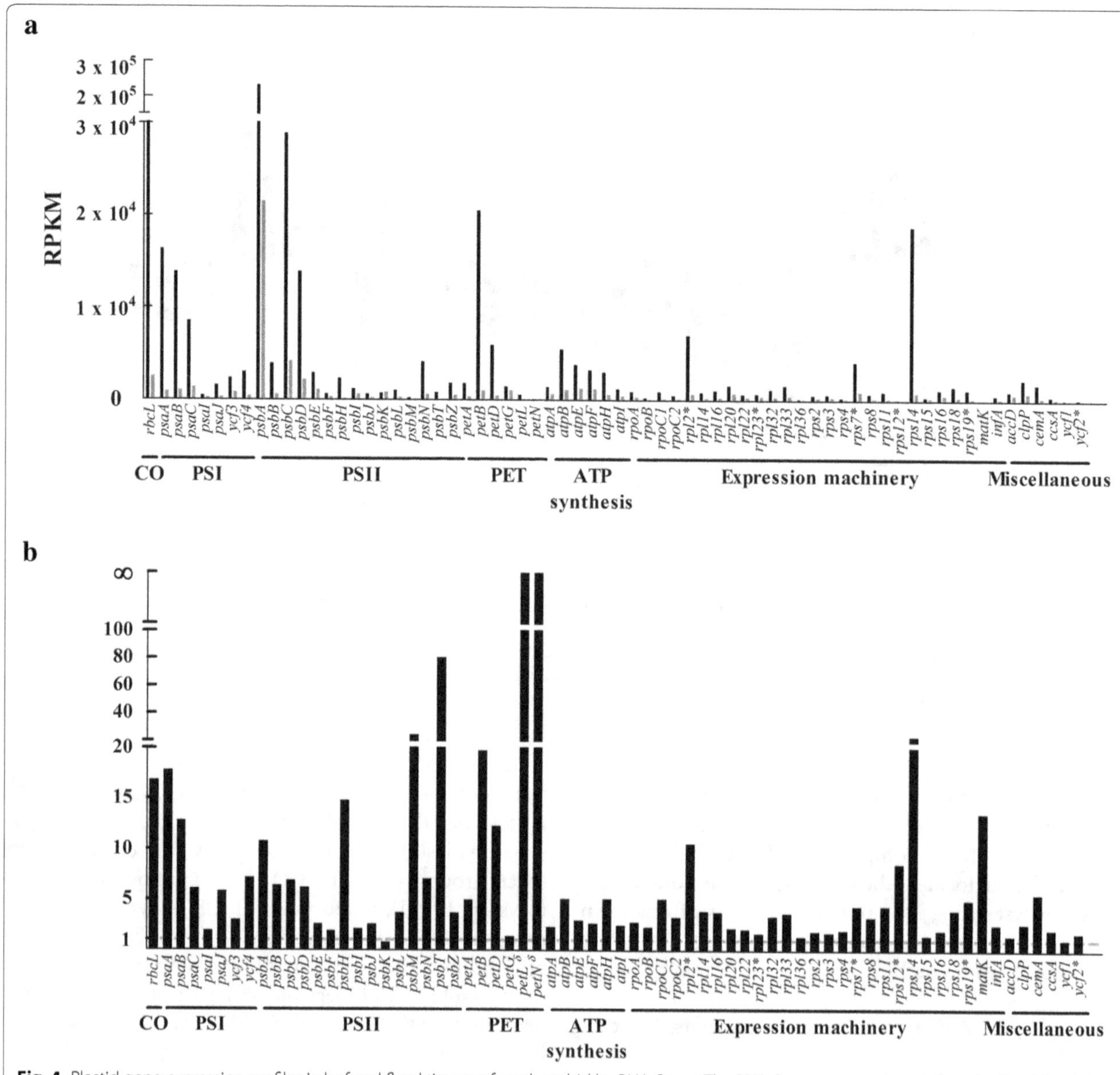

Fig. 4 Plastid gene expression profiles in leaf and floral tissues of moth orchid by RNA-Seq. **a** The RNA-Seq sequenced reads from leaf and floral tissues were imported into CLC Genomics Workbench and mapped to the plastid 68 protein-coding genes. The measurement of leaf (black bar) and floral (grey bar) RPKM is described in "Methods". **b** The relative expression of plastid protein-coding genes at the tissue level is shown by dividing the RPKM of leaf tissue to that of floral tissue. The asterisk indicates the genes present in the inverted repeat region of plastid genome. The δ indicate the RPKM of *petL* and *petN* is 0 in floral tissue

2015). For example, with NGS, 13, 133 and 119 new mitochondrial edits were identified from floral and leaf tissue of Arabidopsis and tobacco leaves, respectively (Bentolila et al. 2013; Grimes et al. 2014; Picardi et al. 2010). As well, up to nine new plastid edits were discovered in Arabidopsis (Bentolila et al. 2013), and many more partial edits could be detected in mitochondria of cotton than with the traditional approach (Suzuki et al. 2013). In this study, we investigated the whole plastid RNA editing status from leaf and floral tissues of *P. aphrodite* by RNA-Seq approaches. We identified 137 plastid RNA edits, including 93 newly discovered edits, which represented an average of 0.09% of the nucleotides examined in the plastid genome (Table 1; Additional file 1: Table S2). Furthermore, comparative cpDNA analysis potentially revealed high conservation (128 of 137) of plastid RNA edits between two endemic moth orchids, *P. aphrodite* and *P. equestris* (Additional file 1: Table S2).

In addition, we also conducted a comparative analysis for those edits from flower tissue between two different approaches, CLC Genomics Workbench and RES-scanner (Additional file 1: Table S5). The results showed that both approaches can simultaneously identify the same 84 edits which most edits have similar editing ratio. Five or nine edits with the most of editing ratio <8.7% are specifically identified by either CLC Genomics Workbench or RES-scanner, respectively, which it is probably due to the difference for the parameters used and the algorithms of software.

Plant mitochondrial RNA editing sites are abundant and highly variable, with approximately 189–635 edits in angiosperms (Bentolila et al. 2013; Grimes et al. 2014; Sloan et al. 2010), whereas plastid RNA edits are relatively limited, with 21–44 nucleotide conversions in seed plants (Chen et al. 2011; Lin et al. 2015a, b; Sasaki et al. 2003, 2006; Tsudzuki et al. 2001; Zeng et al. 2007). All plastid edits in seed plants are the C-to-U conversion type with the exception of the oil palm *Elaeis guineensis* Jacq. (Uthaipaisanwong et al. 2012) and Antarctic hairgrass *Deschampsia antarctica* Desv. (Lee et al. 2014). However, recent reports have revealed more plastid edits in some taxa of seed plants (Additional file 1: Table S3). For example, in the gymnosperm, *Cycas taitungensis* (Chen et al. 2011) and *Ginkgo biloba* (He et al. 2016), 85 and 255 edits were identified from protein-coding transcripts, respectively. In dicot, cotton *Gossypium hirsutum*, 54 edits were identified in 27 transcripts (Jiang et al. 2012). In monocot, coconut *Cocos nucifera* L., duckweed *Spirodela polyrhiza* and Tausch's goatgrass *Aegilops tauschii L.*, 75, 66 and 60 plastid RNA edits were identified, respectively (Huang et al. 2013; Wang et al. 2015, 2017). In the early-branching angiosperm *Amborella trichopoda,* 138 edits were revealed (Hein et al. 2016). Here we identified 137 edits with 126 C-to-U and 11 U-to-C conversion from moth orchid, the highest number so far reported in monocot (Table 1), though nearly 57% of edits showed poor-level (5–20%) editing (Additional file 1: Figure S1, Table S2). However, chloroplast RNA editing is highly variable, with more edits in lower taxa of land plants. For instance, in hornwort *Anthoceros formosae,* at least 943 had 509 (54%) C-to-U and 433 (46%) U-to-C conversions (Kugita et al. 2003). Lycophyte showed a large number of plastid RNA editing sites, up to 3415 edits with only C-to-U conversion in the spike moss *Selaginella uncinata* (Oldenkott et al. 2014). In the fern *Adiantum capillus-veneris*, 350 RNA editing sites with 90% C-to-U edits and 10% U-to-C edits were identified (Wolf et al. 2004). Additionally, the plastid transcriptomes analysis from two early diverging species of fern, *Ophioglossum californicum* and *Psilotum nudum,* identified 297 C-to-U and three U-to-C edits in the former plastid transcripts, but only 27 C-to-U and no U-to-C edits in the latter plastid transcripts (Guo et al. 2015). These analyses suggest an independent gain or loss of RNA edits across the taxa of land plants during evolution.

RNA editing occurs most frequently in the *ndh* transcripts, accounting for >40% of the edits in the chloroplast transcripts of most higher plants with the exception of *Pinus thunbergii* and *Phalaenopsis aphrodite* (Chen et al. 2011; Huang et al. 2013; Lin et al. 2015a, b; Wakasugi et al. 1996; Wang et al. 2015; Zeng et al. 2007) because the functional *ndh* genes were lost from the plastids of these two species (Chang et al. 2006; Wakasugi et al. 1994). In Arabidopsis, those *ndh* edits are differentially regulated in distinct tissue types (Tseng et al. 2013). In this study, only a small fraction (5 of 137) of RNA edits were contributed from *ndh* transcripts in moth orchid, and they also showed poor editing efficiency (3–14%), with the exception of the previously identified *ndhB*-1977 edit (Additional file 1: Table S2). In addition, two sites (*ndhD*-528 and -740) in *ndhD* transcript were completely undetectable in floral tissue (Additional file 1: Table S2), which suggested that some tissue-specific factors may regulate the editing process of *ndh* transcripts in moth orchid. Additionally, the remaining five partial edits in *ndh* transcripts are likely an evolutionary remnant from before the complete loss of plastid RNA edits for non-functional *ndh* pseudogenes in moth orchid.

Previous study showed that the *rpo* (*rpoA*, *rpoB*, *rpoC1* and *rpoC2)* transcripts were the most extensively edited (15 sites, up to 34%) among the functional gene groups in moth orchid (Zeng et al. 2007). We discovered two new edits (*rpoB*-55; *rpoC1*-1638) from *rpo* transcripts, although their editing level was poor (~10%) (Additional file 1: Tables S2, S4). In addition, five edits (*rpoB*-55, *-614, -629, -686* and *rpoC1*-1638) were preferentially occurred in floral tissues. Furthermore, three edits (*rpoB*-55, *rpoC1*-203 and *rpoC1*-1638) were unique to moth orchid as compared with 17 other seed plants (Additional file 1: Table S4), although their editing level was poor with the exception of *rpoC1*-203. In contrast, the four nucleotides (*rpoA*-527, *rpoB*-1241, *rpoC1*-243, -656) of *rpo* transcripts were edited or T is present in the corresponding DNA in most of seed plants but are not edited in moth orchid, which caused the divergence of amino acid residues between moth orchid and other seed plants (Additional file 1: Table S4), though the significance is unknown.

Previous study has demonstrated that the pattern and efficiency of some editing sites in organellar transcripts significantly differed among ecotypes, tissue types, developmental stages and environmental conditions (Bentolila et al. 2005; Chateigner-Boutin and Hanson 2003; Howad and Kempken 1997; Ruf and Kossel 1997; Tillich et al.

2005; Tseng et al. 2013). For instance, in Arabidopsis, the editing of *ndhB*-149, *ndhB*-1255, and *ndhD*-2 occurs in leaf but is completely lost in roots and in lincomycin-treated seedlings (Tseng et al. 2013). In this study, at least 32 sites showed significantly differential (≧20%) RNA editing between leaf and floral tissues in moth orchid (Fig. 3; Additional file 1: Table S2). In addition, 19 and 14 edits are unique for leaf and floral tissues, respectively, which RNA editing is completely undetectable in floral tissue or vice versa (Additional file 1: Table S2). This analysis suggested tissue-specific factors might involve in RNA editing process in moth orchid.

Previously, a C-to-U conversion at the −10 position of the *ndhG* 5′UTR in monocot plants was predicted to modify the RNA secondary (stem/loop) structure (Drescher et al. 2002), but was not observed in moth orchid. Instead, a C-to-U substitution at the −30 position of the *psbH* 5′UTR that can form an energetically less stable secondary structure was reported in moth orchid (Zeng et al. 2007). Therefore, edits that occurred in the UTR regions might play an important role in the regulation of gene expression. We found very high efficiency editing (>81%) at the *psbH* 5′UTR (−30) in both leaf and floral tissues (Additional file 1: Table S2). In addition, at least 50% editing efficiency was observed for the edits of IGSs such as *psaI-ycf4* (60,764), *psbM-rpoB* (27,736), *trnD-psbM* (31,129), *rps16-matK* (3095), *clpP-rpl20 (70,142)* and *psbB-psbT* (75,495) located in the 3′ or 5′UTR of genes such as *psaI* (27, 27 nt downstream of stop codon), *rpoB* (−160, 160 nt upstream of start codon), *psbM* (−85), *matK* (−35), *rpl20* (−134), *psbB* (29), respectively. In comparing the structure of edited and unedited transcripts, significant change in free energy was predicted for the former five (Additional file 1: Figure S5D–H), which suggested that some edits occurring at the 3′ or 5′UTR might play a role in the regulation of gene expression in moth orchid.

Previous study reported that some sites of intron editing are a prerequisite for RNA splicing in plant organelles (Borner et al. 1995; Farre et al. 2012; Vogel et al. 1997), and unspliced RNA is often only partially edited (Yang and Mulligan 1991). In this study, 8 RNA edits were identified in seven introns (Additional file 1: Table S2), in particular, edits in *rps12* (100,611) and *ycf3* (45,108) introns showed high level (>73%) of editing. In addition, leaf-specific editing was found in the *clpP* intron (71,815) and flower-specific editing in the *clpP* intron (72,384), though the editing level is poor. Furthermore, the edit *ycf3* (44,389) in intron showed significantly (>43%) differential editing between leaf and floral tissues (Fig. 3; Additional file 1: Table S2). These analyses suggested that editing in some *cis*-elements of primary transcripts might be important for splicing, and unidentified tissue-specific factors might be required for regulating intron editing in moth orchid. Additionally, RNA editing occurs in tRNA in plant organelles and plays an important role in tRNA processing and maturation (Binder et al. 1994; Kunzmann et al. 1998; Marechal-Drouard et al. 1996). The editing of the tRNA precursor might occur in the anticodon stem, acceptor stem, D stem or even anticodon (Janke and Paabo 1993; Marechal-Drouard et al. 1996). In this study, we observed nearly full RNA editing at the D stem of *trnM* transcripts (52,826 position) in leaf, with significant (>29%) differential editing between leaf and floral tissues (Fig. 3; Additional file 1: Table S2), which suggested that tissue-specific tRNA editing factors might be involved. After being edited in moth orchid, the *trnM* sequence conservation among plant phylogeny was increased, with the exception of *Amborella* species, then the standard clover-leaf tRNA structure could be formed (Fig. 2). The corresponding edit might also occur in *Amborella*, although this has not been reported (Hein et al. 2016).

The Arabidopsis editosome consists of at least four nucleus-encoded protein families, PPR, RIPs, ORRM1 and OZ1, required for plant organelle RNA editing (Sun et al. 2015, 2016). The PPR protein family with more than 400 members in angiosperms is involved in RNA metabolism including RNA editing in plant organelles (Grennan 2011). Although no PPR proteins involved in RNA metabolism have been reported in orchids, approximately 254 expressed sequence tags (EST) potentially encoding PPR proteins in moth orchid were annotated in the Orchidbase database (http://orchidbase.itps.ncku.edu.tw/est/home2012.aspx). In addition, at least three RIP homologues, five OZ1 homologues and two ORRM1-like proteins are available in Orchidbase. Our analysis suggests that the editosome of moth orchids might resemble that of Arabidopsis.

Chloroplast transcription change in large-scale in response to environmental or developmental signals (Barkan 2011). In Arabidopsis, the steady state levels of most plastid gene expression were high in green tissues, while low or undetectable in non-green tissues (Tseng et al. 2013). However, appreciable levels of some plastid transcripts such as *clpP* and *rps14* mRNAs could be detected in non-photosynthetic tissue (Tseng et al. 2013). In this study, the differential steady state level of plastid transcripts between leaf and floral tissues could range from 0.8 to 80 fold (Fig. 4b), with most plastid genes having higher expression level in leaf than floral tissue, with the exception of *psbK*. In addition, genes located in the inverted repeat regions did not show higher expression level than those in single copy regions (Fig. 4). Chloroplast gene expression is regulated at multiple steps such as transcription, splicing, editing, processing, stability, translation, and post-translational modification, and no single step is the primary regulated step (Barkan 2011). Rather, each step might contribute to distinct patterns of plastid gene expression under different environmental and developmental conditions. Therefore, the variation of gene copy number and steady state transcript level probably did not reflect to the protein level in the plastids of distinctive tissues in moth orchid. Further studies are required to clarify how plastid gene expression is regulated in moth orchid.

Conclusions

We have identified 137 edits including 93 newly discovered edits in plastid transcripts of moth orchid by RNA-Seq. It is the highest number reported so far in monocots. Overall, 79 edits were involved in protein-coding transcripts, and the consequence of 58 nucleotide conversions caused the non-synonymous substitution which tend to increase the amino acid hydrophobicity as well as conservation among plant phylogeny. RNA editing occurred in non-protein-coding transcripts such as tRNA, introns and untranslated regulatory regions could affect the formation and stability of secondary structure, which might play a regulatory role in gene expression. For instance, RNA editing in *trnM* is required for the formation of a standard clover-leaf structure. The nearest-neighbor bias towards a U_A context immediately before and after RNA edits occurred in both protein-coding and non-protein-coding transcripts in moth orchid. At least 32 edits showed significant (≧20%) differential editing between leaf and floral tissues, which suggested that some unidentified tissue-specific factors might be responsible for the regulation of RNA editing in moth orchid. Furthermore, most plastid genes expressed in higher level in leaf than that in flower, but it might not reflect to the protein level in the plastids of distinctive tissues in moth orchid.

Abbreviations

cpDNA: chloroplast DNA; DOC: depth of coverage; IGS: intergenic spacer; mtDNA: mitochondrial DNA; MORF: multiple organellar RNA editing factor; NGS: next generation sequencing; PPR: pentatricopeptide repeat protein; RHS: Royal Horticultural Society; RIP: RNA editing-interacting protein; RNA-Seq: RNA sequencing; UTR: untranslated regulatory region.

Authors' contributions

The presented research work was carried out in collaboration between authors. TCC and CCC designed the experiments; TCC, CHW and YCL performed the experiments; TCC, YCL, XW, CHH and CCC analyzed the data; TCC, XW and CCC wrote the paper. All authors read and approved the final manuscript.

Author details

[1] Institute of Biotechnology, National Cheng Kung University, Tainan 701, Taiwan. [2] Department of Biotechnology and Bioindustry Sciences, National Cheng Kung University, Tainan 701, Taiwan. [3] Department of Genetics, University of Georgia, Athens, GA 30602, USA.

Competing interests

The authors declare that they have no competing interests.

Funding

This work was financially supported in part by the Grants (MOST 104-2321-B-006-001, MOST 105-2313-B-006-001 and MOST 106-2313-B-006-003-MY3) to C.-C. Chang from the Ministry of Science and Technology, Taiwan.

References

Barkan A (2011) Expression of plastid genes: organelle-specific elaborations on a prokaryotic scaffold. Plant Physiol 155(4):1520–1532. doi:10.1104/pp.110.171231

Barkan A, Small I (2014) Pentatricopeptide repeat proteins in plants. Annu Rev Plant Biol 65:415–442. doi:10.1146/annurev-arplant-050213-040159

Bentolila S, Chateigner-Boutin AL, Hanson MR (2005) Ecotype allelic variation in C-to-U editing extent of a mitochondrial transcript identifies RNA-editing quantitative trait loci in Arabidopsis. Plant Physiol 139(4):2006–2016. doi:10.1104/pp.105.069013

Bentolila S, Heller WP, Sun T, Babina AM, Friso G, van Wijk KJ, Hanson MR (2012) RIP1, a member of an Arabidopsis protein family, interacts with the protein RARE1 and broadly affects RNA editing. Proc Natl Acad Sci USA 109(22):E1453–1461. doi:10.1073/pnas.1121465109

Bentolila S, Oh J, Hanson MR, Bukowski R (2013) Comprehensive high-resolution analysis of the role of an Arabidopsis gene family in RNA editing. PLoS Genet 9(6):e1003584. doi:10.1371/journal.pgen.1003584

Binder S, Marchfelder A, Brennicke A (1994) RNA editing of tRNA(Phe) and tRNA(Cys) in mitochondria of *Oenothera berteriana* is initiated in precursor molecules. Mol Gen Genet 244(1):67–74

Bock R (2000) Sense from nonsense: how the genetic information of chloroplasts is altered by RNA editing. Biochimie 82(6–7):549–557

Bolger A, Lohse M, Usadel B (2014) Trimmomatic: a flexible trimmer for Illumina sequence data. Bioinformatics 30:2114–2120

Borner GV, Morl M, Wissinger B, Brennicke A, Schmelzer C (1995) RNA editing of a group II intron in *Oenothera* as a prerequisite for splicing. Mol Gen Genet 246(6):739–744

Boussardon C, Avon A, Kindgren P, Bond CS, Challenor M, Lurin C, Small I (2014) The cytidine deaminase signature HxE(x)n CxxC of DYW1 binds zinc and is necessary for RNA editing of *ndhD*-1. New Phytol 203(4):1090–1095. doi:10.1111/nph.12928

Chang CC, Lin HC, Lin IP, Chow TY, Chen HH, Chen WH, Cheng CH, Lin CY, Liu SM, Chang CC, Chaw SM (2006) The chloroplast genome of *Phalaenopsis aphrodite* (Orchidaceae): comparative analysis of evolutionary rate with that of grasses and its phylogenetic implications. Mol Biol Evol 23(2):279–291. doi:10.1093/molbev/msj029

Chateigner-Boutin AL, Hanson MR (2003) Developmental co-variation of RNA editing extent of plastid editing sites exhibiting similar cis-elements. Nucleic Acids Res 31(10):2586–2594

Chen H, Deng L, Jiang Y, Lu P, Yu J (2011) RNA editing sites exist in protein-coding genes in the chloroplast genome of Cycas taitungensis. J Integr Plant Biol 53(12):961–970. doi:10.1111/j.1744-7909.2011.01082.x

Christenson EA (2001) Phalaenopsis: a monograph. Timber Press, Portland

Drescher A, Hupfer H, Nickel C, Albertazzi F, Hohmann U, Herrmann RG, Maier RM (2002) C-to-U conversion in the intercistronic *ndhI/ndhG* RNA of plastids from monocot plants: conventional editing in an unconventional small reading frame? Mol Genet Genom 267(2):262–269. doi:10.1007/s00438-002-0662-9

Dressler RL (2005) How many orchid species? Selbyana 26(1/2):155–158

Farre JC, Aknin C, Araya A, Castandet B (2012) RNA editing in mitochondrial trans-introns is required for splicing. PLoS ONE 7(12):e52644. doi:10.1371/journal.pone.0052644

Fiebig A, Stegemann S, Bock R (2004) Rapid evolution of RNA editing sites in a small non-essential plastid gene. Nucleic Acids Res 32(12):3615–3622. doi:10.1093/nar/gkh695

Grennan AK (2011) To thy proteins be true: RNA editing in plants. Plant Physiol 156(2):453–454. doi:10.1104/pp.111.900412

Grimes BT, Sisay AK, Carroll HD, Cahoon AB (2014) Deep sequencing of the tobacco mitochondrial transcriptome reveals expressed ORFs and numerous editing sites outside coding regions. BMC Genom 15:31. doi:10.1186/1471-2164-15-31

Guo W, Grewe F, Mower JP (2015) Variable frequency of plastid RNA editing among ferns and repeated loss of uridine-to-cytidine editing from vascular plants. PLoS ONE 10(1):e0117075. doi:10.1371/journal.pone.0117075

Hammani K, Giege P (2014) RNA metabolism in plant mitochondria. Trends Plant Sci 19(6):380–389. doi:10.1016/j.tplants.2013.12.008

He P, Huang S, Xiao G, Zhang Y, Yu J (2016) Abundant RNA editing sites of chloroplast protein-coding genes in *Ginkgo biloba* and an evolutionary pattern analysis. BMC Plant Biol 16:257

Hein A, Polsakiewicz M, Knoop V (2016) Frequent chloroplast RNA editing in early-branching flowering plants: pilot studies on angiosperm-wide coexistence of editing sites and their nuclear specificity factors. BMC Evol Biol 16:23. doi:10.1186/s12862-016-0589-0

Hirose T, Sugiura M (2001) Involvement of a site-specific trans-acting factor and a common RNA-binding protein in the editing of chloroplast mRNAs: development of a chloroplast in vitro RNA editing system. EMBO J 20(5):1144–1152. doi:10.1093/emboj/20.5.1144

Howad W, Kempken F (1997) Cell type-specific loss of atp6 RNA editing in cytoplasmic male sterile *Sorghum bicolor*. Proc Natl Acad Sci USA 94(20):11090–11095

Hrubec TC, Robinson JM, Donaldson RP (1985) Isolation of mitochondria from soybean leaves on discontinuous percoll gradients. Plant Physiol 77(4):1010–1012

Huang YY, Matzke AJ, Matzke M (2013) Complete sequence and comparative analysis of the chloroplast genome of coconut palm (*Cocos nucifera*). PLoS ONE 8(8):e74736. doi:10.1371/journal.pone.0074736

Janke A, Paabo S (1993) Editing of a tRNA anticodon in marsupial mitochondria changes its codon recognition. Nucleic Acids Res 21(7):1523–1525

Jheng CF, Chen TC, Lin JY, Chen TC, Wu WL, Chang CC (2012) The comparative chloroplast genomic analysis of photosynthetic orchids and developing DNA markers to distinguish *Phalaenopsis* orchids. Plant Sci 190:62–73. doi:10.1016/j.plantsci.2012.04.001

Jiang Y, Fan SL, Song MZ, Yu JN, Yu SX (2012) Identification of RNA editing sites in cotton (*Gossypium hirsutum*) chloroplasts and editing events that affect secondary and three-dimensional protein structures. Genet Mol Res 11(2):987–1001. doi:10.4238/2012.April.19.4

Kotera E, Tasaka M, Shikanai T (2005) A pentatricopeptide repeat protein is essential for RNA editing in chloroplasts. Nature 433(7023):326–330. doi:10.1038/nature03229

Kugita M, Yamamoto Y, Fujikawa T, Matsumoto T, Yoshinaga K (2003) RNA editing in hornwort chloroplasts makes more than half the genes functional. Nucleic Acids Res 31(9):2417–2423

Kunzmann A, Brennicke A, Marchfelder A (1998) 5′ end maturation and RNA editing have to precede tRNA 3′ processing in plant mitochondria. Proc Natl Acad Sci USA 95(1):108–113

Lee J, Kang Y, Shin SC, Park H, Lee H (2014) Combined analysis of the chloroplast genome and transcriptome of the Antarctic vascular plant *Deschampsia antarctica* Desv. PLoS ONE 9(3):e92501. doi:10.1371/journal.pone.0092501

Lin CP, Ko CY, Kuo CI, Liu MS, Schafleitner R, Chen LF (2015a) Transcriptional Slippage and RNA editing increase the diversity of transcripts in chloroplasts: insight from deep sequencing of *Vigna radiata* genome and transcriptome. PLoS ONE 10(6):e0129396. doi:10.1371/journal.pone.0129396

Lin JY, Lin BY, Chang CD, Liao SC, Liu YC, Wu WL, Chang CC (2015b) Evaluation of chloroplast DNA markers for distinguishing *Phalaenopsis* species. Sci Hortic 192:302–310. doi:10.1016/j.scienta.2015.06.019

Lin BY, Chang CD, Huang LLH, Liu YC, Su YY, Chen TC, Lee RH, Huang HJ, Wu WL, Chang CC (2016) The mitochondrial DNA markers for distinguishing *Phalaenopsis* species and revealing maternal phylogeny. Biol Plant 60(1):68–78. doi:10.1007/s10535-015-0566-2

Liu YC, Lin BY, Lin JY, Wu WL, Chang CC (2016) Evaluation of chloroplast DNA markers for intraspecific identification of *Phalaenopsis equestris* cultivars. Sci Hortic 203:86–94. doi:10.1016/j.scienta.2016.03.021

Marechal-Drouard L, Kumar R, Remacle C, Small I (1996) RNA editing of larch mitochondrial tRNA(His) precursors is a prerequisite for processing. Nucleic Acids Res 24(16):3229–3234

Mortazavi A, Williams BA, McCue K, Schaeffer L, Wold B (2008) Mapping and quantifying mammalian transcriptomes by RNA-Seq. Nat Methods 5(7):621–628. doi:10.1038/nmeth.1226

Oldenkott B, Yamaguchi K, Tsuji-Tsukinoki S, Knie N, Knoop V (2014) Chloroplast RNA editing going extreme: more than 3400 events of C-to-U editing in the chloroplast transcriptome of the lycophyte *Selaginella uncinata*. RNA 20(10):1499–1506. doi:10.1261/rna.045575.114

Picardi E, Horner DS, Chiara M, Schiavon R, Valle G, Pesole G (2010) Large-scale detection and analysis of RNA editing in grape mtDNA by RNA deep-sequencing. Nucleic Acids Res 38(14):4755–4767. doi:10.1093/nar/gkq202

Ruf S, Kossel H (1997) Tissue-specific and differential editing of the two *ycf3* editing sites in maize plastids. Curr Genet 32(1):19–23

Ruwe H, Castandet B, Schmitz-Linneweber C, Stern DB (2013) Arabidopsis chloroplast quantitative editotype. FEBS Lett 587(9):1429–1433. doi:10.1016/j.febslet.2013.03.022

Sasaki T, Yukawa Y, Miyamoto T, Obokata J, Sugiura M (2003) Identification of RNA editing sites in chloroplast transcripts from the maternal and paternal progenitors of tobacco (*Nicotiana tabacum*): comparative analysis shows the involvement of distinct trans-factors for *ndhB* editing. Mol Biol Evol 20(7):1028–1035. doi:10.1093/molbev/msg098

Sasaki T, Yukawa Y, Wakasugi T, Yamada K, Sugiura M (2006) A simple in vitro RNA editing assay for chloroplast transcripts using fluorescent dideoxynucleotides: distinct types of sequence elements required for editing of *ndh* transcripts. Plant J 47(5):802–810. doi:10.1111/j.1365-313X.2006.02825.x

Schattner P, Brooks AN, Lowe TM (2005) The tRNAscan-SE, snoscan and snoGPS web servers for the detection of tRNAs and snoRNAs. Nucleic Acids Res 33 (Web Server issue):W686–689. doi:10.1093/nar/gki366

Schwitzguebel JP, Siegenthaler PA (1984) Purification of peroxisomes and mitochondria from spinach leaf by percoll gradient centrifugation. Plant Physiol 75(3):670–674

Sloan DB, MacQueen AH, Alverson AJ, Palmer JD, Taylor DR (2010) Extensive loss of RNA editing sites in rapidly evolving *Silene* mitochondrial genomes: selection vs. retroprocessing as the driving force. Genetics 185(4):1369–1380. doi:10.1534/genetics.110.118000

Sun T, Germain A, Giloteaux L, Hammani K, Barkan A, Hanson MR, Bentolila S (2013) An RNA recognition motif-containing protein is required for plastid RNA editing in Arabidopsis and maize. Proc Natl Acad Sci USA 110(12):E1169–1178. doi:10.1073/pnas.1220162110

Sun T, Shi X, Friso G, Van Wijk K, Bentolila S, Hanson MR (2015) A zinc finger motif-containing protein is essential for chloroplast RNA editing. PLoS Genet 11(3):e1005028. doi:10.1371/journal.pgen.1005028

Sun T, Bentolila S, Hanson MR (2016) The unexpected diversity of plant organelle RNA editosomes. Trends Plant Sci 21:962–973

Suzuki H, Yu J, Ness SA, O'Connell MA, Zhang J (2013) RNA editing events in mitochondrial genes by ultra-deep sequencing methods: a comparison of cytoplasmic male sterile, fertile and restored genotypes in cotton. Mol Genet Genom 288(9):445–457. doi:10.1007/s00438-013-0764-6

Takenaka M, Zehrmann A, Verbitskiy D, Kugelmann M, Hartel B, Brennicke A (2012) Multiple organellar RNA editing factor (MORF) family proteins are required for RNA editing in mitochondria and plastids of plants. Proc Natl Acad Sci USA 109(13):5104–5109. doi:10.1073/pnas.1202452109

Takenaka M, Zehrmann A, Verbitskiy D, Hartel B, Brennicke A (2013) RNA editing in plants and its evolution. Annu Rev Genet 47:335–352. doi:10.1146/annurev-genet-111212-133519

Takenaka M, Verbitskiy D, Zehrmann A, Hartel B, Bayer-Csaszar E, Glass F, Brennicke A (2014) RNA editing in plant mitochondria-connecting RNA target sequences and acting proteins. Mitochondrion 19(Pt B):191–197. doi:10.1016/j.mito.2014.04.005

Tillich M, Funk HT, Schmitz-Linneweber C, Poltnigg P, Sabater B, Martin M, Maier RM (2005) Editing of plastid RNA in *Arabidopsis thaliana* ecotypes. Plant J 43(5):708–715. doi:10.1111/j.1365-313X.2005.02484.x

Tseng CC, Lee CJ, Chung YT, Sung TY, Hsieh MH (2013) Differential regulation of Arabidopsis plastid gene expression and RNA editing in non-photosynthetic tissues. Plant Mol Biol 82(4–5):375–392. doi:10.1007/s11103-013-0069-5

Tsudzuki T, Wakasugi T, Sugiura M (2001) Comparative analysis of RNA editing sites in higher plant chloroplasts. J Mol Evol 53(4–5):327–332. doi:10.1007/s002390010222

Uthaipaisanwong P, Chanprasert J, Shearman JR, Sangsrakru D, Yoocha T, Jomchai N, Jantasuriyarat C, Tragoonrung S, Tangphatsornruang S (2012) Characterization of the chloroplast genome sequence of oil palm (*Elaeis guineensis* Jacq). Gene 500(2):172–180. doi:10.1016/j.gene.2012.03.061

Vogel J, Hubschmann T, Borner T, Hess WR (1997) Splicing and intron-internal RNA editing of *trnK-matK* transcripts in barley plastids: support for MatK as an essential splice factor. J Mol Biol 270(2):179–187. doi:10.1006/jmbi.1997.1115

Wagoner JA, Sun T, Lin L, Hanson MR (2015) Cytidine deaminase motifs within the DYW domain of two pentatricopeptide repeat-containing proteins are required for site-specific chloroplast RNA editing. J Biol Chem 290(5):2957–2968. doi:10.1074/jbc.M114.622084

Wakasugi T, Tsudzuki J, Ito S, Nakashima K, Tsudzuki T, Sugiura M (1994) Loss of all *ndh* genes as determined by sequencing the entire chloroplast genome of the black pine *Pinus thunbergii*. Proc Natl Acad Sci USA 91(21):9794–9798. doi:10.1073/pnas.91.21.9794

Wakasugi T, Hirose T, Horihata M, Tsudzuki T, Kossel H, Sugiura M (1996) Creation of a novel protein-coding region at the RNA level in black pine chloroplasts: the pattern of RNA editing in the gymnosperm chloroplast is different from that in angiosperms. Proc Natl Acad Sci USA 93(16):8766–8770

Wang M, Cui L, Feng K, Deng P, Du X, Wan F, Weining S, Nie X (2015) Comparative analysis of Asteraceae chloroplast genomes: structural organization, RNA editing and evolution. Plant Mol Biol Rep 33(5):1526–1538. doi:10.1007/s11105-015-0853-2

Wang Z, Lian J, Li Q, Zhang P, Zhou Y, Zhan X, Zhang G (2016) RES-Scanner: a software package for genome-wide identification of RNA-editing sites. GigaScience 5:37

Wang M, Liu H, Ge L, Xing G, Wang M, Weining S, Nie X (2017) Identification and analysis of RNA editing sites in the chloroplast transcripts of *Aegilops tauschii* L. Genes 8(1):13. doi:10.3390/genes8010013

Wolf PG, Rowe CA, Hasebe M (2004) High levels of RNA editing in a vascular plant chloroplast genome: analysis of transcripts from the fern *Adiantum capillus*-veneris. Gene 339:89–97. doi:10.1016/j.gene.2004.06.018

Yagi Y, Hayashi S, Kobayashi K, Hirayama T, Nakamura T (2013) Elucidation of the RNA recognition code for pentatricopeptide repeat proteins involved in organelle RNA editing in plants. PLoS ONE 8(3):e57286. doi:10.1371/journal.pone.0057286

Yang AJ, Mulligan RM (1991) RNA editing intermediates of *cox2* transcripts in maize mitochondria. Mol Cell Biol 11(8):4278–4281. doi:10.1128/MCB.11.8.4278

Zeng WH, Liao SC, Chang CC (2007) Identification of RNA editing sites in chloroplast transcripts of *Phalaenopsis aphrodite* and comparative analysis with those of other seed plants. Plant Cell Physiol 48(2):362–368. doi:10.1093/pcp/pcl058

Permissions

All chapters in this book were first published in BS, by Springer; hereby published with permission under the Creative Commons Attribution License or equivalent. Every chapter published in this book has been scrutinized by our experts. Their significance has been extensively debated. The topics covered herein carry significant findings which will fuel the growth of the discipline. They may even be implemented as practical applications or may be referred to as a beginning point for another development.

The contributors of this book come from diverse backgrounds, making this book a truly international effort. This book will bring forth new frontiers with its revolutionizing research information and detailed analysis of the nascent developments around the world.

We would like to thank all the contributing authors for lending their expertise to make the book truly unique. They have played a crucial role in the development of this book. Without their invaluable contributions this book wouldn't have been possible. They have made vital efforts to compile up to date information on the varied aspects of this subject to make this book a valuable addition to the collection of many professionals and students.

This book was conceptualized with the vision of imparting up-to-date information and advanced data in this field. To ensure the same, a matchless editorial board was set up. Every individual on the board went through rigorous rounds of assessment to prove their worth. After which they invested a large part of their time researching and compiling the most relevant data for our readers.

The editorial board has been involved in producing this book since its inception. They have spent rigorous hours researching and exploring the diverse topics which have resulted in the successful publishing of this book. They have passed on their knowledge of decades through this book. To expedite this challenging task, the publisher supported the team at every step. A small team of assistant editors was also appointed to further simplify the editing procedure and attain best results for the readers.

Apart from the editorial board, the designing team has also invested a significant amount of their time in understanding the subject and creating the most relevant covers. They scrutinized every image to scout for the most suitable representation of the subject and create an appropriate cover for the book.

The publishing team has been an ardent support to the editorial, designing and production team. Their endless efforts to recruit the best for this project, has resulted in the accomplishment of this book. They are a veteran in the field of academics and their pool of knowledge is as vast as their experience in printing. Their expertise and guidance has proved useful at every step. Their uncompromising quality standards have made this book an exceptional effort. Their encouragement from time to time has been an inspiration for everyone.

The publisher and the editorial board hope that this book will prove to be a valuable piece of knowledge for researchers, students, practitioners and scholars across the globe.

List of Contributors

Po-chih Shen and Li-yu Daisy Liu
Biometrics Division, Department of Agronomy, National Taiwan University, Taipei, Taiwan

Ai-ling Hour
Department of Life Science, Fu-Jen Catholic University, Xinbei, Taiwan

Zhihong Guo, Yubao Zhang, Yajun Wang, Guo Yang, Ruoyu Wang and Zhongkui Xie
Gaolan Station of the Agricultural and Ecological Experiment, Northwest Institute of Eco-Environment and Resources, Chinese Academy of Sciences, Lanzhou 730000, China

Le Wang and Wenmei Li
Gaolan Station of the Agricultural and Ecological Experiment, Northwest Institute of Eco-Environment and Resources, Chinese Academy of Sciences, Lanzhou 730000, China
University of Chinese Academy of Sciences, Beijing 100049, China

Li Wang
The Forest Tree Seedling Station of the Alxa League, Alxa League 750300, China

Xiao-Lu Jin, Jessica Cruz, Paveena Vichyavichien, Nwadiuto Esiobu and Xing-Hai Zhang
Department of Biological Sciences, Florida Atlantic University, Boca Raton, FL 33431, USA

Abubakar Ibrahim
Department of Biological Sciences, Florida Atlantic University, Boca Raton, FL 33431, USA
Department of Soil Sciences, Modibbo Adama University of Technology, Yola, Nigeria

Yu-Bin Zhang
Department of Biological Sciences, Florida Atlantic University, Boca Raton, FL 33431, USA
Monitoring Center for Marine Resources and Environments, Guangdong Ocean University, Zhanjiang 524088, China

Van-Chung Mai and Ba-Hoanh Nguyen
Department of Plant Physiology, School of Natural Sciences Education, Vinh University, str. Le Duan 182, Vinh, Nghe An Province, Vietnam

Duc-Dien Nguyen and Le-Ai-Vinh Nguyen
Department of Environmental Sciences, School of Chemo-Biology and Environment Technology, Vinh University, str. Le Duan 182, Vinh, Nghe An Province, Vietnam

Meng Wang, Dou Wang, Qing Zhang, Jia Chai, Yong Peng and Xia Cai
Key Laboratory of Resource Biology and Biotechnology in Western China, (Northwest University), Ministry of Education, Xi'an 710069, China

San-Gwang Hwang
Department of Horticulture, National Chung Hsing University, 145 Xingda Road, South District, Taichung 40227, Taiwan, ROC

Chia-Yun Lee
Institute of Plant and Microbial Biology, Academia Sinica, Taipei 11529, Taiwan, ROC

Ching-Shan Tseng
Taiwan Agricultural Research Institute, Council of Agriculture, Executive Yuan, No.189, Zhongzheng Road, Wufeng District, Taichung 41362, Taiwan, ROC

Shun-Wun Ke and Chin-Ying Yang
Department of Agronomy, National Chung Hsing University, Taichung 40227, Taiwan

Guan-Heng Chen, Chung-Tse Chen and Jason T. C. Tzen
Graduate Institute of Biotechnology, National Chung Hsing University, Taichung 40227, Taiwan

Yao Tang, Yan-Li Yin, Peng Feng and Ying Chang
Northeast Agricultural University, Harbin 150030, China

Wei Zhang
Keshan Branch of Heilongjiang Academy of Agricultural Sciences, Keshan 160000, Heilongjiang Province, China

Hong-ling Li
College of Pharmacy, Qiqihar Medical University, Qiqihar 161006, China

Emily Schofield, Edward P. Jones and Viswambharan Sarasan
Natural Capital and Plant Health, Royal Botanic Gardens, Kew, Richmond TW9 3AE, UK

Wei-Hong Liang and Yuh-Chyang Charng
Department of Agronomy, National Taiwan University, Taipei, Taiwan, Republic of China

Tung-Wu Chang
Hualien District Agricultural Research and Extension Station, Council of Agriculture, Executive Yuan, Hualien, Taiwan, Republic of China

Kuan-Fu Lin, Meng-Yu Tsai, Chung-An Lu, Shaw-Jye Wu and Ching-Hui Yeh
Department of Life Sciences, National Central University, Taoyuan 32001, Taiwan

Xin Zheng and Liang Chen
Key Laboratory for Agricultural Water Resources, Center for Agricultural Resources Research, Institute of Genetics and Developmental Biology, Chinese Academy of Sciences, Shijiazhuang 050021, Hebei, People's Republic of China

Xiaofang Li
Key Laboratory for Agricultural Water Resources, Center for Agricultural Resources Research, Institute of Genetics and Developmental Biology, Chinese Academy of Sciences, Shijiazhuang 050021, Hebei, People's Republic of China
CMLR, Sustainable Minerals Institute, The University of Queensland, Brisbane, QLD 4072, Australia

Jiali Song, Yiting Zhang, Shiwei Song, Wei Su, Riyuan Chen, Guangwen Sun, Yanwei Hao and Houcheng Liu
College of Horticulture, South China Agricultural University, Guangzhou 510642, China

Bing-Hong Huang and Pei-Chun Liao
Department of Life Science, National Taiwan Normal University, 88, Ting-Chow Rd, Sec 4, Taipei 116, Taiwan

Chih-Kai Yang
Department of Life Science, National Taiwan Normal University, 88, Ting-Chow Rd, Sec 4, Taipei 116, Taiwan
The Experimental Forest, College of Bio-Resources and Agriculture, National Taiwan University, 12, Sec. 1, Chien-Shan Rd., Nantou 55750, Taiwan

Yu-Chung Chiang
Department of Biological Sciences, National Sun Yat-sen University, 70, Lien-Hai Rd., Kaohsiung 80424, Taiwan

Li-Ping Ju
Botanical Garden Division, Taiwan Forestry Research Institute, 53, Nan-Hai Rd., Taipei 10066, Taiwan

Fei Luo, Xiong Deng and Yue Liu
College of Life Science, Capital Normal University, Beijing 100048, China

Yueming Yan
College of Life Science, Capital Normal University, Beijing 100048, China
Hubei Collaborative Innovation Center for Grain Industry (HCICGI), Yangtze University, Jingzhou 434025, China

Chunbo Liang and Xutang Huang
Heilongjiang Academy of Agricultural Sciences Postdoctoral Programme, Xuefu Road 368, Harbin 150086, People's Republic of China
Institute of Industrial Crops, Heilongjiang Academy of Agricultural Sciences, Harbin 150086, People's Republic of China

Wenjun Wang, Jing Wang, Jun Ma, Cen Li, Fei Zhou, Shuquan Zhang, Ying Yu and Liguo Zhang
Institute of Industrial Crops, Heilongjiang Academy of Agricultural Sciences, Harbin 150086, People's Republic of China

Weizhong Li
Crop Research and Breeding Center of Land-Reclamation of Heilongjiang Province, Harbin 150036, People's Republic of China

Yusong Jiang, Qinhong Liao, Yong Zou, Yiqing Liu and Jianbin Lan
Research Institute for Special Plants, Chongqing University of Arts and Sciences, Chongqing 402160, China

Renjun Mao, Zhigui He and Ruilian Han
State and Local Joint Research Center of TCM Fingerprint and NP Library, College of Life Sciences, Northwest A&F University, Yangling 712100, Shaanxi, China

Zongsuo Liang
State and Local Joint Research Center of TCM Fingerprint and NP Library, College of Life Sciences, Northwest A&F University, Yangling 712100, Shaanxi, China
College of Life Sciences, Zhejiang Sci-Tech University, Hangzhou 310018, Zhejiang, China

Pengguo Xia
College of Life Sciences, Zhejiang Sci-Tech University, Hangzhou 310018, Zhejiang, China

Yan Liu
Tianjin Tasly Modern TCM Resources Co. Ltd., Tianjin 300400, China

Fenghua Liu and Hongguang Zhao
Shaanxi Tasly Plants Pharmaceutical Co. Ltd., Shangluo 726000, Shaanxi, China

Chi-Hsuan Wu
Institute of Biotechnology, National Cheng Kung University, Tainan 701, Taiwan

Ting-Chieh Chen, Yu-Chang Liu, Chih-Hao Huang and Ching-Chun Chang
Institute of Biotechnology, National Cheng Kung University, Tainan 701, Taiwan
Department of Biotechnology and Bioindustry Sciences, National Cheng Kung University, Tainan 701, Taiwan

Xuewen Wang
Department of Genetics, University of Georgia, Athens, GA 30602, USA

Juanjuan Song and Qing Ye
Key Laboratory of Vegetation Restoration and Management of Degraded Ecosystems, South China Botanical Garden, Chinese Academy of Sciences, 723 Xingke Road, Tianhe District, Guangzhou 510650, China
Guangdong Provincial Key Laboratory of Applied Botany, South China Botanical Garden, Chinese Academy of Sciences, 723 Xingke Road, Tianhe District, Guangzhou 510650, Guangdong, China

Sasivimon C. Swangpol and Paweena Traiperm
Department of Plant Science, Faculty of Science, Mahidol University, Rama VI Road, Ratchathewi, Bangkok 10400, Thailand

Valéria Kocová, Vladislav Kolarčik, Albert Rákai and Pavol Mártonfi
Department of Botany, Faculty of Science, Institute of Biology and Ecology, P.J. Šafárik University, Mánesova 23, 04154 Košice, Slovakia

Nikola Straková
Krosnianska 69, 04022 Košice, Slovakia

Index

www.ingramcontent.com/pod-product-compliance
Lightning Source LLC
LaVergne TN
LVHW080712210926
848485LV00003B/451